NUMERICAL MODELING
OF OCEAN DYNAMICS

ADVANCED SERIES ON OCEAN ENGINEERING

Series Editor-in-Chief
Philip L- F Liu (*Cornell University*)

Advanced Series on Ocean Engineering – Volume 5

NUMERICAL MODELING
OF OCEAN DYNAMICS

Z. Kowalik
Institute of Marine Science, University of Alaska Fairbanks
Alaska

T. S. Murty
Institute of Ocean Sciences, Sidney, B.C.
Canada

World Scientific
Singapore • New Jersey • London • Hong Kong

Published by

World Scientific Publishing Co. Pte. Ltd.
5 Toh Tuck Link, Singapore 596224
USA office: 27 Warren Street, Suite 401-402, Hackensack, NJ 07601
UK office: 57 Shelton Street, Covent Garden, London WC2H 9HE

British Library Cataloguing-in-Publication Data
A catalogue record for this book is available from the British Library.

Advanced Series on Ocean Engineering — Vol. 5
NUMERICAL MODELING OF OCEAN DYNAMICS

ISBN-13 978-981-02-1333-6
ISBN-10 981-02-1333-6
ISBN-13 978-981-02-1334-3 (pbk)
ISBN-10 981-02-1334-4 (pbk)

Preface

This book is a result of collaboration between the authors stretching over a decade. While there are several excellent books dealing with numerical analysis and analytical theory, students and faculty in numerical applications to ocean dynamics are forced to sift through literally hundreds of references. This monograph is our attempt to at least partly rectify this situation. While we do not view this as a cookbook recipe type approach, we kept the graduate student and professor in view in preparing this book. Major chapters (II, III and IV) deal first with the basics and then go on to various applications. Ocean dynamics is a vast field and it will be unrealistic on our part to cover this entire field; instead we confined ourselves to transport equations (diffusion and advection), shallow water phenomena – tides, storm surges and tsunamis, three-dimensional time dependent oceanic motion, natural oscillations and steady state phenomena. The aim of this book is two-fold: it gives an introduction to the application of finite-difference methods to ocean dynamics and it also reviews more complex methods.

A brief introduction to the basic equations, parametrization of coefficients and boundary conditions is given in Chapter I.

The basic numerical methods in the finite-difference approach for one-dimensional problem are depicted in Chapter II. The equation of transport for the advective and diffusive processes is taken as a prototype. Approximation, stability, computational errors and application of explicit versus implicit marching in time have been explained there. The reader is advised to start reading the book from Sections 1 through 8 of this chapter.

Chapter III discusses two-dimensional models. Again a basic approach is described up to Section 7, which can be strengthened by the review given in the second part of the chapter. Particular emphasis is given to space-staggered schemes, problems with open boundaries, nonlinear advective terms and grid refinement.

Chapter IV describes three-dimensional models. The problems encountered in modelling three-dimensional and time dependent motion, i.e., economy of computation, approximation and stability are especially severe. Starting from fjord models, a few approaches to the time integration are delineated: explicit method, split mode method and semi-implicit method. The first approach serves for the shallow water problems, the two latter approaches aim at saving computer time and can be used to study the open ocean dynamics.

Chapter V thoroughly delineates computation of the natural modes. Simple numerical methods are given in Section 10.

Chapter VI depicts basic methods used in the steady state processes. An analogy between iterative processes and time stepping in the time dependent processes is explored.

vi

The first author wishes to thank I. Bang, K.B. Gilkey, D. Glover, R. Honrath and C. Merlin, students from the University of Alaska, Fairbanks for checking some of the numerical methods given in Chapter II and III. He is also grateful to the University of Alaska for sabbatical leave to complete this book. The second author would like to thank the Management at the Institute of Ocean Sciences, Sidney, B.C. for enabling him to do this work. He also thanks Ms. R. Rutka for assistance in preparing the final manuscript.

Table of Contents

CHAPTER I
FORMULATION OF THE GENERAL EQUATIONS

Z. Kowalik

CHAPTER II
TRANSPORT EQUATIONS

Z. Kowalik

CHAPTER III
TWO–DIMENSIONAL NUMERICAL MODELS

Z. Kowalik

CHAPTER IV
THREE-DIMENSIONAL TIME-DEPENDENT MOTION

Z. Kowalik

T.S. Murty

CHAPTER V
NORMAL MODES

T.S. Murty

CHAPTER VI
STEADY STATE PROCESSES

Z. Kowalik

CHAPTER I

FORMULATION OF THE GENERAL EQUATIONS

1. Equations of motion, continuity and diffusion

1.1 General equations

Numerical models describing the fields of velocity and density in the ocean are based on the system of hydrodynamical–thermodynamical equations which incorporate the law of conservation of momentum, mass and energy. These equations will be derived in the Cartesian coordinate system although at times a spherical coordinate system will be used. A right-handed rectangular coordinate system with the origin located at the undisturbed level of the free surface is introduced.

The coordinate system is such that the x axis points towards east, the y axis points towards north, and the z axis points upwards, towards zenith. The components of velocity along these three axes will be respectively denoted by u, v, and w. For convenience and brevity we will use also a tensor summation notation, in which the axes are denoted by x_1, x_2, and x_3 and the velocity components by u_1, u_2, and u_3. In this notation, an index appearing twice in a term implies summation over all three index values.

The equation expressing the conservation of momentum is written in tensor notation as

$$\rho \frac{\partial u_i}{\partial t} + \rho u_j \frac{\partial u_i}{\partial x_j} + 2\rho \epsilon_{ijk} \Omega_j u_k = -\frac{\partial p}{\partial x_i} - g\rho \delta_{3i} + \frac{\partial \sigma_{ij}}{\partial x_j} \qquad (1.1)$$

where ρ is the density of water, t time, p pressure, Ω_j is the component of the Earth's angular velocity, and σ_{ij} are the components of stress due to molecular viscosity. Also note that $\epsilon_{ijk} = +1$ if i, j, k are in cyclic order, $\epsilon_{ijk} = -1$ if i, j, k are in anticyclic order, and $\epsilon_{ijk} = 0$ if any pair or all three indices have the same value; $\delta_{3i} = 1$ for $i = 3$ and $\delta_{3i} = 0$ for otherwise.

If μ is the molecular viscosity, then the stress tensor, σ_{ij}, can be expressed in terms of the rate of deformation of a fluid element by the motion

$$\sigma_{ij} = \mu \left(\frac{\partial u_i}{\partial x_j} + \frac{\partial u_j}{\partial x_i} \right) \qquad (1.2)$$

The conservation of mass can be expressed through the continuity equation

$$\frac{\partial \rho}{\partial t} + \frac{\partial \rho u_i}{\partial x_i} = 0, \tag{1.3}$$

which for an incompressible fluid is

$$\frac{\partial u_i}{\partial x_i} = 0 \tag{1.3a}$$

For any other property of the fluid (e.g., salinity, temperature, etc.), the general form of the conservation equation holds,

$$\frac{\partial \rho \Phi}{\partial t} + \frac{\partial \rho \Phi u_i}{\partial x_i} = -\frac{\partial F_i}{\partial x_i} + q \tag{1.4}$$

Here, F_i are the components of the flux of the property Φ due to internal forces or pressures and q is the total internal source of the property Φ. Note that eq.(1.1) and (1.3) are special forms of eq.(1.4).

1.2 Separation of the flow into an average and variation around an average state

The above equation can not be solved exactly, except in a few simple cases. Even if such a solution can be derived, to understand particular phenomenon we have to understand which terms in the general equations are responsible for its occurrence. One way to simplify the above set and to obtain the equations controlling the particular phenomenon is to separate these equations into two sets: one set expressing the average motion and the other describing the departures from the average. This approach usually is used to separate the motion into the average and turbulent parts (Hinze, 1975). The meaning of an average and variation from it will depend on the averaging period; by changing the averaging time one can separate particular phenomenon. Because of the nonlinear nature of the above equations, this separation can not be achieved as two independent equations, and one has to contend with two sets, in each of which terms expressing interactions between average and variable motion appear.

Multiplying eq.(1.3) by u_i and adding to eq.(1.1) and using eq.(1.2) after ignoring compressibility of the water, we obtain

$$\frac{\partial \rho u_i}{\partial t} + \frac{\partial \rho u_j u_i}{\partial x_j} + 2\rho \epsilon_{ijk}\Omega_j u_k = -\frac{\partial p}{\partial x_i} - g\rho\delta_{3i} + \mu\frac{\partial^2 u_i}{\partial x_j \partial x_j} \tag{1.5}$$

To achieve the above mentioned separation, an average with respect to time is used and the averaging process is denoted by a bar:

$$\bar{\Phi} = \frac{1}{T}\int_0^T \Phi dt \tag{1.6}$$

The next step is to express the velocity, pressure and density fields as an average and a variation:

$$u_i = \bar{u}_i + u'_i \quad p = \bar{p} + p' \quad \rho = \bar{\rho} + \rho' \tag{1.7}$$

The averaging process of eq.(1.5) and eq.(1.3) gives

$$\frac{\partial \overline{\rho u_i}}{\partial t} + \frac{\partial \overline{\rho u_j u_i}}{\partial x_j} + 2\epsilon_{ijk}\Omega_j\overline{\rho u_k} = -\frac{\partial \bar{p}}{\partial x_i} - g\bar{\rho}\delta_{3i} + \mu\frac{\partial^2 \bar{u}_i}{\partial x_j \partial x_j} \tag{1.8}$$

$$\frac{\partial \bar{\rho}}{\partial t} + \frac{\partial \overline{\rho u_i}}{\partial x_i} = 0 \tag{1.9}$$

Subtracting eq.(1.8) from eq.(1.5) and eq.(1.9) from eq.(1.3) the equations of motion and continuity for the variations are derived

$$\frac{\partial(\rho u_i - \overline{\rho u_i})}{\partial t} + \frac{\partial(\rho u_j u_i - \overline{\rho u_j u_i})}{\partial x_j} + 2\epsilon_{ijk}\Omega_j(\rho u_k - \overline{\rho u_k})$$

$$= -\frac{\partial(p - \bar{p})}{\partial x_i} - g(\rho - \bar{\rho})\delta_{3i} + \mu\frac{\partial^2(u_i - \bar{u}_i)}{\partial x_j \partial x_j} \tag{1.10}$$

$$\frac{\partial(\rho - \bar{\rho})}{\partial t} + \frac{\partial(\rho u_i - \overline{\rho u_i})}{\partial x_i} = 0 \tag{1.11}$$

Noting that the variations in time have zero mean by definition, the above equations can be considered again. The nonlinear interactions between the average and variation are expressed by term $\overline{\rho u_i u_j}$ in eq.(1.8). This term becomes

$$\overline{\rho u_i u_j} = \overline{(\bar{\rho} + \rho')(\bar{u}_i + u'_i)(\bar{u}_j + u'_j)}$$

$$= \bar{\rho}\bar{u}_i\bar{u}_j + \bar{\rho}\overline{u'_i u'_j} + \bar{u}_i\overline{\rho'u'_j} + \overline{\rho'u'_i}\bar{u}_j + \overline{\rho'u'_i u'_j} \tag{1.12}$$

In water bodies, generally speaking, variations in density are very small in proportion to the average density, whereas the fluctuations in velocity can be of the same order as the mean velocity. Hence, the terms containing correlations between fluctuations of density and velocity are small compared with the terms containing correlations between velocity components. We will therefore, ignore the density fluctuations in the acceleration terms. However, the correlation between density and velocity fluctuations can be significant in the other equations where velocity correlations do not appear. Especially in the continuity equation, the correlation between density and velocity fluctuations may be important. Accordingly, the eq.(1.12) contains only two terms:

$$\overline{\rho u_i u_j} = \bar{\rho}\bar{u}_i\bar{u}_j + \bar{\rho}\overline{u'_i u'_j} \tag{1.13}$$

Thus, the only additional term in the average equation of motion is correlation between the velocity fluctuations. It is called Reynolds stress tensor, we shall define it as

$$R_{ij} = -\bar{\rho}\overline{u'_i u'_j} \tag{1.14}$$

Then equation of motion and continuity for the average motion can be written as

$$\frac{\partial \bar{\rho}\bar{u}_i}{\partial t} + \frac{\partial \bar{\rho}\bar{u}_j\bar{u}_i}{\partial x_j} - \frac{\partial R_{ij}}{\partial x_j} + 2\epsilon_{ijk}\Omega_j\bar{\rho}\bar{u}_k = -\frac{\partial \bar{p}}{\partial x_i} - g\bar{\rho}\delta_{3i} + \mu\frac{\partial^2 \bar{u}_i}{\partial x_j\partial x_j} \tag{1.15}$$

$$\frac{\partial \bar{\rho}}{\partial t} + \frac{\partial \bar{\rho}\bar{u}_i}{\partial x_i} - \frac{\partial Q_i}{\partial x_i} = 0 \tag{1.16}$$

Here Q_i expresses transport of mass due to fluctuating motion

$$Q_i = -\overline{\rho' u'_i} \tag{1.17}$$

The equations for the variable (turbulent, eddy) motion are

$$\bar{\rho}\frac{\partial u'_i}{\partial t} + \frac{\partial \bar{\rho}\bar{u}_j u'_i}{\partial x_j} + \frac{\partial \bar{\rho}u'_j\bar{u}_i}{\partial x_j} - \frac{\partial R'_{ij}}{\partial x_j} + 2\epsilon_{ijk}\Omega_j\bar{\rho}u'_k$$

$$= -\frac{\partial p'}{\partial x_i} - g\rho'\delta_{3i} + \mu\frac{\partial^2 u'_i}{\partial x_j\partial x_j} \tag{1.18}$$

$$\frac{\partial \rho'}{\partial t} + \bar{u}_i\frac{\partial \rho'}{\partial x_i} + u'_i\frac{\partial \bar{\rho}}{\partial x_i} - \frac{\partial Q'_i}{\partial x_i} = 0 \tag{1.19}$$

In the above equations R'_{ij} and Q'_i are defined as

$$R'_{ij} = -\bar{\rho}(u'_i u'_j - \overline{u'_i u'_j}) \tag{1.20}$$

$$Q'_j = -(\rho' u'_j - \overline{\rho' u'_j}) \tag{1.21}$$

One can see from eq.(1.15) that the influence of the turbulent flow upon the average motion is expressed completely by the Reynolds stress tensor, R_{ij}. However, in the turbulent equations (1.18), the interactions are contained in many terms and are of three different types: (a) transport of fluctuating momentum by the average flow, (b) transport of the average momentum by the fluctuating flow, and (c) divergence of the momentum flux variations of the purely fluctuating flow. The above demonstrated procedure separates slow motion of period longer than T from high-frequency turbulence. Reynolds stress in the average equations (1.15) can be interpreted as a dissipating mechanism. The divergence of R_{ij} is a source of the momentum and longer period motions principally contribute to this source. This means that the motion of a prescribed time scale can receive momentum mainly from the longer time scales and lose momentum to shorter time scales.

Turbulence plays an important role in the dissipative processes in water bodies. However, it can not be easily incorporated into the momentum equations. Usually the dissipative action due to turbulence is taken into account by assuming that the Reynolds stresses are proportional to the strain rate of the mean flow. In analogy to molecular viscosity, this proportionality constant is referred to as eddy viscosity. Considering the fact that the horizontal dimensions of the water body are usually much larger than its vertical dimensions, eddy viscosities of widely different values might be necessary for the horizontal and vertical directions. Sometimes, in the horizontal plane, different eddy viscosity values might be used in the x and y directions (e.g., narrow water bodies). Equations (1.14) can be written as

$$R_{ij} = -\bar{\rho}\overline{u_i' u_j'} = \bar{\rho}N(j)\frac{\partial \bar{u}_i}{\partial x_j} + \bar{\rho}N(i)\frac{\partial \bar{u}_j}{\partial x_i} \tag{1.22}$$

Here, $N(i) = N_h$ for $i, j \neq 3$ and $N(j) = N_z$ for $i, j = 3$. The parameters N_h and N_z are referred to as horizontal and vertical eddy viscosities, respectively. Whenever temperature or salinity display variable distribution, exchange of momentum in the turbulent flow is associated with exchange (or transport) of mass. This transport is described by the eddy diffusivity coefficient, from eq.(1.17)

$$Q_i = -\overline{\rho' u_i'} = D(\rho)\frac{\partial \bar{\rho}}{\partial x_i} \tag{1.23}$$

Eddy diffusivity is usually different along the horizontal and vertical directions. The horizontal eddy diffusivity D_h and vertical eddy diffusivity D_z are respectively introduced to describe this difference in behavior.

The utility of the above parametrization is due to its simplicity, however, the magnitudes of the eddy viscosities or eddy diffusivities vary with variable averaging time in eq.(1.14) and in eq.(1.17). Therefore, the scale of the motion defines the behavior of the turbulent transfer of momentum and mass. The formal approach to this problem was delineated by Hinze(1975). In the equations of the large–scale ocean motion the influence of the mesoscale motion is taken into account through the variable eddy coefficients, but the formulas used for the eddy coefficients reflect the contemporary understanding of the interaction of the mesoscale motion with the large scale motion (Smagorinsky, 1963; Voronov and Ivchenko, 1978).

1.3 Energy equations

The conservation of energy of a mechanical system makes energy one of the best parameters to study the time history of the motion, both from the analytical and numerical points of view. Behavior of the various numerical schemes can be easily observed by invoking the energy of the system. We shall define here two types of energy, i.e., kinetic and potential;

$$E_k = \frac{1}{2}\rho u_i^2 = \frac{1}{2}\rho(u^2 + v^2 + w^2); \qquad E_p = \rho g z \qquad (1.24)$$

The above expressions define the energy in a unit volume of the fluid (Gill, 1982).

We shall start by constructing the kinetic energy equation for the average flow. Multiplying eq.(1.15) by $\bar{u}_i$ and taking into account equation of continuity for incompressible fluid (1.3a), we arrive at

$$\frac{\partial \bar{\rho}\bar{u}_i^2/2}{\partial t} + \frac{\partial \bar{u}_j(\bar{p} + \bar{\rho}\bar{u}_i^2/2)}{\partial x_j} - \bar{u}_i\frac{\partial R_{ij}}{\partial x_j} = -g\bar{\rho}\bar{w} + \bar{u}_i\mu\frac{\partial^2 \bar{u}_i}{\partial x_j \partial x_j} \qquad (1.25)$$

This equation expresses the balance of the kinetic energy in a unit volume of fluid. The first term on the left-hand-side (LHS) describes the change in time of the kinetic energy; the second term expresses a transport of kinetic energy and pressure by an average flow and the third term gives the transfer of energy from the average flow to the variable flow (and vice versa) by the Reynolds stresses. The first term on the right-hand-side (RHS) containing an average vertical component of velocity gives the change of the potential energy in time. The second term on the RHS describes the work done by the viscous stresses and total dissipation caused by the molecular forces. We shall not consider the last term in detail because, usually in oceanic flow molecular friction is not important. The term related to the Coriolis force is also rather small in the average motion and is neglected in the above equation.

The role of dissipation in eq.(1.25) is given by the term containing the Reynolds stresses. Applying the notion of eddy viscosity through eq.(1.22) one can obtain an eddy dissipation similar to the molecular friction term.

To derive the kinetic energy equation for the turbulent flow, eq.(1.18) is multiplied by u_i':

$$\bar{\rho}\frac{\partial u_i'^2/2}{\partial t} + \frac{\partial \bar{\rho}\bar{u}_j u_i'^2/2}{\partial x_j} + u_j'u_i'\frac{\partial \bar{\rho}\bar{u}_i}{\partial x_j} - u_i'\frac{\partial R_{ij}'}{\partial x_j} + 2\epsilon_{ijk}\Omega_j\bar{\rho}u_i'u_k'$$

$$= -\frac{\partial u_i'p'}{\partial x_i} - g\rho'w' + u_i'\mu\frac{\partial^2 u_i'}{\partial x_j \partial x_j} \qquad (1.26)$$

This equation can be averaged again so that an average value for the eddy kinetic energy is obtained. Additionally we delete two terms from the ensuing consideration: molecular viscous term and Coriolis term. Thus,

$$\bar{\rho}\frac{\overline{\partial u_i'^2/2}}{\partial t} + \frac{\partial \bar{\rho}\bar{u}_j\overline{u_i'^2/2}}{\partial x_j} + \overline{u_j'u_i'}\frac{\partial \bar{\rho}\bar{u}_i}{\partial x_j} - \overline{u_i'\frac{\partial R_{ij}'}{\partial x_j}} = -\frac{\overline{\partial u_i'p'}}{\partial x_i} - g\overline{\rho'w'} \qquad (1.27)$$

Some terms in the above equation are of special importance. The last term on LHS involves triple correlation and therefore requires further assumption or closure hypotheses similar to the eddy coefficients. The third term on LHS can be represented as

$$\overline{u'_j u'_i} \frac{\partial \bar{\rho} \bar{u}_i}{\partial x_j} = \frac{\partial \bar{\rho} \bar{u}_i \overline{u'_j u'_i}}{\partial x_j} - \bar{u}_i \frac{\partial \bar{\rho} \overline{u'_j u'_i}}{\partial x_j} = -\frac{\partial \bar{u}_i R_{ij}}{\partial x_j} + \bar{u}_i \frac{\partial R_{ij}}{\partial x_j} \qquad (1.28)$$

This transformation involves definition (1.14). In eq.(1.28) the term $-\dfrac{\partial \bar{u}_i R_{ij}}{\partial x_j}$ equals exactly (but with opposite sign) the term occurring in eq.(1.25). It describes the transfer of kinetic energy from the average motion (larger scale) to the eddy motion (smaller scale) by the Reynolds stresses. Such transfer is essential for maintaining the turbulent flow.

The notion of potential energy can be introduced in two different ways. One way is to use the definition given by (1.24), the other way is to construct the square of the density variations in a similar manner to the eddy kinetic energy – u'^2_i. Let us first multiply eq.(1.9) by gdz and afterwards integrate it over the whole water body to obtain

$$\int gz(\frac{\partial \bar{\rho}}{\partial t} + \frac{\partial \bar{\rho} u_i}{\partial x_i})dxdydz = 0 \qquad (1.29)$$

The first term represents the change in time of the average potential energy. The second term can be rewritten as

$$\int gz\frac{\partial \bar{\rho} u_i}{\partial x_i}dxdydz = \int gz\frac{\partial(\bar{\rho} \bar{u}_i + \overline{\rho' u'_i})}{\partial x_i}dxdydz \qquad (1.30)$$

Along the vertical direction ($i = 3$) it becomes quite similar to the expressions for the eddy potential energy and average potential energy as shown in eq.(1.25) and (1.27).

The change of potential energy in time is often related to the notion of the available potential energy in the fluid body,

$$E_{p,av} = \int \rho gz dx dy dz|_t - \int \rho gz dx dy dz|_{t=t_o} \qquad (1.31)$$

The available potential energy denotes the difference between potential energy at any given time and the potential energy at the initial moment.

As we pointed out the expression for the eddy potential energy can be constructed in a similar way to the kinetic energy by considering the square of the density variation. Multiplying eq.(1.19) by ρ' and g^2 and dividing by $\bar{\rho}^2$, we arrive at

$$\frac{g^2}{2\bar{\rho}^2}\frac{D\rho'^2}{Dt} + \frac{g^2}{\bar{\rho}^2}\rho' u'_i \frac{\partial \bar{\rho}}{\partial x_i} - \frac{g^2}{\bar{\rho}^2}\rho'\frac{\partial Q'_i}{\partial x_i} = 0 \qquad (1.32)$$

Here a full derivative denotes

$$\frac{D}{Dt} = \frac{\partial}{\partial t} + \bar{u}_i \frac{\partial}{\partial x_i} \tag{1.33}$$

Introducing Väisälä–Brunt parameter $N^2 = -\frac{g}{\bar{\rho}} \frac{\partial \bar{\rho}}{\partial z}$ and averaging eq.(1.32) in time, the eddy potential energy is

$$\frac{g^2}{2\bar{\rho}^2 N^2} \frac{D\overline{\rho'^2}}{Dt} + \frac{g^2}{\bar{\rho}^2 N^2} \overline{\rho' u'_k} \frac{\partial \bar{\rho}}{\partial x_k} - \frac{g}{\bar{\rho}} \overline{w' \rho'} - \frac{g^2}{\bar{\rho}^2 N^2} \overline{\rho' \frac{\partial Q'_i}{\partial x_i}} = 0 \tag{1.34}$$

Here index k runs only through $k = 1, 2$, i.e. summation takes place along x and y coordinates. The term $g\overline{w' \rho'}$ appears in the eddy kinetic energy eq.(1.27) as well. Therefore, this term describes energy transfer from kinetic to potential energy and vice versa. Such energy exchange is intricately related to the vertical stratification through the N^2 parameter. In order to identify the important aspects of the energy transfer, let us assume that the fluid is initially stratified only along the horizontal direction, therefore $N^2 = 0$ and the energy exchange term disappears from the potential energy equation. At the same time, potential energy may change in time because of the existence of horizontal stratification: $\dfrac{\partial \bar{\rho}}{\partial x_k} \neq 0$. This energy can be transformed into kinetic energy only when vertical stratification occurs in the density field.

Equation (1.34) describes the energy balance at a given location. Although it is quite illuminating, but from a computational point of view it displays unwanted properties when the vertical stratification is small $N^2 \to 0$. In such case the integral form for the potential energy (1.30) which is free from these limitations, can be used.

1.4 Rectangular system of coordinates

The effects of the Earth's curvature on the motion of a fluid at the relatively small distances of the order of 1000 km may be neglected. To describe this motion a rectangular system of coordinates will be employed, thus making the equations and discussion much simpler. Usually in such studies a set of averaged equations (1.15) is employed, this can be recognized by observing the presence of the Reynolds stresses or the eddy coefficient introduced through definition (1.22). We shall rewrite the set of averaged equations of motion as

$$\frac{Du}{Dt} - fv = -\frac{1}{\rho} \frac{\partial p}{\partial x} + \frac{\partial}{\partial z} N_z \frac{\partial u}{\partial z} + N_h \Delta u \tag{1.35}$$

$$\frac{Dv}{Dt} + fu = -\frac{1}{\rho} \frac{\partial p}{\partial y} + \frac{\partial}{\partial z} N_z \frac{\partial v}{\partial z} + N_h \Delta v \tag{1.36}$$

$$\frac{Dw}{Dt} = -\frac{1}{\rho}\frac{\partial p}{\partial z} - g + \frac{\partial}{\partial z}N_z\frac{\partial w}{\partial z} + N_h\Delta w \qquad (1.37)$$

Here the operator $\frac{D}{Dt}$ is

$$\frac{D}{Dt} = \frac{\partial}{\partial t} + u\frac{\partial}{\partial x} + v\frac{\partial}{\partial y} + w\frac{\partial}{\partial z} \qquad (1.38)$$

In these equations Coriolis parameter $f = 2\Omega\sin\phi$ is a function of the Earth's angular velocity $\Omega = 7.29\times10^{-5}s^{-1}$ and the latitude — ϕ. Two-dimensional Laplace operator Δ is defined as $\Delta = \frac{\partial^2}{\partial x^2} + \frac{\partial^2}{\partial y^2}$. Also notice that for the sake of brevity and clarity we have deleted the bar notation for the time average values.

Above, for the five unknown variables (u, v, w, p, ρ) only three equations have been introduced, therefore two additional equations are required — these will be the equation of continuity and the equation of state. For the problems we intend to solve the sea water can be regarded as an incompressible fluid, and in this case the equation of continuity expresses the conservation of volume (eq.(1.3a)),

$$\frac{\partial u}{\partial x} + \frac{\partial v}{\partial y} + \frac{\partial w}{\partial z} = 0 \qquad (1.39; 1.3a)$$

The density of sea water can be defined from the equation of state as a function of the water salinity (S) and temperature (θ),

$$\rho = \rho(S, \theta) \qquad (1.40)$$

To find the density as a function of temperature and salinity we shall use an expression proposed by Mamayev (1975).

Considering the density as the sum of two terms

$$\rho = 1 + \sigma_\theta 10^{-3} \qquad (1.41)$$

the value σ_θ was given by Mamayev (1975) in the following way

$$\sigma_\theta = (\rho - 1)10^3 = 28.152 + 0.0735\theta - 0.004691\theta^2 + (0.802 - 0.002\theta)(S - 35) \quad (1.42)$$

For a crude estimation of the density dependence on temperature and salinity often a linear equation is used

$$\rho = \rho(35°/_{oo}, 0°) - \beta_\theta\theta + \beta_S(S - 35°/_{oo}) \qquad (1.42a)$$

Here the coefficients are defined as $\beta_\theta = 0.07 \times 10^{-3}$ g/cm^3 deg; $\beta_S = 0.8 \times 10^{-3}$ g/cm^3 °/$_{oo}$; $\rho(35°/_{oo}, 0)$ – density in g/cm^3 at S=35°/$_{oo}$ and $\theta = 0$°C.

Equation (1.42) has been constructed to fit the wide range of temperatures and salinities, and therefore it does not have sufficient accuracy for the specific purposes. An approximation to the equation of state over smaller ranges of temperature and salinity but with the higher accuracy was proposed Bryan and Cox (1972) and Friedrich and Levitus (1972).

Although the density of water is finally defined through expression (1.42) this approach has introduced two new variables, i.e., salinity and temperature. These variables will be defined through the equations of conservation, better known as equations of diffusion:

$$\frac{\partial \theta}{\partial t} + u\frac{\partial \theta}{\partial x} + v\frac{\partial \theta}{\partial y} + w\frac{\partial \theta}{\partial z} = \frac{\partial}{\partial z}D_z\frac{\partial \theta}{\partial z} + D_h\Delta\theta \tag{1.43}$$

$$\frac{\partial S}{\partial t} + u\frac{\partial S}{\partial x} + v\frac{\partial S}{\partial y} + w\frac{\partial S}{\partial z} = \frac{\partial}{\partial z}D_z\frac{\partial S}{\partial z} + D_h\Delta S \tag{1.44}$$

The vertical eddy diffusivity for the turbulent mixing of heat and salt is denoted as D_z, and the horizontal eddy diffusion is D_h.

In summary, seven dependent variables $(u, v, w, p, \rho, S, \theta)$ are defined by seven equations (1.35), (1.36), (1.37), (1.39), (1.40), (1.43), (1.44). To derive a unique solution to this set of equations, suitable boundary and initial conditions ought to be specified. Additionally, the coefficients of the turbulent exchange of momentum (N_z, N_h) and turbulent diffusion (D_z, D_h) will be defined by invoking basic knowledge of the turbulent mixing in the ocean.

1.5 Spherical system of coordinates

Whenever fluid motion is considered along large distances on the globe the equation of motion and continuity can be written in the spherical polar coordinates λ, ϕ and R, defined as longitude, latitude and distance from the Earth's center. If the origin of the system is located on the ocean surface, it is more suitable to introduce a vertical coordinate $z = R - R_0$. Here R_0 is the radius of Earth and is equal 6370km.

Because Earth does not exactly has a spherical shape, the equation given below will better describe the large scale motion relative to the geopotential and not to the spherical surfaces. For further discussion of this problem see Gill (1982).

The equations of motion in the spherical system are

$$\frac{Du}{Dt} - (2\Omega + \frac{u}{R\cos\phi})(v\sin\phi - w\cos\phi) = -\frac{1}{\rho R\cos\phi}\frac{\partial p}{\partial \lambda} + A_\lambda \tag{1.45}$$

$$\frac{Dv}{Dt} + \frac{wv}{R} + (2\Omega + \frac{u}{R\cos\phi})u\sin\phi = -\frac{1}{\rho R}\frac{\partial p}{\partial \phi} + A_\phi \tag{1.46}$$

$$\frac{Dw}{Dt} - \frac{v^2}{R} - (2\Omega + \frac{u}{R\cos\phi})u\cos\phi = -\frac{1}{\rho}\frac{\partial p}{\partial z} - g + A_z \qquad (1.47)$$

where A_λ, A_ϕ, and A_z are the components of the viscous force, and the time operator is expressed as

$$\frac{D}{Dt} = \frac{\partial}{\partial t} + \frac{u}{R\cos\phi}\frac{\partial}{\partial\lambda} + \frac{v}{R}\frac{\partial}{\partial\phi} + w\frac{\partial}{\partial z} \qquad (1.48)$$

The frictional forces are written in a somewhat complicated form:

$$A_\lambda = A_1 u - \frac{N_h}{R^2\cos^2\phi}(u + 2\frac{\partial}{\partial\lambda}(v\sin\phi - w\cos\phi)) \qquad (1.49)$$

$$A_\phi = A_1 v + N_h(-\frac{v}{R^2\cos^2\phi} + \frac{2\sin\phi}{R^2\cos^2\phi}\frac{\partial u}{\partial\lambda} + \frac{2}{R^2}\frac{\partial w}{\partial\phi}) \qquad (1.50)$$

$$A_z = A_1 w + N_h(-\frac{2w}{R^2} - \frac{2}{R^2\cos^2\phi}\frac{\partial u}{\partial\lambda} - \frac{2}{R^2\cos\phi}\frac{\partial}{\partial\lambda}(v\cos\phi)) \qquad (1.51)$$

where the operator A_1 has the following form,

$$A_1 = N_h(\frac{1}{R^2\cos^2\phi}\frac{\partial^2}{\partial\lambda^2} + \frac{1}{R^2\cos\phi}\frac{\partial}{\partial\phi}(\cos\phi\frac{\partial}{\partial\phi})) + \frac{\partial}{\partial z}(N_z\frac{\partial}{\partial z}) \qquad (1.52)$$

The equation of continuity in the spherical system is

$$\frac{1}{R\cos\phi}\frac{\partial u}{\partial\lambda} + \frac{1}{R\cos\phi}\frac{\partial}{\partial\phi}(v\cos\phi) + \frac{\partial w}{\partial z} = 0 \qquad (1.53)$$

Finally the equation of diffusion can be expressed in the following way:

$$\frac{Dc}{Dt} = A_2 c \qquad (1.54)$$

where operator A_2 is

$$A_2 = D_h(\frac{1}{R^2\cos^2\phi}\frac{\partial^2}{\partial\lambda^2} + \frac{1}{R^2\cos\phi}\frac{\partial}{\partial\phi}(\cos\phi\frac{\partial}{\partial\phi})) + \frac{\partial}{\partial z}(D_z\frac{\partial}{\partial z}) \qquad (1.55)$$

Variable c stands for concentration, and it can denote salinity, temperature or any passive admixture in the sea water.

1.6 Boussinesq and hydrostatic approximations

The above derived equations of motion describe a complete spectrum of motion, but we shall study mainly the meso and large scale motion such as long waves, wind and density-driven currents. To obtain from a general set of equations the specific set which will describe only a certain class of motion, traditionally the so-called dimensional analysis has been applied, see, e.g., Pedlosky (1982). This approach assumes that certain information related to the temporal and spatial scale of motion is known *a priori*.

The hydrostatic approximation is only related to the vertical component of equation of motion (1.37) or (1.47). Because the horizontal extent of the ocean is much larger than the vertical extent, the vertical component of the motion is much weaker than the horizontal velocity. Typical vertical velocities in the oceans are of the order 10^{-2} cm/s, whereas the horizontal velocities are 1000 times greater. If the flow is predominantly horizontal, and the vertical acceleration is small compared to the gravity acceleration, the equation of vertical motion can be reduced to the simple hydrostatic law (Proudman, 1953). Rigorous justification of this approximation has never been given, and it is possible that locally (where the depth changes abruptly) large vertical currents may occur. The hydrostatic assumption simplifies equations (1.37) and (1.47) to

$$-\frac{1}{\rho}\frac{\partial p}{\partial z} - g = 0 \qquad (1.56)$$

Boussinesq (1903) approximation describes the way the density variations enter into equations of motion. Essentially water density will be considered as constant and the variations of density will lead to a new term in the equation of motion which characterize Archimedian (buoyancy) force — see Landau and Lifshitz (1959).

The Boussinesq approximation can be introduced by assuming that the basic state of the fluid is hydrostatic (the state of no motion), defined by pressure p_o and density ρ_o. Motion will arise due only to the density ρ' or pressure p' variations, thus,

$$p = p_o(z) + p'(x, y, z, t)$$
$$\rho = \rho_o(z) + \rho'(x, y, z, t)$$
$$u = u'(x, y, z, t)$$
$$v = v'(x, y, z, t)$$
$$w = w'(x, y, z, t) \qquad (1.57)$$

where $\rho' \ll \rho_o$, and $p' \ll p_o$.

Let us first introduce (1.57) into equation of motion (1.35),

$$(\rho_o + \rho')(\frac{Du}{Dt} - fv) = -\frac{\partial(p_o + p')}{\partial x} + \text{Fric.terms} \qquad (1.58)$$

from which a simplified equation in the Boussinesq form follows:

$$\rho_o \frac{Du}{Dt} - \rho_o fv = -\frac{\partial p'}{\partial x} + \text{Fric.terms} \qquad (1.59)$$

The equation of motion along the vertical direction (1.56) with the Boussinesq approximation is given by

$$\frac{\partial p}{\partial z} = -g(\rho_o + \rho') \qquad (1.60)$$

and can now be integrated from any depth z to the free surface $z = \zeta(x, y, t)$,

$$\int_z^\zeta dp = -\int_z^\zeta g(\rho_o + \rho')dz$$

Noting that the pressure at the free surface $p(\zeta)$ is equal to the atmospheric pressure (p_a) and pressure at any depth $p(z)$ is equal $p_o + p'$, we write the final result as

$$p' = p_a - p_o + \int_z^0 g\rho_o dz + \int_0^\zeta g\rho_o dz + g\int_z^\zeta \rho' dz$$

$$= p_a - p_o + \int_z^0 g\rho_o dz + g\rho_o(\zeta - z) + g\int_z^\zeta \rho' dz \qquad (1.61)$$

Thus we were able, by integrating the equation of motion along the vertical coordinate, to express unknown pressure variations as a sum of terms related to the atmospheric pressure p_a, hydrostatic pressure p_o, sea level changes (ζ), and density stratification. Introducing pressure variation into the equation of motion (1.59), and applying similar considerations to the equation of motion along the y coordinate, the system of equations with the Boussinesq and hydrostatic approximation is obtained:

$$\frac{Du}{Dt} - fv = -\frac{1}{\rho_o}\frac{\partial p_a}{\partial x} - g\frac{\partial \zeta}{\partial x} - \frac{g}{\rho_o}\frac{\partial}{\partial x}\int_z^\zeta \rho' dz + \frac{\partial}{\partial z}N_z\frac{\partial u}{\partial z} + N_h\Delta u \qquad (1.62)$$

$$\frac{Dv}{Dt} + fu = -\frac{1}{\rho_o}\frac{\partial p_a}{\partial y} - g\frac{\partial \zeta}{\partial y} - \frac{g}{\rho_o}\frac{\partial}{\partial y}\int_z^\zeta \rho' dz + \frac{\partial}{\partial z}N_z\frac{\partial v}{\partial z} + N_h\Delta v \qquad (1.63)$$

In the above equations the horizontal components of acceleration have been related to the atmospheric pressure (p_a), sea level (ζ) and the density distribution (ρ'). The motion is modified by the Coriolis force and the friction. A similar approach can be taken to derive equations in the spherical system of coordinates.

Although the vertical acceleration was assumed to be negligible and the vertical equation has been reduced to the equation of hydrostatic state the vertical velocity still can be calculated from the equation of continuity (1.39) or (1.53).

2. Two-dimensional equations

2.1 Vertically integrated equations

The pressure due to the sea level slope is constant throughout the water depth, it is therefore feasible to derive a new set of equations from (1.62) and (1.63) by vertical integration. The new set may be used to find the unknown sea slope components or it can be applied to solve problems which do not depend on the vertical distribution. The vertically averaged equations have been applied to predict the sea surface variations in time or the averaged current in tides, storm surges and tsunamis (Murty 1984).

Integration over the vertical coordinate will take place from the bottom $z = -H(x, y)$ to the free surface $z = \zeta(x, y, t)$. Differentiating these expressions in time one can obtain the boundary conditions:

$$\frac{\partial \zeta}{\partial t} + u\frac{\partial \zeta}{\partial x} + v\frac{\partial \zeta}{\partial y} = w \quad \text{at} \quad z = \zeta(x, y, t) \tag{1.64}$$

$$w = -\frac{DH}{Dt} = -u\frac{\partial H}{\partial x} - v\frac{\partial H}{\partial y} \quad \text{at} \quad z = -H(x, y) \tag{1.65a}$$

A kinematic condition at the bottom requires the flow to be parallel to the bottom contour. Velocity in the frictional flow decays at the bottom. To express this fact a no-slip condition will be imposed:

$$u = v = w = 0 \quad \text{at} \quad z = -H(x, y) \tag{1.65b}$$

Now let us analyze term by term in eq.(1.62). The vertical component of the frictional force renders

$$\rho_0 \int_{-H}^{\zeta} \frac{\partial}{\partial z} N_z \frac{\partial u}{\partial z} dz = \tau_x^s - \tau_x^b \tag{1.66}$$

and

$$\rho_0 \int_{-H}^{\zeta} \frac{\partial}{\partial z} N_z \frac{\partial v}{\partial z} dz = \tau_y^s - \tau_y^b \tag{1.67}$$

Surface (τ^s) and bottom (τ^b) stresses are introduced based on the stress definition

$$\tau_x = \rho_0 N_z \frac{\partial u}{\partial z} \quad \text{and} \quad \tau_y = \rho_0 N_z \frac{\partial v}{\partial z} \tag{1.68}$$

Generally, stress is defined as a tangential force acting on a unit surface and it has the dimension of pressure. To proceed further the vertically averaged velocity is defined:

$$\tilde{u} = \frac{1}{D} \int_{-H}^{\zeta} u \, dz \quad \text{and} \quad \tilde{v} = \frac{1}{D} \int_{-H}^{\zeta} v \, dz \tag{1.69}$$

Here $D = H + \zeta$ denotes a total depth. The horizontal acceleration term in eq.(1.62) is transformed as

$$\int_{-H}^{\zeta} \frac{\partial u}{\partial t} dz = \frac{\partial}{\partial t} \int_{-H}^{\zeta} u\, dz - [u\frac{\partial \zeta}{\partial t}]_{z=\zeta}$$

$$\int_{-H}^{\zeta} u\frac{\partial u}{\partial x} dz = \frac{\partial}{\partial x} \int_{-H}^{\zeta} u^2\, dz - \int_{-H}^{\zeta} u\frac{\partial u}{\partial x} dz - [u^2\frac{\partial \zeta}{\partial x}]_{z=\zeta} - [u^2\frac{\partial H}{\partial x}]_{z=-H}$$

$$\int_{-H}^{\zeta} v\frac{\partial u}{\partial y} dz = \frac{\partial}{\partial y} \int_{-H}^{\zeta} uv\, dz - \int_{-H}^{\zeta} u\frac{\partial v}{\partial y} dz - [uv\frac{\partial \zeta}{\partial y}]_{z=\zeta} - [uv\frac{\partial H}{\partial y}]_{z=-H}$$

$$\int_{-H}^{\zeta} w\frac{\partial u}{\partial z} dz = [uw]_{z=\zeta} - [uw]_{z=-H} - \int_{-H}^{\zeta} u\frac{\partial w}{\partial z} dz \qquad (1.70)$$

Now let us combine the terms whose sum is equal to zero,

$$-[u(\frac{\partial \zeta}{\partial t} + u\frac{\partial \zeta}{\partial x} + v\frac{\partial \zeta}{\partial y})]_{z=\zeta} -$$

$$\int_{-H}^{\zeta} u(\frac{\partial u}{\partial x} + \frac{\partial v}{\partial y} + \frac{\partial w}{\partial z})dz + [uw]_{z=\zeta} - [uw]_{z=-H}$$

The first term is equal to the third term, the second term is equal to zero because it comprises equation of continuity (1.39). Finally, the last term vanishes due to the bottom boundary condition (1.65b). The integration of Coriolis term and pressure terms in (1.62) is straightforward, but the derivation of the horizontal friction term is rather cumbersome. The final result is

$$\frac{\partial}{\partial t} \int_{-H}^{\zeta} u\, dz + \frac{\partial}{\partial x} \int_{-H}^{\zeta} u^2\, dz + \frac{\partial}{\partial y} \int_{-H}^{\zeta} uv\, dz - f \int_{-H}^{\zeta} v\, dz$$

$$= -\frac{H+\zeta}{\rho_0}\frac{\partial p_a}{\partial x} - g(H+\zeta)\frac{\partial \zeta}{\partial x} - \frac{g}{\rho_0} \int_{-H}^{\zeta}\int_{z}^{\zeta} \frac{\partial \rho'}{\partial x} dz dz + (\tau_x^s - \tau_x^b)/\rho_0 + \text{Hor.Fric.} \qquad (1.71)$$

A similar equation can be written along the y direction. These equations will serve as a starting point for a few different approximations.

Shallow water approximation, was probably applied for the first time by Hansen (1949) to study numerically propagation of tides. It assumes that the horizontal velocities are constant along the vertical direction, also the motion is usually considered to be barotropic i.e., density stratification is neglected. The nonlinear and horizontal terms in (1.71) can be written as

$$\frac{\partial}{\partial x} \int_{-H}^{\zeta} u^2\, dz + \frac{\partial}{\partial y} \int_{-H}^{\zeta} uv\, dz \approx \frac{\partial}{\partial x} \tilde{u}\tilde{u}D + \frac{\partial}{\partial y} \tilde{u}\tilde{v}D$$

$$\int_{-H}^{\zeta} (N_h \Delta u) dz \approx D N_h \Delta \tilde{u} \qquad (1.72)$$

With the above simplifications taken into consideration, the shallow water equations along E-W and N-S directions become

$$\rho_0 \left(\frac{\partial \tilde{u}}{\partial t} + \tilde{u} \frac{\partial \tilde{u}}{\partial x} + \tilde{v} \frac{\partial \tilde{u}}{\partial y} - f \tilde{v} \right) = -\frac{\partial p_a}{\partial x} - g \rho_0 \frac{\partial \zeta}{\partial x} + \tau_x^s / D - \tau_x^b / D + \rho_0 N_h \Delta \tilde{u} \quad (1.73)$$

$$\rho_0 \left(\frac{\partial \tilde{v}}{\partial t} + \tilde{u} \frac{\partial \tilde{v}}{\partial x} + \tilde{v} \frac{\partial \tilde{v}}{\partial y} + f \tilde{u} \right) = -\frac{\partial p_a}{\partial y} - g \rho_0 \frac{\partial \zeta}{\partial y} + \tau_y^s / D - \tau_y^b / D + \rho_0 N_h \Delta \tilde{v} \quad (1.74)$$

One observation is important; the exchange of momentum in the vertical direction in eq.(1.62) and (1.63) has been described by means of the eddy viscosity coefficient, in eq.(1.73) and (1.74) only stresses at the surface and at the bottom ought to be defined. This makes the shallow water problem much simpler, because the eddy viscosity coefficients are often difficult to define.

Deep water approximation. The deep water motion is considered to be rather slow. Comparison of the various terms in the equations of motion shows that certain terms, like nonlinear and horizontal friction terms are small. Let us begin by comparing the nonlinear term to the Coriolis term,

$$u \frac{\partial u}{\partial x} = f v \qquad (1.75)$$

Now suppose that we know the characteristic (typical) value of the horizontal components of the current and both are equal u_c. The characteristic value of the horizontal distance is denoted as L. Applying to eq.(1.75) a dimensional analysis based on the characteristic values, we can define distance over which the Coriolis term begins to dominate over the nonlinear term,

$$u_c u_c / L = f u_c \qquad \text{thus} \qquad L = u_c / f \qquad (1.76)$$

Introducing typical values $u_c = 10$ cm/s and $f = 10^{-4} \text{s}^{-1}$ into above formulae, the horizontal distance $L = 1$ km is obtained. Therefore, at distances larger than 1 km the nonlinear terms can be disregarded. A similar result may be inferred for the horizontal friction term through a comparison to the Coriolis term,

$$N_h u_c / L^2 = f u_c \qquad \text{thus} \qquad L = \sqrt{N_h / f} \qquad (1.77)$$

For $N_h = 10^8 \text{cm}^2 \text{s}^{-1}$ the typical distance equals to $L = 10$ km. Here the conclusion is that at scales less than 10 km horizontal friction play a significant role in oceanic flow.

It is important to stress that in both cases the characteristic values (i.e., velocity and horizontal eddy diffusivity) were assumed to be known *a priori*. Applying all these approximations we shall write the vertically averaged equations for the deep water motion (also since $\zeta \ll H$, we take $D \simeq H$)

$$\rho_o \frac{\partial \bar{u}}{\partial t} - \rho_o f \bar{v} = -\frac{\partial p_a}{\partial x} - g \rho_o \frac{\partial \zeta}{\partial x} - \frac{g}{H} \int_{-H}^{\zeta} \int_{z}^{\zeta} \frac{\partial \rho'}{\partial x} dz dz + \tau_z^a/H - \tau_z^b/H + \rho_o N_h \Delta \bar{u}$$

$$(1.78)$$

$$\rho_o \frac{\partial \bar{v}}{\partial t} + \rho_o f \bar{u} = -\frac{\partial p_a}{\partial y} - g \rho_o \frac{\partial \zeta}{\partial y} - \frac{g}{H} \int_{-H}^{\zeta} \int_{z}^{\zeta} \frac{\partial \rho'}{\partial y} dz dz + \tau_y^a/H - \tau_y^b/H + \rho_o N_h \Delta \bar{v}$$

$$(1.79)$$

In the above equations pressure changes due to the density stratification are again included because density-driven currents are the major component of the deep water motion.

Equation of continuity. Integrating the equation of continuity (1.39) along the vertical direction and applying the relations

$$\int_{-H}^{\zeta} \frac{\partial u}{\partial x} dz = \frac{\partial}{\partial z} \int_{-H}^{\zeta} u dz - \left[u \frac{\partial \zeta}{\partial x} \right]_{z=\zeta} - \left[u \frac{\partial H}{\partial x} \right]_{z=-H}$$

$$\int_{-H}^{\zeta} \frac{\partial v}{\partial x} dz = \frac{\partial}{\partial z} \int_{-H}^{\zeta} v dz - \left[v \frac{\partial \zeta}{\partial x} \right]_{z=\zeta} - \left[v \frac{\partial H}{\partial x} \right]_{z=-H}$$

and also taking into account the boundary conditions (1.64) and (1.65a), the equation of continuity for the vertically averaged flow is obtained:

$$\frac{\partial}{\partial x} \bar{u} D + \frac{\partial}{\partial y} \bar{v} D + \frac{\partial \zeta}{\partial t} = 0 \qquad (1.80)$$

Transport equations. Vertical integration defined by (1.69) admits possibility of introducing a different set of variables. We define here the components of the horizontal transport as

$$M_x = D\bar{u} = \int_{-H}^{\zeta} u dz \quad \text{and} \quad M_y = D\bar{v} = \int_{-H}^{\zeta} v dz \qquad (1.81)$$

Using this definition we proceed to eq.(1.71) to rewrite equation of motion in the transport variables. However simple it may seem the nonlinear terms and frictional terms defined by eq.(1.72) ought to be redefined with the proper approximation. Thus starting from eq.(1.72)

$$\frac{\partial}{\partial x}\int_{-H}^{\zeta} u^2 dz + \frac{\partial}{\partial y}\int_{-H}^{\zeta} uv dz \approx \frac{\partial}{\partial x}\bar{u}\bar{u}D + \frac{\partial}{\partial y}\bar{u}\bar{v}D = \frac{\partial}{\partial x}(\frac{M_x}{D}M_x) + \frac{\partial}{\partial y}(\frac{M_x}{D}M_y)$$

$$\int_{-H}^{\zeta}(N_h\Delta u)dz \approx DN_h\Delta\bar{u} \approx N_h\Delta M_x \tag{1.82}$$

With these simplifications the equation of motion (1.71) under the shallow water approximation is written as

$$\rho_o[\frac{\partial M_x}{\partial t} + \frac{\partial}{\partial x}(\frac{M_x}{D}M_x) + \frac{\partial}{\partial y}(\frac{M_x}{D}M_y) - fM_y]$$

$$= -D\frac{\partial p_a}{\partial x} - gD\rho_o\frac{\partial\zeta}{\partial x} + \tau_x^s - \tau_x^b + \rho_o N_h\Delta M_x \tag{1.83}$$

Similarly, the motion along the y coordinate is described by the equation

$$\rho_o[\frac{\partial M_y}{\partial t} + \frac{\partial}{\partial x}(\frac{M_y}{D}M_x) + \frac{\partial}{\partial y}(\frac{M_y}{D}M_y) + fM_x]$$

$$= -D\frac{\partial p_a}{\partial y} - gD\rho_o\frac{\partial\zeta}{\partial y} + \tau_y^s - \tau_y^b + \rho_o N_h\Delta M_y \tag{1.84}$$

Notice that in the above equations the density stratification is neglected.

Equation of continuity (1.80) for the transport variables is

$$\frac{\partial M_x}{\partial x} + \frac{\partial M_y}{\partial y} + \frac{\partial\zeta}{\partial t} = 0 \tag{1.85}$$

2.2 Vertically integrated equations on the sphere

Whenever an integration domain stretches over 1000 km the use of a spherical system is advisable. For the shallow water equations of motion the assumptions made in the previous section namely the absence of the vertical motion and the hydrostatic and Boussinesq approximations are valid here as well. The technique used to derive a set of vertically integrated equations in the previous section can be repeated here for the spherical coordinates. Equations of motions (1.45) and (1.46) can be written as

$$\frac{Du}{Dt} - (2\Omega + \frac{u}{R\cos\phi})v\sin\phi = -\frac{1}{\rho_o R\cos\phi}\frac{\partial p_a}{\partial\lambda} - \frac{g}{R\cos\phi}\frac{\partial\zeta}{\partial\lambda} + A'_\lambda \tag{1.86}$$

$$\frac{Dv}{Dt} + (2\Omega + \frac{u}{R\cos\phi})u\sin\phi = -\frac{1}{\rho_o R}\frac{\partial p_a}{\partial\phi} - \frac{g}{R}\frac{\partial\zeta}{\partial\phi} + A'_\phi \tag{1.87}$$

Here

$$\frac{D}{Dt} = \frac{\partial}{\partial t} + \frac{u}{R\cos\phi}\frac{\partial}{\partial\lambda} + \frac{v}{R}\frac{\partial}{\partial\phi} \tag{1.88}$$

$$A'_\lambda = A_1 u - \frac{N_h}{R^2\cos^2\phi}(u + 2\frac{\partial}{\partial\lambda}(v\sin\phi)) \tag{1.89}$$

$$A'_\phi = A_1 v + N_h(-\frac{v}{R^2\cos^2\phi} + \frac{2\sin\phi}{R^2\cos^2\phi}\frac{\partial u}{\partial\lambda}) \tag{1.90}$$

where operator A_1 has the following form

$$A_1 = N_h(\frac{1}{R^2\cos^2\phi}\frac{\partial^2}{\partial\lambda^2} + \frac{1}{R^2\cos\phi}\frac{\partial}{\partial\phi}(\cos\phi\frac{\partial}{\partial\phi})) + \frac{\partial}{\partial z}(N_z\frac{\partial}{\partial z}) \tag{1.91}$$

In the above equations density stratification is neglected, although these terms can be easily introduced into the system by comparison to eqs.(1.62) and (1.63). Now, integrating eqs.(1.86) and (1.87) along the vertical direction and taking into account the discussion from the previous section we arrive at

$$\frac{D\tilde{u}}{Dt} - (2\Omega + \frac{\tilde{u}}{R\cos\phi})\tilde{v}\sin\phi = -\frac{1}{\rho_o R\cos\phi}\frac{\partial p_a}{\partial\lambda} - \frac{g}{R\cos\phi}\frac{\partial\zeta}{\partial\lambda} + \frac{\tau^a_\lambda - \tau^b_\lambda}{\rho_o D} + A'_1\tilde{u} \tag{1.92}$$

$$\frac{D\tilde{v}}{Dt} + (2\Omega + \frac{\tilde{u}}{R\cos\phi})\tilde{u}\sin\phi = -\frac{1}{\rho_o R}\frac{\partial p_a}{\partial\phi} - \frac{g}{R}\frac{\partial\zeta}{\partial\phi} + \frac{\tau^a_\phi - \tau^b_\phi}{\rho_o D} + A'_1\tilde{v} \tag{1.93}$$

Here tilde above the variables denotes vertical averaging the way it is defined in formulas (1.69). The horizontal friction term has been simplified in (1.92) and (1.93) so that the operator A'_1 is

$$A'_1 = N_h(\frac{1}{R^2\cos^2\phi}\frac{\partial^2}{\partial\lambda^2} + \frac{1}{R^2\cos\phi}\frac{\partial}{\partial\phi}(\cos\phi\frac{\partial}{\partial\phi})) \tag{1.94}$$

Finally integrating equation of continuity (1.53) along the vertical direction we arrive at equation (which is very similar to eq.(1.80))

$$\frac{\partial\zeta}{\partial t} + \frac{1}{R\cos\phi}\frac{\partial\tilde{u}D}{\partial\lambda} + \frac{1}{R\cos\phi}\frac{\partial}{\partial\phi}(D\tilde{v}\cos\phi) = 0 \tag{1.95}$$

2.3 Vertically integrated equations on a stereographic map

When studying the oceanic motion in a large area where the Earth's curvature is of importance the spherical system of coordinate is used. Difficulties arise if integration is performed in the region near the pole. In spherical system of coordinates

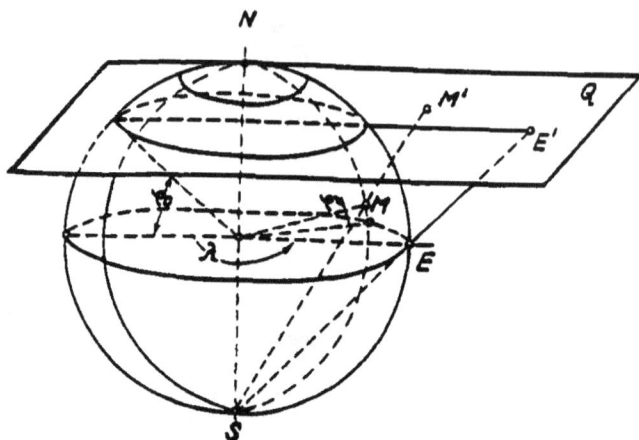

Fig. 1.1
Polar stereographic projection. (Kowalik and Nguyen Bich Hung, 1977)

at the pole $\cos\phi \to 0$ and to avoid this obstacle a stereographic polar coordinate system is introduced approximating spherical coordinates by means of a scale factor m (Kowalik and Bich Hung, 1977; Gjevik and Straume, 1989). Stereographic projection is obtained by establishing a plane Q passing through the parallel of latitude ϕ_o (usually $\phi_o = 60°$) — Fig.1.1.

That portion of the sphere above latitude ϕ_o is projected onto the surface Q. The new system is written in the plane Q. It has its origin at the axis of rotation (north pole). The x axis points in direction of a 0^0 meridian and y axis points in direction of a 90^0 meridian. This new system is related to the spherical system by the following coordinate transformation,

$$x = mR\cos\lambda \qquad \text{and} \qquad y = mR\sin\lambda \tag{1.96}$$

Scale factor m for a map projection relates a surface area on the stereographic map to its image on a sphere.

$$m = \frac{\delta S(\text{map})}{\delta S(\text{sphere})} = \frac{1 + \sin\phi_o}{1 + \sin\phi} \tag{1.97}$$

Usually $\phi_o = 60°$, therefore the magnitude of m is close to unity ($m = 0.93$ at the pole).

Transforming the set of equation (1.92), (1.93) and (1.95) to the Cartesian coordinates in the plane Q we arrive at

$$\frac{\partial \tilde{u}}{\partial t} - f\tilde{v} = -m\frac{1}{\rho_o}\frac{\partial p_a}{\partial x} - mg\frac{\partial \zeta}{\partial x} + \frac{\tau_z^s - \tau_z^b}{D\rho_o} + N_h m^2 \Delta\tilde{u} \tag{1.98}$$

$$\frac{\partial \tilde{v}}{\partial t} + f\tilde{u} = -m\frac{1}{\rho_o}\frac{\partial p_a}{\partial y} - mg\frac{\partial \zeta}{\partial y} + \frac{\tau_y^s - \tau_y^b}{D\rho_o} + N_h m^2 \Delta \tilde{v} \qquad (1.99)$$

$$\frac{\partial \zeta}{\partial t} = -m^2 \left[\frac{\partial}{\partial x}(\frac{\tilde{u}D}{m}) + \frac{\partial}{\partial y}(\frac{\tilde{v}D}{m}) \right] \qquad (1.100)$$

Notice that only the linear form of the equation is transformed and all the nonlinear terms are excluded.

Equation of continuity (1.100) can be simplified if one assumes that the relative change of velocity is greater than the relative change in the scale factor

$$|\frac{1}{\tilde{u}}\frac{\partial \tilde{u}}{\partial x}| \gg |\frac{1}{m}\frac{\partial m}{\partial x}| \qquad \text{and} \qquad |\frac{1}{\tilde{v}}\frac{\partial \tilde{v}}{\partial y}| \gg |\frac{1}{m}\frac{\partial m}{\partial y}|$$

These assumptions allow us to rewrite eq.(1.100) as

$$\frac{\partial \zeta}{\partial t} = -m \left[\frac{\partial}{\partial x}(\tilde{u}D) + \frac{\partial}{\partial y}(\tilde{v}D) \right] \qquad (1.100a)$$

2.4 Laterally integrated equations

Estuaries or fjords can be idealized as elongated channels with the dominant (horizontal) component of velocity along the axis of the channel. The cross-channel motion can be neglected by averaging the equations of motion and continuity along a cross-channel coordinate. Directing the x coordinate along the channel, z coordinate upward, and y coordinate across the channel, the average velocity and density is defined through the following set of equations. Equation of continuity,

$$\frac{\partial}{\partial x}bu + \frac{\partial}{\partial z}bw = 0 \qquad (1.101)$$

is obtained through the integration of eq.(1.39). Equation of motion

$$\frac{\partial u}{\partial t} + \frac{u}{b}\frac{\partial}{\partial x}ub + w\frac{\partial u}{\partial z} = -\frac{1}{\rho_o}\frac{\partial p_a}{\partial x} - g\frac{\partial \zeta}{\partial x} - \frac{g}{\rho_o}\frac{\partial}{\partial x}\int_z^\zeta \rho' dz + \frac{\partial}{\partial z}N_z\frac{\partial u}{\partial z} + \frac{N_h}{b}\frac{\partial}{\partial x}(b\frac{\partial u}{\partial x}) \qquad (1.102)$$

is derived through integration of (1.62).

Integration of eq.(1.101) over depth yields an additional equation for the unknown sea level,

$$\frac{\partial \zeta}{\partial t} + \frac{1}{b}\frac{\partial}{\partial x}(b\int_{-H}^\zeta udz) = 0 \qquad (1.103)$$

Equation of salt or heat transport (diffusion) is obtained by integration (1.43) or (1.44) along the width of the channel (b),

$$\frac{\partial S}{\partial t} + \frac{u}{b}\frac{\partial}{\partial x}Sb + w\frac{\partial S}{\partial z} = \frac{\partial}{\partial z}D_z\frac{\partial S}{\partial z} + \frac{D_h}{b}\frac{\partial}{\partial x}(b\frac{\partial S}{\partial x}) \qquad (1.104)$$

Solution of (1.104) will define distribution of the salinity and temperature, therefore equation of state (1.42) can be used to find density and to estimate density driven currents through the integral in (1.102). Although we did not write tilde above the variables in the equations the dependent variables like velocity or salinity, are actually averaged over the channel width b.

In the equation of motion (1.102) we have neglected the term due to Earth's rotation. One way to justify this approach is to compare the properties of the long waves when rotation is included and when it is neglected. Without rotation both internal and surface long waves propagate along the channel as plane waves. Rotation, on other hand, will produce a Kelvin wave with a cross-stream variation in the wave amplitude. The length scale which characterizes the effect of rotation on long wave propagation is the Rossby radius of deformation. When the width of the fjord is small compared to the Rossby radius, cross-stream effects may be neglected (Huppert, 1980). The Rossby radii (r_i) relates the long wave speed (C_i) to the Coriolis parameter (f) in a continuously stratified fluid, $r_i = C_i/f$. The Rossby radius for the surface waves is usually quite large, because $C = \sqrt{gH}$, but for the internal waves it strongly depends on the stratification and wave mode number. For a channel of 500 m depth, a typical Rossby radii for the first three modes range from 2 km to 15 km.

3. Application of the stream function

3.1 Steady state Ekman problem

We shall start with the equation of motion in the spherical system of coordinates (1.86), (1.87), (1.47), and the density stratification is included as well

$$\rho_o \frac{Du}{Dt} - \rho_o(2\Omega + \frac{u}{R\cos\phi})v\sin\phi$$

$$= -\frac{1}{R\cos\phi}\frac{\partial p_a}{\partial\lambda} - \frac{\rho_o g}{R\cos\phi}\frac{\partial\zeta}{\partial\lambda} - \frac{g}{R\cos\phi}\int_z^\zeta \frac{\partial\rho'}{\partial\lambda}dz + \rho_o A'_\lambda \qquad (1.105)$$

$$\rho_o \frac{Dv}{Dt} + \rho_o(2\Omega + \frac{u}{R\cos\phi})u\sin\phi$$

$$= -\frac{1}{R}\frac{\partial p_a}{\partial\lambda} - \frac{\rho_o g}{R}\frac{\partial\zeta}{\partial\lambda} - \frac{g}{R}\int_z^\zeta \frac{\partial\rho'}{\partial\phi}dz + \rho_o A'_\phi \qquad (1.106)$$

We shall study the steady state deep water motion, therefore nonlinear terms, horizontal friction and time dependence are discarded from the above equations. The linear set has been applied by Ekman to study wind driven currents. In the set below both the wind and density currents are present:

$$-fv\rho_o = -\frac{1}{R\cos\phi}\frac{\partial p_a}{\partial\lambda} - \frac{\rho_o g}{R\cos\phi}\frac{\partial\zeta}{\partial\lambda} - \frac{g}{R\cos\phi}\int_z^\zeta \frac{\partial\rho'}{\partial\lambda}dz + \rho_o \frac{\partial}{\partial z}N_z\frac{\partial u}{\partial z} \qquad (1.107)$$

$$fu\rho_o = -\frac{1}{R}\frac{\partial p_a}{\partial \phi} - \frac{\rho_o g}{R}\frac{\partial \zeta}{\partial \phi} - \frac{g}{R}\int_z^\zeta \frac{\partial \rho'}{\partial \phi}dz + \rho_o \frac{\partial}{\partial z}N_z\frac{\partial v}{\partial z} \qquad (1.108)$$

Density is assumed to be known either from observations or from the computations via equation of diffusion. To define an unknown sea level slope an additional equation based on the vertically integrated equations of motion and continuity is constructed. Integrating equations (1.107) and (1.108) from the bottom $z = -H(x,y)$ to the free surface $z = \zeta(x,y)$,

$$-f(H+\zeta)\tilde{v}\rho_o = -\frac{H+\zeta}{R\cos\phi}\frac{\partial p_a}{\partial\lambda} - \frac{\rho_o g(H+\zeta)}{R\cos\phi}\frac{\partial\zeta}{\partial\lambda} - \frac{g}{R\cos\phi}\int_{-H}^\zeta\int_z^\zeta \frac{\partial\rho'}{\partial\lambda}dzdz + \tau_\lambda^s - \tau_\lambda^b$$
$$(1.109)$$

$$f(H+\zeta)\tilde{u}\rho_o = -\frac{H+\zeta}{R}\frac{\partial p_a}{\partial\phi} - \frac{\rho_o g(H+\zeta)}{R}\frac{\partial\zeta}{\partial\phi} - \frac{g}{R}\int_{-H}^\zeta\int_z^\zeta \frac{\partial\rho'}{\partial\phi}dzdz + \tau_\phi^s - \tau_\phi^b$$
$$(1.110)$$

In the above, the components of the average velocity are defined as in (1.69). The surface and bottom stresses are introduced based on the stress definition (1.68). In the ensuing considerations we shall assume that $H \approx H + \zeta$ and the bottom stresses τ^b as a linear function of the average velocity

$$\tau_\lambda^b = r\tilde{u} \qquad \text{and} \qquad \tau_\phi^b = r\tilde{v} \qquad (1.111)$$

The stream function will be introduced through the equation of continuity (1.95) as

$$\frac{1}{R\cos\phi}(\frac{\partial\tilde{u}H}{\partial\lambda} + \frac{\partial}{\partial\phi}(\tilde{v}H\cos\phi)) + \frac{\partial\zeta}{\partial t} = 0 \qquad (1.112)$$

Because in the steady state the sea level is constant in time, the equation of continuity

$$\frac{\partial\tilde{u}H}{\partial\lambda} + \frac{\partial}{\partial\phi}(\tilde{v}H\cos\phi) = 0 \qquad (1.113)$$

permits to introduce a stream function defined in the following way:

$$\tilde{v}HR\cos\phi = \frac{\partial\Psi}{\partial\lambda} \qquad \text{and} \qquad \tilde{u}HR = -\frac{\partial\Psi}{\partial\phi} \qquad (1.114)$$

By crossdifferentiation of (1.109) and (1.110) and applying equation (1.114) the equation for the stream function is derived,

$$\frac{\partial}{\partial\lambda}k_\lambda\frac{\partial\Psi}{\partial\lambda} + \frac{\partial}{\partial\phi}k_\phi\frac{\partial\Psi}{\partial\phi} + \frac{\partial\alpha}{\partial\phi}\frac{\partial\Psi}{\partial\lambda} - \frac{\partial\alpha}{\partial\lambda}\frac{\partial\Psi}{\partial\phi}$$

$$= \frac{\partial}{\partial\lambda}(R\tau_\phi^s/H) - \frac{\partial}{\partial\phi}(R\tau_\lambda^s\cos\phi/H) + \frac{\partial B_\lambda}{\partial\phi} - \frac{\partial B_\phi}{\partial\lambda} \qquad (1.115)$$

In (1.115) the following notation has been introduced, $\alpha = f\rho_o/H$,

$$k_\lambda = r/(H^2 \cos\phi) \quad \text{and} \quad k_\phi = r\cos\phi/H^2$$

$$B_\lambda = \frac{g}{H} \int_{-H}^{\zeta} \int_z^{\zeta} \frac{\partial \rho'}{\partial\lambda} dz dz \quad \text{and} \quad B_\phi = \frac{g}{H} \int_{-H}^{\zeta} \int_z^{\zeta} \frac{\partial \rho'}{\partial\phi} dz dz$$

In conclusion, the solution of the system (1.107) and (1.108) (i.e. three-dimensional distribution of the current) could be easily derived if the sea level slope components were known. To find these components first the equations for the vertically averaged current (1.109) and (1.110) were constructed, afterwards based on the equation of continuity (1.113) the stream function is introduced and eq.(1.109) and (1.110) are reduced to one equation (1.115) of the second order. The actual solution will proceed backward, first equation for the stream function (1.115) will be solved with a suitable boundary condition, next the average velocity will be computed from (1.114) and finally eq.(1.109) and (1.110) will be solved for the sea level slope components $\frac{\partial\zeta}{\partial\lambda}$ and $\frac{\partial\zeta}{\partial\phi}$. Given the density, wind and atmospheric pressure from measurements, the above method renders three-dimensional distribution of the current. This information is not always available and the fields of density and atmospheric forcing are not always adjusted so that they can be related through eq.(1.115). This method finds nonetheless many applications as a first approach to describe the interaction of the fields of density and currents.

3.2 Nonsteady motion

The equations to study general oceanic circulation incorporate the so-called rigid lid assumption, Bryan (1969). It allows a more efficient calculation since the time step of the numerical calculation can be substantially increased. The rigid lid condition

$$w = \frac{\partial\zeta}{\partial t} = 0 \quad \text{at} \quad z = 0 \tag{1.116}$$

requires the vertical velocity at the free surface to be zero. It follows from (1.116) that the equations of continuity (1.113) is still valid and therefore the stream function can be introduced as in (1.114). The temporal motion is governed by the equations (1.105) and (1.106). Integrating this set of equations along the vertical direction allows us to introduce the vertically averaged velocity and stream function

$$-\frac{1}{HR}\frac{\partial}{\partial\phi}\frac{\partial\Psi}{\partial t} - \frac{f}{HR\cos\phi}\frac{\partial\Psi}{\partial\lambda}$$

$$= -\frac{1}{\rho_o R\cos\phi}\frac{\partial p_a}{\partial\lambda} - \frac{g}{R\cos\phi}\frac{\partial\zeta}{\partial\lambda} - \frac{g}{\rho_o H R\cos\phi}\int_{-H}^{0}\int_z^{0}\frac{\partial\rho'}{\partial\lambda}dz dz + \frac{1}{H}\int_{-H}^{0}G^\lambda dz \tag{1.117}$$

$$\frac{1}{HR\cos\phi}\frac{\partial}{\partial\lambda}\frac{\partial\Psi}{\partial t} - \frac{f}{HR}\frac{\partial\Psi}{\partial\phi}$$

$$= -\frac{1}{R\rho_o}\frac{\partial p_a}{\partial\phi} - \frac{g}{R}\frac{\partial\zeta}{\partial\phi} - \frac{g}{HR\rho_o}\int_{-H}^{0}\int_{z}^{0}\frac{\partial\rho'}{\partial\phi}dzdz + \frac{1}{H}\int_{-H}^{0}G^{\phi}dz \qquad (1.118)$$

By cross-differentiation of the above equations and subtracting on either side, the equation for the stream function is obtained:

$$\frac{\partial}{\partial\lambda}\left(\frac{1}{H\cos\phi}\frac{\partial^2\Psi}{\partial\lambda\partial t}\right) + \frac{\partial}{\partial\phi}\left(\frac{\cos\phi}{H}\frac{\partial^2\Psi}{\partial\phi\partial t}\right) + \frac{\partial}{\partial\phi}\left(\frac{f}{H}\frac{\partial\Psi}{\partial\lambda}\right) - \frac{\partial}{\partial\lambda}\left(\frac{f}{H}\frac{\partial\Psi}{\partial\phi}\right)$$

$$= \frac{\partial}{\partial\phi}\left(\frac{g}{H\rho_o}\int_{-H}^{0}\int_{z}^{0}\frac{\partial\rho'}{\partial\lambda}dzdz\right) - \frac{\partial}{\partial\lambda}\left(\frac{g}{H\rho_o}\int_{-H}^{0}\int_{z}^{0}\frac{\partial\rho'}{\partial\phi}dzdz\right)$$

$$+ \frac{\partial}{\partial\lambda}\left(\frac{R}{H}\int_{-H}^{0}G^{\phi}dz\right) - \frac{\partial}{\partial\phi}\left(\frac{R\cos\phi}{H}\int_{-H}^{0}G^{\lambda}dz\right) \qquad (1.119)$$

It remains to explain that G^{λ} and G^{ϕ} represent nonlinear and viscous terms in equations (1.105) and (1.106).

3.3 Two-dimensional motion

Assuming a transversal homogeneity (for example along y axis) of the flow and density one may neglect all the terms containing derivatives with respect to y. The equation of continuity (1.39) will reduce to

$$\frac{\partial u}{\partial x} + \frac{\partial w}{\partial z} = 0 \qquad (1.120)$$

With this approximation, a flow can be studied only with the Boussinesq approximation allowing for the vertical motion to be resolved, consequently from eq.(1.35) and (1.37) with (1.57) taken into account:

$$\frac{\partial u}{\partial t} + u\frac{\partial u}{\partial x} + w\frac{\partial u}{\partial z} = -\frac{1}{\rho_o}\frac{\partial p'}{\partial x} + N_z\frac{\partial^2 u}{\partial z^2} + N_h\Delta u \qquad (1.121)$$

$$\frac{\partial w}{\partial t} + u\frac{\partial w}{\partial x} + w\frac{\partial w}{\partial z} = -\frac{1}{\rho_o}\frac{\partial p'}{\partial z} - g\frac{\rho'}{\rho_o} + N_z\frac{\partial^2 w}{\partial z^2} + N_h\Delta w \qquad (1.122)$$

For the sake of simplicity the eddy viscosity coefficients both along the horizontal and vertical directions are taken as constant and the density variation is assumed to be known. To solve (1.120), (1.121) and (1.122) two approaches are feasible: (1) primary variables, i.e. velocities u, w and pressure p' are searched, or (2) pressure is eliminated by cross-differentiation and the new variables, i.e., vorticity η and stream function Ψ are introduced. To proceed either way, additional equations are required. Both pressure (in case 1), and stream function (in case 2) are governed by elliptical equation of Poisson type (Fromm, 1964). Our preference is with the second approach because for the vorticity and the stream function the boundary conditions are easily derived. Endoh (1977) found this approach very useful in studying thermohaline circulation, Kowalik and Matthews (1983) applied it to the brine rejection circulation in the Arctic lagoons. Introducing the vorticity component η normal to the $x - z$ plane,

$$\eta = \frac{\partial w}{\partial x} - \frac{\partial u}{\partial z} \tag{1.123}$$

by cross-differentiation of (1.121) and (1.122) and subtracting on either side, the following vorticity equation is obtained:

$$\frac{\partial \eta}{\partial t} + \frac{\partial}{\partial x}(u\eta) + \frac{\partial}{\partial z}(w\eta) = -\frac{g}{\rho_o}\frac{\partial \rho'}{\partial x} + N_z\frac{\partial^2 \eta}{\partial z^2} + N_h\frac{\partial^2 \eta}{\partial x^2} \tag{1.124}$$

Through the equation of continuity (1.120) a stream function Ψ is defined as

$$u = -\frac{\partial \Psi}{\partial z} \quad \text{and} \quad w = \frac{\partial \Psi}{\partial x} \tag{1.125}$$

thus the vorticity can be related to the stream function:

$$\frac{\partial^2 \Psi}{\partial x^2} + \frac{\partial^2 \Psi}{\partial z^2} = \eta \tag{1.126}$$

4. Viscosity in the turbulent flow

Exchange of momentum between various water particles is generally caused by the normal (pressure) and tangential stresses. As we have seen above the stress is expressed through the eddy viscosity and derivative of velocity along the normal direction to the flow. We shall not discuss the validity of this assumption for the turbulent flow, but we shall try to express the eddy viscosity as a function of the flow itself. In a few instances the tangential stress can be easily derived from the experiments.

4.1 Tangential stresses at the free surface and at the bottom

The definition of the tangential stress along a horizontal direction is (see eq.(1.68))

$$\tau_x = \rho N_z \frac{\partial u}{\partial z} \quad \text{and} \quad \tau_y = \rho N_z \frac{\partial v}{\partial z} \tag{1.127}$$

The magnitude of this stress at the free surface is equal to the tangential stress over the ocean i.e., to the wind stress. The experimental value of the wind stress τ^a

includes the wind drag coefficient C_{10}, the air density ρ_a, and the wind speed (W) at the altitude 10m above the sea level. The component of the wind stress along x and y directions are:

$$\tau_x^a = C_{10}\rho_a|W|W_x \quad \text{and} \quad \tau_y^a = C_{10}\rho_a|W|W_y \qquad (1.128)$$

where $|W| = \left(W_x^2 + W_y^2\right)^{1/2}$.

Garratt (1977) analyzed almost all measured data and found that C_{10} under a neutral atmospheric stability depends linearly on the wind velocity

$$C_{10} = (0.75 + 0.067 \times 10^{-2}W) \times 10^{-3} \qquad (1.129)$$

where wind velocity is expressed in centimeters per second.

In practice in the storm surge computations (Henry and Heaps, 1976) the wind drag coefficient is usually set as constant and as large as 2.7×10^{-3}. Since the average density of the air is about 1.2×10^{-3} g/cm^3, the coefficient in expression (1.128), $C_{10}\rho = 3.2 \times 10^{-6}$. The value of the drag coefficient used in the storm surge calculations equals the value derived from the Garratt's expression when the wind speed of 29 m/s is introduced into (1.129). The reason for the use of a such a large value is not clear, most probably the wind used in the calculation is produced from the geostrophic method which usually gives smaller than observed wind speed.

At the bottom of the ocean the quadratic dependence of the bottom stress on the velocity is also well recognized:

$$\tau_x^b = \rho r|V|u \quad \text{and} \quad \tau_y^b = \rho r|V|v \qquad (1.130)$$

where $V = \sqrt{u^2 + v^2}$. The bottom drag coefficient (r) is a function of the bottom roughness and the properties of the bottom boundary layer (Komar, 1976). This coefficient is usually taken in the range $(2 - 4) \times 10^{-3}$. In our considerations $r = 3 \times 10^{-3}$. The stress, drag coefficient and the speed of the current are taken at the altitude of 1m above the bottom.

The linearized equations of motion include the linear expression for the bottom stress. Such a form is deduced from eq.(1.130) by taking $\rho r|V| = r_1$ as constant. Thus the linear form of eq.(1.130) is

$$\tau_x^b = r_1 u \quad \text{and} \quad \tau_y^b = r_1 v \qquad (1.130a)$$

This coefficient's magnitude is related to the average flow condition.

In the polar seas where pack ice is present the interaction of the atmosphere, ice and water is expressed through the wind stress, similar to expression (1.128), but the interaction of the ice and water will introduce a new stress. This water stress (τ^w) is sensitive both to the relative motion of the water and ice and to the magnitude of the drag coefficient (r_w),

$$\tau_x^w = \rho r_w |V_i - V|(u_i - u)$$

$$\tau_y^w = \rho r_w |V_i - V|(v_i - v) \tag{1.131}$$

where u_i and v_i components of the ice velocity along the x and y directions respectively; $|V_i - V| = \left((u_i - u)^2 + (v_i - v)^2\right)^{1/2}$. The water drag coefficient is a function of the hydrodynamic properties of the ice-water interface and the relative motion, its magnitude ranges from 3×10^{-3} to 5.5×10^{-3}. For pack ice drift in summer as a result of wind stress, McPhee (1980) estimated the water drag magnitude to be from 4×10^{-3} to 5.5×10^{-3}. He has also postulated that the ratio of the water drag coefficient to the wind drag coefficient is close to 2 ($r_w/C_{10} \approx 2$).

4.2 Vertical exchange

Equations of motion and continuity serve to describe an average distribution of velocity and density. The motion is generally irregular and turbulent. The interaction of the turbulent motion with the average motion is described by the vertical and horizontal eddy viscosities and the exchange of mass by the vertical and horizontal eddy diffusivity. Since the motion in the ocean takes place along the two basic directions the intensity of the turbulent motion is different for these directions and therefore eddy coefficients differ as well. An average eddy viscosity coefficient along the horizontal direction (N_h) ranges from $10^5 \text{cm}^2/\text{s}$ to $10^8 \text{cm}^2/\text{s}$, while the magnitude of the vertical eddy viscosity (N_z) is $1 - 10^3 \text{cm}^2/\text{s}$. This strong anisotropy is apparently imposed by the disparity of the vertical and horizontal scales.

The generation of turbulence in the ocean is usually related to the local Richardson number (Turner, 1973),

$$Ri = -\frac{g}{\rho}\frac{\partial \rho}{\partial z} \bigg/ \left[(\frac{\partial u}{\partial z})^2 + (\frac{\partial v}{\partial z})^2\right] \tag{1.132}$$

The transition from the laminar motion to the turbulent one occurs at Ri=1/4. If $Ri > 1/4$ the laminar motion will dominate if on the other hand $Ri < 1/4$ the instabilities in fluid due to the velocity shear and density stratification set forth the turbulent motion. This limit is not very strict for the geophysical turbulence which is a superposition of the various scales of motion and is generated through the various sources of instability. In the upper portion of the ocean due to wind forcing and low stratification turbulence is almost always present. A second turbulent layer appears in close proximity to the bottom in the region of strong velocity shear.

The dependence of turbulence on the Richardson number leads to the assumption that the vertical eddy coefficients should be a function of the same number.

Munk and Anderson (1948) analyzed the available distributions of the current and recommended the following expressions:

$$N_z = N_{z,o}(1 + 10Ri)^{-0.5} \tag{1.133}$$

$$D_z = D_{z,o}(1. + 3.33)^{-3/2} \tag{1.134}$$

The magnitudes $N_{z,o}$ and $D_{z,o}$ are the eddy viscosity and eddy diffusivity coefficients defined for the homogeneous fluid ($\frac{\partial \rho}{\partial z} = 0$).

In order to alleviate deficiencies present in the above formulas, Mamayev (1958) proposed to use an exponential dependence on the Richardson number,

$$N_z = N_{z,o}e^{-1.5Ri} \tag{1.135}$$

$$D_z = D_{z,o}e^{-3Ri} \tag{1.136}$$

The actual values of the exponents were recommended by Leendertse and Liu (1975). They have also conducted an extensive comparison of the various expressions for the eddy coefficients.

Although the above formulas describe quite well the turbulent processes generated by the wind at the ocean surface, they are unsatisfactory for bottom turbulence and for the time dependent behavior. To characterize the near-bottom (or near-wall) turbulence, Prandtl introduced the concept of the turbulence scale, also called, length scale or mixing length. The mixing length (l) is assumed to grow linearly with the distance from the bottom

$$l = \kappa(z + z_o + H) \tag{1.137}$$

and at the bottom ($z = -H$),

$$l = \kappa z_o \tag{1.138}$$

where z_o is roughness length, $\kappa = 0.4$ is von Karman's constant.

Along with the mixing length the following formula to express eddy viscosity has also been proposed:

$$N_z = l^2\left(\left(\frac{\partial u}{\partial z}\right)^2 + \left(\frac{\partial v}{\partial z}\right)^2\right)^{1/2} \tag{1.139}$$

This formula, of course, leads to the well-known logarithmic velocity distribution in the boundary layer. Consider near-bottom flow along the x direction only, than from eq.(1.127),

$$\tau_z^b = \rho N_z \frac{\partial u}{\partial z}$$

Inserting eq.(1.139) into this expression, the relation between stress and mixing length is established as

$$\tau_z^b = \rho l^2 (\frac{\partial u}{\partial z})^2 \tag{1.140}$$

According to numerous observations, the stress in the bottom boundary layer is close to be constant. Therefore, starting from eq.(1.140), introducing length scale from eq.(1.137), and integrating along z direction, the logarithmic distribution of velocity follows:

$$u = \frac{1}{\kappa} \sqrt{\tau_z^b/\rho} \ln(z + z_o + H) \tag{1.141}$$

The basic tool which is applied at the present time to derive the vertical eddy coefficients is an equation for the kinetic turbulent energy (E), see Philips (1966), Monin and Yaglom (1971). This is eq.(1.27) taken in the boundary layer of the ocean under the assumption that the basic parameters. i.e., velocity and density are functions of time and the vertical coordinate (z),

$$\frac{\partial E}{\partial t} + u \frac{\partial E}{\partial x} + v \frac{\partial E}{\partial y} + w \frac{\partial E}{\partial z}$$

$$= N_z [(\frac{\partial u}{\partial z})^2 + (\frac{\partial v}{\partial z})^2] + \frac{\partial}{\partial z}(N_z \frac{\partial E}{\partial z}) - \frac{g}{\rho} D_z \frac{\partial \rho}{\partial z} - \epsilon \tag{1.142}$$

The energy equation has been constructed from a general equation of transport for the quantity E,

$$\frac{DE}{Dt} = \text{Production}(E) - \text{Dissipation}(E) + \text{Diffusion}(E) \tag{1.143}$$

Here $E = \overline{u_i'^2}/2$ — see eq.(1.27). The relationship among the unknown eddy viscosity, eddy diffusivity and dissipation (ϵ) follows from the Kolmogorov's hypothesis of similarity
(Monin and Yaglom, 1971).

$$N_z = c^{1/4} l E^{1/2}; \qquad \epsilon = c^{3/4} E^{3/2}/l = N_z^3/l^4; \qquad D_z/N_z = \sigma \tag{1.144}$$

Constant $c \approx 0.08$, was arrived at through matching the Prandtl model and results from the energy equation. The nondimensional number σ ranges from 0.1 to 0.5. The details are analyzed by Launder and Spalding (1972).

The energy equation (1.142) includes Prandtl's and Karman's formulas. Consider eq.(1.142) for the steady state without nonlinear and diffusion terms

$$N_z [(\frac{\partial u}{\partial z})^2 + (\frac{\partial v}{\partial z})^2] - \frac{g}{\rho} D_z \frac{\partial \rho}{\partial z} - \epsilon = 0 \tag{1.145}$$

Introducing dissipation term ϵ from eq.(1.144) we arrive at

$$N_z[(\frac{\partial u}{\partial z})^2 + (\frac{\partial v}{\partial z})^2] - \frac{g}{\rho}D_z\frac{\partial \rho}{\partial z} - N_z^3/l^4 = 0 \qquad (1.145a)$$

Taking into account definition of the Richardson number (1.132) and definition of σ from eq. (1.144), an expression for the eddy viscosity is derived,

$$N_z = l^2\left[(\frac{\partial u}{\partial z})^2 + (\frac{\partial v}{\partial z})^2\right]^{1/2}\sqrt{1 + \sigma\,Ri} \qquad (1.146)$$

This expression defines eddy viscosity as a function of the length scale and Richardson number. The old formula for the eddy viscosity (1.139) follows for the nonstratified fluid ($Ri = 0$).

Since the mixing length enters practically into every definition of the turbulent properties let us study its behavior starting from the Prandtl expression (1.137). According to this formula the mixing length grows linearly without any limits. But since the turbulence is decaying well above the bottom the mixing length should reach some asymptotic value (l_o). Blackadar (1962) through comparison with observed wind profiles, introduced (l_o) into the definition of the length scale, as

$$l = \frac{\kappa z}{(1 + \kappa z/l_o)} \qquad (1.147)$$

This result when applied to the bottom layer gives

$$l = \frac{\kappa(z + H)}{1 + \kappa(z + H)/l_o} \qquad (1.148)$$

A similar approach can be used to the turbulence generated at the sea surface and the mixing length will be

$$l = -\frac{\kappa z}{(1 - \kappa z/l_o)} \qquad (1.149)$$

The asymptotic value of the length scale was determined from the turbulent energy distribution by Mellor and Yamada (1974),

$$l_o = \alpha \int_{-H}^{z+h} zE\,dz \Big/ \int_{-H}^{z+h} E\,dz \qquad (1.150)$$

In eq.(1.150) h is the level above the bottom where the bottom generated turbulence is completely suppressed. A similar expression can be easily constructed for the surface boundary turbulent layer. The value of α was derived for the bottom and surface turbulence in the range from 0.1 to 0.3.

In the Russian literature the closure of the system is often done by assuming the length scale to be a function of the energy distribution (Marchuk and Kagan, 1984)

$$l = \kappa \frac{\sqrt{E}}{l} \left(\frac{\partial}{\partial z} \frac{\sqrt{E}}{l} \right)^{-1} \tag{1.151}$$

In shallow water with a depth of the order of 100 m or less, Leendertse and Liu (1975) recommended the length scale as

$$l = -\kappa z (1 + \kappa z/H) \tag{1.152}$$

To obtain a solution to the equation of the turbulent energy the relevant boundary condition should be added. At the bottom, the flux of turbulent energy across the bed is zero:

$$\frac{\partial E}{\partial z} = 0 \qquad \text{at } z = -H \tag{1.153}$$

At the surface the turbulent energy is generated by the wind stress, and a simplified equation is deduced from (1.145) as

$$0 = N_z \left[\left(\frac{\partial u}{\partial z} \right)^2 + \left(\frac{\partial v}{\partial z} \right)^2 \right] - \epsilon$$

Multiplying both sides by the eddy viscosity given in eq.(1.144), we arrive at

$$E = c^{-1/2} \sqrt{(\tau_x^s)^2 + (\tau_y^s)^2} \tag{1.154}$$

More complicated models and relations derived from the various models were summarized by Mellor and Yamada (1982).

New methods of defining eddy viscosity will be developed through the data assimilation and solution of the inverse problem. A simple and very useful inverse method was constructed by Felzenbaum (Kielmann and Kowalik, 1980) some 30 years ago now. Steady state linear Ekman problem described in Sec.3.1 with assumptions of the constant eddy coefficient and flat bottom domain was solved analytically. At the next step by introducing additional information from the measurements an inverse problem was solved to define eddy viscosity from the given analytical solution. In this case additional information is an empirical coefficient relating the current velocity at the surface and the wind velocity. This coefficient, called β, usually ranges from 0.015 to 0.030. The eddy viscosity coefficient through this inverse approach is

$$N_z = \gamma W H/(4\beta) \qquad \text{for the shallow sea}$$

$$N_z = (\gamma/\beta)^2 W^2/f \qquad \text{for the deep sea} \tag{1.155}$$

The depth which differentiates between a shallow and deep sea is defined as

$$H_d = 4(\gamma/\beta)W/f \qquad (1.156)$$

In the above formulas: $\gamma = C_{10}\rho_a$; W is the wind speed; f denotes Coriolis parameter and H the total depth.

These are useful formulas giving a quick method to estimate the eddy viscosity magnitude. From eq.(1.155) we conclude that in the shallow water domain the eddy viscosity is a function of the wind and depth; in the deep water domain it is a function of the wind and Coriolis parameter only.

4.3 Large scale motion — horizontal exchange

At the present time, the role of horizontal eddy viscosity in the exchange of momentum is not well understood. Constant eddy viscosity coefficient or various exchange models are used, although the validity of given type of representation is very difficult to establish from measurements. A general approach results from the procedure of calculating an average large scale motion through the Reynolds equation (1.15) while small scale dissipation (subgrid dissipation) is described by the Reynolds stresses and usually is parametrized by the eddy viscosity coefficients. Turbulent energy cascading from the large scale to the numerical grid scale is dissipated by the eddy viscosity term and often the same term serves to preserve numerical stability. Frequently, it is difficult to differentiate between the physical and numerical role of the horizontal eddy viscosity coefficient.

In the averaged equations the horizontal friction (eq.(1.35), (1.36)) is expressed by a constant coefficient and the horizontal Laplace operator

$$N_h\Delta\chi = N_h\left(\frac{\partial^2\chi}{\partial x^2} + \frac{\partial^2\chi}{\partial y^2}\right) \qquad (1.157)$$

Presently, this term is often replaced by the biharmonic operator (O'Brien, 1989)

$$-B\Delta(\Delta\chi) = -B\left[\frac{\partial^2}{\partial x^2}(\Delta\chi) + \frac{\partial^2}{\partial y^2}(\Delta\chi)\right] \qquad (1.158)$$

Here B is a biharmonic friction coefficient. Its value, for the scale of about 100 km equals -8×10^{19} (CGS). This friction term is used to better resolve eddies in the numerical models. Biharmonic friction has the advantage over the Laplacian friction as it dissipates small scales more quickly than the large scale motions so that the frictional effects on the mesoscale eddies and large scale circulation can be lower than in Laplacian term. For this reason, it is called scale-selective friction. Here we are arrive at the essence of the large scale-subgrid scale exchange scheme. Since geophysical turbulence is in the domain of vigorous investigations at present, new measurements will influence representation of the energy flow in the numerical models. Recent measurements show that mesoscale eddies are always present in the ocean and they interact with the large scale motion. The large scale oceanic models only partly resolve the mesoscale eddies. Some eddies, not resolved by the models, will be located in the subgid motion.

We shall consider briefly what kind of parametrization should be introduced into the set of equations of motion to take into account mesoscale eddies. Importance of this effect is evident from interaction of the mesoscale eddies with the bottom topography (Holloway, 1987), resulting in an additional bottom stress. Such stress has strong effect on the large scale ocean circulation.

Investigation of the eddy properties will be performed through the vorticity equation in which the mesoscale eddies can be identified through the application of proper filtering technique. For the derivation of a vorticity equation consider eq.(1.1) in the form,

$$\frac{\partial u_i}{\partial t} + u_j \frac{\partial u_i}{\partial x_j} + 2\epsilon_{ijk}\Omega_j u_k = -\frac{1}{\rho}\frac{\partial p}{\partial x_i} + F_i \qquad (1.159)$$

Here

$$F_i = -g\delta_{3i} + \frac{1}{\rho}\frac{\partial \sigma_{ij}}{\partial x_j}$$

Application of the *curl* operation to eq.(1.159) will give the vorticity equation. In the tensor notation the i component of the vorticity vector is defined as

$$\omega_i = \epsilon_{ijk}\frac{\partial u_k}{\partial x_j} \qquad (1.160)$$

Thus for example the component of vorticity along x_3 (z) axis is

$$\omega_3 = \epsilon_{321}\frac{\partial u_1}{\partial x_2} + \epsilon_{312}\frac{\partial u_2}{\partial x_1} = -\frac{\partial u_1}{\partial x_2} + \frac{\partial u_2}{\partial x_1} = \frac{\partial v}{\partial x} - \frac{\partial u}{\partial y}$$

For the sign of the antisymmetric tensor ϵ reader should refer to Sec.1 of this chapter. Applying *curl* operator $\epsilon_{ijk}\dfrac{\partial}{\partial x_j}$ to eq.(1.159) results in the following vorticity equation,

$$\frac{\partial \omega_m}{\partial t} + \frac{\partial}{\partial x_j}(\omega_m u_j - \omega_j u_m) + 2\frac{\partial}{\partial x_j}(\Omega_m u_j - \Omega_j u_m) = \epsilon_{mji}\frac{1}{\rho^2}\frac{\partial p}{\partial x_i}\frac{\partial \rho}{\partial x_j} + \epsilon_{mji}\frac{\partial F_i}{\partial x_j}$$

$$(1.161)$$

Besides a friction term, which is buried in F_i, this equation incorporates a baroclinic source term $\epsilon_{mji}\dfrac{1}{\rho^2}\dfrac{\partial p}{\partial x_i}\dfrac{\partial \rho}{\partial x_j}$. Mesoscale eddies are somewhat similar to the large vortex tubes oriented along the x_3 (z) axis and therefore we are only seeking the time history of ω_3 component, from eq.(1.161),

$$\frac{\partial \omega_3}{\partial t} + \frac{\partial}{\partial x_j}(\omega_3 u_j - \omega_j u_3) - 2\Omega_3 \frac{\partial u_3}{\partial x_3} + \beta u_2 = W \qquad (1.162)$$

Here

$$W = \epsilon_{3ji} \frac{1}{\rho^2} \frac{\partial p}{\partial x_i} \frac{\partial \rho}{\partial x_j} + \epsilon_{3ji} \frac{\partial F_i}{\partial x_j} \quad \text{and} \quad \beta = 2\frac{\partial \Omega_3}{\partial x_2}$$

To understand the above result it is useful to remember that the vector of the earth's angular velocity $\vec{\Omega} = (0, 0, \Omega_3)$.

Considering now an eddy – "turbulent" motion, one can proceed from here in the customary manner, i.e., by introducing into eq.(1.161) an average and changes around the average. In deriving the equations for the Reynolds stresses (1.14), (1.15) time averaging is used. In recent studies, volume and surface averages have been used with the space filters which separates the scale of interest (Nikolayevski, 1970; Leonard, 1974). The result of application of such filter as shown by Leonard (1974) are exchange processes defined by Reynolds stresses supplemented by an additional term which depends on the averaging scale (Reynolds, 1976). The development of this concept to the interaction of mesoscale eddies with the large scale oceanic motion was done by Ivchenko (1977) and Voronov and Ivchenko (1978). Circumventing all the details of the space averaging let us tackle in a straightforward manner eq.(1.162). Denoting averaging procedure by a bar, the equation for the average vorticity is

$$\frac{\partial \overline{\omega}_3}{\partial t} + \frac{\partial}{\partial x_j}\overline{(\omega_3 u_j - \omega_j u_3)} - 2\Omega_3 \frac{\partial \overline{u}_3}{\partial x_3} + \beta \overline{u}_2 = \overline{W} \qquad (1.163)$$

Here a Pandora's box is open with the new eddy viscosity coefficients to define turbulent exchange of the vorticity. The nonlinear term at the left-hand side of eq.(1.163) when expressed by average and fluctuating vorticity will introduce new terms like $\overline{\omega_3' u_1'}$ or $\overline{\omega_3' u_3'}$. They can be transformed by the same method which served to derive eddy viscosity (1.22). In analogy to molecular exchange by assuming proportionality to mean vorticity field,

$$\overline{\omega_3' u_1'} = N_h^\omega \frac{\partial \overline{\omega}_3}{\partial x_1} \quad \text{or} \quad \overline{\omega_3' u_3'} = N_z^\omega \frac{\partial \overline{\omega}_3}{\partial x_3} \qquad (1.164)$$

Including these expressions into eq.(1.163) we arrive at

$$\frac{\partial \overline{\omega}_3}{\partial t} + \frac{\partial}{\partial x_1}(\overline{\omega}_3 u_1) + \frac{\partial}{\partial x_2}(\overline{\omega}_3 u_2) + u_3 \frac{\partial \overline{\omega}_3}{\partial x_3} - N_h^\omega \Delta\omega_3 - N_z^\omega \frac{\partial^2 \overline{\omega}_3}{\partial x_3^2} - 2\Omega_3 \frac{\partial \overline{u}_3}{\partial x_3} + \beta \overline{u}_2 = \overline{W}$$
$$(1.165)$$

Here $\Delta = \dfrac{\partial^2}{\partial x_1^2} + \dfrac{\partial^2}{\partial x_2^2}$. Averaging operation in the given volume results in an average vorticity or an average angular velocity of this volume

$$\overline{\omega}_3 = \epsilon_{3kj}\overline{\dfrac{\partial u_j}{\partial x_k}} = \alpha_3 + \chi_3 \quad \text{and} \quad \alpha_3 = \epsilon_{3kj}\dfrac{\partial \bar{u}_j}{\partial x_k} \tag{1.166}$$

Here χ_3 is an average vorticity of the mesoscale velocity. It describes explicitly angular velocity of the mesoscale eddies. The average over small scale motion is equal to zero. Whenever the mesoscale eddies are not present in the averaging volume, $\chi_3 = 0$. Now we need to bring into the scope an averaged equation of motion. For this purpose consider eq.(1.15)

$$\dfrac{\partial \bar{\rho}\bar{u}_i}{\partial t} + \dfrac{\partial \bar{\rho}\bar{u}_j\bar{u}_i}{\partial x_j} - \dfrac{\partial R_{ij}}{\partial x_j} + 2\epsilon_{ijk}\Omega_j\bar{\rho}\bar{u}_k = -\dfrac{\partial \bar{p}}{\partial x_i} - g\bar{\rho}\delta_{3i} + \mu\dfrac{\partial^2 \bar{u}_i}{\partial x_j\partial x_j} \tag{1.167}$$

In eq.(1.15) tensor R_{ij} is symmetrical, meaning $R_{ij} = R_{ji}$, this property does not hold in case of the volume averaging. An addition to the Reynolds stress for the mesoscale–large scale interactions is asymmetrical. Because this asymmetry is brought about by mesoscale vorticity one can assume

$$R_{ij} - R_{ji} = 2\gamma\epsilon_{ij3}\chi_3 \tag{1.168}$$

The symmetrical part of the tensor is defined in the old fashion by eq.(1.22)

$$R_{ij} + R_{ji} = 2\bar{\rho}N(j)\dfrac{\partial \bar{u}_i}{\partial x_j} + \bar{\rho}N(i)\dfrac{\partial \bar{u}_j}{\partial x_i} \tag{1.169}$$

Combining eqs.(1.168) and (1.169) the tensor R_{ij} is defined and it can be introduced into eq.(1.167). We shall write here equations of motion along the x and y directions and to simplify the notation, the nonlinear terms and molecular friction are omitted, thus

$$\dfrac{\partial u}{\partial t} - 2\Omega v = -\dfrac{1}{\rho}\dfrac{\partial p}{\partial x} + \gamma\dfrac{\partial \chi_3}{\partial y} + N_z\dfrac{\partial^2 u}{\partial z^2} + N_h\Delta u \tag{1.170}$$

$$\dfrac{\partial v}{\partial t} + 2\Omega u = -\dfrac{1}{\rho}\dfrac{\partial p}{\partial y} - \gamma\dfrac{\partial \chi_3}{\partial x} + N_z\dfrac{\partial^2 v}{\partial z^2} + N_h\Delta v \tag{1.171}$$

New (and asymmetrical) terms due to the mesoscale vorticity describe interaction of the mesoscale and large scale domain. The mesoscale vorticity is defined through solution of eq.(1.165).

CHAPTER II

TRANSPORT EQUATIONS

1. Mathematical rudiments

1.1 Taylor's theorem application

To solve numerically any partial differential equation a continuous space–time domain has to be replaced by a discrete grid domain. We shall apply Taylor's theorem to define dependent variables and their derivatives in the discrete domain. Suppose function $c(x)$ and its derivatives are finite and continuous; than if $c(x)$ is given at the point x, it can be defined at the point $x + h$ by the series

$$c(x + h) = c(x) + h\frac{\partial c}{\partial x} + \frac{h^2}{2}\frac{\partial^2 c}{\partial x^2} + \frac{h^3}{3!}\frac{\partial^3 c}{\partial x^3} + O(h^4) \qquad (2.1)$$

and at the point $x - h$ as

$$c(x - h) = c(x) - h\frac{\partial c}{\partial x} + \frac{h^2}{2}\frac{\partial^2 c}{\partial x^2} - \frac{h^3}{3!}\frac{\partial^3 c}{\partial x^3} + O(h^4) \qquad (2.2)$$

From (2.1) the first derivative of $c(x)$ at the point x is

$$\frac{\partial c}{\partial x} = \frac{c(x + h) - c(x)}{h} + O(h) \qquad (2.3)$$

and from (2.2),

$$\frac{\partial c}{\partial x} = \frac{c(x) - c(x - h)}{h} + O(h) \qquad (2.4)$$

Derivatives (2.3) and (2.4) are called forward and backward derivatives, respectively. Subtraction of the two series gives

$$c(x + h) - c(x - h) = 2h\frac{\partial c}{\partial x} + \frac{2h^3}{3!}\frac{\partial^3 c}{\partial x^3} + O(h^5) \qquad (2.5)$$

and thus the first derivative is

$$\frac{\partial c}{\partial x} = \frac{c(x + h) - c(x - h)}{2h} + O(h^2) \qquad (2.6)$$

This numerical form is called the central derivative. All three expressions not only differ in the form but also in the approximation or in the order of accuracy. Each derivative is expressed by the difference form: thus $\dfrac{\partial c}{\partial x}$ is the limit of the difference ratio as $h \to 0$, the higher order of accuracy is achieved through the central difference form because the error of approximation in (2.6) is proportional to h^2.

The difference formulae for the second derivative can be introduced by adding series (2.1) and (2.2),

$$c(x+h) + c(x-h) = 2c(x) + h^2\frac{\partial^2 c}{\partial x^2} + O(h^4) \qquad (2.7)$$

and from (2.7),

$$\frac{\partial^2 c}{\partial x^2} = \frac{c(x+h) + c(x-h) - 2c(x)}{h^2} + O(h^2) \qquad (2.8)$$

One possible approach to increase the order of accuracy in (2.3), (2.4) and (2.6) is to construct the difference formula based not on two or three grid points but on a larger number of grid points (Fox, 1962; Mitchell and Griffiths, 1980). Let us take five points $x - 2h, x - h, x, x + h, x + 2h$ along the x axis and consider expression (2.3) which has an approximation error of the order of h,

$$\frac{\partial c}{\partial x} = \frac{c(h+x) - c(x)}{h} - \frac{h}{2}\frac{\partial^2 c}{\partial x^2} + O(h^2) \qquad (2.9)$$

If the second derivative can be inserted explicitly into (2.9) the order of approximation of $\dfrac{\partial c}{\partial x}$ will increase to $O(h^2)$. We shall construct the second derivative on the lattice $x, x + h, x + 2h$ with $x + h$ as a central point:

$$\frac{\partial^2 c}{\partial x^2} = \frac{c(x+2h) + c(x) - 2c(x+h)}{h^2} + O(h^2) \qquad (2.10)$$

and introducing (2.10) into (2.9) the finite difference expression for the first derivative with the second order of approximation is derived,

$$\frac{\partial c}{\partial x} = \frac{-3c(x) + 4c(x+h) - c(x+2h)}{2h} + O(h^2) \qquad (2.11)$$

A similar approach can be taken to increase accuracy of the backward derivative (2.4),

$$\frac{\partial c}{\partial x} = \frac{c(x) - c(x-h)}{h} + \frac{h}{2}\frac{\partial^2 c}{\partial x^2} + O(h^2) \qquad (2.12)$$

with the second derivative formed on the lattice $x - 2h, x - h, x$, this will result in

$$\frac{\partial c}{\partial x} = \frac{3c(x) - 4c(x - h) + c(x - 2h)}{2h} + O(h^2) \qquad (2.13)$$

By combining (2.11) and (2.13) a new expression for the central derivative of order h^3 is obtained:

$$\frac{\partial c}{\partial x} = \frac{1}{2}((2.11) + (2.13))$$

$$= \frac{2(c(x + h) - c(x - h))}{2h} - \frac{c(x + 2h) - c(x - 2h)}{4h} + O(h^3) \qquad (2.14)$$

In (2.14) the first derivative is defined on the lattice with mesh size h and afterwards a "correction" is introduced on the lattice with double mesh size $2h$. An interesting corollary is that the accuracy of the numerical formulas (or numerical solutions) can be improved by combining the results from the different lattices.

In the near-surface layer of the ocean, temperature or current often exhibit a large gradient. To resolve properly such local variations a lattice with a variable space grid is used. The above expressions for the derivatives will require some change. For this purpose let us consider the grid distribution depicted in Fig.2.1.

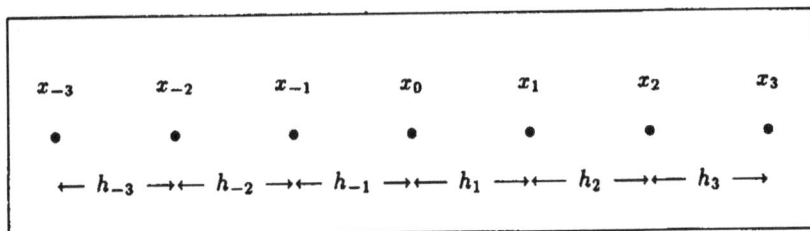

Fig. 2.1

Irregular grid: h denotes space step, x is coordinate.

The basic series expressing the values of the function $c(x)$ at the points $x_1 = x_0 + h_1$ and $x_{-1} = x_0 - h_{-1}$ correspond to (2.1) and (2.2):

$$c(x_0 + h_1) = c(x_0) + h_1 \frac{\partial c}{\partial x} + \frac{h_1^2}{2} \frac{\partial^2 c}{\partial x^2} + \frac{h_1^3}{3!} \frac{\partial^3 c}{\partial x^3} + O(h^4) \qquad (2.15)$$

$$c(x_0 - h_{-1}) = c(x_0) - h_{-1} \frac{\partial c}{\partial x} + \frac{h_{-1}^2}{2} \frac{\partial^2 c}{\partial x^2} - \frac{h_{-1}^3}{3!} \frac{\partial^3 c}{\partial x^3} + O(h^4) \qquad (2.16)$$

The construction method for the backward and forward derivatives from the above series is similar to (2.3) and (2.4), for example,

$$\frac{\partial c}{\partial x} = \frac{c(x + h_1) - c(x)}{h_1} + O(h_1) \qquad (2.15a)$$

The central derivative is obtained by subtracting (2.16) from (2.15):

$$\frac{\partial c}{\partial x} = \frac{c(x_0 + h_1) - c(x_0 - h_{-1})}{h_1 + h_{-1}} - \frac{(h_1 - h_{-1})}{2}\frac{\partial^2 c}{\partial x^2} - \frac{h_1^3 + h_{-1}^3}{3!(h_1 + h_{-1})}\frac{\partial^3 c}{\partial x^3} \quad (2.17)$$

Only if $(h_1 - h_{-1}) \to 0$ (i.e. both space steps are equal) the second order of accuracy is achieved in (2.17), but generally this derivative is approximated with the first order of accuracy. Unequal mesh size tends to diminish the accuracy of approximation. There is also another possible way to construct the central derivative. For this, eq.(2.15) is divided by h_1 and eq.(2.16) by h_{-1} and afterwards the resulting equations are subtracted on both sides as

$$\frac{c(x_0 + h_1)}{h_1} - \frac{c(x_0 - h_{-1})}{h_{-1}} = \frac{c(x_0)}{h_1} - \frac{c(x_0)}{h_{-1}} + 2\frac{\partial c}{\partial x} + \frac{\partial^2 c}{\partial x^2}\frac{(h_1 - h_{-1})}{2} + \frac{\partial^3 c}{\partial x^3}\frac{(h_1^2 + h_{-1}^2)}{3!} \quad (2.18)$$

From (2.18),

$$\frac{\partial c}{\partial x} = \frac{1}{2}\Big(\frac{c(x_0 + h_1) - c(x_0)}{h_1} + \frac{c(x_0) - c(x_0 - h_{-1})}{h_{-1}}\Big) - \frac{h_1 - h_{-1}}{4}\frac{\partial^2 c}{\partial x^2} - \frac{\partial^3 c}{\partial x^3}\frac{h_1^2 + h_{-1}^2}{2\,3!} \quad (2.19)$$

Expression (2.19) is a somewhat better approximation than (2.17) since an approximation error has been diminished by factor of 2.

The second derivative on the variable lattice is constructed from (2.15) and (2.16) by adding these expressions:

$$\frac{\partial^2 c}{\partial x^2} = \Big(\frac{c(x_0 + h_1) - c(x_0)}{h_1} + \frac{c(x_0 - h_{-1}) - c(x_0)}{h_{-1}}\Big)/0.5(h_1 + h_{-1}) - \frac{h_1 - h_{-1}}{3(h_1 + h_{-1})}\frac{\partial^3 c}{\partial x^3} \quad (2.20)$$

To improve the accuracy of the first derivative the approach previously delineated can be employed again. Considering the series (2.15) and numerical lattice given in Fig.2.1, we shall write the first derivative as,

$$\frac{\partial c}{\partial x} = \frac{c(x_0 + h_1) - c(x_0)}{h_1} - \frac{h_1}{2}\frac{\partial^2 c}{\partial x^2} + O(h^2) \quad (2.21)$$

The second derivative in (2.21) is taken as

$$\frac{\partial^2 c}{\partial x^2} = \Big(\frac{c(x_2) - c(x_1)}{h_2} + \frac{c(x_0) - c(x_1)}{h_1}\Big)/0.5(h_1 + h_2) \quad (2.22)$$

with the central grid point located at x_1. Introducing (2.22) into (2.21) the forward difference formula for the first derivative with the second order of accuracy on the variable mesh is obtained:

$$\frac{\partial c}{\partial x} = \frac{c(x_1) - c(x_0)}{h_1}\Big(1 + \frac{h_1}{h_1 + h_2}\Big) - \frac{c(x_2) - c(x_1)}{h_2}\frac{h_1}{h_1 + h_2} \quad (2.23)$$

Employing series (2.16) and constructing a second derivative around point x_{-1}, the backward numerical expression for the first derivative is obtained,

$$\frac{\partial c}{\partial x} = \frac{c(x_0) - c(x_{-1})}{h_{-1}}\left(1 + \frac{h_{-1}}{h_{-1} + h_{-2}}\right) + \frac{c(x_{-2}) - c(x_{-1})}{h_{-2}} \frac{h_{-1}}{h_{-1} + h_{-2}} \qquad (2.24)$$

It is important to remember that unequal mesh size will diminish the accuracy. Even a symmetrical expression for the second derivative (2.20) has an error proportional to the first power of the space step! A good way to compare the uniform and nonuniform grids is to subtract (2.3) from (2.15a). The difference is

$$Diff = \frac{h_1 - h}{2}\frac{\partial^2 c}{\partial x^2} + \dots + \frac{h_1^{n-1} - h^{n-1}}{(n-1)!}\frac{\partial^{n-1} c}{\partial x^{n-1}}$$

and it results from selecting the nonuniform grid. If $h = h_1$, than $Diff = 0$, and thus a uniform grid minimizes the error. The nonuniform grid is often a source of the approximation errors.

1.2 Numerical notation and numerical grid

The first step towards the construction of the numerical algorithms is to replace a continuous domain by the discrete domain. The points of the numerical lattice are defined through the coordinates in the following manner:

$$x_j = jh, j = 1, 2, 3, \dots$$

$$y_k = kh, k = 1, 2, 3, \dots$$

$$z_l = lh, l = 1, 2, 3, \dots$$

$$t_m = mT, m = 1, 2, 3, \dots \qquad (2.25)$$

Two parameters of the numerical lattice are space (h) and time (T) steps. Both are constant throughout the whole domain of integration. Previously we have dealt with a continuous function $c(x)$ at the discrete points x_j, in the ensuing considerations a discrete function $c_{j,k,l}^m$ will be defined on the discrete space. A simple lattice in the $x - t$ plane is shown in Fig.2.2 to explain the basic notation.

The coordinates of the point P in $x - t$ plane are jh and mT but to shorten the notation we shall write these coordinates as j and m.

A similar short notation is adopted for the derivatives, for example (2.3) takes the form

$$\frac{\partial c}{\partial x} = \frac{c_{j+1} - c_j}{h} \qquad (2.26)$$

and (2.8) is

$$\frac{\partial^2 c}{\partial x^2} = \frac{c_{j+1} + c_{j-1} - 2c_j}{h^2} \qquad (2.27)$$

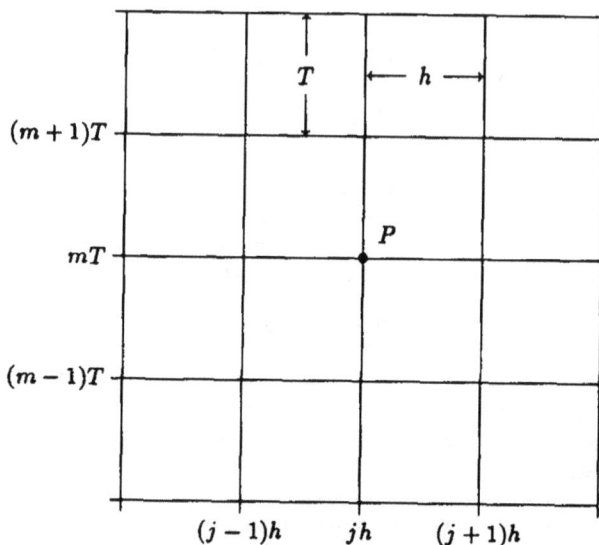

Fig. 2.2
Numerical grid in the $x - t$ plane.

For the nonuniform grid the space step is different along the x, y, z directions and advancing in time is also done through the variable time step. The set of coordinates is defined as

$$x_J = \sum_{j=0}^{J} h_{x,j}; \quad J = 1, 2, 3, \ldots$$

$$y_K = \sum_{k=0}^{K} h_{y,k}; \quad K = 1, 2, 3, \ldots$$

$$z_L = \sum_{l=0}^{L} h_{z,l}; \quad L = 1, 2, 3, \ldots$$

$$t_M = \sum_{m=0}^{M} T_m; \quad M = 1, 2, 3, \ldots \tag{2.28}$$

With the above notation, the central derivative (2.19) becomes

$$\frac{\partial c}{\partial x} = \frac{1}{2}\left(\frac{c_{j+1} - c_j}{h_{x,j+1}} + \frac{c_j - c_{j-1}}{h_{x,j}}\right) \tag{2.29}$$

and the forward derivative (2.23) is

$$\frac{\partial c}{\partial x} = \frac{c_{j+1} - c_j}{h_{x,j+1}}(1 + \frac{h_{x,j+1}}{h_{x,j+1} + h_{x,j+2}}) - \frac{c_{j+2} - c_{j+1}}{h_{x,j+2}} \frac{h_{x,j+1}}{h_{x,j+1} + h_{x,j+2}} \tag{2.30}$$

2. Boundary and initial conditions

The diffusion equation

$$\frac{Dc}{Dt} = A_2 c \tag{2.31}$$

has been introduced in Ch.I, both in Cartesian (eq.(1.43), (1.44)) and spherical (eq.(1.54), (1.55)) system of coordinates. Assuming that the velocity distribution is given in the domain Ω and along boundary Γ, one may ask what are the necessary boundary and initial conditions for the concentration to obtain a unique solution to the diffusion equation. Generally, since eq.(2.31) is a second order partial differential equation of the parabolic type, one initial and two boundary conditions (for each coordinate axes) of the type

$$\alpha \frac{\partial c}{\partial n} + \beta c = \gamma \tag{2.32}$$

ought to be specified. In eq.(2.32) α, β, γ are known functions, and $\frac{\partial}{\partial n}$ denotes the derivative along the normal direction to the boundary Γ.

Let us consider (2.31) in the Cartesian system of coordinates, along the x axis only

$$\frac{\partial c}{\partial t} + u \frac{\partial c}{\partial x} = \frac{\partial}{\partial x} D_h \frac{\partial c}{\partial x} \tag{2.33a}$$

or

$$\frac{\partial c}{\partial t} = \frac{\partial}{\partial x}(D_h \frac{\partial c}{\partial x} - uc) \tag{2.33b}$$

The term in parenthesis (on the right-hand side of eq.(2.33b)) represents flux of the substance through a surface perpendicular to the x axis. Boundary condition (2.32) can be rewritten for this special case as

$$D_h \frac{\partial c}{\partial x} - uc = \gamma \tag{2.34}$$

If the diffusive flux in (2.34) is negligible, the boundary condition simplifies to

$$-uc = \gamma \tag{2.35}$$

Concentration c should remain positive in the domain Ω, this may require negative velocity at the boundary Γ or the flux should be directed into the domain of integration.

Generally speaking two types of the boundary conditions are encountered in ocean models. In the first case, when the entire boundary consists of the shoreline, the normal velocities and normal transports at the boundary are equal to zero. In the second case, when the boundary is a division line between two water bodies, the concentration (or fluxes) and velocities specified at the boundary should lead to a unique solution. This is a difficult problem, because often we are unable to specify a set of boundary conditions for a unique solution, or if we know how to specify the conditions, the high quality measurements that are needed are difficult to obtain. Finally it ought to be pointed out that the eddy diffusion coefficients and the velocity fields are highly variable and have complicated dependence on the time and space scales (Ozmidov, 1986; Ramming and Kowalik, 1980). This creates a formidable obstacle in studying diffusion through high accuracy numerical formulas.

3. Basic numerical properties

3.1 Simple numerical solution of the advective problem

Let us start with an elementary problem. In the domain $0 \leq x \leq \infty$, $t \geq 0$ we shall search a solution to the equation

$$\frac{\partial c}{\partial t} + u\frac{\partial c}{\partial x} = 0 \tag{2.36}$$

subject to the initial condition

$$c(x,0) = c_j^0 \tag{2.37}$$

Eq.(2.36) describes variations of concentration along the x axis due to advection by velocity u.

There are several possible ways of constructing a difference equation from the differential form (2.36). Time integration of (2.36) over one time step

$$c^{m+1} - c^m = -\int_{mT}^{(m+1)T} (u\frac{\partial c}{\partial x})dt \tag{2.38}$$

can be carried out in several different ways as

$$c^{m+1} - c^m = -(u\frac{\partial c}{\partial x})^m T \tag{2.38a}$$

or

$$c^{m+1} - c^m = -(u\frac{\partial c}{\partial x})^{m+1} T \tag{2.38b}$$

or

$$c^{m+1} - c^m = -\alpha(u\frac{\partial c}{\partial x})^m T - \beta(u\frac{\partial c}{\partial x})^{m+1} T \tag{2.38c}$$

Here, $\alpha \leq 1$, $\beta \leq 1$ and $\alpha + \beta = 1$. The first formula (2.38a) is explicit in time, because the advective term is calculated at the old (m) time step. The second expression is implicit, here both the unknown function (c^{m+1}) and the advective term are taken at the m+1 time step. The last formula combines the explicit and implicit approaches. The coefficients α and β indicate the relative weighting to be applied to the explicit terms and to the implicit terms, respectively.

To construct a numerical counterpart of (2.36) we shall use a simple grid given in Fig.2.2,

$$\frac{c_j^{m+1} - c_j^m}{T} = -u\frac{c_{j+1}^m - c_j^m}{h} \tag{2.39}$$

This is explicit in time, because all the concentrations at the right-hand side of (2.39) are known from the time level m, and the initial values are specified everywhere by (2.37). Does this process actually lead to a solution which is close to the exact (analytical) solution — $C(x,t)$ of the original problem formulated by equation (2.36)? To answer this question we shall investigate the error e_j^m between the exact and numerical solutions,

$$e_j^m = C(m,j) - c_j^m \tag{2.40}$$

Introducing c_j^m from (2.40) into (2.39),

$$\frac{C(m+1,j) - C(m,j)}{T} - \frac{e_j^{m+1} - e_j^m}{T} + u\frac{C(m,j+1) - C(m,j)}{h} - u\frac{e_{j+1}^m - e_j^m}{h} = 0 \tag{2.41}$$

and next developing function C(x,t) with the help of (2.1) and (2.2) into a Taylor's series, we arrive at

$$\frac{C(m+1,j) - C(m,j)}{T} + u\frac{C(m,j+1) - C(m,j)}{h} = \frac{\partial C}{\partial t} + u\frac{\partial C}{\partial x} = O(T,h) \tag{2.42}$$

Through comparison of (2.36), (2.41), and (2.42) the equation for the error is obtained

$$e_j^{m+1} = (\frac{uT}{h} + 1)e_j^m - \frac{Tu}{h}e_{j+1}^m + T \times O(T,h) \tag{2.43}$$

If $0 \leq -Tu/h \leq 1$, the coefficients in (2.43) are positive (velocity is negative !) and less than 1, thus

$$|e_j^{m+1}| \leq (\frac{uT}{h} + 1)|e_j^m| - \frac{Tu}{h}|e_{j+1}^m| + T \times O(T,h) \tag{2.43a}$$

Denoting

$$R_j^m = Max(|e_j^m|, |e_{j+1}^m|)$$

the inequality (2.43a) yields

$$R_j^{m+1} \leq (1 + \frac{uT}{h})R_j^m - \frac{Tu}{h}R_j^m + T \times O(T,h) = R_j^m + T \times O(T,h) \tag{2.44}$$

The error is, therefore, limited and during one time step it will increase by $T \times O(T, h)$. If the integration is extended to $m + 1$ time step then

$$R_j^{m+1} \leq R_j^0 + (m + 1)TO(T, h) \tag{2.45}$$

Initial conditions, both for the analytical $C(x, t)$ and numerical c_j^m solutions, are given by (2.37), thus $R_j^0 = 0$, and if $T \to 0$ and $O(T, h) \to 0$ then $R_j^{m+1} \to 0$. This proves convergence of the numerical solution derived through (2.39) to the analytical solution of the original problem given by (2.36). To prove convergence we have assumed that the coefficients are in the range $0 \leq -Tu/h \leq 1$, otherwise if $-Tu/h \geq 1$, the inequality (2.45) will not converge. The example presented above was rather simple, usually the calculations necessary to prove convergence are difficult to carry out. At the present time it is customary to study the properties of numerical schemes by means of stability.

3.2 Stability

Stability requires that the numerical solution remains bounded as the integration in time takes place. For the linear difference equation the convergence follows from the approximation and stability. Stability of the linear difference equation is usually studied by the Fourier method (Sod, 1985) or by investigating the eigenvalues of the coefficient matrix (Smith, 1978). The latter method, also more rigorous (it investigates stability related to the equations and boundary conditions), because of the cumbersome algebra, has been rarely used.

Let us assume that the error δc is associated with solution $c(x, t)$. The error may be caused by an error of approximation or by a rounding off error due to computer arithmetic. This error can be described as a harmonic wave or Fourier component,

$$\delta c = c^* e^{i\omega t} e^{i\kappa_1 x} e^{i\kappa_2 y} e^{i\kappa_3 z} \tag{2.46}$$

where: c^* is amplitude (usually unit amplitude is used $c^*=1$), ω is frequency of the oscillation, $\kappa_1, \kappa_2, \kappa_3$, are the wave numbers along the x, y, z axes, respectively. Introducing space–time lattice defined by (2.25), the above expression can be rewritten as

$$\delta c = c^* e^{i\omega m T} e^{i\kappa_1 jh} e^{i\kappa_2 kh} e^{i\kappa_3 lh} \tag{2.47}$$

or

$$\delta c = \lambda^m e^{i\kappa_1 jh} e^{i\kappa_2 kh} e^{i\kappa_3 lh} \tag{2.48}$$

where

$$\lambda = c^* e^{i\omega T} \tag{2.49}$$

The coefficient λ, called amplification factor or stability parameter, describes the change of amplitude in time for the one time step increment ($m = 1$). The condition established by von Neumann (Richtmayer and Morton, 1967) requires $|\lambda| \leq 1$ for the stability of a numerical scheme. To determine the stability of a numerical scheme we shall introduce (2.48) into a difference equation and analyze the magnitude of λ. This can be done because the difference equation for an error is of the same form as a difference equation for the dependent function (compare 2.39 and 2.43).

Let us apply this approach to equation (2.39). For the sake of simplicity κ_1 is denoted as κ,

$$\frac{\lambda^{m+1}e^{i\kappa jh} - \lambda^m e^{i\kappa jh}}{T} = -u\frac{\lambda^m e^{i\kappa(j+1)h} - \lambda^m e^{i\kappa jh}}{h} \tag{2.50}$$

This equation can be simplified through division of the both sides by λ^m and $e^{i\kappa jh}$,

$$\frac{\lambda - 1}{T} = -\frac{u}{h}(e^{i\kappa h} - 1) \tag{2.51}$$

Denoting $a = \cos \kappa h$ and $q = -uT/h$ (q is often called the Courant number), the absolute value of the stability parameter takes the following form

$$|\lambda|^2 = 1 + 2q(q-1)(1-a) \leq 1 \tag{2.52}$$

As $-1 \leq a \leq 1$, from (2.52) it follows that $|\lambda| \leq 1$ when

$$-\frac{uT}{h} \leq 1 \tag{2.53}$$

Inequality (2.53) is necessary condition for the stability of the numerical scheme (2.39). Condition (2.53) is exactly the same as the one used to prove convergence of the (2.45); both conditions are valid for the negative velocities only.

Amplification factor is a complex number, therefore it can be represented as a point in the plane by means of polar coordinates (ρ, ϕ), where $\rho = |\lambda|$ is the distance of the point from the coordinate origin, and $\phi = \omega T$ is the angle which the radius vector of the given point makes with the positive direction of the axis of abscissas. Referring to the amplification factor we define the following numerical solutions:

$$|\lambda| = \begin{cases} > 1 & \text{unstable;} \\ = 1 & \text{neutral;} \\ < 1 & \text{damping.} \end{cases}$$

The change of phase of a numerical solution is easily deduced through comparison with the change of phase of the analytical solution. We shall also perform a comparison of the propagation velocity, calculated from an analytical and a numerical solution. In the case of the diffusive process, a propagation velocity is the advective velocity; for the long waves the phase or group velocity is considered. The comparison will be done through a velocity ratio (VR),

$$VR = \text{velocity} \quad \text{ratio} = \frac{\text{numerical} \quad \text{speed}}{\text{analytical} \quad \text{speed}}$$

This relative change in velocity defines a numerical solution as

$$VR = \begin{cases} > 1 & \text{accelerating;} \\ = 1 & \text{nochange;} \\ < 1 & \text{decelerating.} \end{cases}$$

We shall analyze the behavior of the amplification factor in the ensuing sections, now let us notice that in (2.51) and (2.52) the amplification factor has different values for the different wave length (L). The parameter which controls stability $a = \cos \kappa h$ varies when wave number $(\kappa = 2\pi/L)$ is changing. Hence, we conclude that the signal with the different wave length will be differently amplified by the numerical scheme(2.39), or differently reproduced, or some part of the wave spectra will be reproduced with less error. Numerical scheme (2.39) strongly damps the signal in the close proximity to the range of the shortest waves $L = 2h$ (Mesinger and Arakawa, 1976). For example, if $q = -uT/h = 0.5$, and $\kappa = 2\pi/(2h)$, then from (2.52) $|\lambda| = 0$.

4. Explicit versus implicit numerical schemes

We continue with the numerical scheme (2.39), in order to develop similar formula for the positive transport velocities and to generalize this formula for both the negative and positive velocities. In (2.39) the space derivative is taken at time level m, therefore, this scheme is uncentered in time (the centered scheme will be defined at time level $m + 1/2$). This scheme incorporates two space points, let us now consider three points along the x direction, i.e., points $j - 1, j$ and $j + 1$ (see Fig.2.3).

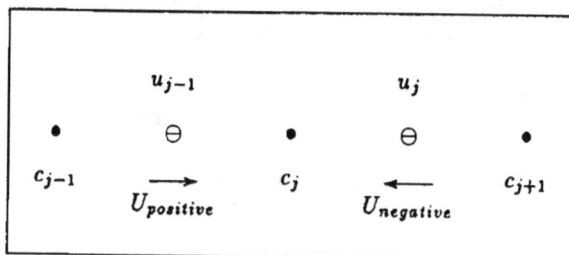

Fig. 2.3
Stencil for the staggered numerical grid;
u and c grid points are located a half space step apart.

We have just found that for an uncentered derivative between points $j+1$ and j, the difference scheme is stable for the negative transport velocities. It is obvious that a positive velocity will require a different space derivative based on the points j and $j-1$. The numerical scheme (2.39) can be generalized for the both negative and positive velocities in the following manner:

$$\frac{c_j^{m+1} - c_j^m}{T} + \frac{u - |u|}{2} \frac{c_{j+1}^m - c_j^m}{h} + \frac{u + |u|}{2} \frac{c_j^m - c_{j-1}^m}{h} + O(h, T) = 0 \qquad (2.54)$$

Here $u = 0.5(u_{j-1} + u_j)$. This difference equation is stable when

$$0 \le \frac{T|u|}{h} \le 1 \qquad (2.55)$$

For the obvious reason this type of numerical scheme is called an upwind (space) difference.

The constraints imposed by the inequality (2.55) or by other stability requirements necessitate quite short time step (T). Sometimes a large scale oceanographic process has to be considered for the multi-year time span and it can be very difficult, or even impossible, to march with the short time step permitted by the stability criteria. One possible way to increase the time step of the numerical integration is to introduce an implicit (in time) numerical scheme. For this purpose we shall rewrite (2.54) as

$$\frac{c_j^{m+1} - c_j^m}{T} + \frac{u - |u|}{2} \frac{c_{j+1}^{m+1} - c_j^{m+1}}{h} + \frac{u + |u|}{2} \frac{c_j^{m+1} - c_{j-1}^{m+1}}{h} + O(h, T) = 0 \quad (2.56)$$

Space differencing in (2.56) remains the same as in (2.54) but the variables are considered at the time step $m+1$. The answer to the most important question, i.e., how to solve this implicit equation, will be dealt with later on. Now, let us apply the von Neumann method to find out whether (2.56) is really stable. Introducing (2.48) into (2.56) leads to

$$\frac{\lambda - 1}{T} + \frac{u - |u|}{2} \frac{\lambda(e^{i\kappa h} - 1)}{h} + \frac{u + |u|}{2} \frac{\lambda(1 - e^{-i\kappa h})}{h} = 0 \qquad (2.57)$$

The absolute value of amplification parameter is

$$|\lambda| = \frac{1}{\left(1 + 2q(1 + q)(1 - \cos \kappa h)\right)^{1/2}} \le 1 \qquad (2.58)$$

because $-1 \le \cos \kappa h \le 1$, and parameter $q = |u|T/h$ is always positive.

It should be stressed that the actual time step of calculation can not be chosen only to satisfy a stability condition or to speed up the computation. For example, if

12 hours tide will be resolved by 24 hours time step, the result will be unacceptable. Additionally, a very long time step may result in quite large rounding off errors. There is no unique answer as to how the time step of an implicit scheme has to be chosen. Comparison of the explicit and implicit numerical equations used to describe the same phenomenon gives a tentative indication that the time step of an implicit numerical scheme can be chosen 10-20 times larger than the time step of an explicit scheme (Kowalik, 1975).

The amplification factor (2.58) is different for the signals of different wave number and also changes when relation of the time step to the space step is not constant. The amplification factor is always less than one, thus the difference equation (2.56) tends to damp a signal. This damping is a function of a parameter q, (the longer the time step, the stronger the damping will be), and the wave number κ. For the shortest wavelength $L = 2h$, $\kappa h = 2\pi h/2h = \pi$, and

$$|\lambda| = \frac{1}{(1 + 4q(1 + q))^{1/2}} \tag{2.59}$$

the damping is the strongest. For the longer wave length the damping will decrease and finally when $L \to \infty$, $\lambda \to 1$, the damping will disappear. Fig.2.4 illustrates this behavior of the amplification factor for the implicit numerical equation (2.56) as a function of the parameter q and wavelength of the signal.

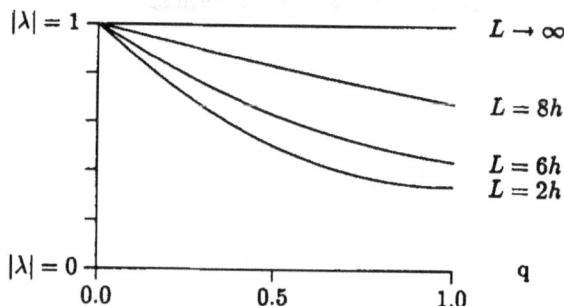

Fig. 2.4
Amplification factor $|\lambda|$ as a function of wave length L and q parameter.

5. Centered Numerical Schemes

5.1 Second-order space differencing

One way to improve the order of approximation of the advective equation (2.36) is to construct centered in space and time schemes of the second-order approximation

$$\frac{c_j^{m+1} - c_j^{m-1}}{2T} + u\frac{c_{j+1}^m - c_{j-1}^m}{2h} + O(h^2, T^2) = 0 \tag{2.60}$$

To find stability of the above equation we introduce (2.48) into (2.60) and obtain for the amplification factor,

$$\lambda - \lambda^{-1} + \frac{u2Ti}{h} \sin \kappa h = 0 \qquad (2.61)$$

From (2.61) it follows that $|\lambda| = 1$ when $|u|T/h \leq 1$. Therefore (2.60) is stable when the above inequality is satisfied. Since $|\lambda| = 1$, the signal is neither damped nor amplified by this numerical scheme. Although the difference scheme (2.60) possesses second order of approximation it has some other drawbacks which make this scheme difficult to apply.

A different scheme, centered in time and space, was developed by Crank and Nicholson (1947),

$$\frac{c_j^{m+1} - c_j^m}{2T} + u\frac{c_{j+1}^{m+1/2} - c_{j-1}^{m+1/2}}{2h} + O(h^2, T^2) = 0 \qquad (2.62)$$

where;

$$c_j^{m+1/2} = \frac{c_j^m + c_j^{m+1}}{2}$$

This second-order approximation scheme is absolutely stable because $|\lambda| = 1$ for any $q = uT/h$. It is implicit in time and thus requires a special method to arrive at the solution.

5.2 Angular derivative

The angular derivative method introduced by Roberts and Weiss (1966), is illustrated in Fig.2.5. Let us assume that the calculation of the new value of the concentration at the time step $m + 1$ is performed along the line, starting from the left end towards the right end. When the calculation is about to begin at the point with coordinates $j, m + 1$, the values of the concentration in the points depicted as bullets, are already known. Basing on this observation the following difference formula for eq.(2.36) may be constructed:

$$\frac{c_j^{m+1} - c_j^m}{T} + \frac{u}{h}\left(\frac{c_j^{m+1} + c_{j+1}^m}{2} - \frac{c_{j-1}^{m+1} + c_j^m}{2}\right) + O(h^2, T^2) = 0 \qquad (2.63)$$

The angular derivative is defined at the center of the time–space lattice. The sweep in this calculation starts from the left boundary and is carried to the right boundary.

The reverse sweep is possible as well, i.e., from the right to the left. The numerical expression for this case is

$$\frac{c_j^{m+1} - c_j^m}{T} + \frac{u}{h}\left(\frac{c_{j+1}^{m+1} + c_j^m}{2} - \frac{c_j^{m+1} + c_{j-1}^m}{2}\right) + O(h^2, T^2) = 0 \qquad (2.64)$$

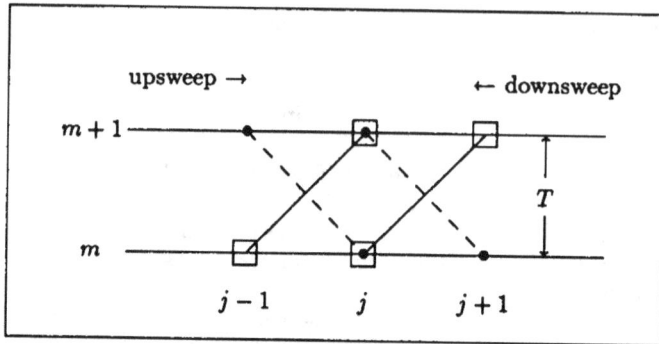

Fig. 2.5
The construction of the first derivative by an angular derivative method at the $j, m + 1$ grid point. Bullets are used in upsweep and squares in the downsweep.

The grid points used to derive this expression are denoted in Fig.2.5 by the square boxes. Both (2.63) and (2.64) seem to be implicit in time but in reality c_{j-1}^{m+1} in (2.63) and c_{j+1}^{m+1} in (2.64) are known, because they were calculated just one grid point before the present grid point. Both schemes display absolute stability. We shall prove this property only for the first difference equation. Introducing (2.48) into (2.63),

$$\lambda - 1 + q[\lambda(1 - e^{-i\kappa h}) + (e^{i\kappa h} - 1)] = 0 \qquad (2.65)$$

where $q = uT/h$. From (2.65), a few algebraic transformation leads to $|\lambda| = 1$. The angular derivative method has many plausible applications and it can serve to construct the difference form both for the first derivative and for the second derivative as well.

6. Computational errors: diffusion and dispersion

The difference equations constructed to solve the advection problem, even if they fulfill the required stability, often do not reproduce the advective processes correctly. The principal errors are due to the so-called numerical diffusion (dissipation) and dispersion. In Fig.2.6 the process of advecting of the concentration jump (front) along the x axis is illustrated. The calculation are done through the explicit upwind numerical scheme (2.54), explicit central numerical scheme (2.60), and also an ideal solution is given.

In the upwind numerical solution the vertical front is not only advected with velocity u but also it slowly changes shape as if through a strong diffusion process. A central numerical scheme better reproduces the vertical front, so it is less diffusive. Unfortunately it displays small ripples generated at the upper portion of the graph, as a result of the dispersive processes. One can also notice that the concentration is sometimes greater than the initial value (i.e., greater than 1) which denies the basic conservation law. These and other problems spurred the efforts to improve and to develop new algorithms.

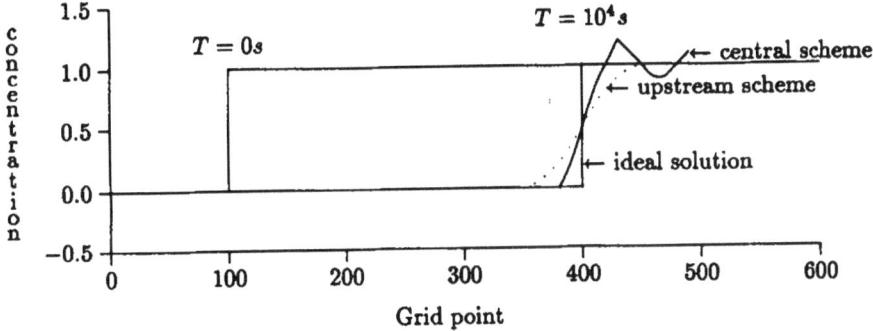

Fig. 2.6

Advection of the concentration jump from an initial moment till $T = 10^4$s, reproduced by an ideal solver, upstream difference scheme and central difference scheme.

Before we consider more complicated numerical expressions we shall describe a few basic reasons for the diffusion and dispersion errors. Numerical diffusion (or in the case of an equation of motion — numerical dissipation), is closely related to the order of approximation of a differential equation by a difference equation. We shall illustrate this problem through the explicit equation (2.39). By the definition of the Taylor series (2.1) the derivative in (2.39) are

the time derivative:

$$\frac{c_j^{m+1} - c_j^m}{T} = \frac{\partial c}{\partial t} + \frac{T}{2}\frac{\partial^2 c}{\partial t^2} + O(T^2) \tag{2.66}$$

the space derivative:

$$\frac{c_{j+1}^m - c_j^m}{h} = \frac{\partial c}{\partial x} + \frac{h}{2}\frac{\partial^2 c}{\partial x^2} + O(h^2) \tag{2.67}$$

Introducing the above formulas into (2.39) we arrive at

$$\frac{\partial c}{\partial t} + \frac{T}{2}\frac{\partial^2 c}{\partial t^2} + u\frac{\partial c}{\partial x} + \frac{uh}{2}\frac{\partial^2 c}{\partial x^2} + O(h^2, T^2) = 0 \tag{2.68}$$

The original advective eq.(2.36) is embedded into (2.68) but the additional terms appear in (2.68) as well because we have included all the terms up to the second order of approximation. These additional terms,

$$\frac{T}{2}\frac{\partial^2 c}{\partial t^2} + \frac{uh}{2}\frac{\partial^2 c}{\partial x^2} \tag{2.69}$$

represent a numerical diffusion process. They can always be transformed into the time or space domain only, by using equation (2.36) (and by assuming u as constant),

$$\frac{T}{2}\frac{\partial^2 c}{\partial t^2} = \frac{Tu^2}{2}\frac{\partial^2 c}{\partial x^2} \qquad (2.70)$$

Hence, eq.(2.69) describes the diffusion process with the numerical eddy coefficient equal to

$$-\frac{uh}{2} - \frac{Tu^2}{2} \qquad (2.71)$$

We conclude from the above considerations that the numerical friction enters into the equation of advection due only to the first order of approximation of the differential equation by the difference equation. When turbulent or chemical diffusion is also present all precautions should be taken so that the numerical diffusion will not dominate the physical or chemical processes.

The signal reproduced through the difference equation is not only altered by the numerical diffusion (dissipation) but, as we have seen in Fig.2.6, it will also be distorted by the dispersion. To illustrate the dispersive properties of the equation (2.39) we shall compare the amplitude and phase of the wave reproduced by the differential equation (2.36) and by the difference equation (2.39). Let us consider a wave of unit amplitude in the form

$$c = e^{i\kappa x}e^{i\omega t} \qquad (2.72)$$

Introducing (2.72) into (2.36) the phase velocity is defined as

$$\frac{\omega}{\kappa} = -u \qquad (2.73)$$

The same wave introduced into the difference equation (2.39) defines the following relation,

$$\lambda = e^{i\omega T} = q(e^{i\kappa h} - 1) + 1 \qquad (2.74)$$

where $q = -uT/h$.

In the complex variable equation the real values on the left side are equal to the real values on the right side and the same rule holds for the imaginary values. Basing on this rule, we first obtain from (2.74) two equations for $\sin \omega T$ and $\cos \omega T$. Next, we arrive at an equation which defines the phase change of the signal over one time step,

$$\tan \omega T = \frac{q \sin \kappa h}{1 - q(1 - \cos \kappa h)} \qquad (2.75a)$$

From (2.74) the change of amplitude during one time step can be found as well:

$$|\lambda|^2 = 1 + 2q(q - 1)(1 - a) \qquad (2.75b)$$

Here $a = \cos \kappa h$.

Eq.(2.75b) describes the amplification factor defined previously by (2.49).

To shorten the notation, the right side of (2.75a) is denoted as Q. Assuming Q to be small, the left side of (2.75a) can be developed into a series ($\omega T = \arctan Q$):

$$\omega T = Q - \frac{Q^3}{3} + \frac{Q^5}{5} - \ldots \tag{2.76}$$

Since $-uT/h \ll 1$ and $\kappa h \ll 1$, the right side of (2.75a) can also be developed into a series. Retaining only a few terms from the series, we obtain

$$\omega T = -\frac{uT}{h}\kappa h + \frac{uT}{h}\frac{(\kappa h)^3}{6} \tag{2.77}$$

From eq.(2.77) the phase velocity of the wave resolved by the finite difference scheme is

$$C = \frac{\omega}{\kappa} = -u + u\frac{\kappa^2 h^2}{6} \tag{2.78a}$$

The phase velocity is no longer a constant value equal to $-u$ but is a function of the wave number κ, thus the numerically calculated waves display a dispersive character and only if $\kappa h = 0$ the nondispersive wave occurs again. The (2.77) may also serve to define group velocity as

$$C_{gr} = \frac{\partial \omega}{\partial \kappa} = -u + u\frac{\kappa^2 h^2}{2} \tag{2.78b}$$

From the imaginary part of (2.74) we know that $\sin \omega T = q \sin \kappa h$, upon differentiation we obtain an exact formula for the group velocity,

$$C_{gr} = -u\frac{\cos \kappa h}{\cos \omega T} \tag{2.78c}$$

The recent stability theories for the hyperbolic problem (such a problem is described by the time dependent advection) established a stability criterion which can be expressed by the group velocity (Trefethen, 1983). This criterion takes into account not only the internal portion of domain but the boundary as well. In this respect the stability parameter we study is a weaker criterion, because the boundary is left out from the formulation.

Above simple considerations are quite helpful in understanding how actually the numerical scheme works and what it does to the wave. The change of amplitude, phase lag and the phase velocity carry information about an error introduced by the finite-difference scheme over one time step (T). To show how the formulas given above can be used, we shall carry out a few simple computations varying parameters $q = -uT/h$ and L/h. The latter shows how well the wave is resolved by the numerical grid and we shall call it parts per wavelength.

The velocity ratio or relative computed velocity $c/(-u)$ as a function of the resolution is given in Table 2.1.

Table 2.1

Relative computed velocity as a function of L/h.

L/h→	4	6	10	100
c/(-u)→	.6	.81	.93	.999

Because the velocity ratio is less that unity the signal calculated by the numerical scheme (2.39) is decelerating. According to our expectation the relative velocity is approaching unity for the finer resolution.

The numerical solution also does not follow exactly the prototype wave (2.72), it lags behind the analytical solution with the phase lag given in Table 2.2.

Table 2.2

Phase lag of the computed wave.

q ↓	parts per wavelength(L/h)			
	4	6	10	100
0.1	6.3°	5.2°	3.4°	0.35°
0.5	45°	30°	18°	1.8°
0.9	83.7°	54.8°	32.5°	3.2°

The phase lag diminishes for the high grid resolution but it also depends strongly on parameter q.

On other hand the computed amplitude given in Table 2.3 approaches the prototype amplitude for the high resolution grid but its dependence on parameter q is not unique.

Table 2.3

Amplitude of the computed wave.

q ↓	parts per wavelength(L/h)			
	4	6	10	100
0.1	0.9	0.95	0.98	0.999
0.5	0.71	0.87	0.95	0.999
0.9	0.9	0.95	0.98	0.999

Above considerations set an estimate for the error over one time step. If we are interested to know the errors over one period we can compute the number of time step as $m = L/(uT)$ required for one period and sum up the errors.

A simplified and very useful approach to investigate the numerical properties is to limit the consideration to the space domain only, and search a solution in the time domain as a simple harmonic wave. Suppose in eq.(2.60) the space derivative remains of the second order of approximation but the time derivative is in the differential form, leading to the following difference–differential equation:

$$\frac{\partial c_j}{\partial t} + u \frac{c_{j+1} - c_{j-1}}{2h} = 0 \tag{2.79}$$

Searching a solution to this equation in the form of $\exp(imt)\,\exp(ij\kappa h)$ we shall find

$$i\omega + i\kappa \left[\frac{u\sin(\kappa h)}{\kappa h} \right] = 0 \tag{2.80}$$

In this case, waves propagate with the phase speed

$$C = \frac{\omega}{\kappa} = -\frac{u\sin(\kappa h)}{\kappa h} \tag{2.81}$$

which is a function of the wave number, and thus the numerical scheme gave rise to numerical dispersion. It can be seen that, as argument κh increases, the phase velocity decreases monotonically from u, and C becomes zero for the shortest resolvable wavelength $2h$, when $kh = \pi$. Thus the propagation speed for all waves is less than the true speed, u, and departure of C from u increases as the wavelength decreases. For the differential equation (2.36) the phase velocity (2.73) is constant and the group velocity can be defined as

$$C_{gr} = \frac{\partial \omega}{\partial \kappa} = -\frac{\partial(u\kappa)}{\partial \kappa} = -u \tag{2.82a}$$

Hence, the group velocity is equal to the phase velocity. However, for the difference-differential equation (2.79) the group velocity is

$$C_{gr} = \frac{\partial(C\kappa)}{\partial \kappa} = -u\cos(\kappa h) \tag{2.82b}$$

Hence, as κh increases from zero the group velocity decreases monotonically, and C_{gr} is equal u (not $-u$) at the shortest resolvable wavelength, $2h$. To summarize, central difference of the space derivative causes both the phase and group velocity to decrease as the wave number increases. Waves having length less than $4h$ have the opposite group velocity; this motion is inconsistent with the expected behavior of the advection equation, and therefore such waves are referred to as parasitic waves.

One may ask the question whether an improvement is possible through the central finite-difference scheme but of a higher order of approximation. Using a Taylor series, one can approximate the term $\frac{\partial c}{\partial x}$ by

$$\frac{c_{j+1} - c_{j-1}}{2h} = \frac{\partial c}{\partial x} + \frac{1}{3!}\frac{\partial^3 c}{\partial x^3}h^2 + O(h^4) \tag{2.83}$$

One can replace h in the above equation with $2h$ and write

$$\frac{c_{j+2} - c_{j-2}}{4h} = \frac{\partial}{\partial cx} + \frac{4}{3!}\frac{\partial^3 c}{\partial x^3}h^2 + O(h^4) \tag{2.84}$$

A fourth-order scheme, which is formed by a linear combination of eq.(2.83) and eq.(2.84) and in which the truncation error due to eq.(2.83) cancels those due to eq.(2.84), is the following:

$$\frac{4}{3}\frac{c_{j+1} - c_{j-1}}{2h} - \frac{1}{3}\frac{c_{j+2} - c_{j-2}}{4h} = \frac{\partial c}{\partial x} + O(h^4) \tag{2.85}$$

Advection equation (2.36) can rewritten using eq.(2.85) for the first derivative in the following difference–differential form

$$\frac{\partial c_j}{\partial t} + u\left[\frac{4}{3}\frac{c_{j+1} - c_{j-1}}{2h} - \frac{1}{3}\frac{c_{j+2} - c_{j-2}}{4h}\right] = 0 \tag{2.86}$$

Proceeding as in eq.(2.81), the phase velocity can be written as

$$C^{IV} = -u\left[\frac{4}{3}\frac{\sin(\kappa h)}{\kappa h} - \frac{1}{3}\frac{\sin(2\kappa h)}{2\kappa h}\right] \tag{2.87}$$

$$C_{gr}^{IV} = -u\left[\frac{4}{3}\cos(\kappa h) - \frac{1}{3}\cos(2\kappa h)\right] \tag{2.88}$$

For the small value of κ, eqs.(2.81) and (2.87) can be developed into a Taylor series

$$C = -u[1 - \frac{1}{3!}(\kappa h)^2 + \ldots]$$

$$C^{IV} = -u[1 - \frac{4}{5!}(\kappa h)^4 + \ldots] \tag{2.89}$$

It can be seen that the fourth-order scheme is more accurate, because the advection speed is better represented by this scheme. The group velocity at the shortest wave ($L = 2h$) for the second-order scheme is equal to u, but for the fourth-order scheme it increases to $u5/3$, thus propagating more energy towards parasitic short waves. Hence, we conclude that the higher–order scheme is more accurate but it also causes more numerical dispersion. The main drawback of higher order differencing scheme is the requirement of additional grid rows, which leads to computational modes in space, similar to computational modes in time, and makes the prescription of boundary conditions rather difficult. Also, the computational effort increases considerably.

7. Computational and physical modes of the numerical solution

We again scrutinize and compare two explicit numerical schemes to show certain "not very desirable" properties of the centered numerical scheme (2.60). The simple explicit scheme (2.39) is based on two-time levels while the centered scheme (2.60) is based on three-time levels. The latter scheme requires, therefore, two independent initial conditions to start a computational process, the former scheme requires only one initial condition. The additional condition for (2.60) has to be found, let us say, by application of explicit scheme (2.39). This difference between (2.39) and (2.60) is even more profound; the centered numerical scheme actually represents two independent solutions. One is strictly related to the physics of the process described by the differential equation and the other is purely a numerical solution, not related to the physics but to the structure of the numerical equation. To facilitate further analysis let us notice that we rendered an error of the solution in the form of the harmonic wave (2.48), actually, because the linear equations are studied, the solution of these equations can be written in the same form:

$$c_j^m = \lambda^m e^{i\kappa j h} \tag{2.90}$$

The solution of (2.39) with the help of (2.90) can be also written as

$$c_j^m = \lambda^m e^{i\kappa j h} \tag{2.91}$$

where

$$\lambda = 1 - \frac{uT}{h}(e^{i\kappa h} - 1)$$

As we know from previous consideration this solution is stable when $-uT/h \leq 1$ and both the numerical and analytical solutions represent a wave propagating with the phase velocity close to $\omega/\kappa = -u$.

The centered scheme has two roots for λ (2.61):

$$\lambda_{1,2} = -iq_1 \pm (1 - q_1^2)^{1/2} \tag{2.92}$$

where

$$q_1 = \frac{uT}{h}\sin \kappa h$$

Because $|\lambda| = 1$, the two roots can be represented as

$$\lambda_1 = e^{-i\Phi} \quad \text{and} \quad \lambda_2 = -e^{i\Phi} \tag{2.92a}$$

where

$$\Phi = \arctan \frac{q_1}{(1 - q_1^2)^{1/2}} \tag{2.93}$$

Introducing (2.92a) into (2.90) two solutions are obtained. The physical solution,

$$c_j^m = e^{-i\Phi m} e^{i\kappa jh} \tag{2.94}$$

and the computational solution,

$$c_j^m = (-e^{i\Phi})^m e^{i\kappa jh} \tag{2.95}$$

The (2.95) alternates sign on every time step. This feature is also the easiest way to recognize the computational mode in practical computations. It will look like a saw-tooth curve. The source of the computational solution is probably related to two time steps required to calculate a new value by the symmetrical numerical scheme. To start computations by this scheme two initial values have to be prescribed. Since only one can be derived from the observations the second one has to be computed by a noncentered scheme similar to (2.39).

Certain properties of the centered scheme can be deduced from the numerical dispersion of this scheme. Since $\lambda = e^{i\omega T}$ and also it is given by (2.92), two results follow:

$$\omega T = -\Phi \quad \text{and} \quad \omega T = \Phi \tag{2.96}$$

Assuming as usual that

$$\frac{uT}{h} \quad \text{and} \quad \kappa h$$

are small values,

$$q_1 = \frac{uT}{h} \sin \kappa h \approx uT\kappa \quad \text{and} \quad q_1 \ll 1 \tag{2.97}$$

From (2.93) Φ can be obtained and introduced into (2.96) giving two values of the phase velocity

$$\frac{\omega}{\kappa} = -u \quad \text{and} \quad \frac{\omega}{\kappa} = u \tag{2.98}$$

The phase velocity of the physical solution equals to the phase velocity of the analytical solution, but the phase velocity of the numerical mode is opposite to the analytical solution. The numerical wave gives rise to two group velocities as well. These results are valid to the first order of approximation only. Inclusion into numerical solution of the higher order terms (as we did in eq.(2.77)) will somewhat diminish the magnitude of numerical velocities.

8. Diffusive processes

8.1 Explicit and implicit schemes

We have made a few attempts to construct a numerical solution to the advective part of the transport equation, now we turn our attention to the diffusive part described by the simple one dimensional problem

$$\frac{\partial c}{\partial t} = \frac{\partial}{\partial x}\left(D\frac{\partial c}{\partial x}\right) \tag{2.99}$$

where $D_h = D$. Some of the techniques developed for the advective problem can be carried over to the diffusion problem as well. Let us start with the explicit numerical scheme and define both space and time steps as constant. Using expression (2.3) for the time derivative and eq.(2.8) for the space derivative, the following difference equation is constructed:

$$\frac{c_j^{m+1} - c_j^m}{T} = \frac{D}{h^2}(c_{j+1}^m + c_{j-1}^m - 2c_j^m) + O(T, h^2) \tag{2.100}$$

It is obvious that (2.100) has first order approximation in time and second order in space. The difference equation is written for the constant eddy diffusivity (D). For the variable coefficient the grid plotted in the Fig.2.7 is quite useful. Space points related to the concentration and to the eddy diffusivity are located a half grid space apart. They form a staggered grid net but two nearby points have the same index, as illustrated in Fig.2.7.

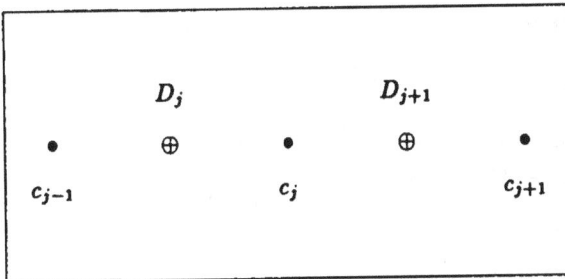

Fig. 2.7
Staggered grid for the diffusion equation
with the variable coefficient of diffusion – D.

A difference equation corresponding to (2.99) for the case of a variable eddy diffusivity has the following form:

$$\frac{c_j^{m+1} - c_j^m}{T} = \frac{D_{j+1}(c_{j+1}^m - c_j^m) - D_j(c_j^m - c_{j-1}^m)}{h^2} + O(T, h^2) \tag{2.101}$$

We shall find only the stability conditions of the above schemes for the simpler case of constant coefficient. Introducing (2.48) into (2.100) we arrive at

$$\frac{\lambda - 1}{T} = \frac{2D}{h^2}\left(\frac{e^{i\kappa h} - e^{-i\kappa h}}{2} - 1\right) \tag{2.102}$$

and because $\cos a = \cos^2(a/2) - \sin^2(a/2)$, the expression in parenthesis simplifies to

$$\cos \kappa h - 1 = -2\sin^2(\kappa h/2) \tag{2.103}$$

Finally,

$$\lambda = 1 - \frac{4DT}{h^2}\sin^2(\kappa h/2) \tag{2.104}$$

Stability requirement $|\lambda| \leq 1$, or $-1 \leq \lambda \leq 1$, provides two inequalities,

$$0 \leq \frac{2DT}{h^2}\sin^2(\kappa h/2) \leq 1 \tag{2.105}$$

From (2.105) one can deduce that for the fixed values of the space and time step the increase of the eddy diffusivity (D) may cause instability. The trigonometric term in (2.105) can play a similar, but not such a strong role. It is dependent on the wavelength of the signal. The shortest wavelength $L = 2h$ sets the argument of the trigonometric function to: $\kappa h/2 = \pi/2$, and therefore $\sin^2(\kappa h/2) = 1$. Since the longer wavelength is related to the smaller values of trigonometric function, it follows from (2.105) that the shortest wave is the plausible region of instabilities. Aside from the numerical requirements the physics of the process may cause an instability as well. In the generally recognized models of turbulence the energy flow is usually directed from the large scale towards the smaller scale and finally the energy is dissipated, at the very small scale, into heat. This phenomenon can not be reproduced exactly by numerical means, mainly because the shortest scale resolved by the numerical grid is still much longer than the dissipation scale. Physical processes reproduced by the numerical solutions have no mechanism to dissipate energy accumulated at the shortest wavelength ($L = 2h$). This may bring about an instability in the short wave range.

Along with the time explicit numerical schemes which are governed by the stability conditions, one can construct an unconditionally stable implicit scheme:

$$\frac{c_j^{m+1} - c_j^m}{T} = \frac{D}{h^2}(c_{j+1}^{m+1} + c_{j-1}^{m+1} - 2c_j^{m+1}) + O(T, h^2) \tag{2.106}$$

Stability parameter for (2.106) is equal to

$$\lambda = 1/[1 + \frac{4DT}{h^2}\sin^2(\kappa h/2)] \tag{2.107}$$

and therefore this scheme is unconditionally stable for any value of time and space step, wavelength, and eddy diffusivity. But before starting any computation it will be wise to check what the numerical scheme actually does to the unit wave. Stability parameter (2.107), which is also amplitude of the computed wave, shows that an amplitude may be severely distorted when time step is growing.

8.2 Centered numerical schemes

We shall start by briefly analyzing a centered space–time numerical scheme,

$$\frac{c_j^{m+1} - c_j^{m-1}}{2T} = \frac{D}{h^2}(c_{j+1}^m + c_{j-1}^m - 2c_j^m) + O(T^2, h^2) \qquad (2.108)$$

It is based on three-time levels and on second-order approximation. Stability parameter is defined by a quadratic equation

$$\lambda - 1/\lambda = -P \qquad (2.109)$$

Two roots of (2.109) are

$$\lambda_{1,2} = -\frac{P}{2} \pm (\frac{P^2}{4} + 1)^{1/2} \qquad (2.110)$$

where:

$$P = -\frac{8TD}{h^2} \sin^2(\kappa h/2)$$

This scheme is always unstable because the absolute value one of these roots is always greater than 1.

The centered numerical scheme although unstable is of second order approximation, therefore several stable schemes have been constructed to emulate properties of the centered numerical scheme.

Crank-Nicholson centered numerical scheme

We shall illustrate this method through the following difference–differential equation,

$$\frac{c_j^{m+1} - c_j^m}{T} = \frac{D}{2}\frac{\partial^2 c^m}{\partial x^2} + \frac{D}{2}\frac{\partial^2 c^{m+1}}{\partial x^2} + O(h^2, T^2) \qquad (2.111)$$

In (2.111) diffusion term is split into two parts taken at different time steps m and $m+1$. Difference scheme is of the second order of approximation and it is centered around the point with coordinates $m+1/2, j$. To prove approximation a Taylor series has to be developed around this central point. Although Crank–Nicholson's method is stable it requires a special method to solve the one dimensional implicit equation.

Split-up method (Fractional step method)

Instead of a time step T let us introduce a half-time step $T/2$ and write the difference–differential equation for each substep:

$$\frac{1}{2}\frac{c_j^{m+1/2} - c_j^m}{T/2} = \frac{c_j^{m+1/2} - c_j^m}{T} = \frac{D}{2}\frac{\partial^2 c^m}{\partial x^2} = \frac{D}{2}\frac{\partial^2 c^{m+1/2}}{\partial x^2} \qquad (2.112a)$$

and

$$\frac{1}{2}\frac{c_j^{m+1} - c_j^{m+1/2}}{T/2} = \frac{c_j^{m+1} - c_j^{m+1/2}}{T} = \frac{D}{2}\frac{\partial^2 c^{m+1}}{\partial x^2} = \frac{D}{2}\frac{\partial^2 c^{m+1/2}}{\partial x^2} \qquad (2.112b)$$

If both (2.112) are added on either side, the Crank–Nicholson equation follows, but the split-up method is more general and is well suited for the multidimensional equations.

The above equations represent an implicit approach; now we shall introduce the time explicit integration methods.

DuFort–Frankel numerical scheme

In the DuFort–Frankel method (1953) three time levels are used and a second order approximation is obtained by centering diffusion term in time,

$$\frac{c_j^{m+1} - c_j^{m-1}}{2T} = \frac{D}{h^2}\left[(c_{j+1}^m - c_j^{m+1}) - (c_j^{m-1} - c_{j-1}^m)\right] \qquad (2.113)$$

The points used to construct the diffusion term are depicted in Fig.2.8.

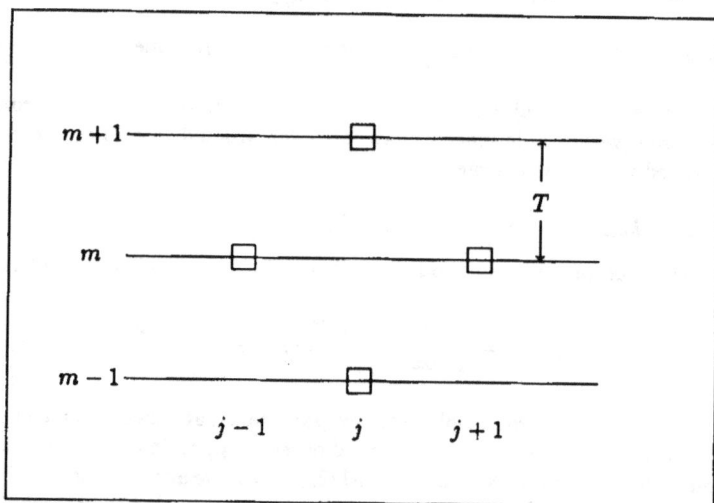

Fig. 2.8
Grid points for the diffusion term in the DuFort–Frankel method.

Saul'ev method (1957)

This method has been developed long ago and it is closely related to the angle derivative method. Only two time levels are used and two solutions are constructed to apply with an alternating direction sweep method at alternate time steps – see Fig. 2.9.

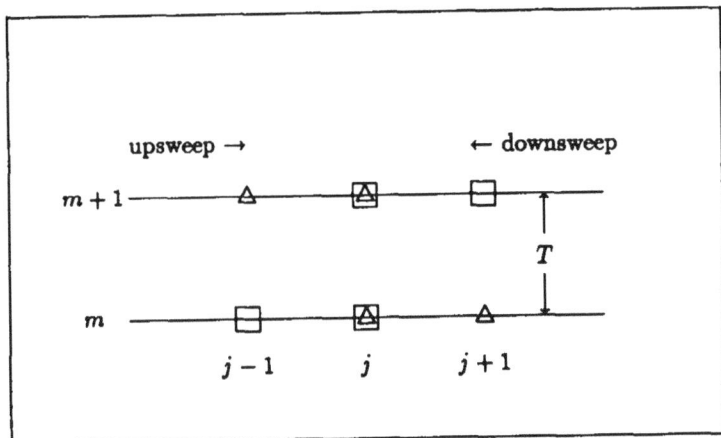

Fig. 2.9
Numerical grid for the diffusion term in Saul'ev method.
Triangles serve for the upsweep and squares for the downsweep.

The points to construct formula for the upsweep, i.e., for the sweep from j low to j high, are denoted in Fig.2.9 by triangles. The difference equation is the upsweep is

$$\frac{c_j^{m+1} - c_j^m}{T} = \frac{D}{h^2}\left[(c_{j+1}^m - c_j^{m+1}) - (c_j^m - c_{j-1}^{m+1})\right] \tag{2.114}$$

Downsweep formula (from j high to j low) is based on the points denoted by the squares in Fig.2.9,

$$\frac{c_j^{m+1} - c_j^m}{T} = \frac{D}{h^2}\left[(c_{j+1}^{m+1} - c_j^m) - (c_j^{m+1} - c_{j-1}^m)\right] \tag{2.115}$$

It is interesting to notice that both the above formulas can be derived from the different geometrical premises by not invoking the angle derivatives:

$$\frac{c_j^{m+1} - c_j^m}{T} = \frac{D}{h^2}\left[(c_{j+1}^m - c_j^m) - (c_j^{m+1} - c_{j-1}^{m+1})\right] \tag{2.114a}$$

In the Saul'ev and DuFort–Frankel methods stability has been achieved by introducing into Laplacian an implicit term c_j^{m+1}. Although implicit in the notation both methods are explicit in the realization (Richtmayer and Morton, 1967).

8.3 Multidimensional problem

A solution to the diffusion equation in the two-dimensional (2D) space,

$$\frac{\partial c}{\partial t} = D(\frac{\partial^2 c}{\partial x^2} + \frac{\partial^2 c}{\partial y^2}) \qquad (2.116)$$

can be searched by the simple explicit method (2.100). Given the space steps along the x and y axis are equal, the inequality (2.105) can be written for 2D case as

$$0 \le \frac{4DT}{h^2} \sin^2(\kappa h/2) \le 1 \qquad (2.117)$$

Assuming that D and h are prescribed parameters and $\sin^2(\kappa h/2) \le 1$, the time step of numerical computations can be deduced:
  for 1D case from (2.105),

$$0 \le T \le \frac{h^2}{2D} \qquad (2.118a)$$

for 2D case from (2.117),

$$0 \le T \le \frac{h^2}{4D} \qquad (2.118b)$$

and by way of analogy for 3D case,

$$0 \le T \le \frac{h^2}{6D} \qquad (2.118c)$$

It is obvious that the multidimensional case will require a shorter time step than the one dimensional problem.

  An implicit approach in 2D or 3D space requires the solution of two- or three-dimensional implicit equations. This rather formidable task can be easily circumvented by an application of the split-up method (Yanenko, 1971). Through equations (2.112) splitting along one direction has been demonstrated. We shall apply the same idea to split equation (2.116) along different directions so that a 2D problem will be replaced by the set of 1D equations

$$\frac{1}{2}\frac{\partial c}{\partial t} = D\frac{\partial^2 c}{\partial x^2} \qquad (2.119a)$$

$$\frac{1}{2}\frac{\partial c}{\partial t} = D\frac{\partial^2 c}{\partial y^2} \qquad (2.119b)$$

Splitting in the spatial dimensions allows us to construct very efficient numerical schemes which can be solved by 1D numerical algorithms. To obtain a numerical scheme fully centered in time and space we shall integrate (2.119a) on the first time substep by an explicit method,

$$\frac{c_{j,k}^{m+1/2} - c_{j,k}^m}{T} = \frac{D}{h^2}(c_{j+1,k}^m + c_{j-1,k}^m - 2c_{j,k}^m) \qquad (2.120a)$$

and eq.(2.119b) on the second substep by an implicit method,

$$\frac{c_{j,k}^{m+1} - c_{j,k}^{m+1/2}}{T} = \frac{D}{h^2}(c_{j,k+1}^{m+1} + c_{j,k-1}^{m+1} - 2c_{j,k}^{m+1}) \qquad (2.120b)$$

The first equation is solved explicitly along the x direction and second equation is solved along the y direction by the factorization method (see Appendix 1).

The splitting above shown is by no means unique — there are other plausible ways. To illustrate this point we shall again split (2.116) using two fractional steps but this time in the following way:

$$\frac{c_{j,k}^{m+1/2} - c_{j,k}^m}{T} = \frac{D}{2}(\frac{\partial^2 c^{m+1/2}}{\partial x^2} + \frac{\partial^2 c^m}{\partial y^2}) \qquad (2.121a)$$

$$\frac{c_{j,k}^{m+1} - c_{j,k}^{m+1/2}}{T} = \frac{D}{2}(\frac{\partial^2 c^{m+1/2}}{\partial x^2} + \frac{\partial^2 c^{m+1}}{\partial y^2}) \qquad (2.121b)$$

This form displays better symmetry from the point of view of geometry and numerical solution. In (2.120) one direction was solved by the explicit method the other direction by the implicit method. In (2.121) the x and y directions are solved by the same implicit method and also the x and y portions of the two-dimensional diffusion term are kept together at each substep.

The splitting-up method is usually devised to take into account a numerical method which will be used later on to solve a difference equation. For example, if one feels uneasy with the implicit methods (which sometimes may be quite cumbersome) the split-up equations can be solved by Saul'ev or DuFort–Frankel time explicit methods.

Stability criteria of the above systems can be searched by the standard approach through (2.48) or by scrutinizing solution in the form (2.120). By way of analogy to (2.90) a linear solution of (2.120) can be written as

$$c_{j,k}^m = \lambda^m e^{i\kappa_1 jh} e^{i\kappa_2 kh} \qquad (2.122)$$

Inserting (2.122) into (2.120) we obtain from (2.120a),

$$\lambda_1 = \lambda^{1/2} = 1 - \frac{4DT}{h^2}\sin^2(\kappa_1 h/2) \qquad (2.123a)$$

and from (2.120b),

$$\lambda_2 = \lambda^{1/2} = 1/\left[(1 + \frac{4DT}{h^2} \sin^2(\kappa_2 h/2)\right] \tag{2.123b}$$

By this means the stability parameter is defined for one time step as

$$\lambda = \lambda_1 \lambda_2 = \frac{h^2 - 4DT \sin^2(\kappa_1 h/2)}{h^2 + 4DT \sin^2(\kappa_2 h/2)} \tag{2.124}$$

Although the stability criterion which follows from (2.124) appears to be different from the one for the explicit scheme (2.117), we have to remember that κ_1 and κ_2 are the wave numbers along the different axes. When $\kappa_2 \to 0$, κ_1 may as well be equal to $2\pi/2h$, which will set exactly the same stability criterion as for the explicit scheme.

The set of implicit equations (2.121) is unconditionally stable. Searching for the stability parameter by the method delineated in (2.122)–(2.124), we obtain

$$\lambda = \frac{\left[h^2 - 2DT \sin^2(\kappa_1 h/2)\right] \left[h^2 - 2DT \sin^2(\kappa_2 h/2)\right]}{\left[h^2 + 2DT \sin^2(\kappa_1 h/2)\right] \left[h^2 + 2DT \sin^2(\kappa_2 h/2)\right]} \tag{2.125}$$

Modulus of (2.125) is always less than 1, which proves the absolute stability of the system (2.121).

To construct difference schemes for solving multidimensional problems, very often a partitioning of the complicated problem into a few simpler problems has been used. The aim of the split method (often called fractional-step method) is to obtain a set of simple equations which can be solved by a simple and efficient numerical method (Marchuk, 1976; Yanenko, 1971) Usually the line factorization method is applied in conjunction with the split method (see Appendix 1). It is understood that along with employing this method to the equation, the boundary condition must be split as well. This statement may seem to be superfluous, but often when a new equation is derived it is not clear what kind of boundary condition is suitable to solve the problem. Finally, when constructing a numerical scheme by fractional-step approach, caution is suggested, too many splittings usually generate a numerical (parasitic) solution.

We shall employ two kinds of splittings — one along the spatial directions and another in terms of physical or chemical processes described by the equation. It is also plausible to apply splitting to derive better approximation. Let us consider a set of equations of motion, continuity and transport as one vector equation which defines a time dependent problem in the following way:

$$\frac{\partial \Psi}{\partial t} = L\Psi + \mathbf{f} \tag{2.126}$$

where L is space differential operator. It is a positive definite operator and is expressed as

$$L = \sum_{i=1}^{n} L_i \qquad (2.127)$$

and all L_i are also positive definite.

Splitting of the above set of equations is performed in the following way. The numerical solution is advanced in time from time level m to the time level $m + 1$ by time step T. Now we split this time step into n substeps, and n equations will be constructed as

$$\frac{1}{n} \frac{\Psi^{m+1/n} - \Psi^m}{T/n} = L_1 \Psi^{m+1/n} + f/n$$

$$\frac{1}{n} \frac{\Psi^{m+2/n} - \Psi^{m+1/n}}{T/n} = L_2 \Psi^{m+2/n} + f/n$$

$$\vdots$$

$$\frac{1}{n} \frac{\Psi^{m+1} - \Psi^{m+(n-1)/n}}{T/n} = L_n \Psi^{m+1} + f/n \qquad (2.128)$$

Let us briefly summarize the reasons for applying the splitting procedure. First of all it allows us to organize the process of computations into a simple and repeatable algorithm. The procedure is often based on the time implicit scheme and therefore the time step can be increased so that the computations become more efficient. The strong stability properties allows us also to learn more about the physical properties of the equations, because the various coefficients (like eddy diffusivity) may be changed over wide ranges without violating stability conditions.

9. Application of the higher order computational schemes to the advective equation

The high order of approximation and stability do not provide completely satisfying numerical schemes because often the numerical errors cause high numerical diffusivity and dispersion. Split-up method or Saul'ev method applied to the diffusive part of the transport equation gave quite reasonable results but numerical schemes for the advective portion provided rather poor results, often a numerical diffusion/dispersion smears the physics and chemistry of the process. A survey of the second-order steady state difference schemes (Gupta, 1983) showed that even if a scheme has $O(h^2)$ truncation error its behavior is not universally of the second order, especially when convection dominates diffusion. Although a great number of algorithms for the advection were developed (Rood, 1982) they can not serve every purpose. Sometimes the coding difficulties and storage requirements are the principal obstacles. Higher order approximation schemes usually require longer computational time and substantial programming effort. A scheme of this kind called a flux-corrected transport method (Boris and Book, 1973, 1975) displays very good results but the large computer time requirements prohibit application to the oceanic processes. Oceanic diffusion and dynamics display a different intensity along the vertical and horizontal directions. Vertical processes take place over shorter distance and time scales. It seems appropriate to apply a splitting method along the different directions to separate the vertical and horizontal processes. After the horizontal and vertical processes have been separated one can split, let us say, the vertical transport equation

$$\frac{\partial c}{\partial t} + w\frac{\partial c}{\partial z} = \frac{\partial}{\partial z}\left(D_z\frac{\partial c}{\partial z}\right) \tag{2.129}$$

but this time in terms of the physical processes i.e., advection and diffusion,

$$\frac{1}{2}\frac{\partial c}{\partial t} + w\frac{\partial c}{\partial z} = 0 \tag{2.130a}$$

$$\frac{1}{2}\frac{\partial c}{\partial t} = \frac{\partial}{\partial z}\left(D_z\frac{\partial c}{\partial z}\right) \tag{2.130b}$$

In the ensuing sections we shall analyze a few general methods, their applications is not limited to the transport equation only.

9.1 Weighted differencing scheme for the advection–diffusion problem

The idea behind this approach is very interesting indeed. Why not use for the construction of a numerical scheme an analytical local solution? The method was developed by Allen (Allen and Southwell, 1955), Il'in (1969) and Fiadeiro and Veronis (1977). The method is appealing because the formal analysis displays the second-order approximation in space on the three point grid. In actual fact, the order of approximation depends both on the advection velocity and diffusion coefficient. Raithby and Torrance (1974) and Gupta (1983) showed that if $u/D \gg 1$, the method is approaching the first-order upwind differencing scheme. If on other hand $u/D \ll 1$, the method is close to the second-order centered numerical scheme. The results from this model are always positive definite.

Let us solve the advection–diffusion equation

$$u\frac{\partial c}{\partial x} = D\frac{\partial^2 c}{\partial x^2} \tag{2.131}$$

between $j-1$ and j grid points as

$$c = c_{j-1} + (c_j - c_{j-1})\frac{\exp[\frac{u}{D}(x - x_{j-1})] - 1}{\exp[\frac{u}{D}(x_j - x_{j-1})] - 1} \tag{2.132}$$

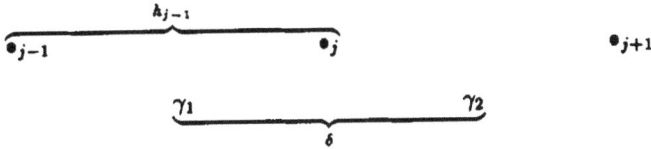

Fig. 2.10
Grid for the weighted differencing scheme.

and denote grid points as $j-1$, j, $j+1$. The distances between the grid points are h_{j-1} and h_j – see Fig.2.10.

First we express eq.(2.131) in the flux form as

$$\frac{\partial}{\partial x}(D\frac{\partial c}{\partial x} - uc) = \frac{\text{Flux}(\gamma_2) - \text{Flux}(\gamma_1)}{\delta} \tag{2.133}$$

Here $\delta = 0.5(h_{j-1} + h_j)$.

To calculate the values of fluxes, the analytical solution (2.132) will be used. Thus,

$$\text{Flux}(\gamma_2) = (D\frac{\partial c}{\partial x} - uc)_{\gamma_2}$$

$$= D(c_{j+1} - c_j)\frac{u}{D}\frac{\exp[(x - x_j)u/D]}{\exp(h_ju/D) - 1}$$

$$-uc_j - u(c_{j+1} - c_j)\frac{\exp[(x - x_j)u/D] - 1}{\exp(h_ju/D) - 1}$$

$$= \frac{u}{\exp(h_ju/D) - 1}c_{j+1} - (u + \frac{u}{\exp(h_ju/D) - 1})c_j$$

$$= (\frac{u}{\exp(h_ju/D) - 1} + \frac{u}{2}\frac{\exp(h_ju/D) - 1}{\exp(h_ju/D) - 1} - \frac{u}{2})c_{j+1}$$

$$-(\frac{u}{\exp(h_ju/D) - 1} + \frac{u}{2}\frac{\exp(h_ju/D) - 1}{\exp(h_ju/D) - 1} + \frac{u}{2})c_j$$

$$= (\frac{u}{2}\frac{\exp(h_ju/D) + 1}{\exp(h_ju/D) - 1} - \frac{u}{2})c_{j+1} - (\frac{u}{2}\frac{\exp(h_ju/D) + 1}{\exp(h_ju/D) - 1} + \frac{u}{2})c_j \tag{2.134}$$

To proceed further we introduce two auxiliary quantities

$$q_j = u_jh_j/(2D_j) \quad \text{and} \quad \alpha_j = \coth(q_j) - \frac{1}{q_j} \tag{2.135}$$

The last line of (2.134) will change to

$$\text{Flux}(\gamma_2) = (\frac{D_j}{h_j}q_j\coth(q_j) - \frac{u}{2})c_{j+1} - (\frac{D_j}{h_j}q_j\coth(q_j) + \frac{u}{2})c_j$$

$$= (\frac{D_j}{h_j} - \frac{u}{2}(1 - \alpha_j))c_{j+1} - (\frac{D_j}{h_j} + \frac{u}{2}(1 + \alpha_j))c_j \qquad (2.136)$$

The last expression is usually used in the computation and construction of the algorithms. For example, an implicit formula for the time integration can be constructed in the following manner:

$$\frac{c_j^{m+1} - c_j^m}{T} = \frac{1}{\delta}\{[\frac{D_j}{h_j} - u_j(1 - \alpha_j)/2]c_{j+1}^{m+1} - [\frac{D_j}{h_j} + u_j(1 + \alpha_j)/2]c_j^{m+1}$$

$$-[\frac{D_{j-1}}{h_{j-1}} - u_j(1 - \alpha_{j-1})/2]c_j^{m+1} + [\frac{D_{j-1}}{h_{j-1}} + u_{j-1}(1 + \alpha_{j-1})/2]c_{j-1}^{m+1}\} \qquad (2.137)$$

In Fig.2.11, a somewhat unusual geometry has been applied to derive the difference equation along the vertical direction. With z axis pointing upward, the l index increases in the opposite direction.

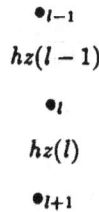

$$\bullet_{l-1}$$
$$hz(l-1)$$
$$\bullet_l$$
$$hz(l)$$
$$\bullet_{l+1}$$

Fig. 2.11
Grid distribution along the vertical direction.

In this geometry the right-hand side of (2.137) is

$$\frac{1}{\delta}\{[\frac{D_{l-1}}{hz(l-1)} - w_j(1 - \alpha_{l-1})]c_{l-1}^{m+1} - [\frac{D_{l-1}}{hz(l-1)} + w_j(1 + \alpha_{l-1})]c_l^{m+1}$$

$$-[\frac{D_l}{hz(l)} - w_j(1 - \alpha_l)]c_l^{m+1} + [\frac{D_l}{hz(l)} + w_j(1 + \alpha_l)]c_{l+1}^{m+1}\} \qquad (2.138)$$

We mentioned above that the scheme is formally of the second order of approximation but real approximation is changed due to the coefficients of the equation. One can demonstrate this property by developing the coefficients into a series. Instead we shall bring in another interesting difference equation through which we will demonstrate this property. Let us write the following numerical equation to solve (2.131) in the staggered grid stencil from Fig.2.7:

$$\frac{1}{h^2}[(\sqrt{D_j} + \frac{uh}{4\sqrt{D_j}})^2(c_{j+1} - c_j) - (\sqrt{D_{j-1}} - \frac{uh}{4\sqrt{D_{j-1}}})^2(c_j - c_{j-1})]$$

$$= \frac{1}{h^2}[D_j(c_{j+1} - c_j) - D_{j-1}(c_j - c_{j-1})] + u\frac{(c_{j+1} - c_{j-1})}{2h} + \frac{u^2}{16}[\frac{c_{j+1} - c_j}{D_j} - \frac{c_j - c_{j-1}}{D_{j-1}}]$$
$$(2.139)$$

The scheme is positive definite and has a second order of approximation. The last term represents the error of approximation and it can be written as

$$\frac{h^2u^2}{16}\frac{\partial}{\partial x}\frac{1}{D}\frac{\partial c}{\partial x}$$

Although this error is proportional to h^2 the approximation will be of the second order if $u^2/(16D) \approx 1$. When advection dominates diffusion eq.(2.139) has less than second order of approximation.

9.2 Positive definite scheme

This method has been proposed by Smolarkiewicz (1983; 1984). To delineate the basic steps of this method let us consider the equation of advection (2.36) and the difference equation (2.54). The difference equation is stable in time if the parameters of this equation are chosen according to the inequality (2.55). The numerical scheme (2.54) is defined as positive definite if from the initial condition $c_j^0 \geq 0$ follows $c_j^m \geq 0$ for all the time and space points. It is important to conserve a positive definite property at each time step because some schemes (see Fig.2.6) do not conserve concentration, or introduce negative values. The upstream advection scheme (2.54) preserves the positive definite property but, as it was shown in Sec. 6 of this chapter, the numerical schemes with the first-order approximation lead to excessive numerical diffusion (2.71). The Smolarkiewicz's idea is to compensate for this effect by introducing an additional time level of numerical computations. Let us consider (2.54) and introduce the expressions for the derivatives from the Taylor series (2.1) and (2.2) then,

$$\frac{\partial c}{\partial t} + \frac{T}{2}\frac{\partial^2 c}{\partial t^2} + u\frac{\partial c}{\partial x} - \frac{|u|h}{2}\frac{\partial^2 c}{\partial x^2} + O(T^2, h^2) = 0 \qquad (2.140)$$

Transformation (2.70) changes the above equation into

$$\frac{\partial c}{\partial t} + u\frac{\partial c}{\partial x} = \frac{|u|h}{2}\frac{\partial^2 c}{\partial x^2} - \frac{Tu^2}{2}\frac{\partial^2 c}{\partial x^2} + O(T^2, h^2) \qquad (2.141)$$

From (2.141) it is obvious that (2.54) approximates with the second order of accuracy not the advection equation (2.36) but the following advection/diffusion equation:

$$\frac{\partial c}{\partial t} + u\frac{\partial c}{\partial x} = D_n\frac{\partial^2 c}{\partial x^2} \qquad (2.142)$$

where $D_n = 0.5(|u|h - Tu^2)$ denotes the numerical diffusion coefficient. The process of numerical diffusion in (2.142) is described by

$$\frac{\partial c}{\partial t} = D_n\frac{\partial^2 c}{\partial x^2} = \frac{\partial}{\partial x}(D_n\frac{\partial c}{\partial x}) \qquad (2.143)$$

The idea is to reverse, at the next time level, the action of this term and to correct for excessive diffusion. Since the reversal of the diffusion processes is impossible an advection equation is introduced as

$$\frac{\partial c}{\partial t} = -u_d\frac{\partial c}{\partial x} = -\frac{\partial}{\partial x}(u_d c) \qquad (2.144)$$

where

$$u_d = \begin{cases} -D_n c^{-1}\frac{\partial c}{\partial x} & \text{if } c > 0 \\ 0 & \text{if } c = 0 \end{cases} \qquad (2.145)$$

Here u_d is a diffusion velocity which has to be reversed to correct for numerical diffusion. The reversed diffusion velocity ($\tilde{u}$) is defined as

$$\tilde{u} = \begin{cases} -u_d & \text{if } c > 0 \\ 0 & \text{if } c = 0 \end{cases} \qquad (2.146)$$

The overall solution will proceed in the following way: on the first time substep, eq.(2.54) is solved and on the second substep the same equation will be solved but with the advection velocity equal to $\tilde{u}$. This solution preserves the positive definite value of concentration but it does display dispersional ripples.

Correction is not completely achieved because on both the substeps the first-order accuracy method has been applied. Smolarkiewicz (1983) has found that if velocity $\tilde{u}$ is increased by as much as 8% the quality of solution is significantly improved. We have described above a method for the time explicit integration. Probably a similar approach can be devised for the time implicit integration based on eq.(2.56).

9.3 Higher order approximation to the upwind (upstream) method

As we have seen, before the upwind (or upstream) method is stable, it conserves positive definite property and is quite simple to program, but it has an excessive diffusion coefficient. It is therefore reasonable to explore the same approach and to find out whether this method can be improved by application of the higher order of approximation to the first derivative. Let us briefly recall the upwind method described in Sec.3 of this chapter. If a three-point stencil is given (Fig.2.3) the expression for the first space derivative can be chosen depending on the direction of the current. If velocity is negative, the advective process is treated by the following difference equation:

$$\frac{c_j^{m+1} - c_j^m}{T} + \frac{u - |u|}{2} \frac{c_{j+1}^m - c_j^m}{h} = 0 \qquad (2.147a)$$

If on the other hand velocity is positive,

$$\frac{c_j^{m+1} - c_j^m}{T} + \frac{u + |u|}{2} \frac{c_j^m - c_{j-1}^m}{h} = 0 \qquad (2.147b)$$

Although the three-point stencil is used the derivative is constructed from two-points only (Fig.2.12a). It would appear that the five-point stencil may bring better approximation and smaller numerical diffusion. Let us consider five-point stencil depicted in Fig.2.12b.

The construction of the first (upwind) derivative can be carried over through various means i.e., using three-points or four-points formulas. Let us first consider three-points formulas. The negative flow can be resolved by the derivative

$$\frac{\partial c}{\partial x} = \frac{-3c_j^m + 4c_{j+1}^m - c_{j+2}^m}{2h} + O(h^2) \qquad (2.148a)$$

based on the function given at the three-points j, $j + 1$ and $j + 2$. To resolve the positive flow,

$$\frac{\partial c}{\partial x} = \frac{3c_j^m - 4c_{j-1}^m + c_{j-2}^m}{2h} + O(h^2) \qquad (2.148b)$$

three points $j - 2$, $j - 1$ and j are applied. A similar expression has been introduced in Sec.1 of this chapter — for the constant grid step: formulas (2.11) and (2.13), and for the variable grid size: formulas (2.23) and (2.24). Let us assume that the computational grid runs from a point $j = JS$ to point $j = JE$. The application of the expressions (2.148) is straightforward, except in the close proximity to the boundaries. At the left boundary ($j = JS$) expression (2.148a) is applied. One point away from the boundary for $j = JS + 1$ the formula (2.148a) is easily applied but to apply (2.148b) we are lacking one point. One possibility is to use the first-order

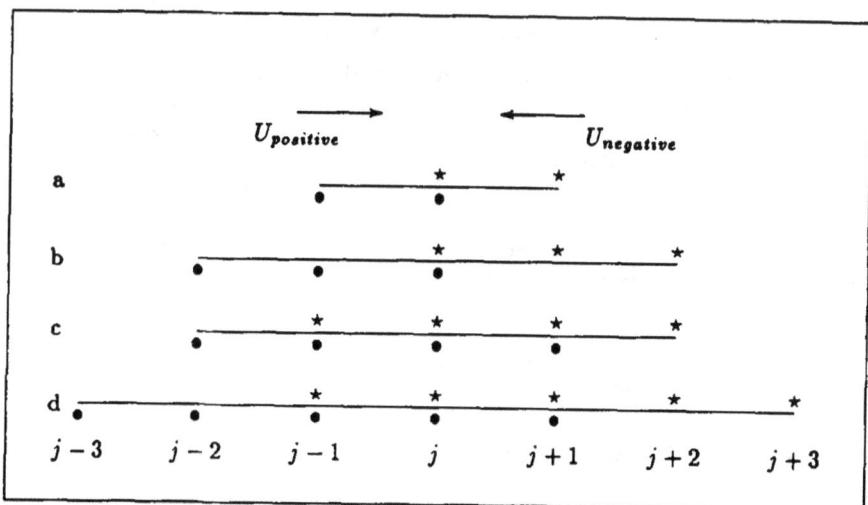

Fig. 2.12

Numerical stencils and grid points used to construct the first derivatives by upstream method. Stars denote points for negative flow and bullets denote points for positive flow.

expression (2.147b), which of course will violate the second-order approximation. A similar procedure can be implemented at the point $j = JE - 1$ by using (2.147a).

The four-point formulas constructed on the five-point stencil presents a more consistent approach both for the interior points and the boundary points. The location of the upwind points is depicted in Fig.2.12c. The derivative for the positive flow is

$$\frac{\partial c}{\partial x} = (c_{j-2}^m - 6c_{j-1}^m + 3c_j^m + 2c_{j+1}^m)/(6h) + O(h^3) \qquad (2.149a)$$

Not only two grid points to the left of the point j are used but also one point to the right from the central point is introduced as well. For negative flow the first derivative

$$\frac{\partial c}{\partial x} = (-2c_{j-1}^m - 3c_j^m + 6c_{j+1}^m - c_{j+2}^m)/(6h) + O(h^3) \qquad (2.149b)$$

is constructed by using two upstream points $j + 1$, $j + 2$ and one downstream point $j - 1$. In the interior of the integration domain from point $JS + 2$ to point $JE - 2$, the formulas (2.149) can be used. For $JS + 1$ point, the expression (2.149b) can be applied and for $JE - 1$ point the expression (2.149a) can be used. Now for the boundary points the four-point upstream formulas can be introduced which use only points to the left or to the right from the central point j. At the left boundary,

$$\frac{\partial c}{\partial x} = (-11c_j^m + 18c_{j+1}^m - 9c_{j+2}^m + 2c_{j+3}^m)/(6h) + O(h^3) \qquad (2.150a)$$

first derivative is constructed from the upstream points $j+1, j+2, j+3$. At the right boundary an asymmetrical formula is used:

$$\frac{\partial c}{\partial x} = (-2c_{j-3}^m + 9c_{j-2}^m - 18c_{j-1}^m + 11c_j^m)/(6h) + O(h^3) \qquad (2.150b)$$

Thus in this approach all the grid points are treated in a consistent manner by the four-point algorithm which has third-order approximation.

In Sec.1 of this chapter we did not introduce four-point formulas for the first derivative like we did for the three-points formulas. The methods to construct multipoint derivatives by means of Lagrange polynomials is usually delineated in the courses on numerical analysis, see, e.g., Burden *et al.*(1981). These derivatives can be also constructed by the method given in Sec.9.8 of this chapter. The application of the third order formulas greatly improves the numerical calculations both diminishing numerical diffusion, observed in the upstream first derivative, and dispersion observed in the symmetrical first derivative. The success of the four-points derivative prompted development of the five-points derivative on the seven-point stencil, see Fig.2.12d. This method encounters again the same problems in the vicinity of the boundary. At the left boundary, a five-point upstream formula is constructed:

$$\frac{\partial c}{\partial x} = (-25c_j^m + 48c_{j+1}^m - 36c_{j+2}^m + 16c_{j+3}^m - 3c_{j+4}^m)/(12h) + O(h^4) \qquad (2.151a)$$

To the point $JS + 1$ the main formula for the negative velocity can be applied:

$$\frac{\partial c}{\partial x} = (-3c_{j-1}^m - 10c_j^m + 18c_{j+1}^m - 6c_{j+2}^m + c_{j+3}^m)/(12h) + O(h^4) \qquad (2.152a)$$

Point $JS + 2$ again has to be treated in a special way:

$$\frac{\partial c}{\partial x} = (c_{j-2}^m - 8c_{j-1}^m + 0c_j^m + 8c_{j+1}^m - c_{j+2}^m)/(12h) + O(h^4) \qquad (2.153)$$

At the right boundary, similar problems will be encountered. For the point $JE - 2$ expression (2.153) is used. For the point $JE - 1$ the principal expression for the positive flow can be applied:

$$\frac{\partial c}{\partial x} = (-c_{j-3}^m + 6c_{j-2}^m - 18c_{j-1}^m + 10c_j^m + 3c_{j+1}^m)/(12h) + O(h^4) \qquad (2.152b)$$

Finally, at the right boundary the following formula is used:

$$\frac{\partial c}{\partial x} = (3c_{j-4}^m - 16c_{j-3}^m + 36c_{j-2}^m - 48c_{j-1}^m + 25c_j^m)/(12h) + O(h^4) \qquad (2.151b)$$

9.4 Quadratic upstream interpolation

The high-order upstream formulas for the first derivative already show a strong improvement when compared against a simple first-order derivative. Leonard (1979a, 1979b) applied an upwind algorithm based on the quadratic interpolation surface. To construct various schemes the control-volume approach is adopted, which considers the budget of the constituents within the control-volume. We shall present two methods. The QUICK method serves to solve problems related to the steady state flows. The QUICKEST is a more complicated version, it includes Lagrangian properties of the flow and serves for the unsteady motion. Our experience shows that another method of a similar strength as QUICKEST is the method constructed by Russell and Lerner (1981).

9.4a QUICK (Quadratic upstream interpolation for convective kinematics)

The stencil of the grid points is shown in Fig.2.13. As we did before for the flow with a positive velocity, a formula for the first derivative will be constructed based on the points $j+1$, j, $j-1$ and $j-2$. The first derivative is defined as a difference between points $j+1/2$ and $j-1/2$. To find the values of a function in these points the quadratic interpolation formulas are used. At the point $j-1$ a local coordinate (α) is introduced and the dependent function is defined as

$$c = c_0 + c_1\alpha + c_2\alpha^2 \qquad (2.154)$$

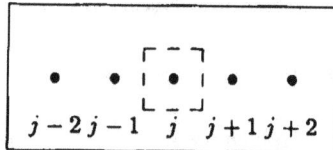

Fig. 2.13
Five-point stencil for the QUICK algorithm.

The unknown coefficients can be defined in terms of the given functions at the nodes. Thus at the point $j - 1$ the local coordinate is $\alpha = 0$, and $c_0 = c_{j-1}$. Similarly we get

$$c_j = c_0 + c_1 h + c_2 h^2 \quad \text{at the point} \quad j$$

and

$$c_{j-2} = c_0 - c_1 h + c_2 h^2 \quad \text{at the point} \quad j - 2$$

Equation (2.154) can be rewritten with all the coefficients defined explicitly:

$$c = c_{j-1} + \frac{c_j - c_{j-2}}{2h} \alpha + \frac{c_j + c_{j-2} - 2c_{j-1}}{2h^2} \alpha^2 \qquad (2.155)$$

To construct a derivative we need to know the values of this function at the points $j + 1/2$ and $j - 1/2$. The above local expression will serve to introduce a value at $j - 1/2$. To derive a value at the point $j + 1/2$, eq.(2.154) ought to be written at the point j. From eq.(2.155) by setting $\alpha = h/2$, we get

$$c_{j-1/2} = (c_j + c_{j-1})/2 - \frac{1}{8}(c_j + c_{j-2} - 2c_{j-1}) \qquad (2.156)$$

As an analogy to the above expression, a formula at the point $j + 1/2$ is derived:

$$c_{j+1/2} = (c_j + c_{j+1})/2 - \frac{1}{8}(c_{j-1} + c_{j+1} - 2c_j) \qquad (2.157)$$

The finite-difference formulation of the advective equation is taken as explicit in time:

$$\frac{c_j^{m+1} - c_j^m}{T} = -u \frac{c_{j+1/2}^m - c_{j-1/2}^m}{h} \qquad (2.158)$$

Introducing (2.156) and (2.157) into (2.158) we obtain an explicit numerical formula for the positive velocity case:

$$c_j^{m+1} = c_j^m - q(\frac{1}{8}c_{j-2}^m - \frac{7}{8}c_{j-1}^m + \frac{3}{8}c_j^m + \frac{3}{8}c_{j+1}^m) \qquad (2.159)$$

where $q = uT/h$.

The negative velocity case can be investigated in a similar manner. The values of the unknown function at the points $j+1/2$ and $j-1/2$ are again searched through the upstream quadratic interpolation formulas but this time based on the points $j - 1, j, j + 1$ and $j + 2$. At the point $j - 1/2$,

$$c_{j-1/2} = 0.5(c_j + c_{j-1}) - \frac{1}{8}(c_{j+1} + c_{j-1} - 2c_j) \qquad (2.160)$$

At the point $j + 1/2$,

$$c_{j+1/2} = 0.5(c_j + c_{j+1}) - \frac{1}{8}(c_j + c_{j+2} - 2c_{j+1}) \qquad (2.161)$$

The advection process is described here by an explicit scheme for the time differencing:

$$c_j^{m+1} = c_j^m - q(-\frac{3}{8}c_{j-1}^m - \frac{3}{8}c_j^m + \frac{7}{8}c_{j+1}^m - \frac{1}{8}c_{j+2}^m) \qquad (2.162)$$

Stability properties of the above numerical schemes were investigated by Leonard (1979b) and shown to be exactly the same as for the simple explicit scheme.

By combining (2.159) and (2.162) one expression for both the positive and negative velocity can be derived. For this purpose (2.159) is rewritten as

$$c_j^{m+1} = c_j^m - qA \quad \text{and (2.162) as} \quad c_j^{m+1} = c_j^m - qB$$

and the combined algorithm is

$$c_j^{m+1} = c_j^m - 0.5(q + |q|)A - 0.5(q - |q|)B \qquad (2.163)$$

For a two-dimensional algorithm, a 13 grid-point stencil is used, see Fig.2.14. The local system of coordinate (α, β) is attached first to the point $j - 1$, k, and a six-point quadratic interpolation formula is used to define a function:

$$c = c_0 + c_1\alpha + c_2\alpha^2 + c_3\beta + c_4\beta^2 + c_5\alpha\beta \qquad (2.164)$$

This time we consider not only four points around the central point but also the control surface at the distant $0.5h$ from the center, see Fig.2.14.

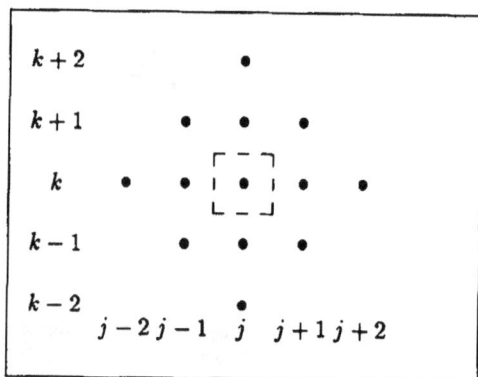

Fig. 2.14
Grid stencil and control surface for the two-dimensional QUICK algorithm.

The unknown coefficients in (2.164) are evaluated in similar way to eq.(2.155), thus considering only the new points,

$$c_{j-1,k+1} = c_0 + c_3 h + c_4 h^2$$

$$c_{j-1,k-1} = c_0 - c_3 h + c_4 h^2$$

This allows us to define the coefficients and the expression (2.164) is rewritten as

$$c = c_{j-1,k} + \frac{c_{j,k} - c_{j-2,k}}{2h}\alpha + \frac{c_{j,k} + c_{j-2,k} - 2c_{j-1,k}}{2h^2}\alpha^2$$

$$+ \frac{c_{j-1,k+1} - c_{j-1,k-1}}{2h}\beta + \frac{c_{j-1,k+1} + c_{j-1,k-1} - 2c_{j-1,k}}{2h^2}\beta^2 \qquad (2.165)$$

The value of a function at the point $j - 1/2, k$ is defined as an average value along β direction from $-h/2$ to $h/2$,

$$c_{j-1/2,k} = \frac{1}{h}\int_{-h/2}^{h/2} c(h/2, \beta)d\beta \qquad (2.166)$$

Taking (2.165) at the point $j - 1/2, k$ and substituting into (2.166) results in

$$c_{j-1/2,k} = 0.5(c_{j-1,k} + c_{j,k}) - \frac{1}{8}(c_{j-2,k} + c_{j,k} - 2c_{j-1,k})$$

$$+ \frac{1}{24}(c_{j-1,k+1} + c_{j-1,k-1} - 2c_{j-1,k}) \qquad (2.167)$$

At the control face which crosses point $j + 1/2, k$ a similar formula results:

$$c_{j+1/2,k} = 0.5(c_{j,k} + c_{j+1,k}) - \frac{1}{8}(c_{j+1,k} + c_{j-1,k} - 2c_{j,k})$$

$$+ \frac{1}{24}(c_{j,k+1} + c_{j,k-1} - 2c_{j,k}) \qquad (2.168)$$

Searching for the solution to the advection equation through the explicit scheme

$$\frac{c_{j,k}^{m+1} - c_{j,k}^m}{T} + u\frac{c_{j+1/2,k}^m - c_{j-1/2,k}^m}{h} = 0 \qquad (2.169)$$

we arrive at

$$c_{j,k}^{m+1} = c_{j,k}^m - q_1\left(\frac{1}{8}c_{j-2,k}^m - \frac{19}{24}c_{j-1,k}^m + \frac{7}{24}c_{j,k}^m + \frac{3}{8}c_{j+1,k}^m\right)$$

$$-q_1(-\frac{1}{24}c^m_{j-1,k+1} - \frac{1}{24}c^m_{j-1,k-1} + \frac{1}{24}c^m_{j,k+1} + \frac{1}{24}c^m_{j,k-1}) \qquad (2.170)$$

where $q_1 = uT/h$.

Equation (2.170) solves only the case of positive velocity; for the negative velocity a similar approach can be taken and we shall write only the final expression for the explicit calculation:

$$c^{m+1}_{j,k} = c^m_{j,k} - q_1(-\frac{3}{8}c^m_{j-1,k} - \frac{7}{24}c^m_{j,k} + \frac{19}{24}c^m_{j+1,k} - \frac{1}{8}c^m_{j+2,k})$$

$$-q_1(-\frac{1}{24}c^m_{j,k+1} - \frac{1}{24}c^m_{j,k-1} + \frac{1}{24}c^m_{j+1,k+1} + \frac{1}{24}c^m_{j+1,k-1}) \qquad (2.171)$$

Again to derive an algorithm for both positive and negative velocities we can combine (2.170) and (2.171) by first rewriting (2.170) as

$$c^{m+1}_{j,k} = c^m_{j,k} - q_1 A_1 \quad \text{and (2.171) as} \quad c^{m+1}_{j,k} = c^m_{j,k} - q_1 B_1$$

and next

$$c^{m+1}_{j,k} = c^m_{j,k} - 0.5(q_1 + |q_1|)A_1 - 0.5(q_1 - |q_1|)B_1 \qquad (2.172)$$

Algorithm (2.172) contains, of course, only half the story because it describes only the advection along one direction. To introduce the algorithm along the y direction we can apply symmetry which exist between the two directions. Thus (2.170) can rewritten as

$$c^{m+1}_{j,k} = c^m_{j,k} - q_2(\frac{1}{8}c^m_{j,k-2} - \frac{19}{24}c^m_{j,k-1} + \frac{7}{24}c^m_{j,k} + \frac{3}{8}c^m_{j,k+1})$$

$$q_2(-\frac{1}{24}c^m_{j+1,k-1} - \frac{1}{24}c^m_{j-1,k-1} + \frac{1}{24}c^m_{j+1,k} + \frac{1}{24}c^m_{j-1,k}) \qquad (2.173)$$

and (2.171) as

$$c^{m+1}_{j,k} = c^m_{j,k} - q_2(-\frac{3}{8}c^m_{j,k-1} - \frac{7}{24}c^m_{j,k} + \frac{19}{24}c^m_{j,k+1} - \frac{1}{8}c^m_{j,k+2})$$

$$q_2(-\frac{1}{24}c^m_{j+1,k} - \frac{1}{24}c^m_{j-1,k} + \frac{1}{24}c^m_{j+1,k+1} + \frac{1}{24}c^m_{j-1,k+1}) \qquad (2.174)$$

From (2.174) and (2.173) a single formula is constructed as

$$c^{m+1}_{j,k} = c^m_{j,k} - 0.5(q_2 + |q_2|)A_2 - 0.5(q_2 - |q_2|)B_2 \qquad (2.175)$$

Now we are ready to write one grand formula for both the directions and for both positive and negative velocities by combining (2.172) and (2.175):

$$c^{m+1}_{j,k} = c^m_{j,k} - 0.5(q_1 + |q_1|)A_1 - 0.5(q_1 - |q_1|)B_1 - 0.5(q_2 + |q_2|)A_2 - 0.5(q_2 - |q_2|)B_2$$
$$(2.176)$$

9.4b QUICKEST (QUICK with estimated streaming term)

Before undertaking the formulation of the QUICKEST algorithm we shall derive Leith's (1965) numerical algorithm for the advection equation. Starting from the advection problem

$$\frac{\partial c}{\partial t} = -u\frac{\partial c}{\partial x}$$

the second derivative is obtained:

$$\frac{\partial^2 c}{\partial t^2} = -u\frac{\partial^2 c}{\partial x \partial t} = -u\frac{\partial}{\partial x}(-u\frac{\partial c}{\partial x}) = u^2\frac{\partial^2 c}{\partial x^2} \tag{2.177}$$

Considering a Taylor series in the time domain for the function $c(x,t)$,

$$c_j^{m+1} = c_j^m + \frac{\partial c}{\partial t}T + \frac{1}{2}\frac{\partial^2 c}{\partial t^2}T^2 \tag{2.178}$$

By means of (2.177) and advection equation, the time derivatives in (2.178) can be replaced by the space derivatives

$$c_j^{m+1} = c_j^m - uT\frac{\partial c}{\partial x} + \frac{1}{2}u^2T^2\frac{\partial^2 c}{\partial x^2} \tag{2.179}$$

In the nonsteady flow the rate of change of the constituent within a cell is equal to the influx minus outflux through the walls per unit time:

$$\int_{-h/2}^{h/2}(c^{m+1} - c^m)dx = u_l\int_0^T c_l dt - u_r\int_0^T c_r dt \tag{2.180}$$

In (2.180) the l and r indices stand for the right and left wall values, respectively.

For the positive velocity flow a particle which is just crossing through the left wall was at the upstream position

$$x = -\frac{h}{2} - \int_0^T u\,dt \tag{2.181}$$

at the previous time step. The integral at the right-hand side of (2.180) can be written as follows (only for the left wall):

$$u\int_0^T c_l(t + \tilde{t})d\tilde{t} \tag{2.182}$$

First developing c_l into a Taylor series, afterwards changing time derivatives into space derivatives (see (2.179)) and finally integrating, we arrive at

$$u_l \int_0^T c_l d\bar{t} = T u_l c_l^m - \frac{1}{2} u_l^2 T^2 \mathrm{grad}_l + u_l^3 \frac{T^3}{6} \mathrm{curv}_l \qquad (2.183)$$

Introducing a nondimensional Courant number $q = uT/h$ the right-hand side of (2.180) yields

$$\mathrm{RHS} = q_l h(c_l^m - \frac{1}{2} q_l h \mathrm{grad}_l + q_l^2 \frac{h^2}{6} \mathrm{curv}_l) - q_r h(c_r^m - \frac{1}{2} q_r h \mathrm{grad}_r + q_r^2 \frac{h^2}{6} \mathrm{curv}_r) \quad (2.184)$$

Assuming that the velocity at the right wall is equal to the velocity at the left wall (2.184) simplifies to

$$\mathrm{RHS} = hq(c_l^m - q \frac{h}{2} \mathrm{grad}_l + q^2 \frac{h^2}{6} \mathrm{curv}_l) - hq(c_r^m - q \frac{h}{2} \mathrm{grad}_r + q^2 \frac{h^2}{6} \mathrm{curv}_r) \quad (2.185)$$

where,

$$\mathrm{grad}_l = (c_j^m - c_{j-1}^m)/h \qquad \mathrm{curv}_l = (c_{j-2}^m + c_j^m - 2c_{j-1}^m)/h^2$$

$$c_l^m = (3) = 0.5(c_j^m + c_{j-1}^m) - \frac{h^2}{8} \mathrm{curv}_l \qquad (2.186)$$

and

$$\mathrm{grad}_r = (c_{j+1}^m - c_j^m)/h \qquad \mathrm{curv}_r = (c_{j-1}^m + c_{j+1}^m - 2c_j^m)/h^2$$

$$c_r^m = (4) = 0.5(c_j^m + c_{j+1}^m) - \frac{h^2}{8} \mathrm{curv}_r \qquad (2.187)$$

For the left side of (2.180) a quadratic approximation of c over $j-1, j$ and $j+1$ points is taken. The integration of c over the cell is expressed as

$$\int_{-h/2}^{h/2} c(\alpha) d\alpha = \int_{-h/2}^{h/2} (c_0 + c_1 \alpha + c_2 \alpha^2) d\alpha$$

$$= c_0 h + \frac{c_2}{12} h^3 = c_j h + \frac{h^3}{24} (\frac{\partial^2 c}{\partial x^2})_{\alpha=0} \qquad (2.188)$$

Applying above result to the LHS of (2.180) yields

$$\mathrm{LHS} = (c_j^{m+1} - c_j^m)h + \frac{h^3}{24}(\frac{\partial^2 c^{m+1}}{\partial x^2} - \frac{\partial^2 c^m}{\partial x^2}) \qquad (2.189)$$

The last term can be written as

$$\frac{h^3}{24} T \frac{\partial}{\partial t} \frac{\partial^2 c}{\partial x^2} = -\frac{h^3}{24} uT \frac{\partial}{\partial x} \frac{\partial^2 c}{\partial x^2}$$

$$= -\frac{h^3}{24}uT(\text{curv}_r - \text{curv}_l)/h \tag{2.190}$$

Equating LHS and RHS and rearranging we obtain an algorithm to advance the solution in time:

$$c_j^{m+1} = c_j^m - q[0.5(c_j^m + c_{j+1}^m) - q\frac{h}{2}\text{grad}_r - \frac{h^2}{6}(1 - q^2)\text{curv}_r$$

$$-0.5(c_{j-1}^m + c_j^m) + q\frac{h}{2}\text{grad}_l + \frac{h^2}{6}(1 - q^2)\text{curv}_l] \tag{2.191}$$

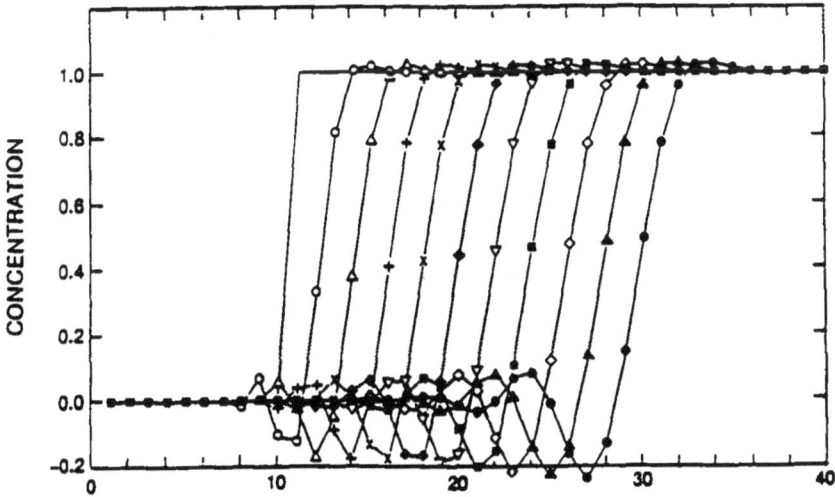

Fig. 2.15

Results of the advective transport computation
by QUICK method. Curves are given every 24 time steps.
Courant number $q = uT/h = 0.07$.

Again this can be simplified by abandoning curv and grad notation,

$$c_j^{m+1} = c_j^m - q[0.5(c_{j+1}^m - c_{j-1}^m) - \frac{1}{2}q(-2c_j^m + c_{j-1}^m + c_{j+1}^m)$$

$$+\frac{1}{6}(1 - q^2)(c_{j-2}^m - 3c_{j-1}^m + 3c_j^m - c_{j+1}^m) \tag{2.192}$$

The algorithm for the negative velocity can be easily deduced; the change is due only to the curvature terms in (2.191). Introducing,

$$\text{curv}_{l,n} = (c_{j+1}^m + c_{j-1}^m - 2c_j^m)/h^2$$

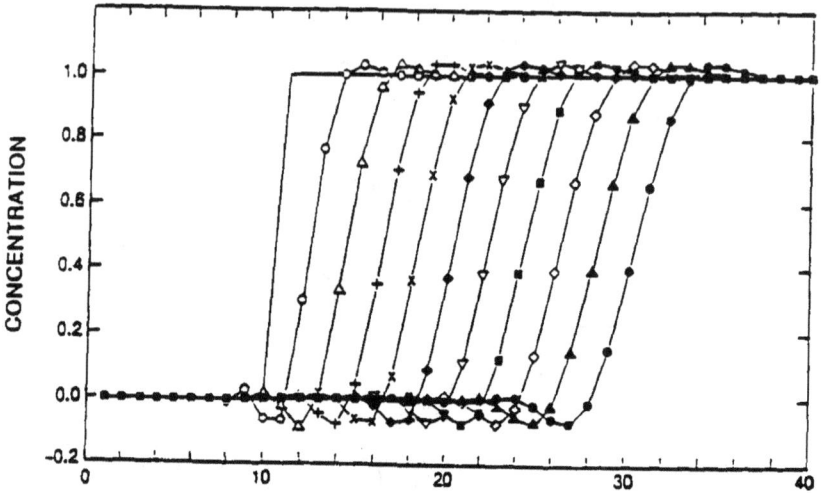

Fig. 2.16

Results of the advective transport computation by QUICKEST method. Curves are given every 24 time steps. Courant number $q = 0.07$.

$$\text{curv}_{r,n} = (c_{j+2}^m + c_j^m - 2c_{j+1}^m)/h^2$$

formula (2.191) can be rewritten for the negative velocity

$$c_j^{m+1} = c_j^m - q[0.5(c_j^m + c_{j+1}^m) - q\frac{h}{2}\text{grad}_r - \frac{h^2}{6}(1 - q^2)\text{curv}_{r,n}$$

$$-0.5(c_{j-1}^m + c_j^m) + q\frac{h}{2}\text{grad}_l + \frac{h^2}{6}(1 - q^2)\text{curv}_{l,n}] \qquad (2.193)$$

The origin of these curvatures is related to expressions (2.160) and (2.161). An algorithm which comprises both the negative and positive cases will be based on two parameters:

$$q_p = (q + |q|)/2 \quad \text{and} \quad q_n = (q - |q|)/2 \qquad (2.194)$$

and on (2.192)

$$c_j^{m+1} = c_j^m - (q_p + q_n)[0.5(c_{j+1}^m - c_{j-1}^m) - 0.5(q_p + q_n)(c_{j+1}^m + c_{j-1}^m - 2c_j^m)]$$

$$-\frac{q_p}{6}(1-q_p^2)(c_{j-2}^m - 3c_{j-1}^m + 3c_j^m - c_{j+1}^m) - \frac{q_n}{6}(1-q_n^2)(c_{j-1}^m - 3c_j^m + 3c_{j+1}^m - c_{j+2}^m) \qquad (2.195)$$

9.4c Results of calculation for 1D advection

We test the QUICK and QUICKEST algorithms for the pure advective transport in which a concentration jump will be advected by a uniform positive velocity. The Courant number in this experiment is equal to 0.07. The results are plotted in Figs. 2.15 and 2.16 every 24 time steps. The QUICK method still possesses the dispersive properties. It generates a quickly decaying ripple. In the QUICKEST the dispersive wave is limited to one small ripple.

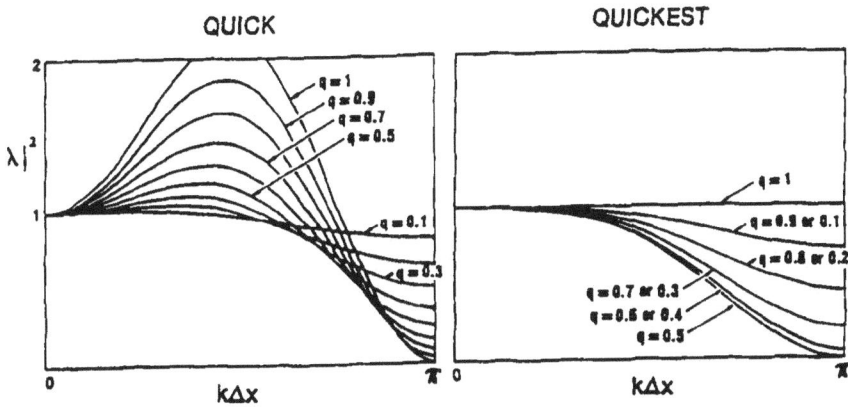

Fig. 2.17

Stability parameter as function of $q = uT/h$ for the QUICK and QUICK-EST.

The dependence of the stability parameter $|\lambda|^2$ on the Courant number is given in Fig.2.17 for the QUICK and for the QUICKEST. The QUICK is rather unstable and only for small values of the Courant number (≤ 0.1) this scheme is stable. The QUICKEST not only reproduces well the concentration jump but is unconditionally stable for $q \leq 1$. This algorithm suppresses the short waves effectively while the amplification factor for the longer waves approaches very fast the value of unity.

It seems that the QUICKEST is so successful, because Leonard (1979b) was able to incorporate into this numerical scheme the Lagrangian properties of the flow.

9.5 Flux-corrected transport

This method employs the equation of transport in the flux form. Such an equation is based on the principle of flux conservation. The difference equations constructed from the differential equations should as well display the same conservative properties. Consider a control surface spanned around grid point j, k – Fig.2.18.

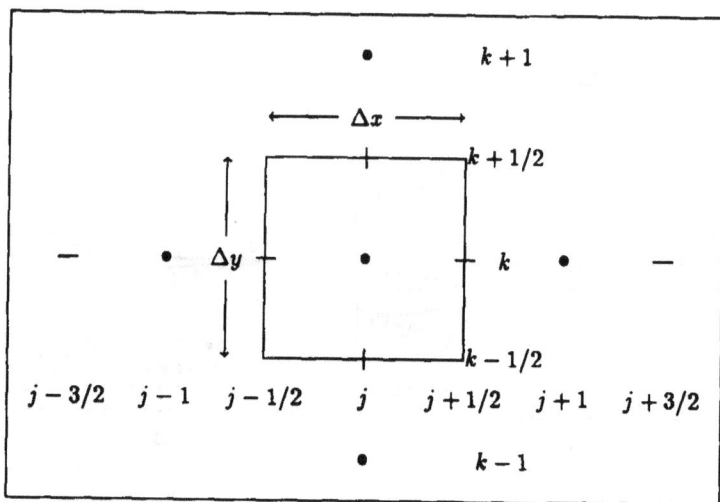

Fig. 2.18

Grid point distribution for the integration of the transport equation. Square box denotes control surface. Dots are located at the concentration point, dashes denote u-component of velocity and vertical bars denote v-component of velocity.

Let us integrate the equation of transport

$$\frac{\partial c}{\partial t} + \frac{\partial uc}{\partial x} + \frac{\partial vc}{\partial y} = 0 \tag{2.196}$$

over a rectangular region $dxdy$ and time dt, then

$$\int_0^T (\int_0^{\Delta x} \int_0^{\Delta y} \frac{\partial c}{\partial t} dxdy)dt = -\int_0^T [\int_0^{\Delta x} \int_0^{\Delta y} (\frac{\partial uc}{\partial x} + \frac{\partial vc}{\partial y})dxdy]dt \tag{2.197}$$

The left-hand side term is easily transformed into a finite-difference form. The surface integral on the right-hand side is transformed into a line integral along the boundary of the control surface. Thus

$$(c_{j,k}^{m+1} - c_{j,k}^m)\Delta x \Delta y = -\{[(uc)_{j+1/2,k}^{m+1/2} - (uc)_{j-1/2,k}^{m+1/2}]\Delta y$$

$$+[(vc)_{j,k+1/2}^{m+1/2} - (vc)_{j,k-1/2}^{m+1/2}]\Delta x\}T \qquad (2.198)$$

or dividing by $\Delta x \Delta y$ and T we arrive at the familiar finite-difference form

$$\frac{c_{j,k}^{m+1} - c_{j,k}^m}{T} = -\frac{(uc)_{j+1/2,k}^{m+1/2} - (uc)_{j-1/2,k}^{m+1/2}}{\Delta x} - \frac{(vc)_{j,k+1/2}^{m+1/2} - (vc)_{j,k-1/2}^{m+1/2}}{\Delta y} \qquad (2.199)$$

In the ensuing considerations we shall review the flux-corrected method only for the one-dimensional problem but both advective and diffusion flux will be included as

$$\frac{\partial c}{\partial t} + \frac{\partial F_x}{\partial x} = 0 \qquad (2.200)$$

Here, $F_x = uc - D_h \frac{\partial c}{\partial x}$.

Various higher order schemes for the transport diffusion equation usually display dispersive waves, although as in the QUICKEST the wave may be limited to one "ripple". Numerical schemes constructed from the first order directional derivatives do not generate dispersive waves but display an excessive numerical diffusion, therefore they can not reproduce abrupt changes in space. We are dealing with two extremities here. On one hand, the high accuracy scheme through the dispersive effects can generate a concentration field which is not positive definite, and on the other hand, the low-order numerical scheme can generate a positive definite value but very smoothed in space. The FCT method employs both numerical schemes to construct a high accuracy scheme without any dispersive effect. The method was developed by Boris and Book (1973), and Book and Boris (1975). In our discussion we shall follow the approach taken by Zalesak (1979). The FCT algorithm constructs the net fluxes point by point as a weighted average of a low-order scheme and a high-order scheme. This weighting is done in such manner as to maximize the use of the high order scheme without overshooting or undershooting. The FCT algorithm can be broken down into six steps.

1. Compute $F_{j+1/2}^L$ – flux by the low-order scheme. The purpose here is to derive a positive definite, or sometimes called monotonic, solution. The monotonic properties of the finite-difference methods were extensively discussed by Sod (1985). For our purpose it is only important that the monotonic solution is ripple-free.

2. Compute $F_{j+1/2}^H$ – flux by the high-order scheme.

3. Introduce an antidiffusive flux as

$$A_{j+1/2} = F_{j+1/2}^H - F_{j+1/2}^L \qquad (2.201)$$

4. Compute low order updated solution

$$c_j^{td} = c_j^m - \frac{T}{h}(F_{j+1/2}^L - F_{j-1/2}^L) \qquad (2.202)$$

5. Limit antidiffusion fluxes in such a way that the new concentration (c^{m+1}) is ripple-free. This is accomplished through limiter $L_{j+1/2}$ as

$$A_{j+1/2}^L = L_{j+1/2} A_{j+1/2}, \quad 0 \le L_{j+1/2} \le 1 \qquad (2.203)$$

6. Use the limited antidiffusive flux to derive concentration c_j^{m+1}

$$c_j^{m+1} = c_j^{td} - \frac{T}{h}(A_{j+1/2}^L - A_{j-1/2}^L) \qquad (2.204)$$

In the construction of the finite-difference formulas we shall use the staggered grid depicted in Fig.2.19.

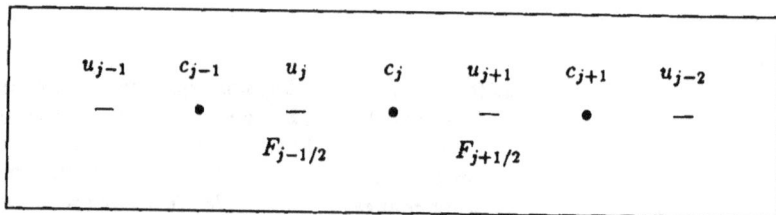

Fig. 2.19
Staggered grid distribution. Dash is velocity point and dot is concentration point.

The advective flux at the grid point $j + 1/2$ is defined as

$$F_{j+1/2}^C = \frac{c_{j+1} + c_j}{2} u_{j+1}$$

and the flux at the grid point $j - 1/2$ is

$$F_{j-1/2}^C = \frac{c_{j-1} + c_j}{2} u_j \qquad (2.205)$$

To perform computations by the low-order scheme we define flux as a function of the velocity direction

$$F_{j+1/2}^D = c_j \frac{u_{j+1} + |u_{j+1}|}{2} + c_{j+1} \frac{u_{j+1} - |u_{j+1}|}{2}$$

and

$$F^D_{j-1/2} = c_{j-1}\frac{u_j + |u_j|}{2} + c_j\frac{u_j - |u_j|}{2} \tag{2.206}$$

We have shown above the basic steps of the FCT, now we shall construct all the formulas pertinent to these steps. Basic numerical method will include both leapfrog and trapezoidal algoritm. Leapfrog algorithm is of second-order approximation, but it also generates a numerical solution. The trapezoidal step is applied to eliminate this parasitic solution. Let us again proceed along the steps of the FCT method.

1. To calculate flux by the low-order scheme we use two steps:

Step A (leapfrog)

$$\tilde{c}_j = c_j^{m-1} - 2T\frac{\partial}{\partial x}[F^D(c^m)]$$

Step B (trapezoidal)

$$c_j^{m+1} = c_j^m - T\frac{\partial}{\partial x}[\frac{F^D(c^m) + F^D(\tilde{c})}{2}] \tag{2.207}$$

The low-order flux $F^L_{j+1/2}$ is defined from (2.207) as

$$F^L_{j+1/2} = \frac{F^D(c^m) + F^D(\tilde{c})}{2}$$

$$= \frac{c_j^m + \tilde{c}_j}{2}(\frac{u_{j+1} + |u_{j+1}|}{2}) + \frac{c_{j+1}^m + \tilde{c}_{j+1}}{2}(\frac{u_{j+1} - |u_{j+1}|}{2}) \tag{2.208}$$

For the ensuing consideration we shall also need flux at the $j - 1/2$ grid point; it can be easily obtained from (2.206) as

$$F^L_{j-1/2} = \frac{c_{j-1}^m + \tilde{c}_{j-1}}{2}(\frac{u_j + |u_j|}{2}) + \frac{c_j^m + \tilde{c}_j}{2}(\frac{u_j - |u_j|}{2}) \tag{2.209}$$

2. To compute the flux by the high-order scheme we shall repeat the above calculations but this time the central derivatives will be employed.

Step A (leapfrog)

$$\hat{c}_j = c_j^{m-1} - 2T\frac{\partial}{\partial x}[F^C(c^m)]$$

Step B (trapezoidal)

$$c_j^{m+1} = c_j^m - T\frac{\partial}{\partial x}[\frac{F^C(c^m) + F^C(\hat{c})}{2}] \tag{2.210}$$

From (2.210) the flux $F^H_{j+1/2}$ follows

$$F^H_{j+1/2} = \frac{F^C(c^m) + F^C(\hat{c})}{2} = \frac{F^C_{j+1/2}(c^m) + F^C_{j+1/2}(\hat{c})}{2}$$

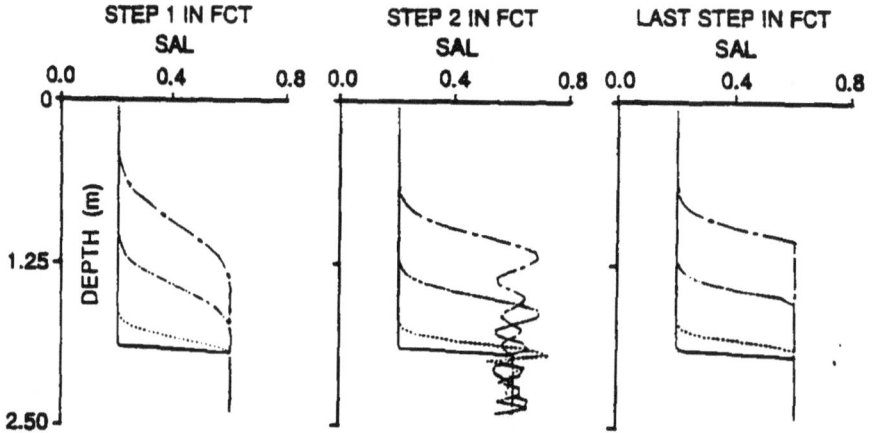

Fig. 2.20

Basic steps of FCT. (1) Result due to the low-order scheme. (2) Results due to the high-order scheme. (last) Results due to the FCT.

$$= 0.25(c_j + c_{j+1} + \widehat{c}_j + \widehat{c}_{j+1})u_{j+1} \qquad (2.211)$$

This result is directly related to (2.205).

3. Calculation of the antidiffusive flux will be done accordingly (2.201) as

$$A_{j+1/2} = F^H_{j+1/2} - F^L_{j+1/2} = (2.210) - (2.208) \qquad (2.212)$$

4. Updated low order solution is given by (2.202) with the low-order fluxes defined by (2.208) and (2.209).

5. This step through a limiter $L_{j+1/2}$ interacts with the antidiffusive flux derived in step 3 so that the resulting solution is ripple-free. The general interaction is expressed by (2.203) The concrete form of the algorithm was originally proposed by Boris and Book (1973)

$$A^L_{j+1/2} = Sgn(A_{j+1/2})\text{Max}\{0, \text{Min}[|A_{j+1/2}|, Sgn(A_{j+1/2})(c^{td}_{j+2} - c^{td}_{j+1})h/T,$$

$$Sgn(A_{j+1/2})(c^{td}_j - c^{td}_{j-1})h/T]\} \qquad (2.213)$$

Here

$$Sgn = \begin{cases} 1, & \text{if } A_{j+1/2} \geq 0; \\ -1, & \text{if } A_{j+1/2} < 0. \end{cases}$$

The generalization of this algorithm and extension to a multidimensional problem have been given by Zalesak (1979).

6. This is the final step and here (2.204) is applied to derive the new value of concentration.

To illustrate the above procedure we have calculated convective transport due to the vertical velocity $w = 10^{-4}$ cm/s of the initial concentration jump located at 1.875 m of depth. A time step of integration is equal $T = 10^3$ s and space step is $h = 5$ cm. The results of the computations are given for the low order numerical scheme in Fig.2.20a, for the high-order numerical scheme in Fig.2.20b, and FCT method is depicted in Fig.2.20c. The concentration change is plotted starting from the initial moment for 100, 500, and 1000 time steps.

9.6 Methods-of-Moments advection scheme

The numerical method developed by Egan and Mahoney (1972) and further improved by Pedersen and Prahm (1974), referred to as the Method-of-Moments scheme, eliminates numerical diffusion almost completely in the advective algorithm. Compared to a simple explicit numerical algorithm it requires larger computer storage and computational time. The increased accuracy is acquired by conserving not only mass (zero moment of concentration), but also the first and second moments (the center of mass and variance, respectively). The mass in each computational cell is followed in a Lagrangian manner over each time step, and the resulting concentration distribution is decomposed onto the Eulerian grid at the end of each time step. The methods simulates the advection of a square distribution exactly; the nonrectangular distribution are also advected with little error. It is a perfect method for the advective process but it does not reproduce well the diffusive processes. In this numerical scheme one has to keep track of three quantities, i.e., the mass, the center of mass and the width of mass. As usually the grid length is given by h. To facilitate the ensuing calculation a non-dimensional distance ζ is given as

$$\zeta = \frac{x - jh}{h} \tag{2.214}$$

it varies from $-1/2$ to $1/2$ over each grid cell. The left wall of the j cell is located at $x = (j-1)h + h/2$ and the right wall is at $x = jh + h/2$.

The total mass, the center of mass and the width in the j cell can be defined as

$$m_j = \int_{-0.5}^{0.5} c_j(\zeta) d\zeta \tag{2.215}$$

$$cm_j = \frac{\int_{-0.5}^{0.5} c_j(\zeta)\zeta d\zeta}{m_j} \tag{2.216}$$

$$w_j^2 = \frac{12 \int_{-0.5}^{0.5} c_j(\zeta)(\zeta - cm_j)^2 d\zeta}{m_j} \tag{2.217}$$

During each time step, the distribution in each cell is advected with a nondimensional velocity (Courant number), $q = uT/h$. It is assumed that $q \leq 1$ so that the concentration moments in a given cell at every time step can be related to the moments at the previous time step in this cell and in the upwind cell only.

To elucidate some of the above introduced parameters let us consider a rectangular distribution shown in Fig.2.21.

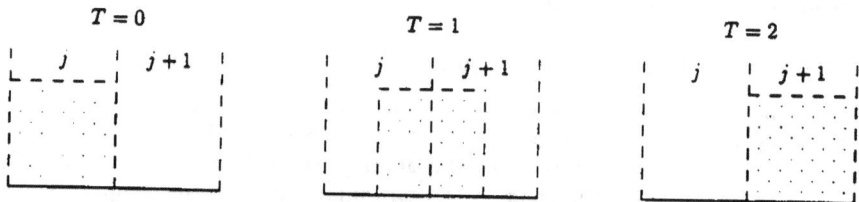

Fig. 2.21

The advection of mass by positive nondimensional velocity $q = 0.5$ from cell j to $j + 1$.

The nondimensional velocity $q = 0.5$. At the initial moment $T = 0$, $c_j^0(\zeta) = 1, m_j^0 = 1, cm_j^0 = 0$, and $w_j^0 = 1$. Now, it also becomes clear why factor 12 has been introduced into (2.217). At the first time step ($T = 1$) with $q = 0.5$,

$$c_j^1(\zeta) = \begin{cases} 0, & \text{if } -0.5 \leq \zeta \leq 0 \\ 1, & \text{if } 0 \leq \zeta \leq 0.5 \end{cases}$$

and $m_j^1 = 0.5, cm_j^1 = 1/4, w_j^1 = 0.5$. In the $(j + 1)$ cell, $m_{j+1}^1 = 0.5, cm_{j+1}^1 = -1/4$ and $w_{j+1}^1 = 1/2$. At the time step ($T = 2$) the mass present in the j and $j + 1$ cell is moved and at the end of this step the total mass in the j cell equals to 0 and in the $j + 1$ cell the mass is 1.

In order to facilitate the ensuing computations the portioning parameter can be introduced,

$$p_j = \frac{2\text{sign}(q_j)(cm_j + q_j) + w_j - 1}{2w_j} \tag{2.218}$$

The value of p_j indicates what fraction of the material in j cell is transported into $(j+1)\text{sign}(q_j)$ cell. If $0 < p_j < 1$ only a fraction p_j is transported; if $p_j < 0$ none of the mass is transferred but if $p_j > 0$ all the mass is transported from j cell to the neighboring cell. The basic three parameters and portioning parameter are shown in Fig.2.22 by considering the advection of a rectangular distribution.

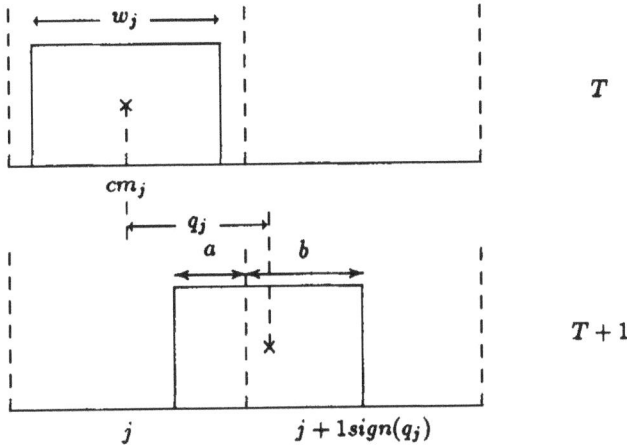

Fig. 2.22

Basic parameters of Methods-of-Moments, w_j — width, cm_j — center of mass, q_j — nondimensional velocity, p_j — portioning parameter, $a = (1 - p_j)w_j$, $b = p_j w_j$.

Let us consider the case of the positive flow in which material is advected into the j cell from the $j-1$ cell, and outflowing to the neighboring $j+1$ cell (Egan and Mahoney, 1972).

The time stepping at the j grid location can be described as

$$m_j^{T+1} = m_a + m_r \tag{2.219}$$

$$cm_j^{T+1} = (m_a cm_a + m_r cm_r)/m_j^{T+1} \tag{2.220}$$

$$(w^2)_j^{T+1} = \left[m_r w_r^2 + m_a w_a^2 + 12 m_r (cm_j^{T+1} - cm_r)^2 + 12 m_a (cm_j^{T+1} - cm_a)^2 \right]/m_j^{T+1} \tag{2.221}$$

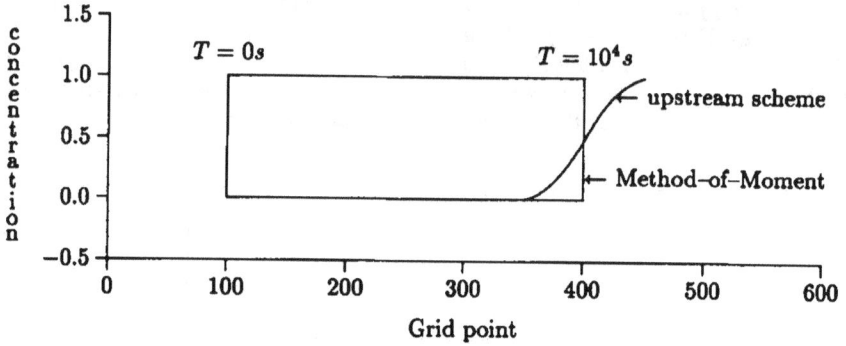

Fig. 2.23

Advection of the concentration jump from initial moment till $T = 10^4$s reproduced by Method-of-Moment and an advective scheme for $q = uT/h = 0.03$.

Here subscripts r and a denotes values remaining and newly-advected into the j cell. To derive these parameters the portioning parameter p_j is used. Contribution to the j cell is governed by the following set of rules:

$$p_j < 0 \qquad m_r = m_j^T, cm_r = m_j^T + q_j, w_r = w_j^T$$

$$p_{j-1} < 0 \qquad m_a = 0, cm_a = 0, w_a = 0$$

$$p_j > 1 \qquad m_r = 0, cm_r = 0, w_r = 0$$

$$p_{j-1} > 1 \qquad m_a = m_{j-1}^T, cm_a = cm_{j-1}^T + q_{j-1} - 1, w_a = w_{j-1}^T$$

$$0 < p_j < 1 \qquad m_r = (1 - p_j)m_j^T, cm_r = (1 - w_j^T + p_j w_j^T)/2, w_r = (1 - p_j)w_j^T$$

$$0 < p_{j-1} < 1 \qquad m_a = p_{j-1}m_{j-1}^T, cm_a = (p_{j-1}w_{j-1}^T - 1)/2, w_a = p_{j-1}w_{j-1}^T \quad (2.222)$$

The accuracy and speed of the method of moments scheme is evaluated by comparing to the results obtained by the explicit in time and the upstream in space difference scheme. The simulation is carried out with the Courant number $q = 0.03$. The computational time was 10 times shorter for the upstream scheme, but the methods of moments reproduces perfectly the advection of a concentration step, see Fig.2.23.

9.7 Convective process — numerical solution on the nonuniform grid

Either a poor space or time approximation may result in the errors which distort the physical solution. Nonuniform grid used to better describe the vertical distributions of temperature or velocity leads often to a finite difference approximation of first order only. Therefore, along with the physical coefficient of diffusion a numerical coefficient will occur having a value close to the physical value. We shall consider this phenomenon for the vertical convection described by

$$\frac{\partial \theta}{\partial t} + \frac{\partial(w\theta)}{\partial z} = 0 \tag{2.223}$$

In space, numerical solution of this problem is carried out on the staggered grid depicted in Fig.2.24. In time, the leapfrog method is applied.

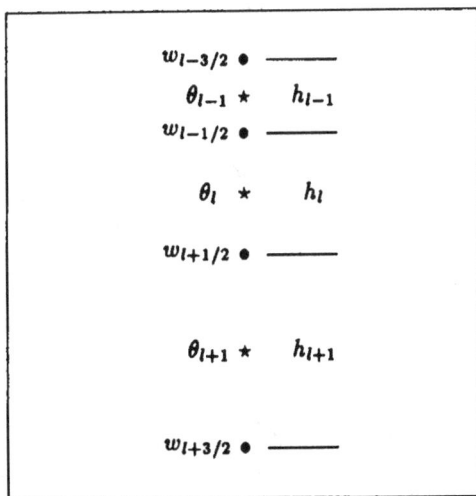

Fig. 2.24
Nonuniform grid distribution, θ – temperature, w, velocity, h_l grid distance.

Let us assume that the grid distance is uniform and equal to h, then space derivative at the point l is

$$\frac{\partial(w\theta)}{\partial z} \simeq \left(\frac{\theta_{l-1} + \theta_l}{2} w_{l-1/2} - \frac{\theta_{l+1} + \theta_l}{2} w_{l+1/2}\right)/h \tag{2.224}$$

To investigate the order of approximation of eq.(2.224) we define with the help of a Taylor series the temperature (θ) at the velocity (w) points (see Sec.1 of this chapter). First the temperature at the grid points $l-1$ and l is defined by the values at the $l-1/2$ point:

$$\theta_{l-1} = \theta_{l-1/2} + h/2\frac{\partial \theta}{\partial z} + \frac{(h/2)^2}{2}\frac{\partial^2 \theta}{\partial z^2} + \frac{(h/2)^3}{3!}\frac{\partial^3 \theta}{\partial z^3} + O(h^4) \tag{2.225a}$$

$$\theta_l = \theta_{l-1/2} - h/2\frac{\partial\theta}{\partial z} + \frac{(h/2)^2}{2}\frac{\partial^2\theta}{\partial z^2} - \frac{(h/2)^3}{3!}\frac{\partial^3\theta}{\partial z^3} + O(h^4) \qquad (2.225b)$$

and then the values at the l and $l+1$ grid points are defined by the values and derivatives at the $l+1/2$ grid point:

$$\theta_l = \theta_{l+1/2} + h/2\frac{\partial\theta}{\partial z} + \frac{(h/2)^2}{2}\frac{\partial^2\theta}{\partial z^2} + \frac{(h/2)^3}{3!}\frac{\partial^3\theta}{\partial z^3} + O(h^4) \qquad (2.225c)$$

$$\theta_{l+1} = \theta_{l+1/2} - h/2\frac{\partial\theta}{\partial z} + \frac{(h/2)^2}{2}\frac{\partial^2\theta}{\partial z^2} - \frac{(h/2)^3}{3!}\frac{\partial^3\theta}{\partial z^3} + O(h^4) \qquad (2.225d)$$

Introducing eqs.(2.225a)–(2.225d) into eq.(2.224) we arrive at

$$\frac{\partial(w\theta)}{\partial z} \simeq (w_{l-1/2}\theta_{l-1/2} - w_{l+1/2}\theta_{l+1/2})/h$$

$$+ \left[w_{l-1/2}\frac{\partial^2\theta}{\partial z^2}\rfloor_{(l-1/2)} - w_{l+1/2}\frac{\partial^2\theta}{\partial z^2}\rfloor_{(l+1/2)}\right] h/8 + O(h^3) \qquad (2.226)$$

By developing the second term on the right-hand side into a Taylor series the second order of approximation follows

$$\frac{\partial(w\theta)}{\partial z} \simeq (w\theta)'_l + \frac{h^2}{8}(w\theta'')'_l \qquad (2.227)$$

In case of a nonuniform spatial grid there are quite a few possibilities to derive a numerical form of the convective term. We shall write it as before setting the average temperature at the velocity points. One approach is to use a similar averaging process as in eq.(2.223). The different approach with the better approximation is to apply the weighted averages,

$$\frac{\partial(w\theta)}{\partial z} = \frac{\partial(w\tilde\theta)}{\partial z} = \frac{w_{l-1/2}\tilde\theta_{l-1/2} - w_{l+1/2}\tilde\theta_{l+1/2}}{h_l} \qquad (2.228)$$

The average temperatures in the above equations are

$$\tilde\theta_{l-1/2} = (h_{l-1}\theta_l + h_l\theta_{l-1})/(h_{l-1} + h_l) \qquad (2.229a)$$

$$\tilde\theta_{l+1/2} = (h_l\theta_{l+1} + h_{l+1}\theta_l)/(h_l + h_{l+1}) \qquad (2.229b)$$

Again, to investigate the order of approximation we shall develop these averages into a Taylor series at the middle points $l-1/2$ and $l+1/2$:

$$\theta_{l-1} = \theta_{l-1/2} + (h_{l-1}/2)\frac{\partial\theta}{\partial z} + \frac{(h_{l-1}/2)^2}{2}\frac{\partial^2\theta}{\partial z^2} + \frac{(h_{l-1}/2)^3}{3!}\frac{\partial^3\theta}{\partial z^3} + O(h_{l-1}^4) \quad (2.230a)$$

$$\theta_l = \theta_{l-1/2} - (h_l/2)\frac{\partial\theta}{\partial z} + \frac{(h_l/2)^2}{2}\frac{\partial^2\theta}{\partial z^2} - \frac{(h_l/2)^3}{3!}\frac{\partial^3\theta}{\partial z^3} + O(h_l^4) \qquad (2.230\text{b})$$

$$\theta_l = \theta_{l+1/2} + (h_l/2)\frac{\partial\theta}{\partial z} + \frac{(h_l/2)^2}{2}\frac{\partial^2\theta}{\partial z^2} + \frac{(h_l/2)^3}{3!}\frac{\partial^3\theta}{\partial z^3} + O(h_l^4) \qquad (2.230\text{c})$$

$$\theta_{l+1} = \theta_{l+1/2} - (h_{l+1}/2)\frac{\partial\theta}{\partial z} + \frac{(h_{l+1}/2)^2}{2}\frac{\partial^2\theta}{\partial z^2} - \frac{(h_{l+1}/2)^3}{3!}\frac{\partial^3\theta}{\partial z^3} + O(h_{l+1}^4) \quad (2.230\text{d})$$

Introducing eqs.(2.230a) and (2.230b) into eq.(2.229a), we arrive at

$$\tilde{\theta}_{l-1/2} = \theta_{l-1/2} + \frac{\partial^2\theta}{\partial z^2}\Big]_{l-1/2}\frac{h_l h_{l-1}}{8} \qquad (2.231\text{a})$$

and in a similar manner introducing eqs.(2.230c) and (2.230d) into eq.(2.229b)

$$\tilde{\theta}_{l+1/2} = \theta_{l+1/2} + \frac{\partial^2\theta}{\partial z^2}\Big]_{l+1/2}\frac{h_l h_{l+1}}{8} \qquad (2.231\text{b})$$

Equations (2.231a) and (2.231b) introduced into eq.(2.227) give the following expression for the first derivative on the nonuniform grid:

$$\frac{\partial(w\theta)}{\partial z} \simeq (w_{l-1/2}\theta_{l-1/2} - w_{l+1/2}\theta_{l+1/2})/h_l$$

$$+ w_{l-1/2}\frac{\partial^2\theta}{\partial z^2}\Big]_{(l-1/2)}\frac{h_{l-1}}{8} - w_{l+1/2}\frac{\partial^2\theta}{\partial z^2}\Big]_{(l+1/2)}\frac{h_{l+1}}{8} \qquad (2.232)$$

We shall again develop the terms containing second derivatives into a Taylor series as

$$w_{l-1/2}\frac{\partial^2\theta}{\partial z^2}\Big]_{(l-1/2)}\frac{h_{l-1}}{8} - w_{l+1/2}\frac{\partial^2\theta}{\partial z^2}\Big]_{(l+1/2)}\frac{h_{l+1}}{8}$$

$$= w_l\frac{\partial^2\theta}{\partial z^2}\Big]_l(\frac{h_{l-1} - h_{l+1}}{8}) + O(h_l h_{l-1} + h_l h_{l+1})$$

Using the identity

$$(w\theta')' = w'\theta' + w\theta''$$

eq.(2.232) can be rewritten in the form

$$\frac{\partial(w\theta)}{\partial z} = (w\theta)'_l + \tilde{\Delta}(w\theta'')_l = (w\theta)'_l - \tilde{\Delta}(w'\theta')_l + \tilde{\Delta}(w\theta')'_l + O(h_l h_{l-1} + h_l h_{l+1}) \quad (2.233)$$

Here $\tilde{\Delta} = (h_{l-1} - h_{l+1})/8$. Thus, the second order of approximation can not be assumed unless we compensate for the two terms in eq.(2.233), namely $-\tilde{\Delta}(w'\theta')_l + \tilde{\Delta}(w\theta')'_l$. These are due to the nonuniform grid distribution, and if the uniform grid is used ($\tilde{\Delta} = 0$) eq.(2.227) will follow. The first term in the above expression according to Yin and Fung (1991) is a small correction to the convective (advective) term. The second term defines numerical diffusion with eddy diffusion coefficient $D_{num} = -\tilde{\Delta}w$. To compensate for the numerical diffusion term we shall reverse the diffusion processss by the convective equation; an approach suggested by Smolarkiewicz (1983) (see Sec.9.2 of this chapter). To reverse in time the effect of the diffusion equation

$$\frac{\partial \theta}{\partial t} = \frac{\partial}{\partial z}(D_{num}\frac{\partial \theta}{\partial z})$$

(2.234)

the convection (advection) equation is introduced

$$\frac{\partial \theta}{\partial t} = -\frac{\partial}{\partial z}(u_d\theta)$$

(2.235)

Here,

$$u_d = \begin{cases} -D_{num}\theta^{-1}\frac{\partial \theta}{\partial z} & \text{if } \theta > 0 \\ 0 & \text{if } \theta = 0 \end{cases}$$

(2.236)

The diffusion velocity defined by this formula has to be reversed to correct for the numerical eddy diffusion. The reversed diffusion velocity ($\tilde{u}$) is defined as

$$\tilde{u} = \begin{cases} -u_d & \text{if } \theta > 0 \\ 0 & \text{if } \theta = 0 \end{cases}$$

(2.237)

The solution will proceed in time in the following way. In the first substep,

$$\frac{\theta^* - \theta^{m-1}}{2T} + \frac{\partial}{\partial z}(w\theta)^m = 0$$

(2.238a)

is solved. Here $\partial(w\theta)/\partial z$ is expressed by eq.(2.228). In the second substep the convection equation with reversed diffusion velocity is solved

$$\frac{\theta^{m+1} - \theta^*}{2T} = -\frac{\partial}{\partial z}(\tilde{u}\theta)^m$$

(2.238b)

In this way an error due to the numerical diffusion is corrected.

9.8 Undefined coefficients method

Here, we shall introduce a method for obtaining the higher order numerical schemes for the advective equation. However, we shall start by considering a method to derive the numerical expressions for the derivatives and equations. The method involves formula with unknown coefficients. The coefficients are to be defined by using a Taylor series and through comparison to an analytical expression. The method is not clearly defined, therefore we may call the whole process the search of the undefined coefficients through the undefined method. However, the results derived are both interesting and useful. We begin by recalling the expressions obtained by application of a Taylor series and defined in Sec.1 of this chapter. Considering the grid points along the x coordinate we can write

$$c_{j-1} = c_j - h\frac{\partial c}{\partial x} + \frac{h^2}{2}\frac{\partial^2 c}{\partial x^2} - \frac{h^3}{3!}\frac{\partial^3 c}{\partial x^3} + O(h^4) \tag{2.239}$$

$$c_{j+1} = c_j + h\frac{\partial c}{\partial x} + \frac{h^2}{2}\frac{\partial^2 c}{\partial x^2} + \frac{h^3}{3!}\frac{\partial^3 c}{\partial x^3} + O(h^4) \tag{2.240}$$

$$c_{j+2} = c_j + 2h\frac{\partial c}{\partial x} + \frac{2^2 h^2}{2}\frac{\partial^2 c}{\partial x^2} + \frac{2^3 h^3}{3!}\frac{\partial^3 c}{\partial x^3} + O(h^4) \tag{2.241}$$

For the time integration we shall use only two-time level schemes, therefore

$$c^{m+1} = c^m + T\frac{\partial c}{\partial t} + \frac{T^2}{2}\frac{\partial^2 c}{\partial t^2} + \frac{T^3}{3!}\frac{\partial^3 c}{\partial t^3} + O(T^3) \tag{2.242}$$

With the above notation let us write the simplest formula for the first derivative as

$$\frac{\partial c}{\partial x} \simeq \alpha c_j + \beta c_{j-1} \tag{2.243}$$

We propose here to construct a numerical expression for the first derivative based on the values of c function at the j and $j-1$ grid points with two unknown coefficients α and β. The c_{j-1} in this equation is developed into a series according to eq.(2.239). Comparing the terms at the left and right side of eq.(2.243), we conclude that at the left side of eq.(2.243) the free term (i.e., term which has no derivatives) is zero, thus,

$$0 = \alpha c_j + \beta c_j \tag{2.243a}$$

The first derivatives are present both at the left and the right side of eq.(2.243), and the equality of the coefficients of these derivatives gives

$$1 = -\beta h \tag{2.243b}$$

Combining eqs.(2.243a) and (2.243b) the unknown coefficients $\alpha = 1/h$ and $\beta = -1/h$ are obtained. Introducing these coefficients into eq.(2.243) the usual expression follows:

$$\frac{\partial c}{\partial x} = \frac{c_j - c_{j-1}}{h} \tag{2.244}$$

Undoubtedly this simple example allows us to understand the approach taken in the method.

Let us derive a more complicated expression for the first derivative, used in the construction of the upwind algorithm. The formula is based on the four grid points $j-1$, j, $j+1$ and $j+2$. Equation with the undefined coefficients is

$$\frac{\partial c}{\partial x} = \alpha c_j + \beta c_{j-1} + \gamma c_{j+1} + \delta c_{j+2} \tag{2.245}$$

Introducing (2.239), (2.240) and (2.241) into the right-hand side of eq.(2.245) the equations for the undefined coefficients are as follows.
Coefficients of the free terms:

$$0 = \alpha + \beta + \gamma + \delta \tag{2.246a}$$

Coefficients of the first derivatives:

$$1 = -\beta h + \gamma h + 2h\delta \tag{2.246b}$$

Coefficients of the second derivatives:

$$0 = \beta + \gamma + 4\delta \tag{2.246c}$$

Coefficients of the third derivatives:

$$0 = -\beta + \gamma + 8\delta \tag{2.246d}$$

Thus we have four equations to define four unknowns. Solving this system we arrive at

$$\alpha = -\frac{3}{6h}; \qquad \beta = -\frac{2}{6h}; \qquad \gamma = \frac{1}{h}; \qquad \delta = -\frac{1}{6h} \tag{2.247}$$

Introducing these coefficients into (2.245) a numerical form for the first derivative is obtained

$$\frac{\partial c}{\partial x} = (-2c_{j-1} - 3c_j + 6c_{j+1} - c_{j+2})/(6h) \tag{2.248}$$

This is exactly formula (2.149b) used in the construction of the upwind algorithm and it is of third order of approximation.

Having learned about the application of the method we shall try constructing not only the derivatives but the numerical forms of the differential equations as well. Let us start from the following suggestion

$$\frac{\partial c}{\partial t} + u \frac{\partial c}{\partial x} = \alpha c_j^{m+1} + \beta c_{j+1}^m + \gamma c_{j-1}^m \tag{2.249}$$

This time we shall also use eq.(2.242) to develop the terms on the right-hand side of eq.(2.249) into a Taylor series. The set of equations for the coefficients follows.

Coefficients of the free term:
$$0 = \alpha + \beta + \gamma \qquad (2.250\text{a})$$

Coefficients of the time derivatives:
$$1 = \alpha T \qquad (2.250\text{b})$$

Coefficients of the space derivatives:
$$u = \beta h - \gamma h \qquad (2.250\text{c})$$

From these equations

$$\alpha = \frac{1}{T}; \qquad \beta = \frac{1}{2}\left(\frac{u}{h} - \frac{1}{T}\right); \qquad \gamma = -\frac{1}{2}\left(\frac{u}{h} + \frac{1}{T}\right) \qquad (2.251)$$

Introducing α, β and γ into eq.(2.249), the construction of the finite difference equation for the advection is achieved

$$\frac{1}{T}(c_j^{m+1} - \frac{c_{j+1}^m + c_{j-1}^m}{2}) + \frac{u}{2h}(c_{j+1}^m - c_{j-1}^m) = 0 \qquad (2.252)$$

This is indeed a very interesting numerical formula. It is of first order of approximation in time and of second order in space and it is stable for $\frac{uT}{h} < 1$. Let us notice that by changing the second term in the time derivative from being a function of the point j only to an average of the two neighboring points $j + 1$ and $j - 1$, we have obtained a conditionally stable numerical scheme. A simple, two-time level numerical scheme in combination with the central first derivative in space leads to an unstable numerical scheme.

Continuing along the same way we shall introduce an equation to derive a two-time level numerical scheme for the advective equation with the second order approximation in time and space.

$$\frac{\partial c}{\partial t} + u\frac{\partial c}{\partial x} = \alpha c_j^{m+1} + \beta c_{j+1}^m + \gamma c_{j-1}^m + \delta c_j^m \qquad (2.253)$$

We shall proceed in a old fashion deriving equations for the coefficients

$$0 = \alpha + \beta + \gamma + \delta \qquad (2.254\text{a})$$

$$1 = \alpha T \qquad (2.254\text{b})$$

$$u = \beta h - \gamma h \qquad (2.254\text{c})$$

At the next step we should construct equations for the coefficients of the second derivatives but the following problem is encountered. If we use the coefficients of the time derivative it will result in the condition $\alpha = 0$, but α is already given by eq.(2.254b), and therefore the problem seems to be not well defined. To make the problem tractable we redefine the second derivative in time through the basic equation

$$\frac{\partial c}{\partial t} + u \frac{\partial c}{\partial x} = 0 \quad \text{as} \quad \frac{\partial^2 c}{\partial t^2} = u^2 \frac{\partial^2 c}{\partial x^2} \tag{2.255}$$

and introduce it into eq.(2.253). The equation for the coefficients of the second derivative (in space) follows,

$$0 = \alpha \frac{u^2 T^2}{2} + \beta \frac{h^2}{2} + \gamma \frac{h^2}{2} \tag{2.254d}$$

Notice that in the derivation of eq.(2.255) an assumption is made that the velocity is constant. It does not mean that this velocity must be constant everywhere in time and space; it is required to be constant only for the space distance equal to h and the time interval equal to T. Using eqs.(2.254) to find coefficients, we arrive at

$$\alpha = \frac{1}{T}; \quad \beta = -\frac{u^2 T}{2h^2} + \frac{u}{2h}$$

$$\gamma = \frac{u^2 T}{2h^2} - \frac{u}{2h}; \quad \delta = -\frac{1}{T} + \frac{u^2 T}{h^2} \tag{2.256}$$

Introducing these coefficients into eq.(2.253), the numerical form for the advective equation is

$$\frac{1}{T}(c_j^{m+1} - c_j^m) + \frac{u}{2h}(c_{j+1}^m - c_{j-1}^m) - \frac{u^2 T}{2h^2}(c_{j+1}^m + c_{j-1}^m - 2c_j^m) = 0 \tag{2.257}$$

Checking the order of approximation of this equation we introduce eqs.(2.239)–(2.242) into eq.(2.257) and obtain

$$\frac{\partial c}{\partial t} + \frac{T}{2} \frac{\partial^2 c}{\partial t^2} + u \frac{\partial c}{\partial x} + O(h^2) - \frac{u^2 T}{2} \frac{\partial^2 c}{\partial x^2} + O(h^2) = 0 \tag{2.258}$$

The first-order terms

$$\frac{T}{2} \frac{\partial^2 c}{\partial t^2} - \frac{u^2 T}{2} \frac{\partial^2 c}{\partial x^2}$$

according to eq.(2.255) compensate each other and bring the approximation up to second order in time and space. This equation is stable for $\frac{uT}{h} < 1$. One observation is very interesting — going back to Sec.9.2 we can see that the above approach is very close to the correction scheme proposed by Smolarkiewicz (1983). The correction in eq.(2.257) is done by incorporating the diffusion term with positive (numerical) viscosity coefficient equal to $\frac{u^2 T}{2}$. The above derived numerical scheme is close to the scheme obtained by Godunov and Riabenki (1977).

CHAPTER III

TWO–DIMENSIONAL NUMERICAL MODELS

1. Basic problems

1.1 Introductory remarks

In this chapter we shall construct numerical algorithms for the partial differential equations describing two-dimensional and time-dependent phenomena, like storm surges, tsunamis, and tides. Let us start from the system of equations for the shallow water introduced in Ch.I. These are equations of motion along E-W and N-S directions,

$$\rho_o\left(\frac{\partial u}{\partial t} + u\frac{\partial u}{\partial x} + v\frac{\partial u}{\partial y} - fv\right) = -\frac{\partial p_a}{\partial x} - g\rho_o\frac{\partial \zeta}{\partial x} + \tau_x^a/D - \tau_x^b/D + \rho_o N_h \Delta u \quad (3.1)$$

$$\rho_o\left(\frac{\partial v}{\partial t} + u\frac{\partial v}{\partial x} + v\frac{\partial v}{\partial y} + fu\right) = -\frac{\partial p_a}{\partial y} - g\rho_o\frac{\partial \zeta}{\partial y} + \tau_y^a/D - \tau_y^b/D + \rho_o N_h \Delta v \quad (3.2)$$

and equation of continuity,

$$\frac{\partial}{\partial x}uD + \frac{\partial}{\partial y}vD + \frac{\partial \zeta}{\partial t} = 0 \qquad (3.3)$$

In the above set of equations, to simplify the notation, the tilde which denotes the vertical averaging has been omitted. The total depth is $D = H + \zeta$. Equations are defined in the space domain Ω, the boundary condition are prescribed at the boundary Γ, initial condition is set at time $t = 0$. The boundary and initial conditions may be defined in various ways, but we hope that the problem posed above has a solution and possibly the derived solution is unique. Before starting numerical considerations let us try to elucidate a few problems associated with the equations, boundary and initial conditions. If we manage to specify to what type of equations our set belongs, we might be able to find out how to specify the boundary conditions and how many boundary conditions will be necessary to derive a unique solution. The prescription of the initial conditions may be difficult if one invokes use of observed data, therefore we shall try to find out how the time varying process depends on the initial condition, hoping that the influence of the initial state will decay in time, and time history of the process will be defined by external forces only.

One has to remember that the construction of a well-behaving numerical algorithm is only the first step towards building a meaningful solution to the problem.

Numerical solution requires high quality measured data for the construction of the initial and boundary conditions. Data are also needed for the verification and validation of the numerical solutions. An important way of establishing the reliability of the numerical results is through comparison against the available analytical solutions.

1.2 Energy equation

We shall start by constructing the energy equation because it is closely related to the mathematical problems we have just mentioned. In order to derive this equation let us multiply (3.1) by Du, (3.2) by Dv, and (3.3) by $g\zeta$. Adding the resulting equation on either side, we arrive at

$$\frac{1}{2}\frac{\partial}{\partial t}\left[D(u^2+v^2)+g\zeta^2\right] + \frac{\partial}{\partial x}\{uD[(u^2+v^2)/2+g\zeta]\} + \frac{\partial}{\partial y}\{vD[(u^2+v^2)/2+g\zeta]\}$$

$$= \frac{1}{\rho_o}(\tau_x^s u + \tau_y^s v - \tau_x^b u - \tau_y^b v) + uDN_h\Delta u + vDN_h\Delta v - \frac{1}{\rho_o}\frac{\partial p_a}{\partial x}uD - \frac{1}{\rho_o}\frac{\partial p_a}{\partial y}vD \quad (3.4)$$

To observe the behavior of this scalar we shall carry integration over the domain Ω. Here we assume that the component of velocity normal to the boundary Γ is equal to zero. This represents a water body closed by coastline. Additionally we shall assume that both the external forces and the frictional forces are also zero. Under these assumptions

$$\frac{1}{2}\frac{\partial}{\partial t}\int\int\left[D(u^2+v^2)+g\zeta^2\right]dxdy = 0 \quad (3.5)$$

the total energy (kinetic and potential)

$$\frac{1}{2}\int\int\left[D(u^2+v^2)+g\zeta^2\right]dxdy = \text{Const} \quad (3.5a)$$

is conserved in time. Because the total energy in the nonfrictional system is conserved in time, it serves well as a tool for testing the behavior of the numerical algorithms. The second and the third terms in (3.4) represent the components of an energy flux vector. This vector characterizes the flux of energy through a unit width surface extended from the surface to the bottom (Kowalik and Untersteiner, 1978). The components of the vector along latitude and longitude are

$$\mathbf{E_h} = \{\rho_o uD[(u^2+v^2)/2+g\zeta],\rho_o vD[(u^2+v^2)/2+g\zeta]\} \quad (3.6)$$

The first order approximation (not valid in very shallow water)

$$\mathbf{E_{h1}} = \{\rho_o gHu\zeta,\rho_o gHv\zeta\} \quad (3.6a)$$

serves often to describe the energy transfer from the source to various locations. Long waves propagating in an oceanic basin are a superposition of traveling and standing waves. The latter usually occurs due to reflection from the coast. Energy flux due to the standing wave when averaged over the wave period is equal to zero; only a traveling wave transports the energy. The magnitude of the energy flux can also serve to characterize energy transfer in the long wave of variable period such as tsunami (Kowalik and Whitmore, 1991).

The remaining terms in (3.4) define the energy sources (here the energy input is caused by the wind stress and surface pressure), and the energy sinks due to the horizontal and bottom friction.

1.3 What type of equation are we dealing with?

The identification of the type of equation will help to define uniqueness of the solution, the boundary and initial conditions. First, we simplify somewhat the basic equation by deleting the horizontal frictional terms; they usually play a secondary role in the shallow water dynamics. Secondly, we rewrite the basic set in the matrix notation because conditions defining the type of equation are best expressed in this notation. Introducing vector $\mathbf{u}(u, v, \zeta)$ and matrices,

$$\mathbf{u} = \begin{pmatrix} u \\ v \\ \zeta \end{pmatrix}; A = \begin{pmatrix} u & 0 & g \\ 0 & u & 0 \\ D & 0 & u \end{pmatrix}; B = \begin{pmatrix} v & 0 & 0 \\ 0 & v & g \\ 0 & D & v \end{pmatrix}; C = \begin{pmatrix} 0 & -f & 0 \\ f & 0 & 0 \\ \partial H/\partial x & \partial H/\partial y & 0 \end{pmatrix}$$
$$(3.7a)$$

we can rewrite the basic set of equations as,

$$\frac{\partial \mathbf{u}}{\partial t} + A\frac{\partial \mathbf{u}}{\partial x} + B\frac{\partial \mathbf{u}}{\partial y} + C\mathbf{u} = \mathbf{F} \tag{3.7}$$

Here vector $\mathbf{F}$ has as components all external forces including bottom stress (which is not really an external force).

For the system (3.7) to be hyperbolic at a point $P(x, y, t)$ of the domain Ω, t the eigenvalues λ of the linear combination $\alpha_1 A + \alpha_2 B$ (given that $\alpha_1^2 + \alpha_2^2 = 1$) should be real and different at the point P (Courant and Hilbert, 1962). The eigenvalues of this linear combination are derived from the determinant

$$\text{Det}(\alpha_1 A + \alpha_2 B - \lambda E) = 0 \tag{3.8}$$

Here E denotes unit matrix. Thus,

$$\lambda_1 = u\alpha_1 + v\alpha_2, \quad \text{and} \quad \lambda_{2,3} = \lambda_1 \pm \sqrt{gD} \tag{3.9}$$

and, therefore, the system is hyperbolic at every point of Ω, t. The number of boundary conditions required for a unique solution of the hyperbolic system equals the number of negative eigenvalues. To deduce conclusions from the above statement for the set (3.7) let us first tackle a simple problem when the nonlinear (advective) terms in the equations of motion are negligible. From (3.8) two eigenvalues follow:

$$\lambda_{1,2} = \pm\sqrt{gD} \qquad (3.10)$$

and one is always negative. In this case, just one boundary condition provides a unique solution. This boundary condition may be posed at the closed boundary (like shoreline), or at the open boundary (any line between two water bodies). Let us choose Ω as rectangular domain, and consider boundary conditions along the boundary $y = $ Const. The normal velocity to this boundary is v, and from the general case described by (3.9), it follows that

$$\lambda_1 = \alpha_2 v \quad \text{and} \quad \lambda_{2,3} = \alpha_2 v \pm \sqrt{gD} \qquad (3.11)$$

Assuming $\sqrt{gD} > \alpha_2 v$, a somewhat disconcerting conclusion follows from (3.11); if the term $\alpha_2 v$ is positive, only one eigenvalue is negative, and only one boundary condition is required; if on the other hand this term is negative two boundary conditions are needed to derive a unique solution. Negative or positive velocity is related to inflow or outflow from the domain Ω. The latter conclusion is also valid for a closed boundary where the normal component of velocity is equal to zero.

The uniqueness of the solution can often be associated with the existence of positive definite operators, again we shall see that the boundary condition may influence the positive definite properties. For further considerations we shall simplify the basic set of equations (3.1)–(3.3) as

$$\frac{\partial u}{\partial t} = -g\frac{\partial \zeta}{\partial x} - Ru \qquad (3.12a)$$

$$\frac{\partial v}{\partial t} = -g\frac{\partial \zeta}{\partial y} - Rv \qquad (3.12b)$$

$$\frac{\partial \zeta}{\partial t} + \frac{\partial}{\partial x}(Hu) + \frac{\partial}{\partial y}(Hv) \qquad (3.12c)$$

In the above set the nonlinear advective terms have been neglected, depth is taken as constant, and $D = H$, the frictional terms are represented by bottom friction, and this friction is a linear function of velocity. By introducing a new variable $\zeta' = \zeta\sqrt{g/H}$ the above set assumes a symmetrical form,

$$\frac{\partial u}{\partial t} + A'\mathbf{u} + B'\mathbf{u} + C'\mathbf{u} = 0 \qquad (3.13)$$

Here $\mathbf{u} = \mathbf{u}(u, v, \zeta')$ and,

$$A' = \begin{pmatrix} 0 & 0 & \sqrt{gH}\partial/\partial x \\ 0 & 0 & 0 \\ \sqrt{gH}\partial/\partial x & 0 & 0 \end{pmatrix} ; \quad B' = \begin{pmatrix} 0 & 0 & 0 \\ 0 & 0 & \sqrt{gH}\partial/\partial y \\ 0 & \sqrt{gH}\partial/\partial y & 0 \end{pmatrix} ;$$

$$C' = \begin{pmatrix} R & 0 & 0 \\ 0 & R & 0 \\ 0 & 0 & 0 \end{pmatrix} \qquad (3.13a)$$

Besides the column vector **u** we shall introduce an adjoint row vector **û**,

$$\mathbf{u} = \begin{pmatrix} u \\ v \\ \zeta' \end{pmatrix} \qquad \hat{\mathbf{u}} = (u \quad v \quad \zeta') \qquad (3.14)$$

This allows us to define a scalar product in the domain Ω, t, as

$$(\hat{\mathbf{u}}, \mathbf{u}) = u^2 + v^2 + \zeta'^2 \qquad (3.15)$$

Multiplying (3.13) by vector **û** we arrive at the following scalar product:

$$(\hat{\mathbf{u}}, \frac{\partial \mathbf{u}}{\partial t}) + (\hat{\mathbf{u}}, A'\mathbf{u}) + (\hat{\mathbf{u}}, B'\mathbf{u}) + (\hat{\mathbf{u}}, C'\mathbf{u})$$

$$= \frac{1}{2}\frac{\partial}{\partial t}(u^2 + v^2 + \zeta'^2) + \sqrt{gH}\frac{\partial}{\partial x}(\zeta'u) + \sqrt{gH}\frac{\partial}{\partial y}(\zeta'v) + R(u^2 + v^2) \qquad (3.16)$$

The first term on the right-hand side of (3.16) expresses the change of total energy in time, and has been introduced by (3.5). The second and third terms describe the energy flux, and the last term is the energy sink caused by bottom friction. Integrating (3.16) over the space and time domain we obtain

$$Q = \int\int\int [\frac{1}{2}\frac{\partial}{\partial t}(\hat{\mathbf{u}}, \mathbf{u}) + (\hat{\mathbf{u}}, A'\mathbf{u}) + (\hat{\mathbf{u}}, B'\mathbf{u}) + (\hat{\mathbf{u}}, C'\mathbf{u})]dxdydt$$

$$= \int\int\int [\frac{1}{2}\frac{\partial}{\partial t}(u^2 + v^2 + \zeta'^2) + \sqrt{gH}\frac{\partial}{\partial x}(\zeta'u) + \sqrt{gH}\frac{\partial}{\partial y}(\zeta'v) + R(u^2 + v^2)]dxdydt$$
$$(3.17)$$

Assuming that the normal component of velocity to the boundary Γ is zero, the energy flux vanishes, and the above scalar product simplifies to

$$Q_1 = \int\int\int [\frac{1}{2}\frac{\partial}{\partial t}(\hat{\mathbf{u}}, \mathbf{u}) + (\hat{\mathbf{u}}, C'\mathbf{u})]dxdydt$$

$$= \int\int\int [\frac{1}{2}\frac{\partial}{\partial t}(u^2 + v^2 + \zeta'^2) + R(u^2 + v^2)]dxdydt \qquad (3.18)$$

The change of energy in time can be written explicitly,

$$\int\int\int [\frac{1}{2}\frac{\partial}{\partial t}(\hat{\mathbf{u}}, \mathbf{u})dxdydt = \int\int\int [\frac{1}{2}\frac{\partial}{\partial t}(u^2 + v^2 + \zeta'^2)]dxdydt$$

$$= \int\int [\frac{1}{2}(u^2 + v^2 + \zeta'^2)]\,dxdy]_{t=t} - \int\int [\frac{1}{2}(u^2 + v^2 + \zeta'^2)]\,dxdy]_{t=0}$$

$$= E(t) - E(0) \tag{3.19}$$

By this means we can rewrite (3.18) as

$$Q_1 = E(t) - E(0) + \int\int R(u^2 + v^2)\,dxdydt \tag{3.18a}$$

The condition for the existence of a unique solution assumes that the scalar product Q or Q_1 is always positive in the domain Ω, t. Generally the matrices (A', B', C'), or operators which render (3.16) positive are called positive definite. Assuming $Q_1 > 0$, and because the last term in (3.18a) is always positive, we can deduce that energy at any instance is always smaller than the energy at initial moment,

$$E(t) \leq E(0) \tag{3.20}$$

This result will be helpful in proving the uniqueness of the solution. The solution to our problem will be not unique if starting from one initial condition the solution will bifurcate in time into several branches. Let us assume that at time (t) two solutions exist, then the energy of these solutions is $E_1(t)$ and $E_2(t)$, respectively. Because at the initial moment $E_1(0) = E_2(0)$, it follows from (3.20) that $E_1(t) = E_2(t)$ and thus uniqueness of solution has been proved. Here again we have to note that this result is due to the assumed boundary condition.

1.4 Practical approach to the open boundary formulation

As we have seen above, the open boundary condition is difficult to formulate so that the uniqueness of the solution is satisfied. Even if such a formulation is clear it is often difficult to derive the high quality data from the measurements to construct the boundary condition. The general set (3.1)–(3.3) constitutes a parabolic problem; because open boundaries often are located far from the shallow water where the horizontal and vertical friction can be neglected and therefore a simpler problem expressed by the hyperbolic set of equations can be formulated. If an integration domain is located in such a way that the dynamical processes occurring inside this domain are not influenced by the processes from outside, one can further simplify the open boundary condition. The open boundary condition should be formulated to allow the signal to propagate outward without any reflection. The first radiating boundary condition for the long waves has been introduced by Reid and Bodine (1968). For the case when advection or wave motion dominates, a general radiation boundary condition was formulated by Orlanski (1976), and it was modified for the water circulation problem by Camerlengo and O'Brien (1980).

To relate the velocity (u) and sea level changes (ζ) at the open boundary Reid and Bodine (1968) used a formula for the long progressive waves:

$$\zeta = \pm u \left(\frac{H}{g}\right)^{1/2} \tag{3.21}$$

For the two-dimensional propagation a more general relation holds:

$$(u^2 + v^2)^{1/2}/c = \pm \zeta / H \tag{3.22}$$

Where, $c = \sqrt{gH}$ is a long–wave velocity and the signs in (3.22) are taken in such a way that u or v are directed outward from the computational domain when the sea level is positive.

When calculating the tide distribution, the currents and sea level can be specified at the open boundary, either from observations or from the encompassing larger scale model. Again, in such case a simple radiation condition is applied

$$u = \hat{u} + \frac{c}{H}(\zeta - \hat{\zeta}) \tag{3.22a}$$

Here, $\hat{\zeta}$ and $\hat{u}$ are the elevation and current prescribed along the open boundary. This condition treats any internally generated deviations from the prescribed sea level and current at the boundary as a free progressive wave propagating outward.

Based on Sommerfeld's (1949) radiation condition Orlanski (1976) proposed to use,

$$\frac{\partial \Phi}{\partial t} + c_p \frac{\partial \Phi}{\partial n} = 0 \tag{3.23}$$

to define the variable Φ on the boundary. The phase velocity (if it is directed outward from the integration domain) is calculated from

$$c_p = -\frac{\partial \Phi}{\partial t} \bigg/ \frac{\partial \Phi}{\partial n} \tag{3.24}$$

In (3.24) the derivatives are assumed to be known from the previous time step of numerical calculation. Actual estimation is performed one grid distance from the boundary. The computation can be summarized as follows. First the phase velocity is calculated based on the values of Φ from the previous time steps in the proximity of the boundary,

$$c_p = \begin{cases} \dfrac{\Phi_{b-1}^m - \Phi_{b-1}^{m-1}}{\Phi_{b-1}^{m-1} - \Phi_{b-2}^{m-1}} \dfrac{h}{T} & \text{for } 0 \leq c_p \leq \frac{h}{T} \\ \dfrac{h}{T} & \text{for } c_p > \frac{h}{T} \\ 0 & \text{for } c_p < 0 \end{cases} \tag{3.25}$$

In the second step the new value of the function is calculated at the boundary:

$$\Phi_b^{m+1} = (1 - \frac{c_p T}{h})\Phi_b^m + c_p \frac{T}{h}\Phi_{b-1}^m \tag{3.26}$$

A radiation condition designed to use at the open boundary for the multidimensional flow was formulated by Raymond and Kuo (1984). Numerous experiments show that Orlanski's condition is far from perfect. The behavior of the condition often depends on the interaction between the numerical solution derived inside the domain and the numerical formulation of the boundary condition. Condition (3.22) often works very well because it does not need additional numerical formulation.

1.5 Initial conditions

Again, formulation of an initial condition from the mathematical point of view is an easy task, but constructing an initial condition from the measurements is often an insurmountable obstacle. We shall rather use the assumption that the dependence of certain processes on the initial condition will decay with time. Therefore, one can set at the initial time all dependent variables as zero:

$$u(x, y, 0) = v(x, y, 0) = \zeta(x, y, 0) \tag{3.27}$$

To show the dependence of the long wave problem on the initial conditions we shall use (3.13) and rewrite in symmetrical form

$$\frac{\partial \mathbf{u}}{\partial t} + A'\mathbf{u} + B'\mathbf{u} + C'\mathbf{u} = \frac{\partial \mathbf{u}}{\partial t} + A_1'\mathbf{u} = 0 \tag{3.28}$$

Here $\mathbf{u} = \mathbf{u}(u, v, \zeta')$, $\zeta' = \zeta\sqrt{g/H}$ and

$$A_1' = \begin{pmatrix} R & 0 & \sqrt{gH}\partial/\partial x \\ 0 & R & \sqrt{gH}\partial/\partial y \\ \sqrt{gH}\partial/\partial x & \sqrt{gH}\partial/\partial y & 0 \end{pmatrix} \tag{3.29}$$

with initial condition

$$\mathbf{u} = \mathbf{u}(x, y, 0) \tag{3.30}$$

The solution to the above problem is searched in the form

$$\mathbf{u} = \mathbf{u}_1 \exp -i(\kappa_1 x + \kappa_2 y) \tag{3.31}$$

This solution assumes space periodicity, therefore it does not depend on the boundary condition. Introducing (3.31) into (3.28) results in

$$\frac{d\mathbf{u}_1}{dt} + A_1'\mathbf{u}_1 = 0 \tag{3.32}$$

with the initial condition

$$u_1 = u_1(0) \tag{3.32a}$$

Here,

$$A_1' = \begin{pmatrix} R & 0 & -i\kappa_1\sqrt{gH} \\ 0 & R & -i\kappa_2\sqrt{gH} \\ -i\kappa_1\sqrt{gH} & -i\kappa_2\sqrt{gH} & 0 \end{pmatrix} \tag{3.32b}$$

Solution of (3.32)

$$u_1 = u_1(0)\exp A_1' t \tag{3.33}$$

will decay in time if the real part of eigenvalues of matrix A_1' is positive. The eigenvalues of A_1' are

$$\lambda_1 = R; \qquad \lambda_{2,3} = R/2 \pm i\sqrt{(\kappa_1^2 + \kappa_2^2)gH - R^2/4} \tag{3.34}$$

Thus all eigenvalues possess a real part. The real parts are positive when, $(\kappa_1^2 + \kappa_2^2)gH \geq R^2/4$. One can conclude therefore, that after a long enough time has elapsed the solution of (3.28) will be independent of the initial conditions due to the influence of bottom friction.

2. Numerical solution of the system of equations

We shall search a numerical solution to the system (3.1)–(3.3) by introducing suitable temporal and spatial grids, and by changing the continuous function into a set of discrete values given at the grid points. Here, as in Ch.II, the time step and space step are denoted as T, and h, respectively. The j index stands for the x coordinate, k index is for the y coordinate and m denotes discrete time. The change of the differential equation into the difference form will be carried out as previously by the Taylor series method. We shall use the same approach we have followed in Ch.II., but it will be applied to the system of equations. When the mesh size (i.e., h and T) approaches zero, the difference equations should approximate the differential equations and solution of the difference equation should converge to the solution of the differential equation. Unfortunately, as we have experienced in Ch.II, the approximation alone does not provide convergence. The errors which are introduced into the computational algorithm (e.g., round off errors) may grow in time in an unbounded manner. Along with the approximation we ought to assure stability of the numerical algorithm.

2.1 Numerical stability of the system of equations

To define stability of the system of equations we shall follow the approach for the one variable given in Ch.II. A set of variables u, v, ζ, will be described by the vector $u(u, v, \zeta)$. An error will be denoted as δu. We shall also assume that the system of equations is linear and both the solution and the error can be represented as a sum of the Fourier components. The nonlinear system of equations can be linearized by "freezing" the variable coefficients.

A convenient form for δu is

$$\delta u = u^* \exp i\omega t \exp i\kappa_1 x \exp i\kappa_2 y \qquad (3.35)$$

Here $u^*(u, v, \zeta)$ is a column vector of the amplitudes determined by the initial conditions. The above expression in discrete form is

$$\delta u = u^* \lambda^m \exp i\kappa_1 jh \exp i\kappa_2 kh \qquad (3.36)$$

Here,

$$\lambda^m = \exp i\omega Tm \qquad (3.36a)$$

ω denotes frequency and κ_1 and κ_2 are the wave numbers along x and y. In (3.36) the space steps along x and y are equal to h. In the linear system the behavior of the error and the solution is similar, therefore in further considerations δ is omitted from (3.36). Substituting (3.36) into numerical equations we shall derive an equation to determine a stability parameter (magnification factor) λ, as a function of the space and time steps and the physical parameters of the problem.

To start with, let us consider long wave propagation in a flat bottom channel. The set (3.1)–(3.3) takes a simpler form:

$$\frac{\partial u}{\partial t} = -g \frac{\partial \zeta}{\partial x} \qquad (3.37a)$$

$$\frac{\partial \zeta}{\partial t} + H \frac{\partial u}{\partial x} = 0 \qquad (3.37b)$$

The difference equations are constructed on the space staggered grid depicted in Fig.3.1. In this figure the velocities and water levels are set at different space locations.

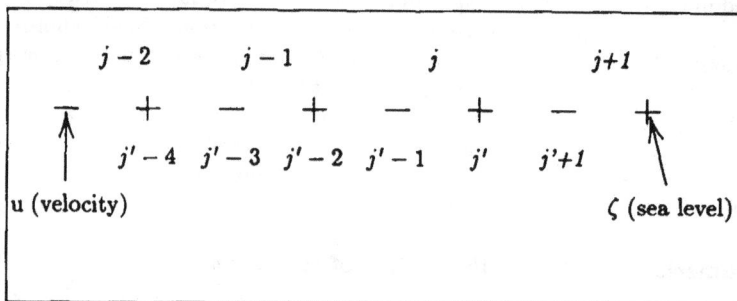

Fig. 3.1

Space staggered grid for the one-dimensional problem.

One can apply an index notation separately for the sea level and velocity points as denoted in Fig.3.1 by the primed indices. A more efficient way of employing index notation is by using one index for the pair of points. Each pair includes one sea level point and one velocity point as shown in the upper portion of Fig.3.1. The grid from Fig.3.1 is quite often used in geophysical fluid dynamics. Advantages and disadvantages of this grid will be discussed later on.

An example of a numerical approximation to eq.(3.37) constructed on the space staggered lattice with an explicit time marching is given by the following set:

$$\frac{u_j^{m+1} - u_j^m}{T} = -g\frac{\zeta_j^m - \zeta_{j-1}^m}{h} \tag{3.38a}$$

$$\frac{\zeta_j^{m+1} - \zeta_j^m}{T} = -H\frac{u_{j+1}^m - u_j^m}{h} \tag{3.38b}$$

Notice that in the above set the j index denotes different locations for the velocity and the sea level. To investigate numerical stability we shall introduce an expression for the error (3.36) into eqs.(3.38), and arrive at

$$\frac{\lambda - 1}{T}u^* + \frac{g}{h}[1 - \exp(-i\kappa_1 h)]\zeta^* = 0 \tag{3.39a}$$

$$\frac{\lambda - 1}{T}\zeta^* + \frac{H}{h}[\exp(i\kappa_1 h) - 1]u^* = 0 \tag{3.39b}$$

The above system has a nontrivial solution if the coefficient determinant is equal to zero,

$$(\lambda - 1)^2 + Q = 0 \tag{3.40}$$

Here $Q = gH(\frac{2T}{h}\sin\frac{\kappa_1 h}{2})^2$
The roots of (3.40) are

$$\lambda_{1,2} = 1 \pm i\sqrt{Q} \tag{3.41}$$

The Neumann necessary condition for the stability states that the magnification factor should satisfy the inequality below (Sod, 1985)

$$|\lambda| \leq 1 \tag{3.42}$$

From (3.41) the absolute value is calculated as

$$|\lambda_{1,2}| = \sqrt{1 + Q} > 1 \tag{3.43}$$

and by this result we can conclude that the magnification factor of (3.38) does not satisfy the inequality (3.42), and the numerical scheme (3.38) will not provide a stable solution.

3. Step by step approach to the construction and analysis
of simple numerical schemes

In order to identify the important steps in the construction and analysis of a numerical scheme we shall consider various approaches to the long wave propagation in a channel. At the left end of the channel a sinusoidal wave is given as

$$\zeta = \zeta_0 \sin(\frac{2\pi}{T_p}t) \qquad (3.44)$$

Here the amplitude ζ_0=100 cm, and the period T_p=600 s.

Propagation of this monochromatic wave towards the right end of the channel will be reproduced. The right end is open and a radiating condition will be used so that the wave can propagate outside without reflection. The channel is 180 km long and 1 km deep. The actual application of the boundary condition is depicted in Fig.3.2.

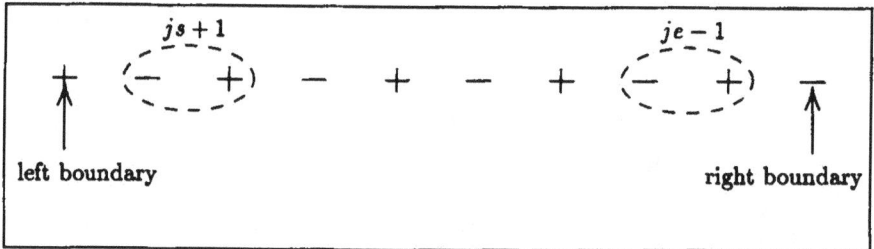

Fig. 3.2

Space staggered grid for the wave propagation problem. Sea level is given at the left boundary as a sinusoidal wave. Radiation condition is used at the right boundary.

The space index runs from $j = js$ at the left-hand boundary to the $j = je$ at the right-hand boundary. As we can see from Fig.3.2 the velocity computation starts at $j = js + 1$, i.e., one grid point away from the left boundary, and proceeds up to the point $j = je$. Sea level computation starts at the left boundary ($j = js$) but it runs only to the $j = je - 1$ point. Sea level at the left boundary is given by (3.44), velocity at the right boundary is specified by (3.21) as

$$u_{je}^m = \zeta_{je-1}^m \sqrt{\frac{g}{H}} \qquad (3.45)$$

3.1 Fischer's (1959) algorithm

The numerical scheme

$$\frac{u_j^{m+1} - u_j^m}{T} = -g\frac{\zeta_j^m - \zeta_{j-1}^m}{h} \tag{3.46a}$$

$$\frac{\zeta_j^{m+1} - \zeta_j^m}{T} = -H\frac{u_{j+1}^{m+1} - u_j^{m+1}}{h} \tag{3.46b}$$

differs from (3.38) because the new value of velocity (u^{m+1}) is used in the continuity equation. The scheme is semi-implicit in time. This approach stabilizes computation in time.

A well-known offspring of the above scheme is a numerical scheme of the second-order approximation in time and in space (Fischer, 1959);

$$\frac{u_j^{m+1} - u_j^{m-1}}{2T} = -g\frac{\zeta_j^m - \zeta_{j-1}^m}{h} \tag{3.47a}$$

$$\frac{\zeta_j^{m+2} - \zeta_j^m}{2T} = -H\frac{u_{j+1}^{m+1} - u_j^{m+1}}{h} \tag{3.47b}$$

In the ensuing consideration we shall study only the first (3.46) set of equations. Stability of this system will be analyzed by applying (3.36) in the form

$$u_j^m = u^*\lambda^m e^{ikjh} \quad \text{and} \quad \zeta_j^m = \zeta^*\lambda^m e^{ikjh} \tag{3.48}$$

Introducing (3.48) into (3.46) we obtain a system of equations from which an equation for the magnification factor follows:

$$\lambda^2 + (Q-2)\lambda + 1 = 0 \tag{3.49}$$

Here

$$Q = 4gH(\frac{T}{h}\sin\frac{\kappa h}{2})^2 \tag{3.50}$$

The roots of (3.49) are

$$\lambda = \frac{1}{2}[(2-Q) \pm \sqrt{Q^2 - 4Q}] \tag{3.51}$$

For stability the absolute value of these roots should be less than 1. It is easy to see that for $Q > 4$ the magnification factor is greater than 1. Let us consider the case when $Q < 4$, then $Q^2 - 4Q < 0$ and from (3.51),

$$\lambda = \frac{1}{2}[(2-Q) \pm i\sqrt{4Q - Q^2}] = a \pm ib \tag{3.52}$$

Since for the complex numbers the following relation holds

$$|\lambda|^2 = (a + ib)(a - ib) = a^2 + b^2 \tag{3.53}$$

it follows from (3.52) that $|\lambda| = 1$ and the algorithm (3.46) is stable for $Q \leq 4$. Introducing into this inequality, the expression for Q, the stability condition is defined as

$$4gH(\frac{T}{h}\sin\frac{\kappa h}{2})^2 \leq 4 \tag{3.54a}$$

or

$$\sqrt{gH} \leq \frac{h}{T} \tag{3.54}$$

This is the so called Courant–Friedrichs–Lewy or CFL stability condition. We can conclude that the numerical scheme (3.46) is stable only if the space step and the time step of a numerical lattice are chosen according to (3.54).

The signal reproduced through this set of equations will be somewhat altered due to the properties of the numerical formulas. As we did in Ch.II. we shall consider a wave of unit amplitude and introduce it into the differential equation (3.37) and the difference equation (3.46). Afterwards, we compare the analytical and numerical solutions to deduce how the numerical scheme alters the amplitude and frequency of the wave. Thus taking the wave

$$u = u^* e^{i\kappa z} e^{i\omega t} \quad \text{and} \quad \zeta = \zeta^* e^{i\kappa z} e^{i\omega t} \tag{3.55}$$

and introducing it into (3.37) we arrive at

$$i\omega u^* + gi\kappa\zeta^* = 0 \tag{3.56a}$$

$$i\omega\zeta^* + Hi\kappa u^* = 0 \tag{3.56b}$$

From this system the well-known expression for the phase velocity (celerity) of the long wave follows:

$$\frac{\omega}{\kappa} = \pm\sqrt{gH} \tag{3.57}$$

Here the positive and negative signs define the direction of propagation. For the numerical system of equations from (3.52) and from the definition (3.36a) it follows that

$$\lambda = e^{i\omega T} = \frac{1}{2}[(2 - Q) \pm i\sqrt{4Q - Q^2}] \tag{3.58}$$

From this, we arrive at the equation which defines the phase change (Φ) of the signal over one time step:

$$\tan\Phi = \tan\omega T = \frac{\sqrt{4Q - Q^2}}{(2 - Q)} \tag{3.59}$$

The phase velocity for the numerical solution then follows:

$$\frac{\omega}{\kappa} = \frac{1}{\kappa T} \arctan \frac{\sqrt{4Q - Q^2}}{(2 - Q)} \qquad (3.60)$$

Parameter

$$Q = 4gH\left(\frac{T}{h}\sin\frac{\kappa h}{2}\right)^2 \qquad (3.50, 3.61)$$

includes both space and time resolution through the nondimensional numbers

$$\frac{T}{h}\sqrt{gH} \qquad \text{and} \qquad \frac{L}{h} \qquad (3.62)$$

The nondimensional number L/h defines a spatial resolution and is connected with the wave number $\kappa = 2\pi/L$. For the well resolved processes the wavelength is covered by many space steps ($\kappa h \to 0$). In such a case $Q \to 0$, and from (3.58)–(3.60) it follows that this numerical scheme neither changes the amplitude ($|\lambda| = 1$) nor the phase ($\Phi = 0$) of the signal. Celerity of the numerical and analytical solution is identical. Actually, this is an ideal situation which rarely can be expected in practical computation. To estimate the phase lag or celerity it is possible to apply an approach we have used in Ch.II, i.e., the above expression can be developed into a series for the small parameter. Instead, we calculate the phase lag and celerity from (3.59) and (3.60) for the various values of nondimensional numbers $\sqrt{gH}T/h$ and L/h. Phase lag (ϕ) follows from (3.59) as

$$\phi = \omega T = \arctan \frac{\sqrt{4Q - Q^2}}{(2 - Q)} \qquad (3.63)$$

Numerically derived celerity is compared against an analytical expression,

$$\text{Celerity} \quad \text{ratio} = \frac{\text{numerical} \quad \text{celerity}}{\text{analytical} \quad \text{celerity}}$$

$$= \frac{\phi}{\kappa T} \frac{1}{\sqrt{gH}} = \frac{\phi}{(2\pi h/L)(\sqrt{gH}T/h)} \qquad (3.64)$$

The results are depicted in Fig.3.3a and Fig.3.3b. Both phase lag and the difference between numerical and analytical celerity diminish for the fine space resolution when L/h is of the order of 10–20. The influence of the time parameter $\sqrt{gH}T/h$ is less important. Decrease of this parameter (less than 1) causes decrease of the phase lag but it increases the difference between the analytical and numerical celerities. Above behavior caused by the space resolution can be observed in the wave propagating along the channel. Since the depth of the channel ($H = 10^5$ cm),

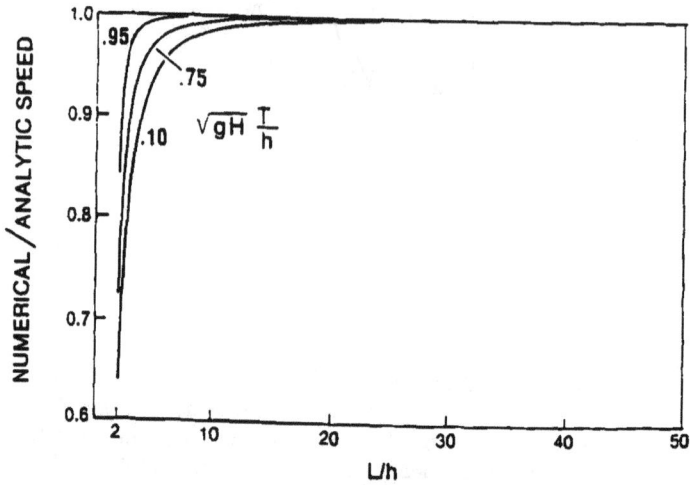

Fig. 3.3a

Ratio of the numerical to analytical long wave speed
for the Fisher's numerical scheme.

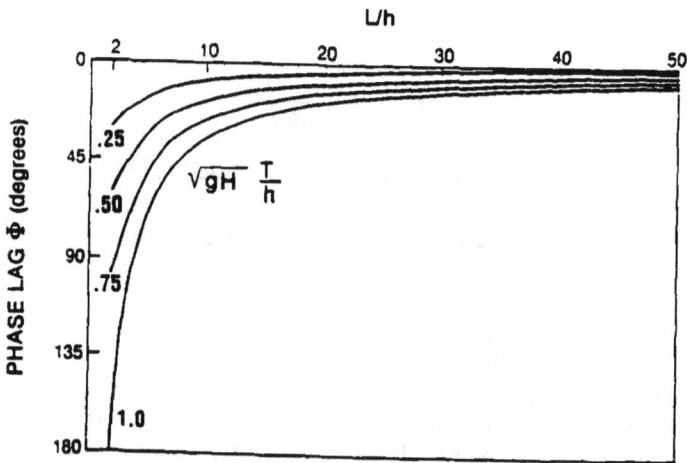

Fig. 3.3b

Phase lag of the Fisher's numerical scheme.

and the period of the wave (T_p =600 s) are known, the wavelength follows from (3.57)

$$L = T_p \sqrt{gH} = 600\sqrt{981 \times 10^5} = 59.43 \text{km} \qquad (3.65)$$

Computation was carried out by numerical scheme (3.46) and two time steps and space steps were used. In the first experiment h=10 km, T=60 s; and in the second experiment h=2 km, T=10 s. The ratio T/h was close in both experiments but L/h changed five times. The result of computations with boundary conditions defined by (3.44) and (3.45) are given in Fig.3.4, after about 1 hr time from the onset of the process. Although in both experiments the space and time steps fulfill stability criteria the derived results for the 60 s time step and 10 km space step, are of poor quality.

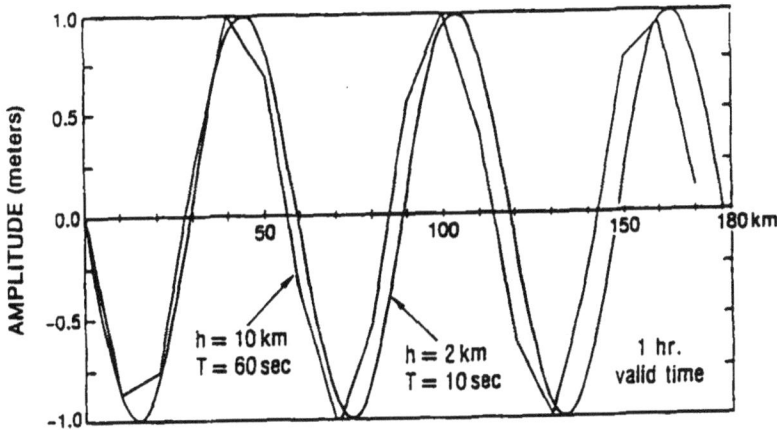

Fig. 3.4

Wave amplitude along the channel, computed by Fisher's algorithm for two different time steps and space steps.

The shape of the wave is poorly reproduced, the wavelength is shorter than 59.43 km and this error is growing over a longer distance of propagation. This spatial error in reproducing the wavelength is analogous to the phase lag error in reproducing the wave period in time domain. The shorter space and time steps used in the second experiment resolved wave propagation quite well. It is also worth mentioning that the radiation condition (3.45) works well and the wave propagates outside without any reflections.

3.2 Leapfrog numerical scheme

Numerical scheme

$$\frac{u_j^{m+1} - u_j^{m-1}}{2T} = -g\frac{\zeta_j^m - \zeta_{j-1}^m}{h} \tag{3.66a}$$

$$\frac{\zeta_j^{m+1} - \zeta_j^{m-1}}{2T} = -H\frac{u_{j+1}^m - u_j^m}{h} \tag{3.66b}$$

possesses the second-order approximation in time and space. Time marching in this method is done with double time step, but because three–time levels are considered ($m-1, m$ and $m+1$) the overall step is equal to one time step; see Fig.3.5.

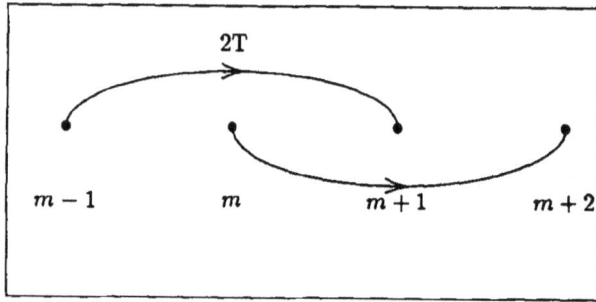

Fig. 3.5

Time stepping in the leapfrog method.

Stability of the system will be analyzed by using expressions (3.48). Introducing (3.48) into (3.66) we obtain the system of equations from which an equation for the magnification factor follows:

$$\lambda^2 - \lambda i\sqrt{Q_1} - 1 = 0 \tag{3.67}$$

Here

$$Q_1 = 16gH(\frac{T}{h}\sin\frac{\kappa h}{2})^2 \tag{3.68}$$

The roots of (3.67) are

$$\lambda = \frac{1}{2}[i\sqrt{Q_1} \pm \sqrt{4 - Q_1}] \tag{3.69}$$

For stability, the absolute value of these roots should be less than 1. It is easy to see that for $Q_1 > 4$ the magnification factor is greater than 1. For $Q_1 \le 4$ from (3.69) it follows that $|\lambda| = 1$, and the algorithm (3.66) is stable in this range. Introducing into above inequality an expression for Q_1, stability condition is defined as

$$16gH(\frac{T}{h}\sin\frac{\kappa h}{2})^2 \le 4 \tag{3.70a}$$

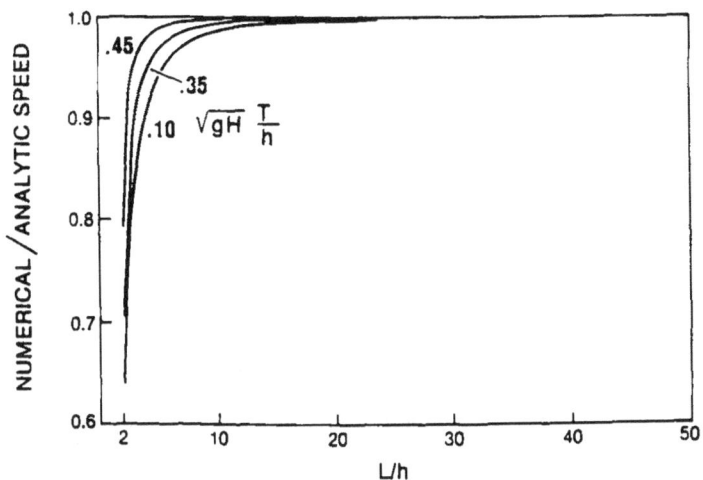

Fig. 3.6a

Ratio of the numerical to analytical long wave speed
for the leapfrog numerical scheme.

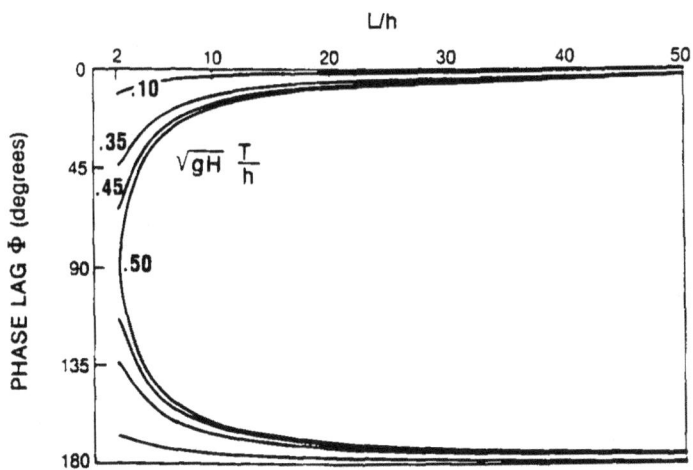

Fig. 3.6b

Phase lag of the leapfrog numerical scheme.

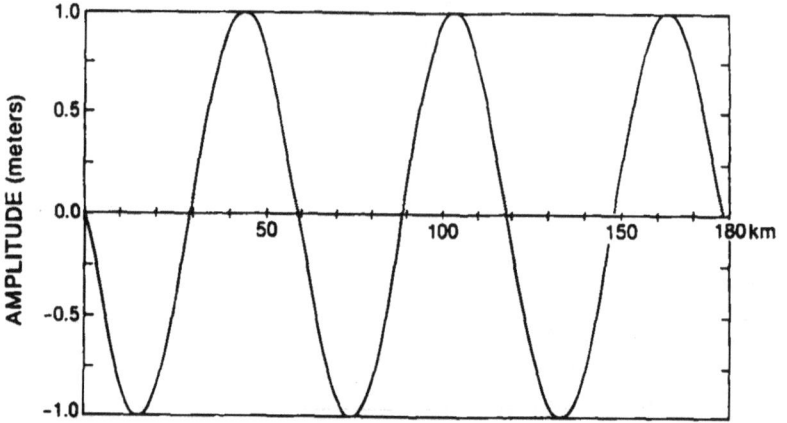

Fig. 3.7a

Wave distribution derived by the application of leapfrog method
with additional application of the first order method every tenth time step.

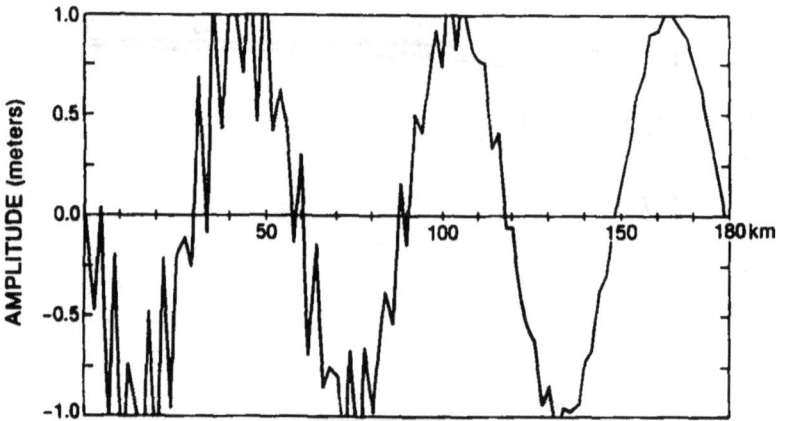

Fig. 3.7b

Wave distribution derived by the application of leapfrog method only.
The solution is distorted by parasitic waves.

or, setting

$$(\sin \frac{\kappa h}{2})^2 = 1$$

we arrive at the stronger condition

$$\sqrt{gH} \leq \frac{h}{2T} \tag{3.70}$$

This condition is quite similar to (3.54) but it allows for the time marching with the double time step.

Again, to estimate the errors in the leapfrog numerical scheme we shall find the phase lag and celerity of the long waves. From (3.69) we arrive at an equation for the phase change over one time step,

$$\phi = \omega T = \arctan \frac{\sqrt{Q_1}}{\pm\sqrt{4 - Q_1}} \tag{3.71}$$

Numerical expression for celerity is compared against an analytical solution

$$\text{Celerity} \quad \text{ratio} = \frac{\text{numerical} \quad \text{celerity}}{\text{analytical} \quad \text{celerity}}$$

$$= \frac{\phi}{\kappa T} \frac{1}{\sqrt{gH}} = \frac{\phi}{(2\pi h/L)(\sqrt{gH}T/h)} \tag{3.72}$$

The phase lag and celerity ratio are depicted in Fig.3.6a and Fig.3.6b. These results differs only slightly from the Fischer's scheme.

The calculations carried out for the wave propagating along the channel by the leapfrog numerical scheme with the boundary condition given by (3.44) and (3.45) are depicted in Fig.3.7a and Fig.3.7b. The result in Fig.3.7b is quite noisy. The reason is (see Ch.II) that the leapfrog method generates two solutions. One solution is defined by the physical properties of the system and the second is a parasitic solution due to the three-time level numerical scheme. The latter solution changes sign at every time step. It can be deleted by application of the first-order method (e.g., Fischer's method) at every 10 or 20 time steps. A smooth plot in Fig.3.7a has been obtained in this way.

The leapfrog method is used quite often in the dynamical problems in oceanography. Here one may notice that the numerical scheme given by (3.47) is also of the second order of approximation. Although (3.47) has no parasitic solution it is becoming unstable when the nonlinear (advective) terms are included, the leap frog method on other hand, can be stable with the nonlinear terms as well.

3.3 Implicit numerical scheme

Now we approach the channel dynamics by constructing an implicit numerical scheme to solve the problem defined by the system (3.37). Before starting the construction we have to know how to solve an implicit equation or the system of equations. Usually for this purpose we shall apply the line inversion method given in Appendix 1.

Let us start by introducing a simple implicit equation,

$$\frac{u^{m+1} - u^m}{T} = -g \frac{\partial \zeta^{m+1}}{\partial x} \tag{3.73a}$$

$$\frac{\zeta^{m+1} - \zeta^{m-1}}{T} = -H \frac{\partial u^{m+1}}{\partial x} \tag{3.73b}$$

These are written in the so called difference–differential form. The assumption here is that the space derivatives are rendered into the numerical form through the space staggered grid. Let us first check the stability requirement to see whether we are on the right track, i.e., did we really obtain a numerical scheme which is stable for the unlimited time or space step? Stability of the system will be analyzed by using expressions (3.48). Introducing (3.48) into (3.73) we obtain a system of equations from which an equation for the magnification factor follows:

$$\lambda^2(1 + Q) - 2\lambda + 1 = 0 \tag{3.74}$$

Here

$$Q = 4gH(\frac{T}{h} \sin \frac{\kappa h}{2})^2 \tag{3.75}$$

The roots of (3.74) are

$$\lambda = \frac{1 \pm i\sqrt{Q}}{1 + Q} \tag{3.76}$$

Absolute value of the stability parameter

$$|\lambda| = \frac{1}{\sqrt{1 + Q}} \tag{3.77}$$

is always less than 1 and the algorithm (3.73) is stable for all time steps (T) and space steps (h). Amplification factor is less than 1 even for the very high resolution of $L/h > 30$. Unfortunately, this method is of first-order approximation in time. The application of this implicit approach will result in high damping of the amplitude. To account for this effect we render equations (3.37) into a symmetrical finite difference form of the second-order approximation in time and space.

$$\frac{u^{m+1} - u^m}{T} = -\frac{g}{2}(\frac{\partial \zeta^{m+1}}{\partial x} + \frac{\partial \zeta^m}{\partial x}) \tag{3.78a}$$

$$\frac{\zeta^{m+1} - \zeta^m}{T} = -\frac{H}{2}(\frac{\partial u^{m+1}}{\partial x} + \frac{\partial u^m}{\partial x}) \tag{3.78b}$$

Introducing (3.48) into (3.78) we obtain an equation for the magnification factor

$$(\lambda - 1) - i\sqrt{Q_2}(\lambda + 1) = 0 \tag{3.79}$$

Here

$$Q_2 = gH(\frac{T}{h}\sin\frac{\kappa h}{2})^2 \tag{3.80}$$

Absolute value of the stability parameter $|\lambda| = 1$, and the algorithm (3.78) is stable for all time steps (T) and space steps (h).

Again, to estimate the errors in the implicit numerical scheme we shall find the phase lag and celerity of the long waves and compare them to the magnitudes derived from the analytical solutions. From (3.79) we arrive at an equation for the phase change over one time step,

$$\phi = \omega T = \arctan\frac{2\sqrt{Q_2}}{1 - Q_2} \tag{3.81}$$

Numerical expression for celerity is compared against an analytical solution

$$\text{Celerity} \quad \text{ratio} = \frac{\text{numerical} \quad \text{celerity}}{\text{analytical} \quad \text{celerity}}$$

$$= \frac{\phi}{\kappa T}\frac{1}{\sqrt{gH}} = \frac{\phi}{(2\pi h/L)(\sqrt{gH}T/h)} \tag{3.82}$$

The phase lag and celerity ratio are depicted in Fig.3.8a and Fig.3.8b. These results differs only slightly from Fischer's or leapfrog scheme.

Before we start the calculation in a channel by the implicit method let us notice that there is no need to solve both (3.78) equations by this method. It is obvious from (3.78) that if a new value of velocity is known, the sea level equation can be solved explicitly, and vice versa, if a new value of the sea level is computed the velocity follows through an explicit computation. Therefore, it is sufficient to solve by the line inversion only one equation from the basic set (3.78). We shall construct a three-point algorithm for the sea level so that the line inversion method (Appendix 1) could be applied. From (3.78a) velocity at the $m + 1$ time level is

$$u^{m+1} = u^m - \frac{gT}{2}(\frac{\partial \zeta^{m+1}}{\partial x} + \frac{\partial \zeta^m}{\partial x}) \tag{3.83}$$

Differentiating (3.83)

$$\frac{\partial u^{m+1}}{\partial x} = \frac{\partial u^m}{\partial x} - \frac{gT}{2}(\frac{\partial^2 \zeta^{m+1}}{\partial x^2} + \frac{\partial^2 \zeta^m}{\partial x^2}) \tag{3.84}$$

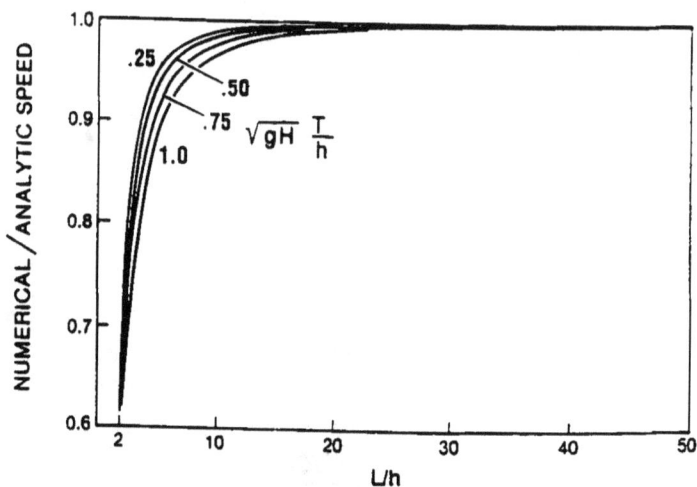

Fig. 3.8a

Ratio of the numerical to analytical long wave speed
for implicit numerical scheme.

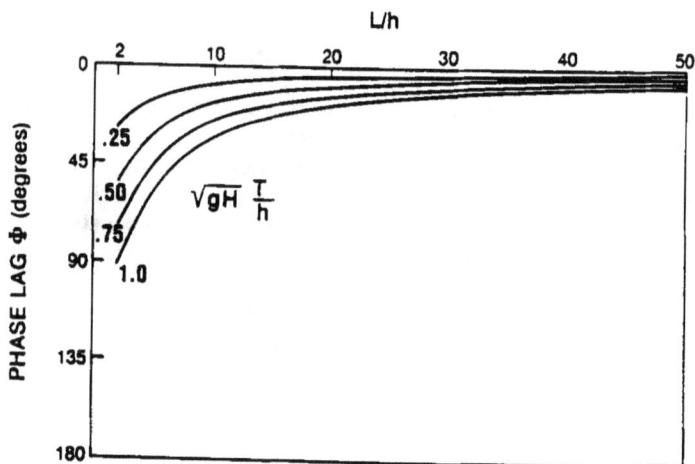

Fig. 3.8b

Phase lag of implicit numerical scheme.

and introducing the above expression to the right side of (3.78b), we obtain

$$\frac{\zeta^{m+1} - \zeta^m}{T} = -\frac{H}{2}\left[\frac{\partial u^m}{\partial x} - \frac{gT}{2}\left(\frac{\partial^2 \zeta^{m+1}}{\partial x^2} + \frac{\partial^2 \zeta^m}{\partial x^2}\right) + \frac{\partial u^m}{\partial x}\right] \qquad (3.85)$$

Expanding the derivatives into finite differences and organizing the resulting equation into a three-point algorithm suitable for the line inversion, we arrive at

$$-a_j \zeta_{j+1}^{m+1} + b_j \zeta_j^{m+1} - c_j \zeta_{j-1}^{m+1} = d_j \qquad (3.86)$$

Here

$$a_j = \frac{gHT^2}{4h^2} \qquad b_j = 1 + 2\frac{gHT^2}{4h^2} \qquad c_j = \frac{gHT^2}{4h^2}$$

and

$$d_j = \zeta^m - \frac{TH}{h}(u_{j+1}^m - u_j^m) + \frac{gHT^2}{4h^2}(\zeta_{j+1}^m - 2\zeta_j^m + \zeta_{j-1}^m)$$

Because a_j, b_j and c_j are positive and $b_j \geq a_j + c_j$, these coefficients fulfill the stability conditions outlined in Appendix 1.

To start the computation by (3.86) the boundary condition ought to be specified and introduced into the line inversion method. The left-hand boundary condition (3.44) is already given, but the right-hand condition has to be specified. We outline here application of the Orlanski (1976) radiating condition given by (3.23). The implicit difference form of this condition is

$$\frac{\zeta_{je}^{m+1} - \zeta_{je}^m}{T} + \frac{c_p}{h}(\zeta_{je}^{m+1} - \zeta_{je-1}^{m+1}) = 0 \qquad (3.87)$$

Here index je stands for the right-hand boundary.

We still need to define the velocity of propagation (c_p) in (3.87). The basic assumption of the radiating boundary condition is that the wave is allowed to leave but not to enter the domain. This requires that the velocity c_p at the right-hand boundary must be nonnegative. Therefore, one way to approach this problem is to look upstream and to find velocity from $je - 1$ grid point and from the previous time step by applying (3.87). There are some additional limits on the value of the velocity, due to stability requirements it can not exceed h/T limit as well. In summary, the following expressions are used to calculate c_p (see 3.25):

$$c_p = \begin{cases} \dfrac{\zeta_{je-1}^m - \zeta_{je-1}^{m-1}}{\zeta_{je-1}^{m-1} - \zeta_{je-2}^{m-1}}\dfrac{h}{T} & \text{for } 0 \leq c_p \leq \dfrac{h}{T} \\ \dfrac{h}{T} & \text{for } c_p > \dfrac{h}{T} \\ 0 & \text{for } c_p < 0 \end{cases} \qquad (3.88)$$

Calculation carried out for the wave propagating along the channel by the implicit numerical scheme with boundary condition given by (3.44) and (3.87) are

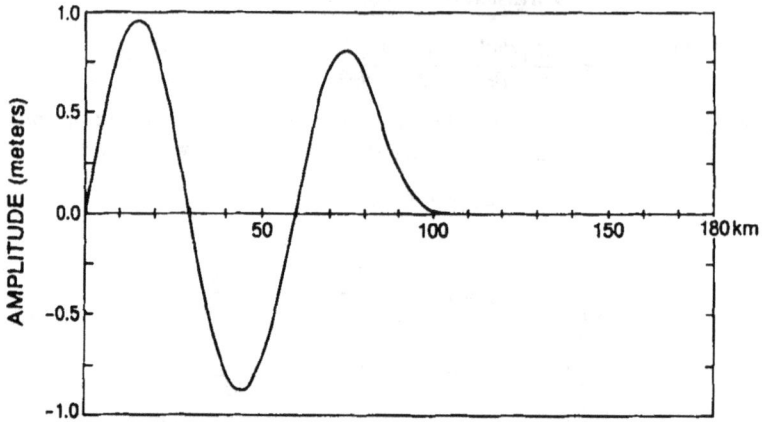

Fig. 3.9a

Initial phase of wave propagation calculated by implicit method.

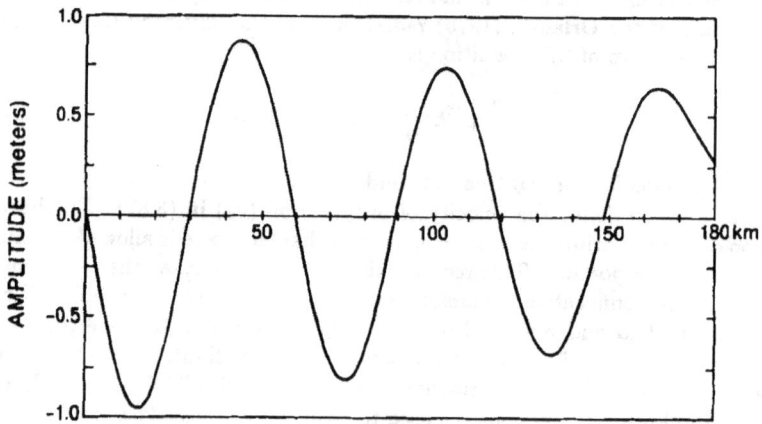

Fig. 3.9b

Wave distribution derived by application of implicit method.

depicted in Figs.3.9a and 3.9b. The result in Fig.3.9a displays an initial phase of propagation before the wave arrives at the right boundary. A plot in Fig.3.9b depicts quite strong damping which is probably due both to the applied radiating boundary condition and the numerical damping of the implicit scheme. In the above algorithm eq.(3.86) and the boundary condition (3.87) are implicit but the phase velocity is calculated by the explicit expression (3.88). This may be one of the reasons of the observed damping.

Finally, a general comment has to be made about the application of the implicit approach. Although this method allows to choose an unrestricted time step or space step, it can not be applied indiscriminately. Due probably to the round off errors the increase of the time step, 10–20 times above an explicit condition, is the practical limit on the application of an implicit method. Possible limitations of the implicit methods can be well established through simple numerical experiments in which both implicit and explicit methods are used to solve the same problem.

4. Two-dimensional models

4.1 Step by step approach

Here, the vertically averaged equations (3.1)–(3.3) will be rendered into numerical form to describe the most common 2D phenomena — storm surges and tides. Before writing any numerical form we need to decide what kind of spatial numerical grid will be adequate for the given problem. We use a staggered grid in which all the points are given in separate locations (see Fig.3.10).

A horizontal bar in this figure is for u (E–W) component of velocity, vertical bar for v (N–S) component of velocity and cross for the sea level – ζ. The shortest distance between the same variables along x or y direction is equal to the space step h. To save computer memory the grid points are organized in the horizontal plane into triples. Each triple contains one point u, one point v and one point ζ (see Fig.3.10). Each triple is located in space by the indices j and k. Actually, eq.(3.1) is solved in the u point, eq.(3.2) in the v point and eq.(3.3) in the ζ point.

Let us write the numerical form for these equations. Since a leapfrog numerical form will be extensively described in Ch.IV, we turn our attention to numerical algorithm based on the Fischer's (1959) formulation:

$$\frac{u_{j,k}^{m+1} - u_{j,k}^{m-1}}{2T} - f\bar{v}^{u,m} = -\frac{g}{h}(\zeta_{j,k}^m - \zeta_{j-1,k}^m) + \frac{1}{\rho_o D_{u,j,k}^m}\tau_{x,j,k}^{s,m}$$

$$-R_{x,j,k}^{m-1}u_{j,k}^{m-1} + \frac{N_h}{h^2}(u_{j+1,k}^{m-1} + u_{j-1,k}^{m-1} + u_{j,k+1}^{m-1} + u_{j,k-1}^{m-1} - 4u_{j,k}^{m-1}) \qquad (3.89)$$

$$\frac{v_{j,k}^{m+1} - v_{j,k}^{m-1}}{2T} + f\bar{u}^{v,m} - \frac{g}{h}(\zeta_{j,k+1}^m - \zeta_{j,k}^m) + \frac{1}{\rho_o D_{v,j,k}^m}\tau_{y,j,k}^{s,m}$$

$$-R_{y,j,k}^{m-1}v_{j,k}^{m-1} + \frac{N_h}{h^2}(v_{j+1,k}^{m-1} + v_{j-1,k}^{m-1} + v_{j,k+1}^{m-1} + v_{j,k-1}^{m-1} - 4v_{j,k}^{m-1}) \qquad (3.90)$$

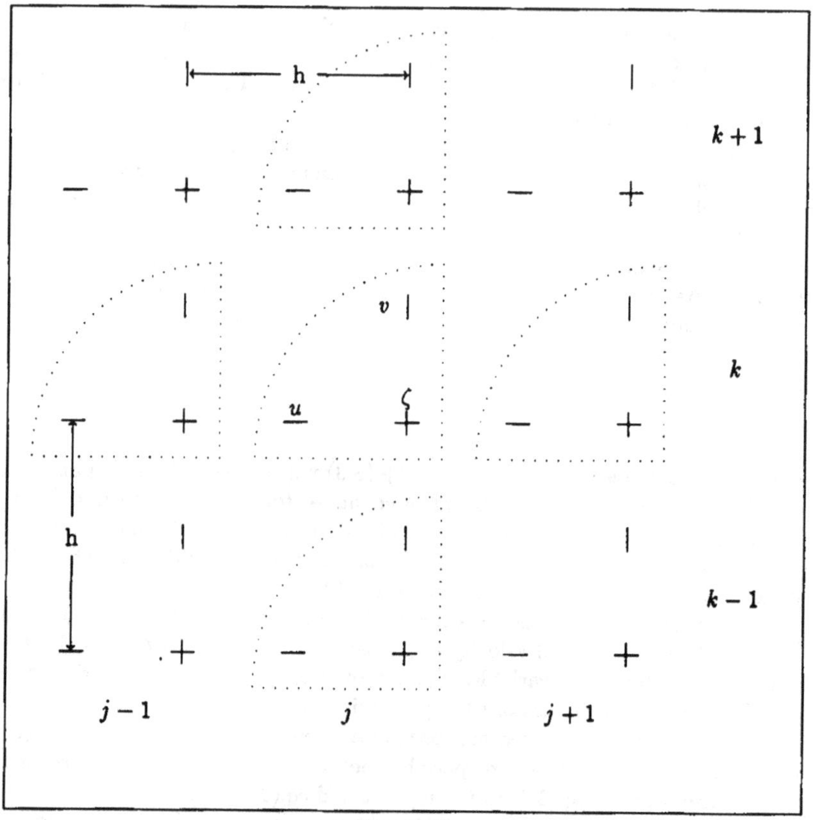

Fig. 3.10
Space-staggered numerical grid.

$$\frac{\zeta_{j,k}^{m+2} - \zeta_{j,k}^m}{2T} = -\frac{1}{h}(u_{j+1,k}^{m+1} D_{u,j+1,k}^m - u_{j,k}^{m+1} D_{u,j,k}^m) - \frac{1}{h}(v_{j,k}^{m+1} D_{v,j,k}^m - v_{j,k-1}^{m+1} D_{v,j,k-1}^m)$$

(3.91)

Here we have the first hint at what actually the staggered grid does. To construct the Coriolis term in (3.89) we need to know how to express v in the point u (and vice versa for eq.(3.90), therefore we take it as an average value of the four v points surrounding the u point. In the second equation a similar approach is taken.

$$\bar{v}^{u,m} = 0.25(v_{j,k}^m + v_{j-1,k}^m + v_{j-1,k-1}^m + v_{j,k-1}^m)$$

(3.92a)

$$\bar{u}^{v,m} = 0.25(u_{j,k}^m + u_{j,k+1}^m + u_{j+1,k+1}^m + u_{j+1,k}^m)$$

(3.92b)

In the equations of motion the nonlinear terms are neglected because they will require a special treatment. To proceed further we ought to notice that in the staggered lattice a certain degree of freedom exists in locating the common variables such as the external forces and depth. We may set the external forces and the depth at every grid point (i.e., u, v and ζ). In the above numerical scheme the depth is prescribed only at the sea level grid points. Afterwards, depth is calculated in the velocity points by averaging. Depth in the u point is defined as $D_{u,j,k} = 0.5(D_{j,k} + D_{j-1,k})$, and depth in the v point, is $D_{v,j,k} = 0.5(D_{j,k} + D_{j,k+1})$. Total depth $D = H + \zeta$ is not only a function of space, since it includes sea level, it depends on time as well.

Frictional forces at the free surface (τ^s) for the storm surge computation are given by the wind stress and for the tides they will be set to zero. At the bottom the nonlinear expression for the stress is applied

$$ru\sqrt{(u^2 + v^2)}/(\rho_o D) \quad \text{or} \quad rv\sqrt{(u^2 + v^2)}/(\rho_o D) \tag{3.93a}$$

Here again, an average v velocity in the u point or an average u velocity in the v point has to be applied. In eqs.(3.89) and (3.90) the following notation is used:

$$R_{x,j,k}^{m-1} = \sqrt{[(u_{j,k}^{m-1})^2 + (\bar{v}^{u,m-1})^2]}/(\rho_o D_{u,j,k}^{m-1})$$

$$R_{y,j,k}^{m-1} = \sqrt{[(\bar{u}^{v,m-1})^2 + (v_{j,k}^{m-1})^2]}/(\rho_o D_{v,j,k}^{m-1}) \tag{3.93b}$$

Bottom friction term has been written at the $m - 1$ time level. The same equation written at the m time level becomes unstable. To improve accuracy and stability of this term a semi-implicit formulation can be taken.

$$0.5R_{x,j,k}^{m-1}(u_{j,k}^{m-1} + u_{j,k}^{m+1})$$

$$0.5R_{y,j,k}^{m-1}(v_{j,k}^{m-1} + v_{j,k}^{m+1}) \tag{3.93c}$$

The bottom friction formulas depend on the total depth D, but this approximation is not valid when $D \to 0$, therefore a minimum depth has to be chosen to prevent instability. The horizontal friction terms also are set at m time level due to stability considerations. It is possible to improve the accuracy (and probably stability) by resorting to the DuFort–Frankel method (see Ch.II). To demonstrate this possibility let us consider only

$$\frac{\partial^2 u}{\partial x^2} \simeq [(u_{j+1}^m - u_j^{m+1}) - (u_j^{m-1} - u_{j-1}^m)]/h^2 \tag{3.94a}$$

The application to other terms is straightforward. A similar result can be obtained by

$$\frac{\partial^2 u}{\partial x^2} \simeq [(u_{j+1}^{m-1} - u_j^{m-1}) - (u_j^{m+1} - u_{j-1}^{m+1})]/h^2 \tag{3.94b}$$

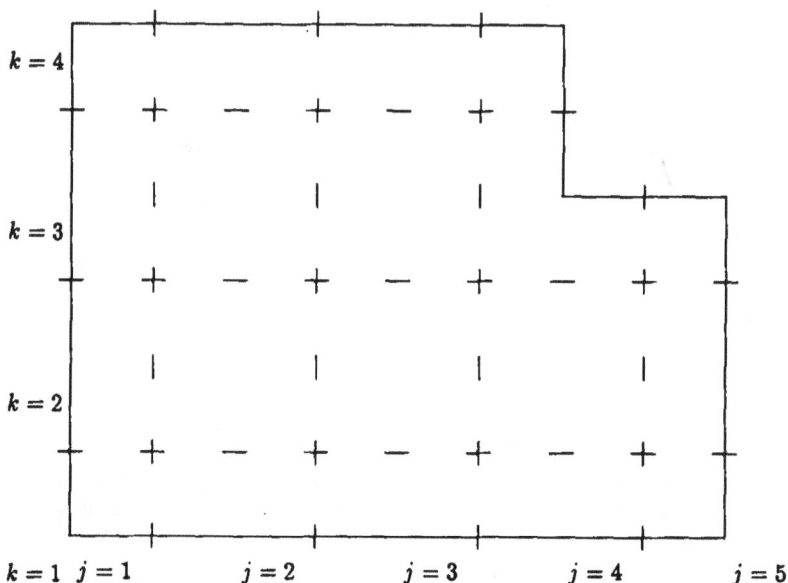

Fig. 3.11
Grid enumeration.

Generally solution of (3.89)–(3.91) proceeds separately for the u, v, and ζ points. A good approach is to organize the solution along horizontal and vertical lines. In Fig.3.11 an example of a computational space domain is given.

Let us analyze first ζ points; we can write the coordinates of every horizontal line as ksu, keu, jsu, jeu. For the first line $ksu = 2, keu = 2, jsu = 1, jeu = 4$. (Only internal points are counted). For the second line $ksu = 3, keu = 3, jsu = 1, jeu = 4$, and for the third line $ksu = 4, keu = 4, jsu = 1, jeu = 3$. Thus the number of the horizontal lines ($nlx = 3$) is 3. One may ask why such a fuss with the counting? Well, if an island is located in the integration domain then the index k does not change but the number of lines nlx does change. For the v equation we shall count the number of vertical lines. Thus $jsv = 1, jev = 1, ksv = 2, kev = 3$, for the first column, and $jsv = 2, jev = 2, ksv = 2, kev = 3$, for the second column, and $jsv = 3, jev = 3, ksv = 2, kev = 3$, for the third column and $jsv = 4, jev = 4, ksv = 2, kev = 2$ for the last column. Again, only the internal points have been counted according to the triple organization depicted in Fig.3.10. Here nly (number of columns) equals 4. Notice that the distribution of the u points and the ζ points is very similar. In every line for an integration at the u points we simply write $jsu + 1$ instead of jsu, the remaining indices are the same.

The time integration is done with the double time step $2T$, see Fig.3.12. This allows us to obtain the second order of approximation in time. The computational process starts by calculating a new velocity (u^{m+1}) from the old sea level (ζ^m), but as soon as the new velocity is available it is used in the calculation of the new sea level (ζ^{m+2}).

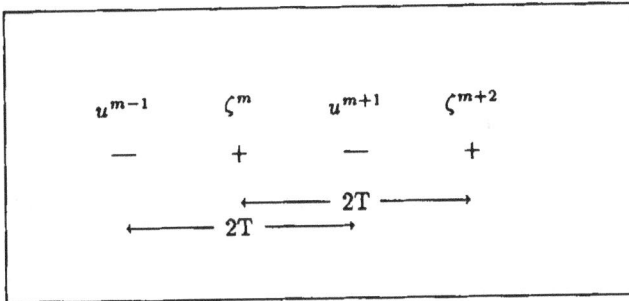

Fig. 3.12

Time integration in the Fischer's numerical scheme.

Finally, we elucidate a few important aspects related to the boundary conditions. If the program is written in FORTRAN all the variables are equal to zero when computation starts. As long as we do not change the values at the boundary they always will stay as zero. Normal component of velocity at the closed boundary is zero, therefore a staggered grid distribution is specially introduced to fulfill this requirement. What shall we do if one of the boundaries is open? For this case, a simple radiation condition for the long wave (3.21) can be applied. Let us assume that the upper boundary in Fig.3.11 is open, thus v along $k = 4$ from $j = 1$ to $j = 3$ is unknown. Invoking (3.21), the following expression can be used for the v component along the open boundary

$$v(j,4) = \zeta(j,4)(g/H(j,4))^{1/2}$$

Sea level $\zeta(j,4)$ is used from the previous time step. If instead of the upper boundary the lower one is open, minus sign will occur in the formulae.

Now we shall briefly discuss the boundary conditions for tide computation. Assuming that the domain depicted in Fig.3.11 is an adjacent sea and tide is entering from outside, and also that the sea level is given from observation at the upper open boundary. How shall we proceed? A glance at $k = 4$ line shows that the best idea will be to delete the v component and start from the sea level because it is given from the observation. But what to do at the u grid points which are located at the line $k = 4$? The actual problem arises when $v(j,3)$ or $u(j,3)$ is calculated by the full equation. We lack some points to calculate horizontal friction at the $u(j,3)$ or Coriolis and horizontal friction terms at $v(j,3)$. Therefore, to start the calculation at the line $k = 3$ the horizontal friction terms ought to be neglected. One can see that in the vicinity of the open boundary the calculations will be carried out through a simpler (hyperbolic) system and the full (parabolic) system of equations is used inside the domain.

To derive the Coriolis term at the open boundary for $v(j, 3)$ component, we choose $u(j, 4) = u(j, 3)$. At the closed boundary the Coriolis term is calculated according to (3.92a) and (3.92b). Jamart and Ozer (1986) showed that this may generate numerical boundary layers. They suggested (for the Coriolis term only) that rather than assuming that the normal velocity at the boundary is zero, it should be taken as equal to the velocity at the neighboring internal point.

The algorithm (3.89)–(3.91) has been constructed by Fischer (1959), but its application was primarily carried out by Hansen (1962) and his collaborators for storm surge computations. Kagan (1970) thoroughly discussed the various features of this numerical scheme.

Besides the second-order approximation in time, quite often computations of long wave propagation can be carried out by the first-order numerical schemes. This two-time level numerical approach is robust and well resolves the wave if the space–time resolution is high enough. The numerical formulation is again based on a space staggered grid.

$$\frac{u_{j,k}^{m+1} - u_{j,k}^m}{T} + UPOS(u_{j,k}^m - u_{j+1,k}^m)/h + UNEG(u_{j+1,k}^m - u_{j,k}^m)/h$$

$$+VAUP(u_{j,k}^m - u_{j,k-1}^m)/h + VAUN(u_{j,k+1}^m - u_{j,k}^m)/h$$

$$-f\bar{v}^{u,m} = -\frac{g}{h}(\zeta_{j,k}^m - \zeta_{j-1,k}^m) + \frac{1}{\rho_o D_{u,j,k}^m}\tau_{x,j,k}^{s,m}$$

$$-R_{x,j,k}^{m-1}u_{j,k}^{m-1} + \frac{N_h}{h^2}(u_{j+1,k}^{m-1} + u_{j-1,k}^{m-1} + u_{j,k+1}^{m-1} + u_{j,k-1}^{m-1} - 4u_{j,k}^{m-1}) \qquad (3.95)$$

$$\frac{v_{j,k}^{m+1} - v_{j,k}^m}{T} + UAVP(v_{j,k}^m - v_{j-1,k}^m)/h + UAVN(v_{j+1,k}^m - v_{j,k}^m)/h$$

$$+VPOS(v_{j,k}^m - v_{j,k-1}^m)/h + VNEG(v_{j,k+1}^m - v^m j,k))/h$$

$$+f\bar{u}^{v,m+1} - \frac{g}{h}(\zeta_{j,k+1}^m - \zeta_{j,k}^m) + \frac{1}{\rho_o D_{v,j,k}^m}\tau_{y,j,k}^{s,m}$$

$$-R_{y,j,k}^{m-1}v_{j,k}^{m-1} + \frac{N_h}{h^2}(v_{j+1,k}^{m-1} + v_{j-1,k}^{m-1} + v_{j,k+1}^{m-1} + v_{j,k-1}^{m-1} - 4v_{j,k}^{m-1}) \qquad (3.96)$$

$$\frac{\zeta_{j,k}^{m+1} - \zeta_{j,k}^m}{T} = -\frac{1}{h}(u_{j+1,k}^{m+1}D_{u,j+1,k}^m - u_{j,k}^{m+1}D_{u,j,k}^m) - \frac{1}{h}(v_{j,k}^{m+1}D_{v,j,k}^m - v_{j,k-1}^{m+1}D_{v,j,k-1}^m)$$

$$\qquad (3.97)$$

Here the nonlinear terms are given according to the upwind–downwind approximation introduced in Ch.II. The following notation is used:

$$UPOS = (u_{j,k}^m + |u_{j,k}^m|)/2 \qquad UNEG = (u_{j,k}^m - |u_{j,k}^m|)/2$$

$$VAUP = (\bar{v}^{u,m} + |\bar{v}^{u,m}|)/2 \qquad VAUN = (\bar{v}^{u,m} - |\bar{v}^{u,m}|)/2$$

$$VPOS = (v_{j,k}^m + |v_{j,k}^m|)/2 \qquad VNEG = (v_{j,k}^m - |v_{j,k}^m|)/2$$

$$UAVP = (\bar{u}^{u,m+1} + |\bar{u}^{u,m+1}|)/2 \qquad UAVN = (\bar{u}^{u,m+1} - |\bar{u}^{u,m+1}|)/2$$

4.2 Numerical stability of the two-dimensional problem

Numerical stability of the system (3.89)–(3.91) can be checked by employing an array of errors (3.36). From the above system a quite complicated set of equations for the stability parameter λ follows:

$$\begin{pmatrix} \lambda^2 - 1 + a_1 + b_1 & c_1\lambda & \lambda d_1 2Tg/h \\ \lambda c_2 & \lambda^2 - 1 + a_2 + b_2 & \lambda d_2 2Tg/h \\ \lambda d_1^* 2TH/h & \lambda d_2^* 2TH/h & \lambda^2 - 1 \end{pmatrix} \begin{pmatrix} u^* \\ v^* \\ \zeta^* \end{pmatrix} = \begin{pmatrix} 0 \\ 0 \\ 0 \end{pmatrix} \qquad (3.98)$$

The stability condition for this system can be obtained by solving the coefficient determinant. The solution can be searched analytically or numerically. The general stability problem is easily simplified by assuming that the instability starts at the shortest resolved space scale. This allows us (see Ch.II) to set the wave number as $\kappa = 2\pi/L = 2\pi/2h$.

In order to identify various stability conditions we shall split the above complicated problem into simple problems.

Two-dimensional motion: only sea level is included

We shall start by neglecting the frictional processes and Coriolis force. In this case only the acceleration and pressure terms due to the sea level remain in the equation, and the determinant of (3.95) is simplified to

$$\det \begin{vmatrix} \lambda^2 - 1 & 0 & \lambda d_1 2Tg/h \\ 0 & \lambda^2 - 1 & \lambda d_2 2Tg/h \\ \lambda d_1^* 2TH/h & \lambda d_2^* 2TH/h & \lambda^2 - 1 \end{vmatrix} = 0 \qquad (3.99)$$

In the above equation $d_1 = 1 - e^{-i\kappa_1 h}$, $d_2 = 1 - e^{-i\kappa_2 h}$ and, d_1^* and d_2^* are complex conjugate to d_1 and d_2. Thus

$$d_1 d_1^* = -4(\sin \frac{\kappa_1 h}{2})^2 \quad \text{and} \quad d_2 d_2^* = -4(\sin \frac{\kappa_2 h}{2})^2$$

From (3.99) the equation for the stability parameter follows:

$$(\lambda^2 - 1)^2 + \lambda^2 4(\frac{2T}{h})^2 gH[(\sin\frac{\kappa_1 h}{2})^2 + (\sin\frac{\kappa_2 h}{2})^2] = 0 \qquad (3.100)$$

The roots of this equation are

$$\lambda^2 = [(2 - a^2) \pm \sqrt{a^4 - 4a^2}]/2 \qquad (3.101)$$

From the above we can single out two cases. If $a^2 > 4$ stability parameter $\lambda^2 > 1$, on other hand if $a^2 \leq 4$, the square root in (3.101) can be set as $ai\sqrt{4 - a^2}$ and the absolute value of $|\lambda^2| = 1$ can be found by multiplying (3.101) by the complex conjugate. Because $a^2 = 4(\frac{2T}{h})^2 gH[(\sin\frac{\kappa_1 h}{2})^2 + (\sin\frac{\kappa_2 h}{2})^2]$, the stability condition is

$$(\frac{2T}{h})^2 gH[(\sin\frac{\kappa_1 h}{2})^2 + (\sin\frac{\kappa_2 h}{2})^2] \leq 1 \quad \text{or} \quad (\frac{2T}{h})^2 gH2 \leq 1 \qquad (3.102)$$

This is the CFL condition for 2D motion when marching in time is done with $2T$ time step. The choice of the space step and time step is dependent on the overall depth. Shallow water will make the problem less severe, because it allows longer time stepping.

Motion due to Coriolis force only

Determinant of (3.99) is simplified to

$$\det\begin{vmatrix} \lambda^2 - 1 & c_1\lambda \\ c_2\lambda & \lambda^2 - 1 \end{vmatrix} = 0 \qquad (3.103)$$

Here $c_1 = -2Tf0.25(1 - e^{-i\kappa_1 h})(1 - e^{-i\kappa_2 h})$ and $c_2 = 2Tf0.25(1 + e^{-i\kappa_1 h})(1 + e^{-i\kappa_2 h})$. The equation for the stability parameter is quite similar to (3.100)

$$(\lambda^2 - 1)^2 + b^2\lambda^2 = 0 \qquad (3.104)$$

Where $b^2 = f^2(2T)^2(\cos\frac{\kappa_1 h}{2})^2(\cos\frac{\kappa_2 h}{2})^2$. By similar computations as before we arrive at the inequality

$$f2T \leq 2 \qquad (3.105)$$

This condition limits the time step to the inertial period, which at the middle latitudes is about 12 hrs.

Horizontal and bottom friction

First we consider the horizontal friction problem. It can be simplified to one equation for the purpose of the stability study as

$$\frac{u_{j,k}^{m+1} - u_{j,k}^{m+1}}{2T} = \frac{N_h}{h^2}(u_{j+1,k}^{m-1} + u_{j-1,k}^{m-1} + u_{j,k+1}^{m-1} + u_{j,k-1}^{m-1} - 4u_{j,k}^{m-1}) \qquad (3.106)$$

Similar approach can be taken to the motion along N–S direction. Introducing (3.36) into (3.106), an equation for the stability parameter follows (see Ch.II),

$$\lambda^2 = 1 - c^2 \qquad (3.107)$$

Here

$$c^2 = \frac{4(2T)N_h}{h^2}[(\sin\frac{\kappa_1 h}{2})^2 + (\sin\frac{\kappa_2 h}{2})^2]$$

In order for this numerical scheme to be stable

$$|\lambda^2| \le 1 \quad \text{or} \quad -1 \le (1 - c^2) \le 1$$

From this two inequalities follow

$$0 \le c^2 \quad \text{and} \quad c^2 \le 2$$

The first one is always fulfilled, the second one defines the stability condition

$$\frac{4(2T)N_h}{h^2} \le 1 \qquad (3.108)$$

Here, we deduce that for the given time and space step there is always a maximum value of the horizontal eddy viscosity N_h for which the numerical scheme is stable. Above this value one encounters an unstable behavior. Experimenting with the various eddy viscosity coefficients one can find that this scheme is unstable not only for the large values but for small magnitude of eddy viscosity as well. This limit is related to the physics of the process and can be found through numerical experiments.

Stability condition for the bottom friction can be searched starting from the equation

$$\frac{u_{j,k}^{m+1} - u_{j,k}^{m+1}}{2T} = -R_{x,j,k}^{m-1}u_{j,k}^{m-1} \qquad (3.109)$$

Here we assume the coefficient R to be constant. Equation for the stability parameter and stability condition follows

$$\lambda^2 = 1 - 2TR \quad \text{and} \quad 2TR \le 2 \qquad (3.110)$$

Bottom friction sets limits on the time step only. The limit is usually less stringent than the CFL condition. The actual value of R_x is

$$ru\sqrt{u^2 + v^2}/(\rho_o D)$$

therefore this term can be large at very small depths.

The nonlinear (advective) terms are not included in (3.89)–(3.91) system but they occur in the (3.95)–(3.97) set. A simplified problem of the time-dependent advective motion in one direction has been analyzed in Ch.II. Sec.4; the stability condition for the upwind–downwind numerical scheme is

$$0 \leq \frac{T|u|}{h} \leq 1 \qquad (3.111)$$

Although the above considerations provide a simple set of stability conditions, the interaction of two or more terms in the equations may result in a more complicated condition. Additionally we should remember that besides the conditions studied above, stability can be influenced by the boundary conditions, external forces, abrupt bottom variations, and so on.

4.3 Errors in the two-dimensional staggered grid

We have found in Ch.II and in Sec.3 of this chapter that the numerical formulation alters space and time properties of the long waves. To study these errors in the two-dimensional models let us start by considering the waves propagating over a flat bottom. The nonlinear, Coriolis and friction terms are neglected in the equations of motion,

$$\frac{\partial u}{\partial t} + g\frac{\partial \zeta}{\partial x} = 0 \qquad (3.112a)$$

$$\frac{\partial v}{\partial t} + g\frac{\partial \zeta}{\partial y} = 0 \qquad (3.112b)$$

$$\frac{\partial \zeta}{\partial t} + H(\frac{\partial u}{\partial x} + \frac{\partial v}{\partial y}) = 0 \qquad (3.112c)$$

Dispersive relation for the long waves, is found by searching for a solution to (3.112) in the following form:

$$(u, v, \zeta) \sim (u^*, v^*, \zeta^*)e^{i\omega t}e^{i\kappa_1 x}e^{i\kappa_2 y} \qquad (3.113)$$

Introducing (3.113) into (3.112) we arrive at

$$\omega^2 = gH(\kappa_1^2 + \kappa_2^2), \qquad (3.114)$$

and from (3.114) the phase velocity for the long wave follows,

$$c = \pm \frac{\omega}{\sqrt{\kappa_1^2 + \kappa_2^2}} = \pm \sqrt{gH} \qquad (3.115)$$

It is useful to notice that long waves are nondispersive i.e., the phase velocity is constant for every wave number.

To ascertain how the numerical formulas developed in the previous section will change the dispersive properties we limit our study to the space domain only. The full study i.e., searching the influence of the space and time numerical formulation may be done by the method delineated in Sec.3 of this chapter. Limiting our formulation only to the space domain allows us to write a difference equation in space and a differential equation in time,

$$\frac{\partial u}{\partial t} = -\frac{g}{h}(\zeta_{j,k} - \zeta_{j-1,k}) \qquad (3.116a)$$

$$\frac{\partial v}{\partial t} = -\frac{g}{h}(\zeta_{j,k+1} - \zeta_{j,k}) \qquad (3.116b)$$

$$\frac{\partial \zeta}{\partial t} - \frac{H}{h}(u_{j+1,k} - u_{j,k} + v_{j,k} - v_{j,k-1}) \qquad (3.116c)$$

These equations has been rendered into the finite difference form through the space-staggered grid given in Fig.3.10. The complete numerical set of equations are given by (3.89)-(3.91), and the numerical stability of the long wave propagation problem analogous to (3.116) is defined by (3.99). Searching, as usual, solution to (3.116) in the form $\exp i\omega t \exp i\kappa_1 h \exp i\kappa_2 h$, we shall find

$$\omega^2 = gH(\frac{2}{h})^2[(\sin\frac{\kappa_1 h}{2})^2 + (\sin\frac{\kappa_2 h}{2})^2] \qquad (3.117)$$

It is useful to notice that the right side of this equation is a portion of (3.100). To simplify ensuing considerations let us assume that the wave propagates along the x direction only. Thus,

$$\omega^2 = gH(\frac{2}{h})^2(\sin\frac{\kappa_1 h}{2})^2 \qquad (3.118)$$

In (3.118) the frequency is a complicated function of the wave number. Only when $\kappa_1 h \to 0$, the analytical result given by (3.114) follows.

Denoting the frequency in (3.114) as ω_a (assuming $\kappa_2 = 0$), and frequency in (3.118) as ω_n, the comparison of both results can done by means of

$$\frac{\omega_n^2}{\omega_a^2} = \frac{(\sin\frac{\pi h}{L})^2}{(\frac{\pi h}{L})^2} \qquad (3.119)$$

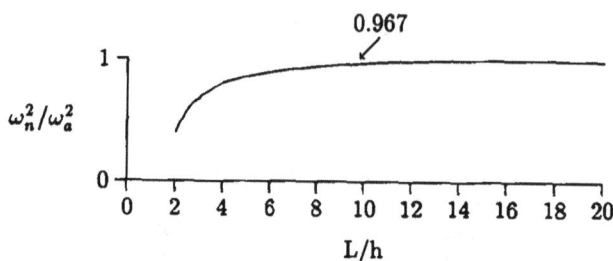

Fig. 3.13

Ratio of square of numerical and analytical frequency
versus resolution for the long gravity waves.

Here $\kappa_1 = 2\pi/L$. In Fig.3.13 this formula is given for various resolutions. The frequency derived from numerical calculation monotonically increases but still for $L/h = 10$ the error between analytical and numerical results is 3.3%.

Dividing both sides of (3.118) by κ_1^2 the formula for the phase velocity is obtained

$$c_N^2 = \frac{\omega^2}{\kappa_1^2} = gH(\frac{2}{\kappa_1^2 h})^2(\sin\frac{\kappa_1 h}{2})^2 \qquad (3.120)$$

Expanding this formula into a series for the small parameter ($\kappa_1 h \to 0$) a simple expression for the velocity follows

$$c_N \simeq \sqrt{gH}[1 - (\kappa_1 h)_2/24] \qquad (3.120a)$$

This phase velocity is smaller than the velocity derived from (3.115) and we conclude that through the numerical approximation an additional friction has been introduced.

Let us continue the above considerations with the Coriolis force included in (3.112) system of equations. Searching for the solution of this system through (3.113) we arrive at the dispersion relation

$$\omega^2 = gH(\kappa_1^2 + \kappa_2^2) + f^2 \qquad (3.121)$$

This formula encompasses both the gravity and inertial waves.

To study the influence of the finite space resolution on the above formula we apply difference–differential equations (3.116) with Coriolis force given by (3.92). Searching solution to this system by the approach given for the system (3.116) we obtain

$$\omega^2 = gH(\frac{2}{h})^2[(\sin\frac{\kappa_1 h}{2})^2 + (\sin\frac{\kappa_2 h}{2})^2] + f^2(\cos\frac{\kappa_1 h}{2})^2(\cos\frac{\kappa_2 h}{2})^2 \qquad (3.122)$$

Combination of gH and f describes an important horizontal scale parameter, the so-called Rossby radius of deformation $\lambda = \sqrt{gH}/f$. Here to deduce the influence of space resolution on the frequency we have to consider both the wave length and Rossby radius. To simplify the formulas we take only one direction assuming $\kappa_2 = 0$. Dividing (3.122) and (3.121) by f^2 we obtain

$$\omega^2/f^2 = (\frac{2\lambda}{h})^2(\sin\frac{\kappa_1 h}{2})^2 + (\cos\frac{\kappa_1 h}{2})^2 \qquad (3.123a)$$

$$\omega^2/f^2 = \lambda^2\kappa_1^2 + 1 \qquad (3.123b)$$

Let us denote again an analytical expression for the frequency (3.123b) as ω_a and numerical expression (3.123a) as ω_n. Next dividing (3.123a) by (3.123b), and denoting $\lambda/h = s$, we arrive at;

$$\omega_n^2/\omega_a^2 = [(2s)^2(\sin\frac{\kappa_1 h}{2})^2 + (\cos\frac{\kappa_1 h}{2})^2]/[s^2\kappa_1^2 h^2 + 1] \qquad (3.124)$$

By assuming first $s > 1$, one finds that the gravity wave solution will dominate in (3.124), therefore this result will be closer to the one previously derived. In Fig.3.14 such a case for $\lambda/h = 2$ is depicted by the continuous line. On the other hand, when $s = 1/2$ both gravitational and inertial waves are of equal importance. This case is depicted by the dashed line in Fig.3.14. The errors for both cases when $\lambda/h = 10$, are equal 6% and 24% respectively.

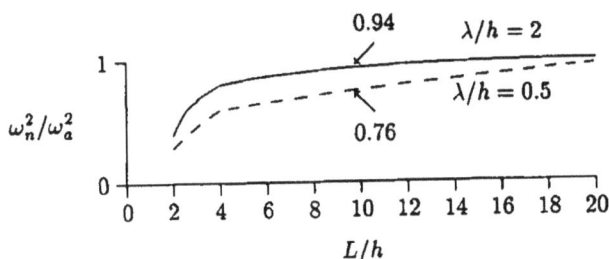

Fig. 3.14

Ratio of square of numerical and analytical frequency versus resolution for the long gravity waves and inertial waves.

The case $\lambda/h = 2$ has been studied by Arakawa (1972), and his results were extended by Batteen and Han (1981). Using Kelvin waves for an illustrative purpose Henry (1981) was able to estimate exactly the errors generated by the staggered grids. This was possible since an analytic solution for Kelvin waves propagating over a uniform depth is well known.

5. Numerical filtering

As we have seen the behavior of the long waves is not well resolved at the small scales when a numerical method is applied. Phillips (1959) has shown that instability in the nonlinear meteorological problems often starts at the shortest wave whose wavelength is equal $2h$. Improving the stability can often be achieved by filtering output over the short wave band. Even a finite differencing schemes constructed through the energy conservation principle (Bryan, 1969), do require some filtering. This is done usually by means of horizontal dissipation process.

The well-known application of the filters to integration in geophysical fluid dynamics problem is due to Shuman (1957). The filtered value $(\overline{F})$ is derived from the three time levels,

$$\overline{F}^m = F^m + 0.5\nu(F^{m-1} - 2F^m + F^{m+1}) \tag{3.125}$$

or,

$$\overline{F}^m = F^m(1 - \nu) + 0.5\nu(F^{m-1} + F^{m+1}) \tag{3.125a}$$

In above formulas ν denotes filter parameter, to be chosen later on.

The filtering process in space is equivalent to (3.125). Instead of three time levels one shall use three grid points. Filter given by (3.125) is centered around level m and it is symmetrical with respect to this level. Let us take a periodic wave,

$$F^m = F^* e^{i\omega t} = F^* e^{i\omega m T} \tag{3.126}$$

and find out how the filter (3.125) will change this wave (i.e., its amplitude, phase or frequency). Introducing (3.126) into the right-hand side of (3.125) the following result is obtained:

$$\overline{F}^m = F^* \left[1 + 0.5\nu(e^{-i\omega T} + e^{i\omega T} - 2)\right] e^{i\omega m T}$$

$$= F^* \left[1 - \nu(1 - \cos\omega T)\right] e^{i\omega m T} = F^* \left[1 - 2\nu(\sin\frac{\omega T}{2})^2\right] e^{i\omega m T} \tag{3.127}$$

By comparing (3.127) and (3.126) one can see that the phase and the frequency of the filtered wave does not change. The amplitude of the wave is changed according to

$$R_c = \frac{\overline{F}^m}{F^m} = 1 - 2\nu(\sin\frac{\omega T}{2})^2 \tag{3.128}$$

Response (R_c) can be chosen by varying the filter parameter ν. The filtered amplitude stays unchanged only for the high resolutions ($\omega T \to 0$) or if $\nu = 0$. The filter will damp the signal when

$$1 - 2\nu(\sin\frac{\omega T}{2})^2 \leq 1 \tag{3.129}$$

and $0 \le \nu \le 0.5$. The shortest wave period $t = 2T$ will be completely removed if

$$1 - 2\nu(\sin\frac{2\pi T}{2t})^2 = 1 - 2\nu(\sin\frac{\pi}{2})^2 = 0 \qquad (3.130)$$

which holds for $\nu = 0.5$. All the other periods will display a diminished amplitude.

For amplification of the signal the response $R_c > 1$ and the filter parameter should be in the range:

$$-0.5 \le \nu < 0 \qquad (3.131)$$

We could, of course, choose in (3.129) $\nu > 0.5$ and in (3.131) $\nu < -0.5$ but this will change sign of the amplitude or equivalently change the phase by $180°$.

Asselin (1972) constructed an interesting and very practical filter. Since the filtering process is usually done consecutively in time, therefore at the time level m the filtered value from the time level $m - 1$ is already available and it can be used in the following manner:

$$\overline{F}^m = F^m + 0.5\nu(\overline{F}^{m-1} - 2F^m + F^{m+1}) \qquad (3.132)$$

This filter has been applied by Asselin (1972) for various purposes. An interesting application is damping of the numerical spurious solution which occurs in the leapfrog method.

Filters can be applied in two dimensions as well. Let us assume that (3.125) is actually applied along x axis then the filtering operation in two dimensions is

$$\overline{F}_{j,k} = \overline{[F_{j,k} + 0.5\nu(F_{j-1,k} - 2F_{j,k} + F_{j+1,k})]}^k \qquad (3.133)$$

To every term inside the square brackets, averaging along the k direction is applied. As an example let us consider the term $F_{j-1,k}$. Using the definition (3.125) but this time to the index k, we obtain,

$$\overline{F}_{j-1,k} = F_{j-1,k} + 0.5\nu(F_{j-1,k-1} - 2F_{j-1,k} + F_{j-1,k+1}) \qquad (3.134)$$

By this means, from (3.133),

$$\overline{F}_{j,k} = F_{j,k} + 0.5\nu(1 - \nu)(F_{j-1,k} + F_{j+1,k} + F_{j,k-1} + F_{j,k+1} - 4F_{j,k})$$

$$+0.25\nu^2(F_{j+1,k-1} + F_{j-1,k-1} + F_{j-1,k+1} + F_{j+1,k+1} - 4F_{j,k}) \qquad (3.135)$$

the nine point filter in two dimensions is constructed.

The basic three element filter (3.125) removes the shortest waves but it also distorts all the other waves. In search for a filter with better characteristics Shapiro (1970) suggested a nine point formulae:

$$F^m = \frac{1}{256}[186F^m + 56(F^{m-1} + F^{m+1}) - 28(F^{m-2} + F^{m+2})$$

$$+8(F^{m-3} + F^{m+3}) - (F^{m-4} + F^{m+4})] \tag{3.136}$$

The horizontal and vertical friction terms in the equation of motion also operate like filters. Considering unidirectional equation of motion in the difference-differential form,

$$\frac{\partial u}{\partial t} = \frac{N_h}{h^2}(u_{j+1} + u_{j-1} - 2u_{j,k}) \tag{3.137}$$

and searching the solution in the form $u = u_1 e^{i\kappa jh}$, we arrive at

$$\frac{\partial u_1}{\partial t} = -u_1 \frac{4N_h}{h^2}(\sin\frac{\kappa h}{2})^2 \tag{3.138}$$

Marching in time as

$$u_1^{m+1} = u_1^m[1 - \frac{4TN_h}{h^2}(\sin\frac{\kappa h}{2})^2] \tag{3.139}$$

one can see that the above expression is the filter with a response function defined as

$$\lambda = \frac{u_1^{m+1}}{u_1^m} = 1 - \frac{4TN_h}{h^2}(\sin\frac{\kappa h}{2})^2 \tag{3.140}$$

The filter's parameter are the eddy viscosity, time and space step. For the shortest wave $\kappa h/2 = 2\pi h/(2L) = \pi/2$, the most effective damping takes place if

$$1 - \frac{4TN_h}{h^2} = 0, \quad \text{or} \quad N_h = \frac{h^2}{4T} \tag{3.141}$$

For the biharmonic friction in the equation of motion,

$$\frac{\partial u}{\partial t} = -N_h^b \frac{\partial^4 u}{\partial x^4} \tag{3.142}$$

and numerical approximation

$$\frac{\partial^4 u}{\partial x^4} \sim \frac{1}{h^4}(u_{j+2} - 4u_{j+1} + 6u_j - 4u_{j-1} + u_{j-2}) \tag{3.143}$$

above considerations lead to

$$\frac{u_1^{m+1}}{u_1^m} = 1 - \frac{TN_h^b}{h^4}(6 - 8\cos\kappa h + 2\cos 2\kappa h) \tag{3.144}$$

The rate of damping in regular (Laplacian) and biharmonic term is different because damping is proportional to h^{-2} in the former case and to h^{-4} in the latter case. Because not all the space–time scales are resolved by the finite–differences, the numerical schemes cause a distortion to the flow of energy in the physical processes. Therefore, filters are used to correct this situation. One interesting and moot problem is the role of the horizontal friction in the calculations and in the physics of the process. Application of three time level scheme constructed by Fischer (1959) without the horizontal friction term resulted in short wave instability. Hansen (1962) damped these instabilities by the five-point space filter related to (3.135). The filter was applied to every variable at every time step:

$$\overline{F}_{j,k} = \nu F_{j,k} + 0.25(1 - \nu)(F_{j-1,k} + F_{j+1,k} + F_{j,k-1} + F_{j,k+1}) \qquad (3.145)$$

with $\nu = 0.99$. Eventually, Brettschneider (1967) achieved stability by applying the horizontal friction term with the eddy viscosity

$$N_h = 0.01\frac{h^2}{T} \qquad (3.146)$$

By means of this coefficient the friction is parameterized for the scales not resolved by the grid or so-called subgrid scales. Can we conclude from the above that the horizontal eddy viscosity is a numerical or a physical parameter? The situation is not clear, on the one hand the processes like barotropic instability depend on friction, on the other hand the presence of friction governs the numerical stability and it is often impossible to distinguish between these usages.

6. Grid refinement

The processes which affect the long waves occur over broad space scales. If all the scales were resolved by a numerical model the computer time required to obtain the solution may be prohibitively large. A solution to this problem is to subdivide the general integration domain into smaller subdomains, where spatial resolution can be adjusted. A simple example of the problem for which variable resolution can be used is tide propagation over deep and shallow areas. Consider M_2 tide propagating (period 12.42 hrs) over a 5 km depth ocean. Its wavelength is about 10^4 km. Over the shallow water of 50 m depth the wavelength decreases to about 10^3 km (assuming the the period remains the same). A numerical space grid of 10^3 km resolves the deep ocean wave quite well but the same space step in the shallow water is too long to resolve the tide wave. In this way we have arrived at a very simple idea of enhancing resolution only at the selected regions. Let us set in the above example over the deep portion of the domain a space step of 10^3 km and over the shallow portion the step is diminished to 10^2 km. Efficient use of the space will lead in turn to computer time saving. Since we have to consider processes in different domains with different space resolution we shall need the means of connecting the coarse resolution domains with the fine resolution domains. Sometimes we shall call this process an embedding of the regional (fine scale) model within the general (coarse scale) model. The way to couple these domains is to match the variables along the common boundary. In our example we have increased space grid 10 times over the boundary from the fine to the coarse grids. One has to wonder if the tide passing through the boundary is reflected or distorted. Such distortion may remove all the advantages of the grid refinement. The important point is to assure a smooth variation on both sides of the boundary and over the boundary. There is no hard rules on how to define the "safe " increase in the resolution over the boundary. This approach has to be tested for various grids and boundaries.

Generally, the coupling of the different domains may be interactive (active) and non-interactive (passive). In the passive approach the variables calculated in the coarser-grid domain are passed to the finer-resolution domain only. The result from the fine grid do not interact with the coarse grid variables. An interactive modeling is implemented by splicing the variables in the both directions. Such interactions may be achieved by various means. The coarse grid may be connected to the fine grid through linear interpolation and the variables from the fine grid are used directly as a boundary condition for the coarse grid. For connecting the coarse and fine grids instead of linear interpolation, a spline interpolation can be used, or information from the fine grid , instead of direct application as the boundary condition for the coarse grid , can be first filtered with the short wave filter so that it is better adjusted to the coarse grid resolution.

A different way to achieve the active interaction is to apply directly the formulas for the derivatives on an irregular grid net. This approach has not been in extensive use and not too much is known about the behavior of such an approach. Let us outline shortly the use of an irregular grid combined with linear interpolation. We shall use Fig.3.15 to construct the basic formulas for the grid interaction. Column j (open circles) is the boundary between the fine grid on the left side (bullets) and the coarse grid on the right side (crosses). To define the first and second derivatives in the open circle points we shall use the formulas given in the Ch.II. Sec.1. The simple application of the given formulas can be implemented at the j, $k + 1$ and j, $k - 2$ points. In constructing the first derivative along the x direction at the j, $k - 1$ and j, k points we shall use a linear approximation. Thus denoting the first derivative as $D1$ one can write for the j, k point:

$$D1(j,k) = \left[(hy_{-2} + hy_{-1})D1(j,k+1) + hy_1 D1(j,k-2)\right]/(hy_{-2} + hy_{-1} + hy_1)$$
(3.147)

and for the j, $k - 1$ point:

$$D1(j,k-1) = \left[hy_{-2}D1(j,k+1) + (hy_1 + hy_{-1})D1(j,k-2)\right]/(hy_{-2} + hy_{-1} + hy_1)$$
(3.148)

Here the derivatives at the j, $k + 1$ and j, $k - 2$ points are calculated through a direct application of the irregular grid formulas. A similar approach is used for the second derivative along the x direction.

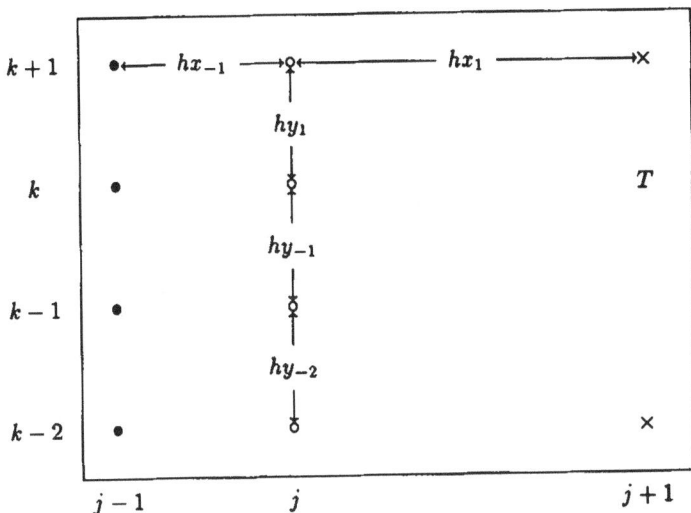

Fig. 3.15
Irregular grid net.

Estimation of the derivatives along the y direction is based on a similar approach. Let us depict the calculation of the second derivative in the j, k point. For this purpose we first calculate the second derivatives in the $j - 1$, k and T points. The second derivative in the $j - 1$, k point is derived by the 3 point formula for the $j - 1$ column and the $k - 1$, k, and $k + 1$ row. The second derivative along the y coordinate in the T point is not ready available. The value of the function in this point has to be obtained first by linear interpolation between the coarse grid points and afterwards the three-point formula can be used to calculate the second derivative in the T point. Finally, linear interpolation between the $j - 1$, k and T points by applying the hx_1 and hx_{-1} distances defines the second derivative in the j, k point.

The second method of the interactive grid splicing involves linear approximation to transfer variables from the coarse grid to the fine grid and the variables from the fine grid are used directly as the boundary condition for the coarse grid.

Fig.3.16 illustrates how this method works. Within the coarse grid, denoted by the capital letter indices, the fine grid is embedded. The usual space staggered grid is used so that the u–component of velocity is denoted by the horizontal bar, the v–component of velocity is denoted by the vertical bar, and the sea level (ζ) location is given by the cross (note that in the coarse grid the bars and crosses

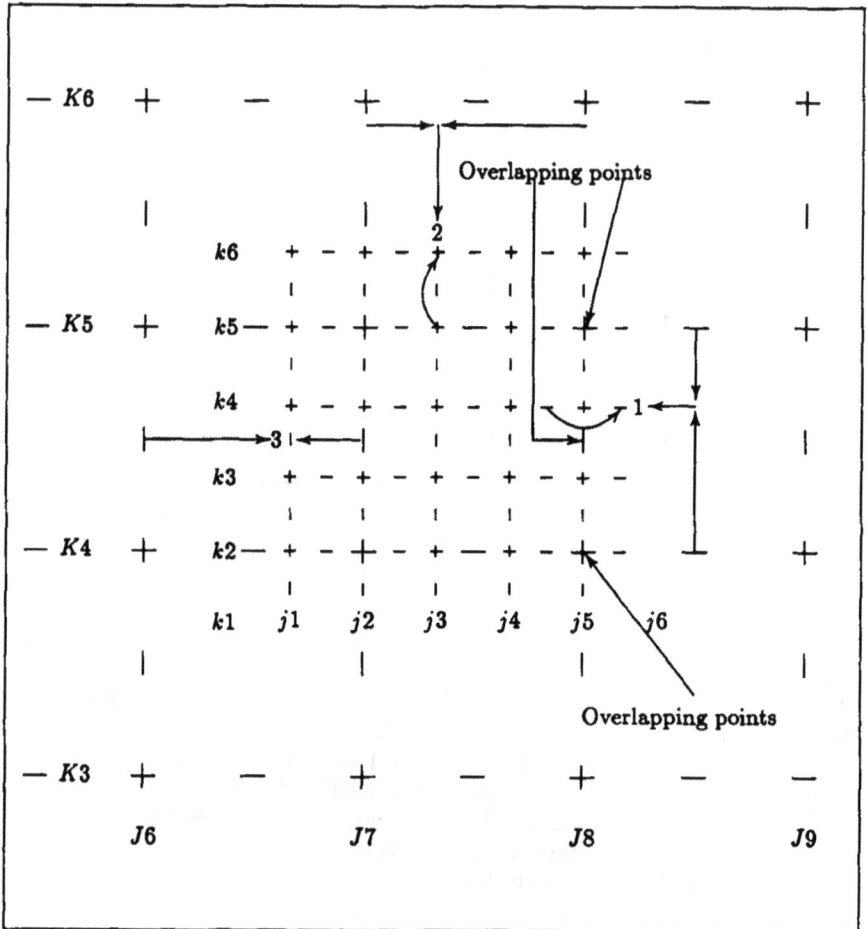

Fig. 3.16
Embedding the fine grid within the coarse grid.

are larger and the capital letters are used). The linear interpolation of the coarse grid variables used in the fine grid calculations as the boundary values is illustrated at the three grid points (1,2,3) in Fig.3.16. Let us consider a vertical boundary located between $j5(J8)$ and $J9$ columns. In the column $j5(J8)$ three points from the coarse grid overlap the fine grid. These are the sea level points at $K4$ and $K5$, and one v–velocity point with coordinate $J8$, $K4$ (see Fig.3.10 for the triple point explanation). Calculation of the sea level in the fine grid point along the $j5$ column require the u–component of velocity to the right from the $j5$ column. The value of this velocity is estimated at the point 1 by the linear interpolation from the coarse grid values $U(J9, K5)$ and $U(J9, K4)$, and from the fine grid value $u(j5, k4)$. Thus velocity at the fine grid point with coordinates $j6, k4$ is

$$u(j6, k4) = [U(J9, K5) * 2/FINC + U(J9, K4) * 1/FINC] * 2/(FINC + 1)$$

$$+u(j5, k4) * 2/(FINC + 1) \qquad (3.149)$$

Here we have assumed that the grid distance from the fine grid to the coarse grid is increased FINC times (in the figure FINC=3)

Another example is depicted in point 2, here the sea level is estimated so that the v-component of velocity at $j3, k5$ point can be calculated. The sea level estimate includes the sea level from the $J7$, $K6$ and $J8$, $K6$ points of the coarse grid and the $j3$, $k5$ point of the fine grid;

$$\zeta(j3, k6) = [\zeta(J7, K6) * 2/FINC + \zeta(J8, K6) * 1/FINC] * 1/FINC$$

$$+\zeta(j3, k5) * 2/FINC \qquad (3.150)$$

Point 3 serves to define the v-component of velocity. This component may serve to calculate the Coriolis or the bottom friction terms. Here,

$$v(j1, k3) = V(J6, K4) * 1/FINC + v(j2, k3) * 2/FINC \qquad (3.151)$$

Thus, in this computational procedure we first solve for the u, v, ζ over the fine grid, and then over the coarse grid. Coarse grid values which overlap the fine grid are then updated with the fine values (An example of the overlapping points is shown in the Fig.3.16). Lastly, the values around the edge of the fine grid, which are necessary for computations within the fine grid, are linearly interpolated between the coarse and fine grids. This technique can be repeated within the same model for more than one boundary, thus achieving a finer resolution at every boundary. The advantage of this approach is that any decrease in space resolution from the coarse to the fine grid, if expressed by an odd whole integer number (FINC=3,5,7,..), can be handled by the same algorithm. The purpose of this technique is to transmit the energy from the fine grid to the coarse grid. A potential problem is that the high

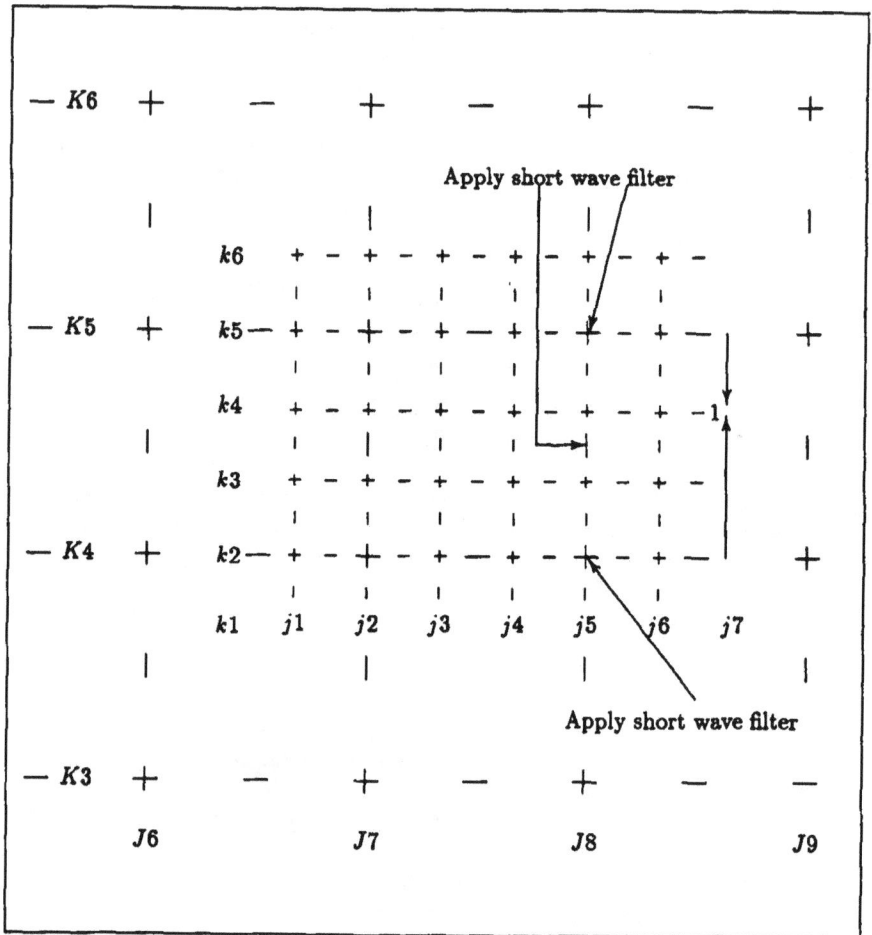

Fig. 3.17
The fine and coarse grid interaction between column $J8$ and $J9$.

frequency waves generated within the fine grid, but not resolved by the coarse grid, will be trapped in the fine grid.

To overcome this problem, which eventually may lead to an instability, we shall construct a more complicated approach to the boundary interaction. The short waves which may be trapped close to the boundary in the fine grid region, will be filtered out by means of Shapiro (1970) filter (see Sec.5 of this Chapter). The new computational procedure repeats the previous procedure, but at the points where the coarse and fine grids overlap, the coarse grid value is not updated by the fine grid value, as before. This fine grid value is filtered first by the nine point space filter and afterwards it is taken as a boundary value for the coarse grid. This technique is demonstrated in Fig.3.17. Here we describe the procedure only for the boundary located between the column $J8(j5)$ and $J9$. To calculate the sea level in the fine grid along column $j5$ we have previously introduced an additional column $j6$, where the u–component of velocity has been estimated. Now, to filter the fine grid sea level and velocity in the points where the coarse grid overlaps the fine grid, an additional column ($j6$) with the sea level is added, but this requires an additional column of the u velocities ($j7$). These velocities are estimated from the coarse and the fine grid values. The velocity approximation which previously was performed in the $j6$ column, now is done in the $j7$ column. In the new construction the fine grid continues into the coarse grid for the two fine grid distances. The Fig.3.17 depicts this approach only for the boundary located between $J8$ and $J9$ columns. Here the linear approximation for the u–velocity in the point 1, column $j7$, and the overlapping points in the $j5$ column have been shown.

A brief review of the literature pertinent to the grid refinement illustrates that the method has been successful for complex topography, and it well reproduces tides, storm surges, and currents (Ramming and Kowalik, 1980; Spall and Holland, 1991; Kowalik and Whitmore, 1991). Chilicka et al. (1983) derived some results by the differencing on the irregular grid, using linear and spline interpolation. The best results were from the irregular grid, the linear and spline interpolation displayed a damped signal in time. This illustrates that the latter two methods do not conserve energy in time. If the conservation property is important (and it should be for the long integration times) a more complicated approach to construct the boundary, proposed by Koss (1971), may be implemented. Koss (1971) also showed that the boundary between the finer and coarse grid can be source of a flow distortion.

The short wave distortion generated at the boundary can be deleted by various means. We have just depicted in Fig.3.17 the filters located close to the boundary. A similar result is achieved by an application of the filter (3.135) or (3.136) all over the integration domain. At the region near the boundary a high viscosity sponge layer (Carton, 1984) brings also the desired results. This layer is constructed by setting around the boundary the horizontal eddy viscosity or the bottom friction to a large value. The signal passing through this region is damped and the short waves are removed. In order to understand how the sponge layer influences the flow, careful numerical experiments are required so that the sponge layer parameters be adjusted to the problem. Actually, all the problems which involve the filters require sensitivity studies so that the dependence of the flow on the filter parameter is elucidated.

7. Simulation of long wave run-up

7.1 Formulation of the problem and brief literature review

The long waves – tides, storm surges and especially tsunamis may develop quite large amplitudes in the near coastal areas. These waves climbing on the shore often cause severe flooding and destruction. In this section we consider the boundary condition and numerical schemes which allow to estimate wave run-up.

The hydrodynamical and mathematical problems connected with discontinuity between wet and dry domains, nonlinearity, friction and computational instability these are the main problems which have to be sorted out in the run-up computation.

To study long wave run-up, the vertically integrated set of equations of motion and continuity is usually used:

$$\frac{\partial u}{\partial t} + u\frac{\partial u}{\partial x} + v\frac{\partial u}{\partial y} = -g\frac{\partial \zeta}{\partial x} - \frac{ru\sqrt{u^2 + v^2}}{D} \tag{3.152}$$

$$\frac{\partial v}{\partial t} + v\frac{\partial v}{\partial y} + u\frac{\partial v}{\partial x} = -g\frac{\partial \zeta}{\partial y} - \frac{rv\sqrt{u^2 + v^2}}{D} \tag{3.153}$$

$$\frac{\partial \zeta}{\partial t} = -\frac{\partial(Du)}{\partial x} - \frac{\partial(Dv)}{\partial y} \tag{3.154}$$

All the notation is standard with u directed along the x direction, and v along the y direction, ζ is the sea level variation and D is the total water depth, $D = H + \zeta$. Numerical models, constructed by Reid and Bodine (1968), Sielecki and Wurtele (1970) and Flather and Heaps (1975), were able to simulate the extent of inundation, but not the actual processes of wave propagation, breaking, and interaction with the coastal structures. This approach was quite successful in predicting flooding due to a major tsunami in Chile (Hebenstreit *et al.*, 1985). A very extensive and thorough study by Lewis and Adams (1983) resulted in a complicated numerical solution for the one-dimensional problem only. Kowalik and Bang (1987) derived a solution using a different numerical algorithm, but again, only to the one-dimensional case. The boundary point between dry and wet domains is a singular point. Its position can be calculated from the equation of motion and continuity by directional derivatives only. One has to suggest a tsunami "predictor" which will move the boundary between dry and wet grid points. The best known approach was proposed by Sielecki and Wurtele (1970). They applied an extrapolation of the sea level to the first dry point based on the continuity equation (3.154). Denoting m as usual, an index of time integration and j as and index of space integration along the x direction, the extrapolated level can be found as:

$$\zeta_j^{m+1} = (\zeta_j^m)_{ext} - \frac{T}{h}\{[(uD)_{j+1}^m]_{ext} - (uD)_j^m\} \qquad (3.155)$$

Here:

$$(\zeta_j^m)_{ext} = 2\zeta_j^m - \zeta_{j-1}^m \qquad (3.155a)$$

$$[(uD)_{j+1}^m]_{ext} = 2(uD)_j^m - (uD)_{j-1}^m, \qquad (3.155b)$$

T is time step and h is space step of numerical integration. Kim and Shimazu (1982), on other hand, applied an equation of transport neglecting friction and inertia. Hibberd and Peregrine (1979) studied the run-up and back-wash by considering the long wave equations together with the wave front condition represented by a bore. Their condition is based on the fact that the total depth equals zero at the boundary and $\zeta = -H$. To find the velocity at the shoreline this condition can be incorporated into the equation of motion. Considering frictionless motion along the x direction only one can write from (3.152),

$$\frac{\partial u_s}{\partial t} + u_s \frac{\partial u_s}{\partial x} = g\frac{\partial H}{\partial x} \qquad (3.156)$$

The next step, which made the wave front prediction more realistic, was an inclusion of the dissipative effects. Matsutomi (1983) proposed a new wave-front condition which depends on the time-varying friction parameter. Mader (1988) suggested a somewhat different and very simple approach to the moving boundary problem. The movement of the boundary follows from the equation of continuity written for the upwind–downwind form. Thus, if the velocity is positive in point j the depth differences are considered between points $j + 1$ and j; for the negative velocity the differences are taken between points j and $j - 1$.

The final choice of the boundary condition will depend on the comparison of analytical or numerical computations and observations. The experiments carried out in a wave tank by Yeh (1987) did show discrepancies between the observed bore front and the proposed analytical or numerical solution. Such experiments are very helpful in searching for a realistic wave-front condition.

Along with the hydrodynamical problems, the mathematical problems connected with discontinuity and nonlinearity make the run-up problem very difficult. Strong nonlinearity and discontinuity are often sources of computational instability which require sophisticated filtering techniques (Lewis and Adams, 1983). A small error near the boundary, due to the approximation of a different order from the general set of equations, will generate short wave oscillations which rapidly propagate into the whole computational domain (Marchuk et al., 1983). To delete the short period oscillations three approaches are feasible:

1. Shapiro's space filter constructed on nine grid points (Shapiro 1970; Lewis and Adams, 1983), see Sec.5 of this chapter.

2. High order upstream derivative method. The high order finite difference formulas constructed on the seven point stencil (Kowalik and Bang, 1987) and applied to the advective term also generate a stable solution, see Ch.II.

3. Filtering by the additional horizontal friction term. By proper tuning this term can be used as a short wave filter without distorting the long period harmonics, see Sec.5 of this chapter.

Along with the line of research described above, a second approach has been evolving. In this approach, a transformation of variables is applied and the independent variables x, t are transformed as

$$X = \frac{x}{l(t)} \quad \text{and} \quad T = t \tag{3.157}$$

where $l(t)$ is the distance from the origin of coordinate to a shore line. Through this transformation the variable region $0 \leq x \leq l(t)$ is transformed into a fixed region $0 \leq X \leq 1$, and no special boundary condition is required. The method has been extensively applied in USSR by L'atkher et al. (1978), in England by Johns (1982) and in Japan by Takeda (1984).

To check numerical results a comparison against the exact solutions can be performed. Here an analytical solution for the wave running up the beach without friction, given by Carrier and Greenspan (1958), may be used. A solution derived by Thacker (1981) for two-dimensional oscillation in parabolic basin is very useful in predicting the extent of inundation.

7.2 Numerical scheme—construction and testing

To construct a numerical scheme, a space staggered grid is applied which requires either sea level or velocity as a boundary condition. The first order scheme is applied in time and in space. Integration is performed along the x and y directions separately. For this purpose the set (3.152)-(3.154) is split in time into two subsets. Denoting m as an index of numerical integration in time, the method can be explained through the following difference–differential equations.

First, these equations are solved along the x direction,

$$\frac{u^{m+1} - u^m}{T} + (u\frac{\partial u}{\partial x})^m + (v\frac{\partial u}{\partial y})^m = -g(\frac{\partial \zeta}{\partial x})^m - (\frac{ru\sqrt{u^2 + v^2}}{D})^m \tag{3.158a}$$

$$\frac{1}{2}\frac{\zeta^* - \zeta^m}{0.5T} = -\frac{\partial(D^m u^{m+1})}{\partial x} \tag{3.158b}$$

and next along the y direction,

$$\frac{v^{m+1} - v^m}{T} + (v\frac{\partial v}{\partial y})^m + (u\frac{\partial v}{\partial x})^m = -g(\frac{\partial \zeta}{\partial y})^m - (\frac{rv\sqrt{u^2 + v^2}}{D})^m \qquad (3.159a)$$

$$\frac{1}{2}\frac{\zeta^{m+1} - \zeta^*}{0.5T} = -\frac{\partial(D^m v^{m+1})}{\partial y} \qquad (3.159b)$$

In this algorithm the calculation of the sea level starts from time step m, and along the x direction a preliminary value ζ^* is obtained. Afterwards, this value is carried over to the y direction to derive the sea level at the $m+1$ time step.

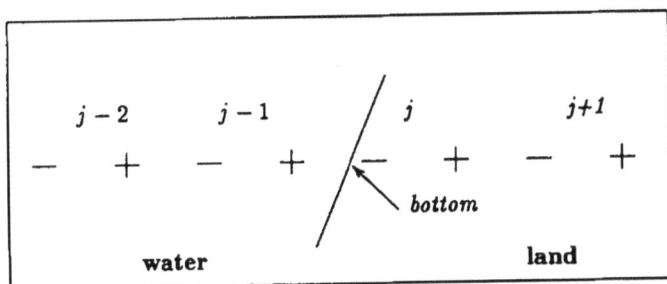

Fig. 3.18
Grid distribution for the run-up problem.

To apply the boundary conditions at the water–land boundary, first the boundary must be located at each time step. This is done by a simple algorithm proposed by Flather and Heaps (1975) for the storm surge computations and explained in Fig.3.18 for the x direction only. To ascertain whether u_j is a dry or wet point (see Fig.3.18), the sea level is tested at this point;

$$\begin{cases} u_j \text{ is wet point} & \text{if } 0.5(D_{j-1} + D_j) \geq 0 \\ u_j \text{ is dry point} & \text{if } 0.5(D_{j-1} + D_j) < 0 \end{cases} \qquad (3.160)$$

It is obvious that the total depth is positive at the wet points and is equal to zero at the boundary because here $\zeta = -H$. The small negative value of the total depth set at the dry points allows to identify the location of all dry regions in the computational process. Next, Sielecki and Wurtele's (1970) extrapolation (3.155)–(3.155b) of the sea level to the first dry point is used. The velocity in the first dry point is extrapolated linearly from the two last points. Linear extrapolation is easy to program but caution should be exercised in the case of a rough beach.

To test the above numerical method, we simulate a case which was solved analytically by Carrier and Greenspan (1958). The wave running up the beach

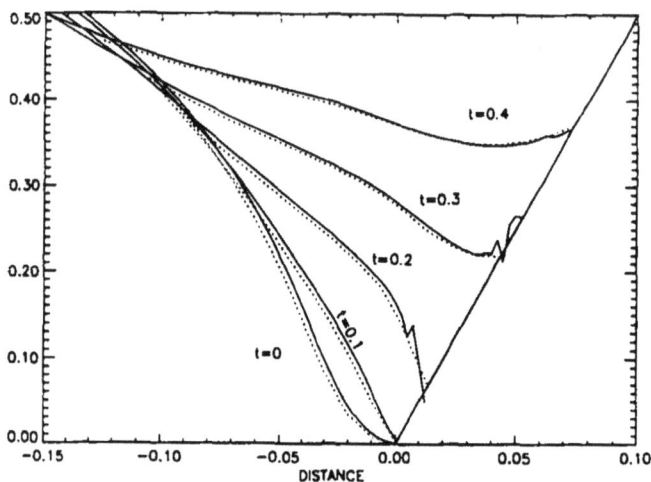

Fig. 3.19a

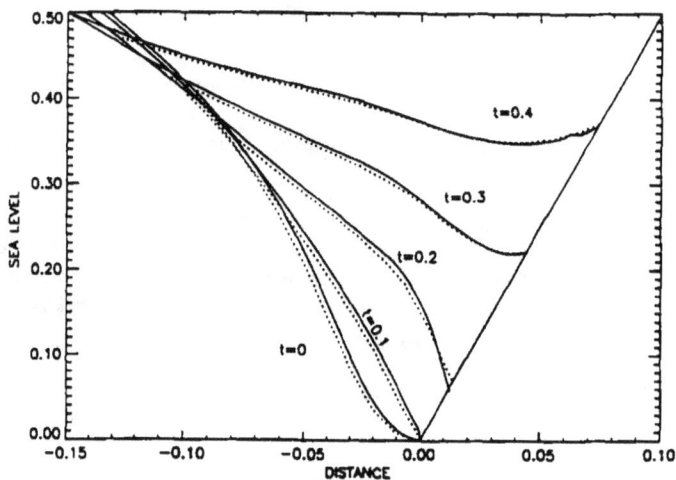

Fig. 3.19b

Wave run-up over sloping beach. Dotted line — analytical solution
and continuous line — numerical solution. All numbers are nondimensional.

without friction was considered and the problem was solved in the dimensionless
coordinates. The distribution of the sea level along the sloping beach at the various
times given both by analytical (dot line) and numerical (continuous line) solutions
is plotted in Fig.3.19a

One can see that the numerical method has reproduced the analytical solution quite well, but for time $t=0.2$ and 0.3 short period oscillations occur in the numerical results. (All the variables in Carrier and Greenspan's (1958) problem are dimensionless.) Small errors near the boundary, due to the different orders of approximation of the general set of equations and boundary conditions, are causing these oscillations. These oscillations are eliminated by a filter (3.136), see Fig.3.19b.

In order to carry out calculation of the wave run-up by the above method one must be sure that the long waves climb a sloping beach without breaking. The condition for the nonbreaking waves to occur was defined by Voltzinger et al. (1989) through a dimensionless parameter (Br),

$$Br = \frac{\zeta_o \omega^2}{g \alpha^2} \qquad (3.161)$$

Here, ζ_o is the amplitude of the wave, and α is the slope of the beach. If $Br < 1$, the nonbreaking waves occur; in case $Br > 1$, the waves will break while climbing the sloping beach.

In this part of the chapter, the earlier simplified approach will be elaborated by reviewing more complex methods. The discussion will center on various time–differencing schemes, application of efficient space grids, open boundary conditions, nonlinear terms, moving boundary models and boundary fitted coordinates.

8. Finite-differencing of the time derivative

The time-derivative terms in the storm surge equations are the time derivatives of the horizontal velocity components u and v in the momentum equations and the time derivative of the free surface height ζ in the continuity equation. Since the terms $\partial u/\partial t$, $\partial v/\partial t$, and $\partial \zeta/\partial t$ all have the same form, discussion will be based on a general relationship of the following form:

$$\frac{DU}{Dt} = F(U,t); \qquad U = U(t) \qquad (3.162)$$

The reason a total time derivative d/dt is used rather than a partial derivative $\partial/\partial t$ is that here, $U = U(t)$ only and its space dependence is not considered since, earlier, the problem of space-differencing was already considered. In this section, liberal use will be made of the works of Mesinger and Arakawa (1976) and Simons (1980).

Several time-differencing schemes are available: two-level schemes without iteration, two-level schemes with iteration, three-level schemes, etc. Discussion will begin with two-level schemes without iteration. In this, three different schemes are well known: the Euler (or forward), backward, and trapezoidal schemes.

In the Euler or forward scheme, the time derivative is approximated as

$$U^{m+1} = U^m + TF^m$$
$$F^m = F(U^m) \tag{3.163}$$

This is a first-order accurate scheme with a truncation error of $O(T)$ and it is an uncentered scheme because F is not centred in time.

In the backward scheme

$$U^{m+1} = U^m + TF^{m+1}$$
$$F^{m+1} = F(U^{m+1}) \tag{3.164}$$

This scheme, as written here, is implicit because F depends on U^{m+1} which must be determined. In the case of partial differential equations, this will require iteration because a set of simultaneous equations (one for each grid point) must be solved. The truncation error of this scheme is also of $O(T)$.

In the trapazoidal scheme

$$U^{m+1} = U^m + \frac{1}{2}T(F^m + F^{m+1}) \tag{3.165}$$

As can be seen, this is also an implicit scheme but its truncation error is of $O(T)^2$. Next, two iterative schemes will be discussed, but still involving two time levels only.

In the Matsuno or Euler backward scheme, the first step is the regular Euler scheme,

$$U^* = U^m + TF^m \tag{3.166}$$

This value of U^* is used to determine F^* through (star indicates an intermediate step between m and $m+1$),

$$F^* = F(U^*) \tag{3.167}$$

This value of F^* is used in a backward step to compute U^{m+1}:

$$U^{m+1} = U^m + TF^* \tag{3.168}$$

As can be seen, this is a first-order accurate scheme and is explicit.

The Heun scheme is a development from the trapezoidal scheme and can be expressed as

$$U^* = U^m + TF^m,$$
$$U^{m+1} = U^m + \frac{T}{2}(F^m + F^*) \tag{3.169}$$

This is also an explicit scheme but is of second-order accuracy.

All the time-differencing schemes introduced thus far can be used for the first time step as well as all the subsequent time steps in the numerical integration. However, the three-level schemes can not be used for the first time step. The most common of the three-level schemes has already been introduced, namely, the leapfrog scheme (also called the midpoint rule or step-over rule). In this scheme

$$U^{m+1} = U^{m-1} + 2TF^m \tag{3.170}$$

with a truncation error of $O(T)^2$.

The Adams–Bashforth scheme,

$$U^{m+1} = U^m + T\left(\frac{3}{2}F^m - \frac{1}{2}F^{m-1}\right) \tag{3.171}$$

is second-order accurate. Another scheme, referred to as the Milne–Simpson scheme, involves fitting a parabola to the values of F^{m-1}, F^m, and F^{m+1} which will lead to an implicit scheme. Young (1968) discussed 13 different time-differencing schemes. For a discussion on the conservation of the energy of low-frequency waves in iterative time integration schemes see Kondo et al. (1982). It is also possible to construct complicated schemes in which space- and time-differencing are treated in a manner that cannot be described separately.

Next, the stability propeties of some of these schemes will be discussed. This can be done only when the form of the function $F(U, t)$ is known. For the oscillation equation,

$$\frac{dU}{dt} = i\omega U; \quad U = U(t) \tag{3.172}$$

$$F = i\omega U \tag{3.173}$$

where $i = \sqrt{-1}$ and ω is the frequency. Note that U could be complex but ω is real.

The general solution of eq.(3.172) is

$$U(t) = U(0)\,e^{i\omega t} \tag{3.174}$$

In the finite-difference form, taking $t = mT$,

$$U(mT) = U(0)\,e^{im\omega T}$$

If the solution is considered in a complex plane, the argument rotates by ωT in each time step T but the amplitude does not change. Introducing stability parameter λ (as we did in Sec.2.1 of this chapter),

$$\lambda = e^{i\omega T}$$

the above equation can be rewritten as

$$U(mT) = U(0)\,e^{im\omega T} = \lambda U(0)e^{i(m-1)\omega T}$$

Therefore the stability parameter is defined through

$$U[(m+1)T] = U^{m+1} = U(0)\,e^{i(m+1)\omega T} = \lambda U^m \tag{3.175}$$

The stability of the various schemes can be ascertained using this equation.

The Euler (or forward) scheme, the backward scheme, and the trapezoidal scheme can be represented by the following equation:

$$U^{m+1} = U^m + T(\alpha F^m + \beta F^{m+1}) \tag{3.176}$$

A consistency condition will be

$$\alpha + \beta = 1 \tag{3.177}$$

and the following is true for each scheme:

$$\begin{aligned}
\text{Euler scheme } \alpha = 1 &\quad \beta = 0 \\
\text{Backward scheme } \alpha = 0 &\quad \beta = 1 \\
\text{Trapezoidal scheme } \alpha = \frac{1}{2} &\quad \beta = \frac{1}{2}
\end{aligned} \tag{3.178}$$

From eqs.(3.173) and (3.176)

$$U^{m+1} = U^m + i\omega T\,(\alpha U^m + \beta U^{m+1}) \tag{3.179}$$

This equation must be solved for U^{m+1} to evaluate λ. Following Mesinger and Arakawa (1976), define

$$P \equiv \omega T \tag{3.180}$$

Then, from eqs.(3.176) and (3.179),

$$\lambda = \frac{U^{m+1}}{U^m} = \frac{1}{1 + \beta^2 P^2}(1 - \alpha\beta P^2 + iP) = \frac{1 + i\alpha P}{1 - i\beta P} \tag{3.181}$$

Using eqs.(3.177) and (3.181), the following is true for each scheme:

$$\text{Euler scheme } \lambda = 1 + iP \tag{3.182}$$

$$\text{Backward scheme } \lambda = \frac{1}{1 + P^2}(1 + iP) \tag{3.183}$$

$$\text{Trapezoidal scheme } \lambda = \frac{1}{1 + \frac{1}{4}P^2}\left(1 - \frac{1}{4}P^2 + iP\right) \tag{3.184}$$

From eq.(3.181) the following is true for each scheme:

$$\text{Euler scheme } |\lambda| = (1 + P^2)^{1/2} \tag{3.185}$$
$$\text{Backward scheme } |\lambda| = (1 + P^2)^{-1/2} \tag{3.186}$$
$$\text{Trapezoidal scheme } |\lambda| = 1 \tag{3.187}$$

From eqs.(3.185)–(3.187), it can be seen that the Euler scheme is unstable and the backward scheme is stable for any T. The trapezoidal scheme is neutral.

Next, two-level iterative schemes will be considered, namely, the Matsuno and the Heun schemes. The finite-difference forms for these schemes can be written as

$$U^* = U^m + TF^m$$
$$U^{m+1} = U^m + T[\alpha F^m + \beta F^*] \tag{3.188}$$
$$\alpha + \beta = 1$$

The following is true for

$$\begin{array}{ccc} \text{Matsuno scheme } \alpha = 0 & \beta = 1 \\ \text{Heun scheme } \alpha = \dfrac{1}{2} & \beta = \dfrac{1}{2} \end{array} \tag{3.189}$$

Applying this to the oscillation equation and proceeding as above gives

$$\lambda = 1 - \beta P^2 + iP \tag{3.190}$$

From eqs.(3.189) and (3.190) the following is true for each scheme:

$$\begin{array}{c} \text{Matsuno scheme } \lambda = 1 - P^2 + iP \\ \text{Heun scheme } \lambda = 1 - \dfrac{1}{2}P^2 + iP \end{array} \tag{3.191}$$

Evaluating $|\lambda|$ gives the following:

$$\begin{array}{c} \text{Matsuno scheme } |\lambda| = (1 - P^2 + P^4)^{1/2} \\ \text{Heun scheme } |\lambda| = \left(1 - \dfrac{1}{4}P^4\right)^{1/2} \end{array} \tag{3.192}$$

Thus, the Matsuno scheme is stable if $|P| \leq 1$. Hence, small T must be used so that

$$T \leq \frac{1}{|\omega|} \tag{3.193}$$

It can be seen that the Heun scheme is always unstable, similar to the Euler scheme. In practice, both these schemes can be used, provided T is not large.

Next, three-level schemes will be considered. The leapfrog scheme, when applied to the oscillation equation, gives

$$U^{m+1} = U^{m-1} + i2\omega T U^m \qquad (3.194)$$

Since this is a three-level scheme, we must specify a computational initial condition U^1 in addition to the physical initial condition U^0. This value of U^1 must be obtained from some two-level scheme.

From eq.(3.175) write

$$U^m = \lambda U^{m-1}$$
$$U^{m+1} = \lambda^2 U^{m-1} \qquad (3.195)$$

From eqs.(3.193) and (3.194), $\lambda^2 - i2P\lambda - 1 = 0$ which is a quadratic equation. The two values of λ are given by

$$\lambda_1 = \sqrt{1 - P^2} + iP$$
$$\lambda_2 = -\sqrt{1 - P^2} + iP \qquad (3.196)$$

Note that as $P \to 0$, $\lambda_1 \to 1$ and $\lambda_2 \to -1$. Hence, solutions associated with λ_1 are the physical modes and those associated with λ_2 are the computational modes (see, Ch.II, Sec.7). Thus, although the leapfrog scheme is a convenient scheme with second-order accuracy and is neutral within the stability range of $|\omega T| \le 1$, its main drawback is the occurrence of a computational mode. This computational mode exhibits a tendency to amplify slowly in the case of nonlinear equations. Lilly (1965) suggested that an occasional use of a two-level scheme, interspersed with a three-level scheme, eliminates the trouble with the computational mode. The time filter constructed by Asselin (1972) when used in conjunction with the leapfrog scheme will remove the computational mode as well.

The Adams–Bashforth scheme applied to the oscillation equation gives

$$U^{m+1} = U^m + i\omega T\left(\frac{3}{2}U^m - \frac{1}{2}U^{m-1}\right) \qquad (3.197)$$

From eq.(3.195)

$$\lambda^2 - \left(1 + i\frac{3}{2}P\right)\lambda + \frac{i}{2}P = 0$$

The solutions are

$$\lambda_1 = \frac{1}{2}\left(1 + i\frac{3}{2}P + \sqrt{1 - \frac{9}{4}P^2 + iP}\right)$$
$$\lambda_2 = \frac{1}{2}\left(1 + i\frac{3}{2}P - \sqrt{1 - \frac{9}{4}P^2 + iP}\right) \qquad (3.198)$$

As $P \to 0$, $\lambda_1 \to 1$ $\lambda_2 \to 0$. Thus, the solution associated with the computational mode is damped (as opposed to neutral for the leapfrog scheme) and hence, there is no trouble from this mode.

It can be shown that the physical mode of the Adams–Bashforth scheme is always unstable. Fortunately, however, as in the Heun scheme, the amplification is only a fourth-order term (i.e., $O(T)^4$) and hence, the scheme can be used with small values of T. The reader is referred to Lilly (1965), Kurihara (1965a), and Young (1968) for the evaluation of some other schemes.

9. Finite-differencing of the space derivative

9.1 Staggered and nonstaggered grid schemes

The staggered grid described above is known as the Richardson lattice. Nonstaggered grids in space and time integration were used in storm surge calculations until the early 1960s. Discussion will begin with the simplest staggered grid schemes associated with central finite-difference.

Away from the boundaries, central-differencing is the most convenient manner of space discretization. However, near (and at) the boundaries, special attention is required; ficticious points can be placed outside the boundary or one-sided difference schemes can be used. One of the simplest central difference schemes is shown in Fig.3.20a. However, this scheme is not convenient for the evaluation of advective terms and the Coriolis terms. For convenient evaluation of these terms, multiple-lattice grids have been used. Simons (1980) suggested coupling the grid in Fig 3.20a with its space supplement shown in Fig.3.20b or its time supplement shown in Fig.3.20c or its space–time supplement shown in Fig.3.20d. Note that the scheme in Fig.3.20d corresponds to the conjugate lattice developed by Platzman (1963).

In double-lattice grids, both components of the horizontal current are defined at the same location, which leads to a combination of the conjugate lattices in Figs.3.20a and 3.20d, as originally proposed by Eliassen (1956). Lilly (1961) used a time interpolation for the Coriolis terms for the space-supplemental lattices in Figs.3.20a and 3.20d.

Single-lattice grids are useful in situations in which Coriolis and nonlinear advective terms are not important. However, for larger bodies of water in which the Earth's rotational effects have to be considered, the truncation errors due to the spatial averaging (that will be required to compute the Coriolis terms) on a single lattice deserve attention.

Double-lattice grids can be formed by combining the space-supplemental lattices in Figs.3.20a and 3.20b or the conjugate lattices in Figs.3.20a and 3.20d (Simons, 1980). The spatial representation of either of these is the same and is shown in Fig.3.21a. The chief advantage of a double-lattice grid over a single-lattice grid is that no spatial averaging will be required for most of the terms in the equations of motion and continuity. The main drawback of a double-lattice grid is that the

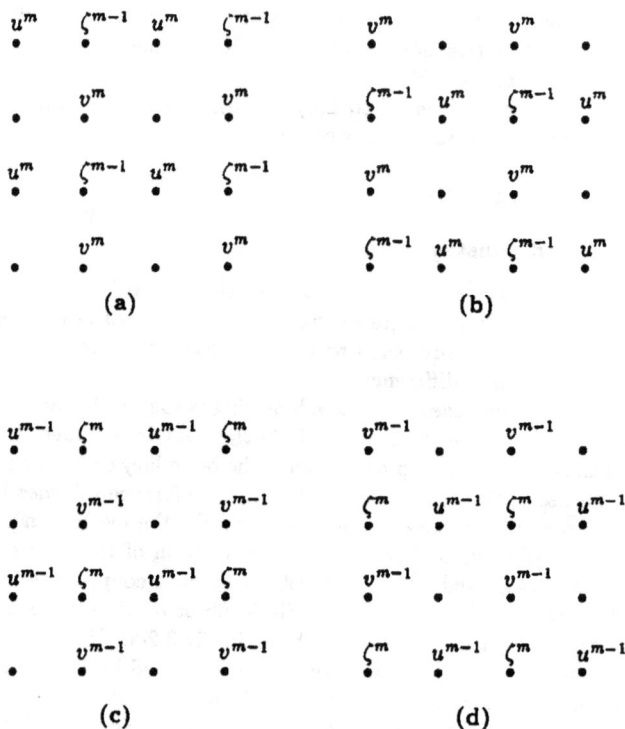

Fig. 3.20
Various staggered grids for the central-difference scheme. (a)Basic scheme; (b)space supplement of (a); (c)time supplement of (a); (d)space-time supplement of (a). Superscript m denotes time stepping.

surface gravity waves travel independently in each lattice and the lattices tend to become decoupled progressively with time, especially for water bodies with irregular boundaries (practically all natural water bodies have irregular boundaries). The Coriolis terms and the nonlinear advective terms will tend to keep the two lattices coupled. However, as was shown by Platzman (1958), some spurious results may be obtained in addition to computational instability.

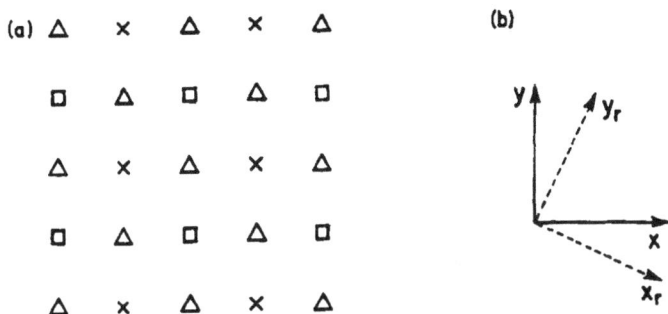

Fig. 3.21

(a)Double-lattice grid. $\triangle$, locations where u and v components are defined; $\times$, sea level belonging to one lattice; $\square$, sea level belonging to a second lattice. (b) Rotated coordinates.

The phenomenon of grid dispersion can become quite a serious hindrance in calculations with double-lattice grids, and various smoothing (in space) operators were developed by Shuman (1957), Harris and Jelesnianski (1964). An alternative to space-smoothing is the introduction of an artificial viscosity (also referred to as pseudo or virtual viscosity) which, in effect, works in a manner similar to a horizontal eddy diffusion of momentum (Obukhov, 1957). Rotation of the basic coordinate system to give a new system x_r and y_r, and then evaluating the Laplacian operator for diffusion along the rotated coordinates has also been used in an attempt to couple the two lattices (e.g., Simons, 1980).

Another approach to keep the two lattices coupled was made by Janjic (1974). Janjic added an artificial diffusion term, not to the momentum equations, as is traditionally done, but to the continuity equation. This term was made proportional to the difference between the two Laplacians of the free surface height field. The first Laplacian is calculated on the x and y coordinates and hence, involves only one lattice. The second Laplacian is evaluated along the rotated coordinates x_r and y_r, and thus, involves both lattices. In principle, if there are no truncation errors, both Laplacians should be identical. However, in practice this will not be so and the artificial diffusion term will be nonzero. This will smooth the surface height field and tend to keep the lattices coupled.

Several authors (e.g., Lauwerier, 1962; Leith, 1965; Heaps, 1969) have used rotated coordinates and evaluated all the derivatives in the relevant equations along these coordinates. However, it should be pointed out (Lauwerier, 1962; Simons, 1973) that any improvements in the elimination of grid dispersion is not only due to the evaluation of all the derivatives along the rotated coordinates but also to the orientation of the grid relative to the boundaries of the water body.

One of the examples of a computation involving multiple lattices is that of Harris and Jelesnianski (1964) in which the two transport components and the surface height field were defined at all the grid points at every time step (this is a combination of eight lattices).

9.2 Some efficient grid schemes

In the literature on numerical weather prediction, the five different grids shown in Fig.3.22 have been used frequently. Messinger and Arakawa (1976) gave a detailed discussion of the properties of these grids. The simplified two-dimensional system for surface gravity waves, assuming the water depth H to be constant, can be written as (in the primitive equation form before vertical integration)

$$
\begin{aligned}
\frac{\partial u}{\partial t} - fv &= -g\,\frac{\partial \zeta}{\partial x} \\
\frac{\partial v}{\partial t} + fu &= -g\,\frac{\partial \zeta}{\partial y} \\
\frac{\partial \zeta}{\partial t} + H\left(\frac{\partial u}{\partial x} + \frac{\partial v}{\partial y}\right) &= 0
\end{aligned}
\tag{3.199}
$$

Use of centred space-differencing with lattice e of Fig.3.23 and the leapfrog scheme for the time-differencing is one way of constructing an efficient scheme.

A space–time combination grid is shown in Fig.3.22. If all the variables were computed at each time level, as explained above, there would be two independent solutions (i.e., the solutions with the variables shown in Fig.3.22 will be independent of the solutions involving the variables not shown on the space-time grid). However, it can be concluded that the space-time lattice obtained by using the lattice shown in Fig.3.23e at every time step is really a superposition of two basic lattices of the type of Fig.3.22. Eliassen (1956) suggested that if eq.(3.199) is solved on only one of the basic lattices, then there will be no computational mode and hence, there will be considerable saving in computer time since the computational effort is reduced to half.

However, Platzman (1963) suggested that the basic Eliassen grid of Fig.3.22 is really a combination of two Richardson lattices (i.e., sea level field calculated at one time level and velocity component calculated at the next time level). It is easy to verify that a single Richardson grid is a time-staggered version of the lattice shown in Fig.3.23c; this can be used for eq.(3.199). However, on an Eliassen grid (i.e.,

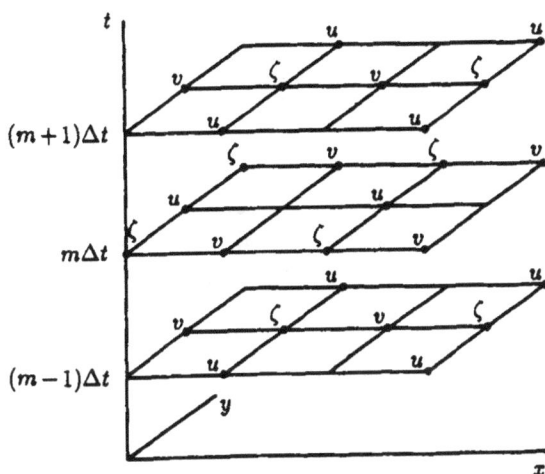

Fig. 3.22

A grid staggered in space and time for use with the leapfrog scheme associated with centered space differencing. (Mesinger and Arakawa, 1976)

Fig.3.22), there are two independent solutions for eq.(3.199) and the only coupling between these solutions is through the Coriolis terms.

Next, two computationally efficient explicit schemes will be compared: (1) the forward-backward scheme and (2) leapfrog time-differencing by the Eliassen grid. In both these schemes, computational effort is reduced by using different integration procedures for the sea level gradient terms in the momentum equations and the divergence term in the continuity equation. Mesinger and Arakawa (1976) referred to these terms as the gravity wave terms.

Economy of computation is achieved in the forward–backward scheme by first integrating the gravity wave terms of either the equations of motion or of continuity forward and the terms of the other equation backward in time. It can be shown that this scheme is stable and allows (from the CFL criterion point of view) twice the size of a time step permitted by the leapfrog scheme.

The second method has just been discussed. Both methods halve the computational effort (as compared with a standard leapfrog method) by avoiding the calculation of the computational mode. Between these two methods, the forward–backward method is superior because all the variables are defined at all grid points at every time step which makes the programming easy. It could also be modified so that the two-grid interval noise (to be defined later) is eliminated.

For the storm surge problems, the forward-backward scheme has been used by Fischer (1959), Sielecki (1968), Lauwerier (1962), and Heaps (1969). Welander

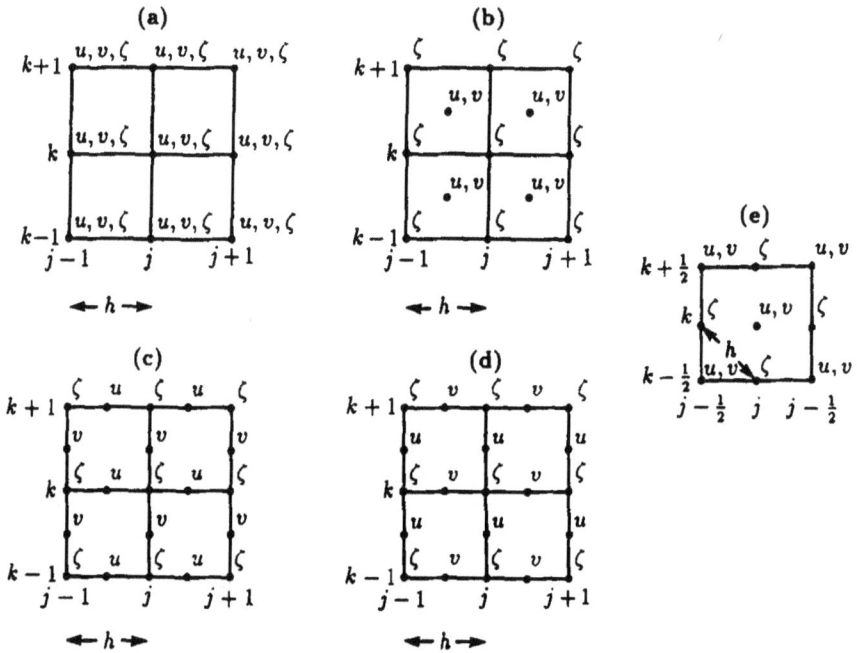

Fig. 3.23
Five different types of grid used in numerical weather prediction models.
(Mesinger and Arakawa, 1976)

(1961) referred to this as the half-implicit scheme. Fischer (1959) showed that the forward time-differencing applied to the Coriolis terms makes this scheme unstable. Later, Fischer (1965a,b,c) showed how to make this scheme stable.

Another computationally economic explicit scheme was that of Shuman *et al.* (1972), referred to as the SBS scheme. Using this scheme in the context of oceanography the sea level field is evaluated at time level $m + 1$ using the leapfrog scheme. Then, the momentum equations are integrated using this sea level averaged over the time interval $2T$ through the trapezoidal rule,

$$\frac{1}{4}\zeta^{m-1} + \frac{1}{2}\zeta^{m} + \frac{1}{4}\zeta^{m+1}$$

This scheme is similar to the forward-backward scheme as far as the stability criterion and the physical solution are concerned. Because of the averaging of the sea level field, time staggering can be used even though this is a three-level scheme.

Although the computational economic schemes described here permit twice the time step that is required by the CFL criterion, the time step is still quite small. For this reason, implicit schemes have been considered as an alternative to the explicit schemes because these schemes are stable for any size of the time step (see Sec.3.3 of this chapter for the construction of one-dimensional implicit scheme). In the simplest implicit scheme (the trapezoidal rule), the finite-difference form for eq.(3.199) with the Coriolis terms omitted could be written as

$$u^{m+1} = u^m - gT\frac{1}{2}(\frac{\partial \zeta^m}{\partial x} + \frac{\partial \zeta^{m+1}}{\partial x})$$

$$v^{m+1} = v^m - gT\frac{1}{2}(\frac{\partial \zeta^m}{\partial y} + \frac{\partial \zeta^{m+1}}{\partial y}) \qquad (3.200)$$

$$\zeta^{m+1} = \zeta^m - HT\frac{1}{2}[(\frac{\partial u}{\partial x} + \frac{\partial v}{\partial y})^m + (\frac{\partial u}{\partial x} + \frac{\partial v}{\partial y})^{m+1}]$$

It can be shown that this scheme is unconditionally stable and neutral.

Although one can use a larger time step in implicit schemes, the main disadvantage of the implicit schemes lies in the necessity of solving a set of simultaneous equations. For example, in eq.(3.200) one can use u^{m+1} and v^{m+1} from the first two equations of eq.(3.200) and substitute them in the third equation to obtain an equation for the sea level. This equation must be solved simultaneously (unlike in the explicit schemes) over the computational region, for instance, by relaxation methods (Crandall, 1956).

An alternative is to use the line inversion method (Appendix 1) also known as the alternating direction implicit (ADI) method. In these schemes, basically in each time step, one first evaluates the derivatives along one horizontal coordinate and then along the second coordinate. This splitting (see Ch.II, Sec.8.3) will permit the solution of simultaneous equations for a single row or column which makes the computation economical. For the gravity wave problems, this has been used by Leendertse (1967) and Abbott *et al.* (1973), Marchuk (1976) and Yanenko (1971).

Several other schemes are available for efficiently inverting large matrices. For example, Noye (1977) described a scheme in which a sparse matrix can be converted into a dense matrix, thereby achieving economy in the computational effort.

Usually Coriolis terms, advection terms, and other terms are omitted more often in implicit schemes than in explicit schemes due to inherent difficulties in the implicit methods. Kwizak and Robert (1971) used a semi-implicit scheme in which these terms are computed explicitly whereas the rest of the integration proceeds through an implicit scheme.

Although the semi-implicit scheme is one way of making an implicit scheme more efficient, there are better methods available which are referred to as the "splitting methods". Splitting methods are preferable to semi-implicit methods in that, with the latter, even though the explicit and implicit parts may be individually stable, there is no assurance that the total scheme will be stable.

To explain the splitting scheme, consider the following simple one-dimensional system

$$\frac{\partial u}{\partial t} + c\frac{\partial u}{\partial x} + g\frac{\partial \zeta}{\partial x} = 0$$
$$\frac{\partial \zeta}{\partial t} + c\frac{\partial \zeta}{\partial x} + D\frac{\partial u}{\partial x} = 0$$

(3.201)

Here, the advection terms are made linear by replacing u with a constant c, and also in the continuity equation the full depth is considered as $D = H + \zeta$, see, eqs.(3.1)–(3.3).

At first by using the general approach delineated in Ch.II, Sec.8.3 and Sec.9, we split the system into advection equation and long wave propagation equation.

Within a given time step, the system of advection equations could first be solved:

$$\frac{1}{2}\frac{\partial u}{\partial t} + c\frac{\partial u}{\partial x} = 0$$
$$\frac{1}{2}\frac{\partial \zeta}{\partial t} + c\frac{\partial \zeta}{\partial x} = 0$$

(3.202)

Let these provisional values (or intermediate values) be represented by u^* and ζ^*. These values then can be used to solve for the equation

$$\frac{1}{2}\frac{\partial u^*}{\partial t} + g\frac{\partial \zeta^*}{\partial x} = 0$$
$$\frac{1}{2}\frac{\partial \zeta^*}{\partial t} + D\frac{\partial u^*}{\partial x} = 0$$

(3.203)

This procedure can be repeated in every time step. The main advantage of the splitting method is that one can use a larger time step for the slower process, such as advection, and a smaller time step for other processes. Its main disadvantage is a greater truncation error than in other schemes. Brown and Pandolfo (1978) devised a scheme to merge two finite-difference schemes with unequal time steps. Basically, this is done by introducing an additional time level and at the expense of increasing the storage requirements.

9.3 Two-grid interval noise

Mesinger and Arakawa (1976) have drawn attention to the important problem of "two-grid interval noise" which is the phenomenon of false stationary waves appearing as neutral solutions of the difference equations. Only with the lattice of Fig.3.22c will this not occur. When the Earth's rotation effects are included, the two-grid interval waves appear with incorrect low frequencies as inertial waves (on the lattice of Fig.3.22d they will appear stationary). Usually by using dissipative schemes with a maximum dissipation for the two-grid interval noise and through additional horizontal diffusion, this is controlled. Arakawa (1972) suggested intermittent use of uncentered space-differencing for the gravity wave terms (defined earlier) alternately on opposite sides of the central point. Mesinger (1973) developed a scheme to suppress two-grid interval noise, which is suited for centered differencing schemes.

Lilly (1965) showed that two-grid interval noise can also occur in time for the leapfrog scheme (i.e., for three time-level schemes). One scheme that was used in numerical weather prediction to suppress two-grid interval noise in time is the intermittent use of a two-level time-differencing scheme. However, this has the disadvantage that the solution that is eliminated is done arbitrarily. An alternative approach is the use of a time filter (Robert, 1966).

10. Treatment of open boundaries

At times, storm surge calculations might have to be performed in a limited region of a large water body. This problem could be tackled in at least two different ways. In one approach, the calculations can be performed for the large water body of which the smaller water body is a part and then use the results for the area of interest. However, this approach is not economical and may not even be possible for certain water bodies. Also, there may be a problem with the resolution since a larger water body has to be modelled. In the second approach, artificial open boundaries can be introduced around the area of interest. However, along these artificial open boundaries, certain conditions have to be introduced and without proper considerations, these conditions might make the results in the interior region inaccurate.

The commonly used practice of putting zero surface elevation at the sea boundary is not at all satisfactory because this amounts to perfect reflection at the sea boundary. A better approximation (Heaps, 1974; Henry and Heaps, 1976) is to assume that all outward travelling waves are normal to the boundary and to calculate the velocity u (or v) from the water level ζ at the nearest interior grid point, i.e., $u = \zeta \sqrt{g/H}$ (see Sec.1.4 of this chapter). This is the so-called radiation condition.

Reid (1975) corrected a misconception, commonly held in applying open boundary conditions that if the derivative of velocity perpendicular to the boundary is set equal to zero, the long waves will pass unimpeded through the artificial boundary.

Reid showed that, although such a condition will permit flow of fluid to or from the system, it will produce total reflection of long waves and not zero reflection.

To show this, following Reid (1975), assume a simple harmonic wave of frequency ω incident at an angle θ relative to the x axis. Owing to this, the transport $x \geq 0$ is

$$M_i = A_i \cos\left[\omega t + \kappa(x\cos\theta + y\sin\theta)\right]$$

where A_i is the local incident amplitude for transport and κ is a local wave number which is determined by the frequency, the local water depth H, and the Coriolis parameter f. In case of reflection, the transport for $x \geq 0$ will be in the form

$$M_r = A_r \cos \left[\omega t + \kappa(-x \cos \theta + y \sin \theta) + \Phi\right]$$

where A_r is the reflected amplitude for transport and Φ is a relative phase angle.
The boundary condition under consideration is

$$\frac{\partial M_x}{\partial x} = 0 \qquad \text{at} \quad x = 0$$

where M_x is the total transport in the x direction, i.e., $(M_i + M_r)$ on the positive
(interior) side of the open boundary. It can be seen that for a nontrivial A_i and
$0 \leq |\theta| < \pi/2$,

$$A_r = A_i \qquad \text{and} \qquad \Phi = 0.$$

Hence, the above boundary condition leads to total reflection with no phase change
for the reflected M_x field.

Reid (1975) suggested that to permit wave transmission through the open
boundary, the normal component of transport has to be forced to be in phase
with the water level deviation. The outward flux of energy J through a unit length
of vertical boundary, from the bottom to the surface, is (see Sec.2.1 of this chapter)

$$J = \rho_o \left(g\zeta + \frac{1}{2}q^2\right) M_n$$

where M_n is the transport through a unit width of boundary taken positive out-
wards, and $q^2 = u^2 + v^2$. For the radiation of energy, it is required that the average
value of J over one wave cycle be positive. If $\zeta \gg q^2/2g$, then the necessary condition is that $\int_0^{T_p} \zeta M_n dt$ be positive at the boundary. Hence, a boundary condition
of the type (used by Reid and Bodine, 1968) $M_n = c\zeta$ where c is a positive coefficient (having dimensions of velocity), permits proper radiation of energy through
the open boundary. The meaning of the above is that the transport and the sea
level are propagating in phase.

Wurtele et al. (1971) developed boundary conditions that will allow disturbances to travel out of the computational region with negligible reflection at the
open boundaries. In developing these boundary conditions, the concept of Riemann
invariants (Abbott, 1966) has to be used. These conditions in one and two space
dimensions have been used to study the flow around a sea mound, island arc, and on
a shelf, and comparison was made with the Sommerfeld's (1949) radiation condition
as described by Vastano and Reid (1966).

Wurtele et al. (1971) coupled the meteorological forcing terms into the single
term (F_x, F_y) and began with the simple system

$$\frac{\partial \zeta}{\partial t} = -\left(\frac{\partial M_x}{\partial x} + \frac{\partial M_y}{\partial y}\right)$$

$$\frac{\partial M_z}{\partial t} = -gD\frac{\partial \zeta}{\partial x} + F_z \tag{3.204}$$

$$\frac{\partial M_y}{\partial t} = -gD\frac{\partial \zeta}{\partial y} + F_y$$

Let δ specify a central–difference operator, e.g. $\delta_j u = u_{j+1} - u_{j-1}$. The finite-difference forms of eq.(3.204) are

$$\zeta_{j,k}^{m+1} = \zeta_{j,k}^{m} - \frac{T}{2h}(\delta_j M_{z,j,k}^{m} + \delta_k M_{y,j,k}^{m+1})$$

$$M_{z,j,k}^{m+1} = M_{z,j,k}^{m} - gD_{j,k}\frac{T}{2h}\delta_j \zeta_{j,k}^{m+1} \tag{3.205}$$

$$M_{y,j,k}^{m+1} = M_{y,j,k}^{m} - gD_{j,k}\frac{T}{2h}\delta_k \zeta_{j,k}^{m+1}$$

For the case of uniform depth and zero external forces, Garabedian (1964) gave the following analytical solution for the water level ζ:

$$\zeta(x, y, t) = \frac{1}{c}\frac{\partial}{\partial t}\frac{1}{2\pi}\int\int \frac{G_1(x + \alpha_1, y + \alpha_2)}{\sqrt{c^2 t^2 - \alpha_1^2 \alpha_2^2}}\, d\alpha_1\, d\alpha_2$$

$$+ \frac{1}{2\pi c}\int\int \frac{G_2(x + \alpha_1, y + \alpha_2)}{\sqrt{c^2 t^2 - \alpha_1^2 \alpha_2^2}}\, d\alpha_1\, d\alpha_2 \tag{3.206}$$

where $\alpha_1^2 \alpha_2^2 < c^2 t^2$, $c = \sqrt{gD}$, $G_1 = \zeta(x, y, 0)$, and $G_2 = \partial \zeta / \partial t(x, y, 0)$. Equation (3.206) gives the solution at any point and time as an area integral over the circular domain of dependence cut from the initial data plane by the characteristic cone

$$(x - x_0)^2 + (y - y_0)^2 - c(t - t_0)^2 = 0$$

and (x_0, y_0, t_0) denotes the initial state. In the one-dimensional case eq.(3.206) reduces to the d'Alembert solution

$$\zeta = \frac{1}{2}[G_1(x + ct) + G_1(x - ct)] + \frac{1}{2}\int_{-ct}^{ct} G_2(x + \xi)\, d\xi \tag{3.207}$$

It can be seen that (in the two-dimensional case) once the disturbance arrives at any point in the computational region, irrespective of how concentrated it might be originally, it will not simply vanish at this point. If this location under consideration is an open boundary, the disturbance outside that boundary will influence the solution in the region interior to the boundary. Thus, the interior solution is quite dependent on the condition applied at the open boundary. Similar results hold for the one-dimensional case. These results are for the case $G_1 \neq 0$, $G_2 \neq 0$. However, if $G_2 = 0$, a concentrated disturbance progresses without change of shape and an arbitrary point returns to its original state after the passage of the disturbance.

For the one-dimensional case, eq.(3.204) reduces to

$$\frac{\partial M_x}{\partial t} = -c^2 \frac{\partial \zeta}{\partial x}, \tag{3.208a}$$

$$\frac{\partial \zeta}{\partial t} = -\frac{\partial M_x}{\partial x} \tag{3.208b}$$

Equation (3.208b) $\times c \pm$ eq.(3.208a) gives

$$\begin{aligned} \frac{\partial}{\partial t}(M_x + c\zeta) + c\frac{\partial}{\partial x}(M_x + c\zeta) = 0 \\ \frac{\partial}{\partial t}(M_x - c\zeta) + c\frac{\partial}{\partial x}(M_x - c\zeta) = 0 \end{aligned} \tag{3.209}$$

Equation (3.209) states that the linear Riemann invariants $M_x \pm c\zeta$ are respectively conserved along the directions

$$\frac{\partial x}{\partial t} = \pm c$$

The characteristics $x \pm ct$ could be drawn to intersect the open boundaries at $x = 0$ and L. The boundary conditions are (see eq.(3.21)),

$$\begin{aligned} M_x + c\zeta = 0 \quad &\text{at} \quad x = 0 \\ M_x - c\zeta = 0 \quad &\text{at} \quad x = L \end{aligned} \tag{3.210}$$

These conditions represent the fact that the wave marches out of the interior region. If ζ is determined at each boundary ($x = 0$ and L) by a one-sided space-difference, then M_x may be determined from eq. (3.210). In the two-dimensional case, the condition becomes

$$(M_x^2 + M_y^2)^{1/2} \pm c\zeta = 0 \tag{3.211}$$

at all the boundaries and the signs are taken in such a way that M_x and M_y are directed outward from the computational domain when sea level is positive. Accuracy of this radiating boundary condition through the calculation of incoming and outgoing wave amplitude was investigated by Foreman (1986).

Next, studies by various authors who have prescribed open boundary conditions for computation of storm surges using shallow water equations, will be considered.

Storm surges are mainly a shallow water coastal phenomena so it is unnecessary to do computations for deep water (Murty and El-Sabh, 1986). Here, a situation will be presented in which conditions will have to be prescribed on three open boundaries as shown in Fig.3.24. Fig.3.25 summarizes the various approaches to this problem described by Harper and Sobey (1983) and Murty and El-Sabh (1986).

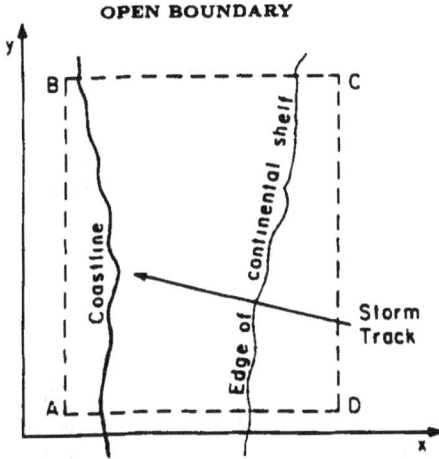

Fig. 3.24
Schematic representation of open boundaries for storm surge calculation.
(from Murty and El-Sabh, 1986)

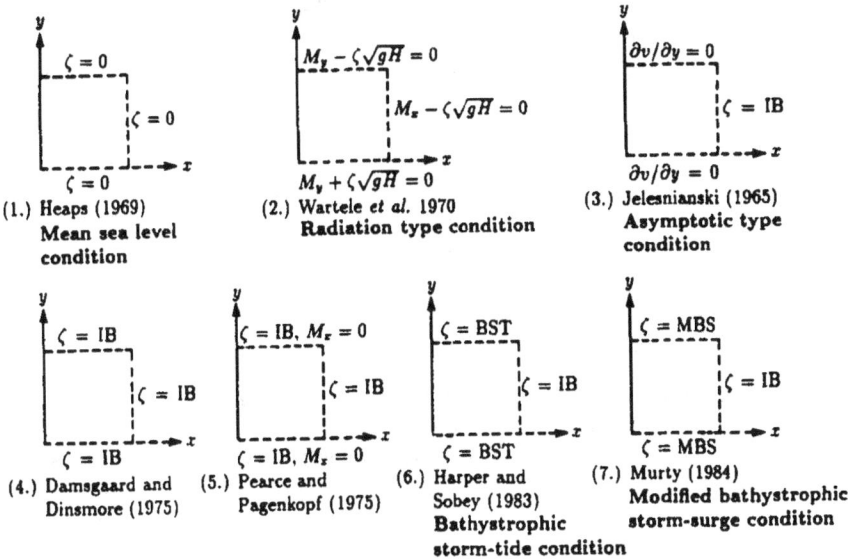

(1.) Heaps (1969)
Mean sea level
condition

(2.) Wartele *et al.* 1970
Radiation type condition

(3.) Jelesnianski (1965)
Asymptotic type
condition

(4.) Damsgaard and
Dinsmore (1975)

(5.) Pearce and
Pagenkopf (1975)

(6.) Harper and
Sobey (1983)
Bathystrophic
storm-tide condition

(7.) Murty (1984)
Modified bathystrophic
storm-surge condition

Fig. 3.25
Open boundary conditions for storm surge models.
(from Murty and El-Sabh, 1986)

In this diagram, equating the storm surge amplitude to zero is referred to as the mean sea level (m.s.l.) condition.

The "inverse barometer effect" (I.B.) refers to the fact that under static conditions approximately a decrease of 1 mb in the atmosphere pressure gives rise to 1 cm increase in the sea level. The sea level determined in this manner is then prescribed at the open boundaries.

The time-dependent two-dimensional storm surge equations (3.1), (3.2) in the bathystrophic (Murty, 1984) (B.S.T.) conditions, are simplified to

$$0 = -g(H + \zeta)\frac{\partial \zeta}{\partial x} - \frac{H + \zeta}{\rho}\frac{\partial p_a}{\partial x} + \frac{\tau_x^s}{\rho} \tag{3.212}$$

Knowing the pressure and wind fields in the storm, and the water depth, eq.(3.213) can be integrated in x (at each time step) to obtain a value for ζ which is then used as the open boundary condition. Murty (1984) modified the B.S.T. condition by including a simple form of bottom stress.

For the review and comparison of various open boundary conditions for a barotropic coastal circulation model, see Chapman (1985) and Røed and Cooper (1986).

Røed and Smedstad (1984), (see also Greatbatch and Otterson, 1991) proposed to split the overall solution near the open boundary into a forced and a free component. For this purpose, considering motion only along x axis, one can write the equation of motion as

$$\frac{\partial u_r}{\partial t} = \frac{\tau_x^s}{\rho_o H} \tag{3.213a}$$

and

$$\frac{\partial u_g}{\partial t} = -g\frac{\partial \zeta}{\partial x} \tag{3.213b}$$

The second equation is solved by the radiation principle. Through the solution of the first equation an additional velocity is added at the boundary by solving the forced problem.

11. Treatment of the nonlinear advective terms

In shallow-water areas and in the computation of the horizontal motion, at times the nonlinear advective terms might have to be included. Charnock and Crease (1957) showed through dimensional analysis that the nonlinear advective terms become important when the free surface height is of the same order of magnitude as the water depth.

Let us first consider a simple advection equation

$$\frac{\partial u}{\partial t} + u\frac{\partial u}{\partial x} = 0 \tag{3.214}$$

This equation is referred as the shock equation by Shuman (1974). Platzman (1964) wrote the general solution as

$$u = f(x - ut) \tag{3.215}$$

where f is an arbitrary function. It will be shown that the finite-difference form of eq.(3.214) will give rise to errors due to the instability of a discrete scheme to resolve wavelengths shorter that $2h$ (or wave number greater than $\kappa_{max} = \pi/h$).

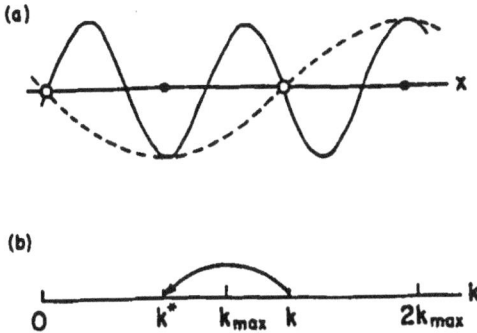

Fig. 3.26

(a) Wave of length $4h/3$ misrepresented by the finite-difference grid as a wave of length $4h$; (b) misrepresentation of a wave number $\kappa > \kappa_{max}$. (Mesinger and Arakawa, 1976)

With the reference to Fig.3.26, consider a function $u(x)$ given by

$$u = \sin(\kappa x) \tag{3.216}$$

for $\kappa < \kappa_{max}$. From eq.(3.216)

$$u\frac{\partial u}{\partial x} = \kappa \sin(\kappa x)\cos(\kappa x) = \frac{1}{2}\kappa \sin(2\kappa x) \tag{3.217}$$

For the wave numbers in the range $\pi/2h < \kappa \le \kappa_{max}$, (wavelength $2h$ to $4h$), the nonlinear term $u(\partial u/\partial x)$ will give rise to a wave number that lies in a range that cannot be resolved by the grid.

The situation for $\kappa > \kappa_{max}$ (Mesinger and Arakawa, 1976) is examined. With reference to Fig.3.26, let wavelength $L = h\,4/3$ (the solid line shows this wave). If the values of the function at the two grid points shown by black dots are known, one cannot distinguish this wave from the wave shown by the broken line. Thus, all wavelengths that cannot be resolved by the grid will be misrepresented (or aliased) as long wavelengths.

Next, the results are generalized to the case when u consists of a number of harmonic components:

$$u = \sum_n u_n \tag{3.218}$$

The term $u(\partial u/\partial x)$ will contain products of harmonics of different wavelengths, such as $\sin(\kappa_1 x)\sin(\kappa_2 x)$. The following trigonometric identity can be easily obtained:

$$\sin(\kappa_1 x)\sin(\kappa_2 x) = \frac{1}{2}[\cos(\kappa_1 - \kappa_2)x - \cos(\kappa_1 + \kappa_2)x] \tag{3.219}$$

Hence, even if initially all wave numbers, κ, are less than or equal to κ_{max}, through nonlinear interaction, wave with $\kappa > \kappa_{max}$ will develop, and aliasing will occur. One can also write

$$\sin(\kappa x) = \sin[2\kappa_{max} - (2\kappa_{max} - \kappa)]x \tag{3.220}$$

It can be seen from Fig.3.26 that the wave number κ will be misinterpreted as a wave number $2\kappa_{max} - \kappa$.

Next, visualization of the consequences of aliasing is attempted. To do this, it can be assumed that the dependent variables of concern here are made up of a series of harmonic components. The energy spectrum of these various components determines the relative contributions of different scales to the dependent variable. Aliasing errors will create a spurious inflow of energy at wave numbers that are not much less than κ_{max} and, with the progress of time, the energy of these components grows in a rapid manner. Phillips (1959) referred to this as nonlinear instability. Miyakoda (1962) showed that similar instability could occur even for linear equations with variable coefficients.

Next, possible schemes for prevention and suppression of nonlinear instability are considered. The simplest approach for suppression of this instability is to tune the horizontal friction so that instability arising from the subgrid scales is minimized. Orszag (1971) showed that aliasing errors could be eliminated by filtering out wave numbers $\kappa > 2/3\kappa_{max}$. Another approach for the suppression of the shortest waves is the Lax and Wendroff (1960) scheme, and this approach was suggested by Richtmayer (1963). Kasahara (1969) suggested that it is sufficient to use the Lax–Wendroff scheme intermittently after long intervals of time integration using other simpler schemes. A third approach for eliminating nonlinear instability is to use a Lagrangian formulation (Leith, 1965; Krishnamurti et al., 1973). Although nonlinear instability might be suppressed, there are several highly undesirable features with Lagrangian schemes (the worst one being the drastic distortion of the basic grid after a few time steps of integration); these schemes will therefore not be considered.

The most sophisticated methods are conservation schemes developed by Arakawa (1966, 1972) and Arakawa and Lamb (1981). These schemes, by retaining the conservation (in the finite-difference forms) of some integral properties of the original differential equations, eliminate nonlinear instability, as well as the spurious inflow of energy to the short waves, rather than artificially suppressing their amplitudes. In these conservation schemes the average values of enstrophy (half the vorticity squared) and kinetic energy do not change, nor does the average wave number.

For the review of the finite-difference schemes and their conservation properties, see Grammelveldt (1969).

The numerical construction of the nonlinear terms in shallow water equations can be approached through the various methods given in Ch.II and Ch.IV. Here one can use algorithms constructed for the advective equation in Ch.II, Secs.4 and 5, including the first order upwind–downwind schemes as wel as the second-order schemes, like Crank–Nicholson and angled derivatives. For the shallow water equations the set of eqs.(3.95)–(3.97) outlines the first order approximation scheme as applied to the nonlinear terms. The higher order of approximation through the leapfrog time stepping is demonstrated in Ch.IV.

Flather and Heaps (1975) developed a model for Morecambe Bay allowing for the inclusion of the nonlinear advective terms. In the depth-averaged form, the equations of motion and continuity are

$$\frac{\partial u}{\partial t} + u\frac{\partial u}{\partial x} + v\frac{\partial u}{\partial y} - fv + \frac{ru(u^2 + v^2)^{1/2}}{D} + g\frac{\partial \zeta}{\partial x} = 0 \qquad (3.221)$$

$$\frac{\partial v}{\partial t} + u\frac{\partial v}{\partial x} + v\frac{\partial u}{\partial y} + fu + \frac{rv(u^2 + v^2)^{1/2}}{D} + g\frac{\partial \zeta}{\partial y} = 0 \qquad (3.222)$$

$$\frac{\partial \zeta}{\partial t} + \frac{\partial}{\partial x}(Du) + \frac{\partial}{\partial y}(Dv) = 0 \qquad (3.223)$$

These are eqs.(3.1)–(3.3) taken to describe the tide propagation, therefore atmospheric pressure, wind stress and horizontal friction terms are neglected. The bottom stress for the x direction is $\tau_x^b = \rho_o ru(u^2 + v^2)^{1/2}$ and similar expression is taken for the y direction. Initially,

$$u = v = 0 \quad \text{and} \quad \zeta = 0 \qquad \text{at } t = 0 \qquad (3.224)$$

The grid is shown in Fig.3.27 and 3.28.

The scheme for the nonlinear advective terms used by Flather and Heaps (1975) is somewhat different from the schemes used by other authors, e.g., Lax and Wendroff (1960), Crowley (1970), and Sielecki and Wurtele (1970). The scheme is based on the angled derivative approach suggested by Roberts and Weiss (1966) and described for one-dimensional approach in Ch.II, Sec.5.2.

To explain this approach, following Flather and Heaps (1975), the simple nonlinear one-dimensional advection equation will be considered:

$$\frac{\partial u}{\partial t} + u\frac{\partial u}{\partial x} = 0 \qquad (3.225)$$

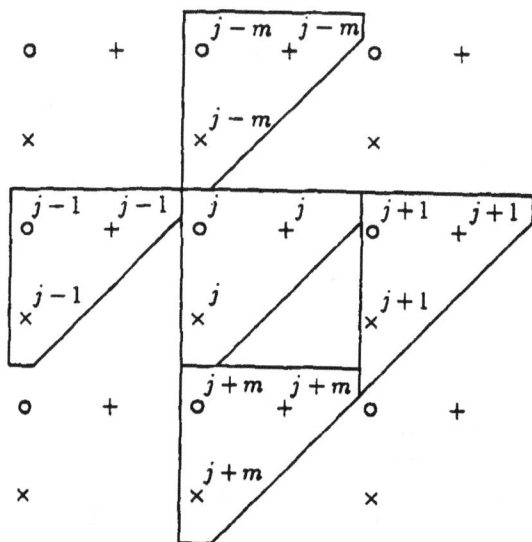

Fig. 3.27

The finite-difference grid used in the numerical solution of the hydro-dynamic equations showing the numbering system for the mesh points; $\bigcirc$, ζ-point; $+$, u-point; $\times$, v-point.

The finite-difference form of this can be written as

$$\frac{u_j(t+T) - u_j(t)}{T}$$
$$+ \overline{u_j}(t)\frac{\frac{1}{2}[u_j(t+T) + u_{j+1}(t)] - \frac{1}{2}[u_{j-1}(t+T) + u_j(t)]}{h} = 0 \qquad (3.226)$$

where

$$\overline{u_j} = \frac{1}{4}(u_{j-1} + 2u_j + u_{j+1}) \qquad (3.227)$$

Here we shall pose for a while with further development of the numerical scheme and discuss shortly the controling mechanism introduced by Flather and Heaps (1975) to suppress instability. The above expression is a filter which will dampen the shortest waves in the numerical grid. The properties of the filter can be investigated by the

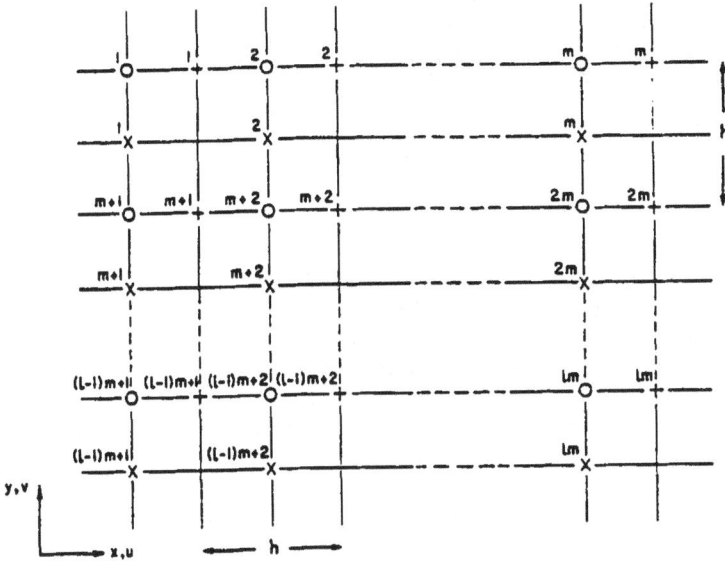

Fig. 3.28
Grid distribution surrounding point j;
m denotes number of columns.

method given in Sec.5 of this chapter. Consider periodical space wave

$$u_j = u^* e^{i\kappa x} = u^* e^{i\kappa j h} \qquad (3.228)$$

and introducing it into eq.(3.227) we arrive at

$$\overline{u_j} = u^*(2e^{i\kappa j h} + e^{i\kappa(j-1)h} + e^{i\kappa(j+1)h} = u^* e^{i\kappa j h} 2[1 + \cos(\kappa h)] = u_j 2[1 + \cos(\kappa h)]$$

This filter response function

$$R = \frac{\overline{u_j}}{u_j} = 2[1 + \cos(\kappa h)] \qquad (3.229)$$

is zero at the shortest wavelength $L = 2h$.

Returning back to the numerical scheme construction we shall introduce h as the distance between successive u values (or successive v or successive ζ values, see Fig.3.27). Assuming that we are located at the point $j1$ in space, and before computing $u_{j1}(t+T)$ by eq.(3.226) the values $u_j(t)$ for $1 \leq j \leq m$ and $u_j(t+T)$ for $1 \leq j \leq j1-1$ are already known. The finite-difference approximation for $\partial u/\partial x$ is called the angled derivative and is correctly centered in space and time.

In the next step, the direction of integration is reversed so that integration is performed by eq.(3.226) in the direction of decreasing j. Then, the finite-difference approximation is

$$\frac{u_j(t+T) - u_j(t)}{T}$$

$$+\overline{u_j}(t)\frac{\frac{1}{2}[u_j(t) + u_{j+1}(t+T)] - \frac{1}{2}[u_{j-1}(t) + u_j(t+T)]}{h} = 0 \qquad (3.230)$$

Again assuming we are just going to perform calculation at the point $j1$ we shall assume that $u_j(t)$ for $1 \leq j \leq m$ and $u_j(t+T)$ for $j1+1 \leq j \leq m$ are known. Equations (3.226) and (3.230) can be rearranged to give explicit forms for $u_j(t+T)$.

Flather and Heaps (1975) described two different ways of extending the angled derivative approach to include all the nonlinear advective terms. In the first method, the variations in the components of the depth-averaged current over one time step due to advection alone are calculated and these are added to the variations caused by other terms.

Before proceeding further notice in Fig.3.27 the numbering for the grid points used by Flather and Heaps (1975). Instead of the two-dimensional enumeration given by indices j, k, one dimensional numbering is used with total number of columns equal to m and number of rows equal to l. The j index will thus assume numbers from 1 to $l \times m$. Also notice that the u, v, and ζ points are organized into triples of the same numeration, with the ζ point being the central, u to the east and v to the south. Fig.3.28 depicts the location of the points in the triples used in numerical schemes. Before writing the finite-difference forms, for convenience, define

$$d_j = \frac{1}{2}(D_j + D_{j+1})$$
$$\mathcal{D}_j = \max(d_j, D_0)$$
$$e_j = \frac{1}{2}(D_j + D_{j+m})$$
$$E_j = \max(e_j, D_0)$$
$$\widetilde{u}_j = \frac{1}{4}(u_{j+1} + u_j + u_{j+m+1} + u_{j+m})$$
$$\widetilde{v}_j = \frac{1}{4}(v_{j-m} + v_{j-m+1} + v_j + v_{j+1})$$

D_0 is a prescribed minimum depth for the denominator in the frictional terms which could otherwise lead to a singularity at $D = 0$. Since $\mathcal{D}_j$ is the maximum value of d_j or D_0, it cannot be smaller than D_0. Similar arguments hold for E_j. The reason for prescribing a minimum depth is (besides avoiding possible division by zero) to cut off frictional damping before it becomes too great (due to very small water depth) and the boundary layer considerations on which the quadratic bottom friction is formulated are no longer applicable.

The finite-difference forms for the tentative values of u and v are denoted by a star. Then,

$$\frac{u_j^*(t+T) - u_j(t)}{T} = f\tilde{v}_j(t) - ku_j^*(t+T)\frac{[u_j^2(t) + \tilde{v}_j^{\,2}(t)]^{1/2}}{D_j(t)}$$
$$- \frac{g[\zeta_{j+1}(t+T) - \zeta_j(t+T)]}{h} \qquad (3.231)$$

$$\frac{v_j^*(t+T) - v_j(t)}{T} = f\tilde{u}_j^{\,*}(t+T) - rv_j^*(t+T)\frac{[\tilde{u}_j^{\,2}(t) + v_j^2(t)]^{1/2}}{E_j(t)}$$
$$- \frac{g[\zeta_j(t+T) - \zeta_{j+m}(t+T)]}{h} \qquad (3.232)$$

For odd time steps with increasing i, the following are the final forms for u_j and v_j:

$$\frac{u_j(t+T) - u_j^*(t+T)}{T} = \frac{1}{2h}\overline{u_j}^{\,*}(t+T)[u_{j+1}^*(t+T) - u_j^*(t+T)$$
$$+ u_j(t+T) - u_{j-1}(t+T)]$$
$$- \frac{1}{4h}\{[v_{j-m}^*(t+T) + v_{j-m+1}^*(t+T)]$$
$$\times [u_{j-m}(t+T) - u_j(t+T)] + [v_j^*(t+T) + v_{j+1}^*(t+T)]$$
$$\times [u_j^*(t+T) - u_{j+m}^*(t+T)]\} \qquad (3.233)$$

and
$$\frac{v_j(t+T) - v_j^*(t+T)}{T} = \frac{1}{4h}\{[u_j^*(t+T) + u_{j+m}^*(t+T)]$$
$$\times [v_{j+1}^*(t+T) - v_j^*(t+T)] + [u_j^*(t+T) + u_{j+m-1}^*(t+T)]$$
$$\times [v_j(t+T) - v_{j-1}(t+T)]\}$$
$$- \frac{1}{2h}\overline{v_j}^{\,*}(t+T)[v_{j-m}(t+T)$$
$$- v_j(t+T) + v_j^*(t+T) - v_{j+m}^*(t+T)] \qquad (3.234)$$

The term for $\overline{u_j}$ has been defined. Define $\overline{v_j}$:

$$\overline{v_j} = \frac{1}{4}(v_{j-m} + 2v_j + v_{j+m}). \qquad (3.235)$$

Equations similar to (3.233) and (3.234) are true for even time steps with decreasing j. The finite-difference form for the continuity equation is

$$\frac{\zeta_j(t+T) - \zeta_j(t)}{T}$$
$$= -\frac{d_j(t)u_j(t) - d_{j-1}(t)u_{j-1}(t) + e_{j-m}(t)v_{j-m}(t) - e_j(t)v_j(t)}{h} \quad (3.236)$$

In the second method, Flather and Heaps (1975) included the advection terms directly and gave the following finite-difference forms:

$$\frac{u_j(t+T) - u_j(t)}{T} = -\frac{1}{2h}\,\overline{u_j}(t)[u_{j+1}(t) - u_j(t) + u_j(t+T) - u_{j-1}(t+T)]$$
$$- \frac{1}{4h}\{[v_{j-m}(t) + v_{j-m+1}(t)][u_{j-m}(t+T) - u_j(t+T)]$$
$$+ [v_j(t) + v_{j+1}(t)][u_j(t) - u_{j+m}(t)]\} + f\widetilde{v_j}(t) - ru_j(t+T)$$
$$\times \frac{[u_j^2(t) + \widetilde{v}_j^{\,2}(t)]^{1/2}}{\mathcal{D}_j(t)} - g\frac{[\zeta_{j+1}(t+T) - \zeta_j(t+T)}{h} \quad (3.237)$$

and

$$\frac{v_j(t+T) - v_j(t)}{T} = -\frac{1}{4h}\{[u_j(t) + u_{j+m}(t)][v_{j+1}(t) - v_j(t)]$$
$$+ [u_{j-1}(t) + u_{j+m-1}(t)][v_j(t+\tau) - v_{j-1}(t+T)]\}$$
$$- \frac{1}{2h}\,\overline{v_j}(t)[v_{j-m}(t+T) - v_j(t+T) + v_j(t) - v_{j+m}(t)]$$
$$- f\widetilde{u_j}(t+T) - kv_j(t+T)\frac{[\widetilde{u}_j^{\,2}(t) + v_j^2]^{1/2}}{E_j(t)}$$
$$- g\frac{h_j(t+T) - h_{j+m}(t+T)}{h}. \quad (3.238)$$

Equations (3.237) and (3.238) are for the odd time steps for increasing j and with u calculated before v. For even time steps, for decreasing j, v will be calculated before u.

One word of caution: since not all the values of the variables required to compute from the above equations are defined adjacent to open boundaries, the advective terms have to be ignored within a distance of h from these boundaries. For a calculation for Morecambe Bay, Flather and Heaps found some instability (in the form of grid scale oscillations) originating from the corner of the open boundary. This instability was removed by omitting the nonlinear advective terms within a distance of $3h$ from the open boundaries. Leendertse (1967) excluded the nonlinear advective terms in a distance of h from land boundaries to suppress instability.

Henry (1982) encountered a similar instability in a calculation of the circulation in Bridport Inlet and suppressed the instability by using a procedure similar to that of Flather and Heaps (1975).

Crean (1978) modified the scheme of Flather and Heaps (1975) in three ways by (1) using the vertically integrated forms of the equations (i.e., transport components) rather than the depth-averaged forms, (2) using the flux forms for the nonlinear advective terms, and (3) including the horizontal frictional terms. Crean et al. (1988) found it necessary to include horizontal friction to control the slight instability due to the inclusion of the nonlinear advective terms (he called this instability noodling). He defined noodling as small grid scale fluctuations in the direction of velocity vectors occurring in the areas where nonlinear advective terms are important. The question of computation of the second derivatives near the boundary (to evaluate the horizontal frictional terms) is avoided by setting the horizontal friction coefficient N_h to zero near the boundaries. This means that in the shallow areas near the coast, bottom stress is dominant over lateral stress.

The forms of the equations of motion used by Crean (1978) are as follows (use Figs.3.27 and 3.28 for grid distribution):

$$\frac{\partial M_x}{\partial t} + \frac{\partial}{\partial x}\left(\frac{M_x^2}{D}\right) + \frac{\partial}{\partial y}\left(\frac{M_x M_y}{D}\right) - f M_y$$

$$+ N_h\left(\frac{\partial^2 M_x}{\partial x^2} + \frac{\partial^2 M_x}{\partial y^2}\right) + g D \frac{\partial \zeta}{\partial x} + \frac{r M_x \sqrt{M_x^2 + M_y^2}}{D^2} = 0 \qquad (3.239\text{a})$$

$$\frac{\partial M_y}{\partial t} + \frac{\partial}{\partial x}\left(\frac{M_x M_y}{D}\right) + \frac{\partial}{\partial y}\left(\frac{M_y^2}{D}\right) + f M_x$$

$$+ N_h\left(\frac{\partial^2 M_y}{\partial x^2} + \frac{\partial^2 M_y}{\partial y^2}\right) + g D \frac{\partial \zeta}{\partial y} + \frac{r M_y \sqrt{M_x^2 + M_y^2}}{D^2} = 0 \qquad (3.239\text{b})$$

$$\frac{\partial \zeta}{\partial t} + \frac{\partial M_x}{\partial x} + \frac{\partial M_y}{\partial y} = 0 \qquad (3.240)$$

where $D = H + \zeta$ is the total water depth. The grid is similar to that used by Flather and Heaps (1975). The finite-difference form using a two time–level explicit scheme (model GF2 from Crean et al. (1988)); for the x-momentum equation is

$$\frac{M_{x,j}^{n+1} - M_{x,j}^n}{T} = f\overline{M}_{y,j}^n - \frac{g\overline{D}_j^x}{h}(\zeta_{j+1}^{n+1} - \zeta_j^{n+1}) - \frac{r M_{x,j}^n\left((M_{x,j}^n)^2 + (\overline{\overline{M}}_{y,j}^n)^2\right)^{1/2}}{(\overline{D}_j^x)^2}$$

$$- \left[\frac{(\overline{M}_{x,j}^{x,n})^2}{\frac{1}{2}(\overline{D}_j^x + \overline{D}_{j+1}^x)} - \frac{(\overline{M}_{x,j-1}^{x,n})^2}{\frac{1}{2}(\overline{D}^x,j-1+\overline{D}_j^x)}\right]\frac{1}{h}$$

$$- \left(\frac{\overline{M}_{x,j}^{y,n}\overline{M}_{y,j-m}^{x,n}}{\overline{\overline{D}}_j} - \frac{\overline{M}_{x,j+m}^y\overline{M}_{y,j}^{x,n}}{\overline{\overline{D}}_{j-m}}\right)\frac{1}{h}$$

$$+\frac{N_{h,j}}{(h)^2}\left(M_{x,j-m}^n + M_{x,j+m}^n + M_{x,j-1}^n + M_{x,j+1}^n - 4M_{x,j}^n\right) \qquad (3.241)$$

The finite-difference form for the y-momentum equation is

$$\frac{M_{yj}^{n+1} - M_{y,j}^n}{T} = -f\overline{M}_{x,j}^n - \frac{g\overline{D}_{j-m}^y}{h}(\zeta_j^{n+1} - \zeta_{j-m}^{n+1}) - \frac{rM_{x,j}^n((\overline{M}_{x,j}^n)^2 + (M_{y,j}^n)^2)^{1/2}}{(\overline{D}_{j+m}^y)^2}$$

$$-\left(\frac{\overline{M}_{x,j+m}^{y,n}\overline{M}_{x,j}^{z,n}}{\overline{\overline{D}}_{j+m}} - \frac{\overline{M}_{x,j+m-1}^{y,n}\overline{M}_{x,j-1}^{z,n}}{\overline{\overline{D}}_{j+m-1}}\right)\frac{1}{h}$$

$$-\left[\frac{(\overline{M}_{yj}^{y,n})^2}{\frac{1}{2}(\overline{D}_j^y + \overline{D}_{j+m}^y)} - \frac{(\overline{M}_{yj+m}^{y,n})^2}{\frac{1}{2}(\overline{D}_{j+m}^y + \overline{D}_{j+2m}^y)}\right]\frac{1}{h}$$

$$+\frac{N_{h,j}}{h^2}\left(M_{y,j-m}^n + M_{y,j+m}^n + M_{y,j-1}^n + M_{y,j+1}^n - 4M_{y,j}^n\right) \qquad (3.242)$$

where

$$\overline{M}_{x,j}^z = \frac{1}{2}(M_{x,j} + M_{x,j+1})$$

$$\overline{M}_{x,j}^y = \frac{1}{2}(M_{x,j} + M_{x,j-m})$$

$$\overline{\overline{D}}_{x_j} = \frac{1}{4}(D_j + D_{j-m} + D_{j-m+1} + D_{j+1}) \qquad (3.243)$$

$$\overline{\overline{M}}_{x,j} = \frac{1}{4}(M_{x,j} + M_{x,j+M} + M_{x,j+m-1} + M_{x,j-1}),$$

$$\overline{\overline{M}}_{y,j} = \frac{1}{4}(M_{y,j} + M_{y,j-m} + M_{y,j-m+1} + M_{y,j+1})$$

The finite-difference form for the continuity equation is

$$\frac{\zeta_j^{n+1} - \zeta_j^n}{T} = -\frac{M_{x,j}^n - M_{x,j-m}^n + M_{y,j-m}^n - M_{y,j}^n}{h} \qquad (3.244)$$

The main stability criterion is the CFL condition

$$T \leq \frac{h}{\sqrt{2gD_{\max}}}$$

The nonlinear instabilities have been often dealt with through the implicit or semi-implicit model constructed by Leendertse (1967), Marchuk et al. (1969), Yeh and Yeh (1976), Falconer (1980) and Ramming and Kowalik (1980). The basic approach to model construction has been delineated in Sec.3.3 of this chapter for the one-dimensional shallow water problem. The numerical schemes given below are in the difference–differential form, so that the time stepping is depicted only, the space differencing will be adjusted to the chosen grid.

The finite-difference equations are expressed in the alternate direction implicit (ADI) form with two successive operations in time during each time step. For the first part from lT to $(l + 1/2)T$, the terms involving u and ζ of the continuity and the x-momentum equation are expressed implicitly whereas the terms involving v are expressed explicitly. For the second part, i.e., during $(l + 1/2)T$ to $(l + 1)T$, the terms involving v and ζ of the continuity equation and the y-momentum equation are expressed implicitly whereas the previous implicit values of u are now represented explicitly.

For the first part of a given time step, the differential-difference equations are

$$\frac{\zeta^{l+1/2} - \zeta^l}{T} = -\frac{1}{2}\frac{\partial}{\partial x}\left[(H + \zeta)u\right]^{l+1/2} - \frac{1}{2}\frac{\partial}{\partial y}\left[(H + \zeta)v\right]^l \qquad (3.245)$$

$$\frac{u^{l+1/2} - u^l}{T} = -(u\frac{\partial u}{\partial x})^l - (v\frac{\partial u}{\partial y})^l + fv^l - g(\frac{\partial \zeta}{\partial x})^l \qquad (3.246)$$

For the second part of the time step,

$$\frac{\zeta^{l+1} - \zeta^{l+1/2}}{T} = -\frac{1}{2}\frac{\partial}{\partial x}\left[(H + \zeta)u\right]^{l+1/2} - \frac{1}{2}\frac{\partial}{\partial y}\left[(H + \zeta)v\right]^{l+1} \qquad (3.247)$$

$$\frac{v^{l+1} - v^{l+1/2}}{T} = -(u\frac{\partial v}{\partial x})^l - (v\frac{\partial v}{\partial y})^l - fu^{l+1/2} - g(\frac{\partial \zeta}{\partial y})^{l+1} \qquad (3.248)$$

These are eqs.(3.1)–(3.3) without frictional terms. In fact the bottom friction can be written in an implicit form to support overall stability, see, e.g., eq.(3.93). In the above system we have split the equation of continuity into two subequations (3.245) and (3.247). The two–step solution involves two unknowns at the first substep, these are $\zeta^{l+1/2}$ and $u^{l+1/2}$. Solution of this system can be obtained by the line inversion method along the x direction, aplying a vector with two components (see, Appendix 1). At the second substep the line inversion is applied along the y direction. In practical applications while marching from step $l + 1$ to step $l + 2$ the whole procedure should be reversed i.e., first equations along y direction should be solved and afterwards equations along x, so that the symmetry in direction is preserved. Only the linear equations, in this semi-implicit system, have been proven to be unconditionally stable. For the nonlinear equations the stability criteria are close to CFL condition. Again, a word of caution against the nondiscriminant use of the implicit equations. When dealing with unconditionally stable equations the time step should not be increased ten times beyond the explicit condition; the approximation errors may render undesirable results.

Book *et al.* (1975) developed "flux-corrected" transport schemes for the proper inclusion of the nonlinear advective terms (see Ch.II, Sec.9.5). In this scheme, any artificial diffusion added to the advection term in the first step is subtracted in the subsequent step. Lam (1977) compared various schemes of this type and showed that a central-difference scheme produces oscillations of great amplitude whereas a one-sided upstream-differencing scheme shows a large false diffusion. However, the one-sided upstream-differencing scheme combined with a flux-corrected transport scheme gave reliable results.

12. Moving boundary models and inclusion of tidal flats

Moving boundary models have been developed to allow for the climbing of the surge on the coastline as well as to include tidal flats, which become submerged during flood and dry during ebb. A simple demonstration of this method is given in Sec.7 of this chapter.

Omitting the nonlinear advective and Coriolis terms, Reid and Bodine (1968) developed a technique for the inclusion of tidal flats. The coastal boundary that follows the grid lines can advance or retreat in discrete steps as the water level rises or falls. To allow for flooding of dry land and to simulate submerged barriers, empirical formulae based on the concept of flow over weirs were used, and application was made to storm surges in Galveston Bay, Texas.

Leendertse (1970) and Leendertse and Gritton (1971) developed an alternating direction implicit technique of allowing for tidal flats, with application to Jamaica Bay, New York. In this model also, the boundary moves along grid lines in discrete steps. However, the condition for a dry area is more stringent than a simple zero local water depth (the stringent condition was used to suppress most of the computational noise due to the movements of the boundary). The programming effort is quite cumbersome, especially due to the implicit scheme used. Other works that dealt with this problem are those of Ramming (1972), Abbott et al. (1973), Backhaus (1976), Runchal (1975), and Wanstrath (1977a, 1977b).

The model of Flather and Heaps (1975) has already been introduced in the section on nonlinear terms. For the calculations in which tidal flats are to be included, they omitted the advective terms and used a simple explicit scheme. The conditions they used depended on an examination of the local water depth and the slope of the water level. Use of the condition on the water level slope especially suppresses the unrealistic movements of the boundary. As in the models of Reid and Bodine (1968) and Leendertse and Gritton (1971), the water–land boundary follows grid lines in discrete time steps.

Before the calculation of currents u and v in the x and y directions at each time step, each grid point was tested to see if it was wet (i.e., positive water depth) or dry (zero water depth). If the point was dry, then the current was prescribed as zero. For wet points, u and v were computed from the relevant equations.

Yeh and Yeh (1976) developed a moving boundary model i.e., the boundary between dry land and the water can move with time using an ADI technique. Since the technique was found to be numerically inefficient, Yeh and Chou (1979) developed an explicit technique. They showed that the moving boundary (MB) model gives storm surge amplitudes that could be 30% smaller than those given by a fixed boundary (FB) model, and observations are in better agreement with the results of the MB model. In other words, FB models that assume a fixed vertical wall at the water–land boundary could overestimate the surge by about 30%.

At the seaward open boundary, water level ζ is taken as the sum of the tide plus the inverse barometer effect. The land–water boundary advances or retreats according to the rise or recession of the surge level, i.e.,

$$\vec{V}\cdot\vec{n} = 0 \quad \text{at} \quad S[x,\, y,\, \zeta(t)] = 0$$

where $\vec{V}$ is the velocity vector, $\vec{n}$ is the unit vector normal to the curve S, and $S[x,\, y,\, \zeta(t)]$ is the water–land interface which is determined by the solution. Note that at the water–land boundary the normal velocity is taken as zero. However, this boundary is not fixed but is allowed to move freely depending on the surge elevation. This is done by varying the boundary lines on units of the grid (thus making discrete changes). However, this may give rise to computational instability problems. Hence, in very shallow areas, the bottom friction was increased (which is physically justified). Yeh and Yeh (1976) stated that the criteria upon which the location of water–land interface are based are relatively simple (see the original paper for details).

The first set of numerical experiments were for the Gulf of Mexico. The surge as calculated by the MB and FB models at Eugene Island is shown in Fig.3.29. Three differences exist between the results for both models. With the MB model (a) the maximum amplitude is about 30% lower, (b) the curve is flatter, and (c) the peak surge occurs later. The observed surge (not shown here) is in better agreement (mainly at the peak) with the MB model results.

A second set of experiments was performed for the Hurricane Carla storm surge of September 1961. A third set of experiments was performed for the south coast of Maine. The surge from a northeaster on 3 February 1972 was simulated. A maximum surge of 1.13 m was observed at Portland, Maine. In this case, both models gave almost identical surges because of the steep slope in the inland areas and the small amplitude of the surge. Lynch (1980) suggested that these models should be referred to as "sequence of fixed boundary simulations". Yeh and Chou (1979) preferred the term "discrete moving boundary".

Tetra Tech Inc. (1978) developed coastal flooding storm surge models which included the nonlinear advective terms, Coriolis terms, wind stress, atmospheric pressure gradients, and bottom stress. Here, discussion will be confined to the treatment of the land–water boundary. Usually, the landslope onshore is much greater than the slope of the ocean floor. In such situations, the coastal surge is assumed to propagate overland to its corresponding contour level (when the distance

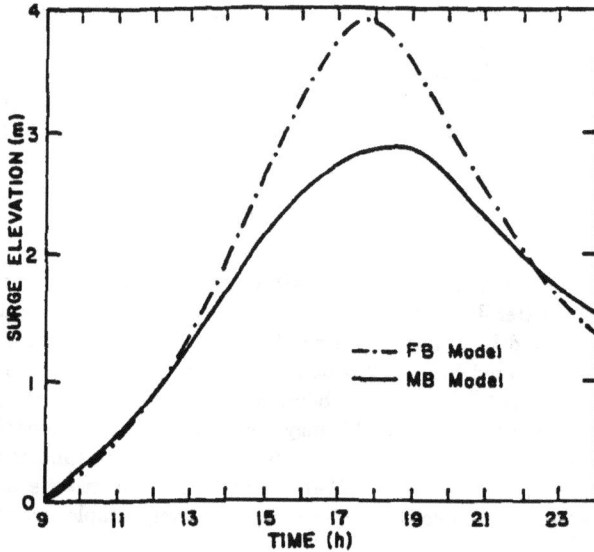

Fig. 3.29
Computed storm surges at Eugene Island, U.S.A., using fixed
(FB) and moving boundary (MB) models. (Yeh and Chou, 1979)

to that contour line is much less than one grid interval). However, there are certain
regions, such as western Florida, where the onshore slope is very small and the
limiting contour interval may be several kilometres inland. For such cases, a one-
dimensional run-up model is used at various traverses.

Sielecki and Wurtele (1970) developed a moving boundary scheme in which the
lateral boundary of the fluid is determined as a part of the solution. They tested
the validity of their scheme by comparing the results of some simple numerical experiments with the results from analytical solutions. Actually, their scheme consists
of three different methods: (1) Lax–Wendroff scheme (Lax and Wendroff, 1960) as
modified by Richtmeyer (Richtmeyer, 1963); (2) using the principle of energy conservation as formulated by Arakawa (1966); (3) using the quasi-implicit character
of the difference equations.

Sielecki and Wurtele (1970) wrote the equation of continuity and the equations
of motion in a symmetrical manner that is somewhat different from the traditional

forms:

$$\frac{\partial D}{\partial t} + \frac{\partial M_x}{\partial x} + \frac{\partial M_y}{\partial y} = 0 \tag{3.249}$$

$$\frac{\partial M_x}{\partial t} + \frac{\partial}{\partial x}(M_x u + GD^2) + \frac{\partial}{\partial y}(M_y u) - fM_y = gD\frac{\partial H}{\partial x} \tag{3.250}$$

$$\frac{\partial M_y}{\partial t} + \frac{\partial}{\partial x}(M_x v) + \frac{\partial}{\partial y}(M_y v + GD^2) + fM_x = gD\frac{\partial H}{\partial y} \tag{3.251}$$

where $(D = H + \zeta)$, $(M_x = Du, M_y = Dv)$, and $G = g/2$.

For use with the Lax–Wendroff scheme, a nonstaggered grid was used. For the finite-differencing the following notation is used,

$$\begin{aligned}
\Delta_j \phi^m_{j,k} &= \phi^m_{j+1,k} - \phi^m_{j-1,k} \\
\Delta_k \phi^m_{j,k} &= \phi^m_{j,k+1} - \phi^m_{j,k-1} \\
\overline{\phi}^m_{j,k} &= \frac{1}{4}(\phi^m_{j+1,k} + \phi^m_{j-1,k} + \phi^m_{j,k+1} + \phi^m_{j,k-1})
\end{aligned} \tag{3.252}$$

and

$$\Delta x = \Delta y = h$$

Let

$$\epsilon = \frac{T}{2h} \tag{3.253}$$

The time integration of scheme I consists of two parts in a given time step. In the first part,

$$\zeta^{m+1}_{j,k} = \overline{\zeta}^m_{j,k} - \epsilon(\Delta_j M^m_{x,j,k} + \Delta_k M^m_{y,j,k}) \tag{3.254}$$

$$M^{m+1}_{x,j,k} = \overline{M}^m_{x,j,k} - \epsilon\left[\Delta_j\left(\frac{(M^m_{x,j,k})^2}{D^m_{j,k}}\right) + G\Delta_j(D^m_{j,k})^2 + \Delta_k\left(\frac{M^m_{x,j,k}M^m_{y,j,k}}{D^m_{j,k}}\right)\right]$$
$$+ fTM^m_{y,j,k} + \epsilon g D^{m-1}_{j,k}\Delta_j H_{j,k} \tag{3.255}$$

$$M^{m+1}_{y,j,k} = \overline{M}^m_{y,j,k} - \epsilon\left[\Delta_j\left(\frac{M^m_{x,j,k}M^m_{y,j,k}}{D^m_{j,k}}\right) + \Delta_k\left(\frac{(M^m_{y,j,k})^2}{D^m_{j,k}}\right) + G\Delta_k(D^m_{j,k})^2\right]$$
$$- fTM^m_{x,j,k} + \epsilon g D^{m-1}_{j,k}\Delta_k H_{j,k} \tag{3.256}$$

In the second part,

$$D_{j,k,n+2} = D^m_{j,k} - 2\epsilon(\Delta_j M^{m+1}_{x,j,k} + \Delta_k M^{m+1}_{y,j,k}) \tag{3.257}$$

$$M^{m+2}_{x,j,k} = M^m_{x,j,k} - 2\epsilon\left[\Delta_j\left(\frac{(M^{m+1}_{x,j,k})^2}{D^{m+1}_{j,k}}\right) + G\Delta_j(D^{m+1}_{j,k})^2 + \Delta_k\left(\frac{M^{m+1}_{x,j,k}M^{m+1}_{y,j,k}}{D^{m+1}_{j,k}}\right)\right]$$

$$+ 2fTM_{y,j,k}^{m+1} + 2\epsilon g D_{j,k}^m \Delta_j H_{j,k}, \qquad (3.258)$$

$$M_{y,j,k}^{m+2} = M_{y,j,k}^m - 2\epsilon \left[\Delta_j \left(\frac{M_{x,j,k}^{m+1} M_{y,j,k}^{m+1}}{D_{j,k}^{m+1}} \right) + \Delta_k \left(\frac{(M_{y,j,k}^{m+1})^2}{D_{j,k}^{m+1}} \right) + G\Delta_k (D_{j,k}^{m+1})^2 \right]$$

$$- 2fTM_{x,j,k}^{m+1} + 2\epsilon g D_{j,k}^m \Delta_k H_{j,k} \qquad (3.259)$$

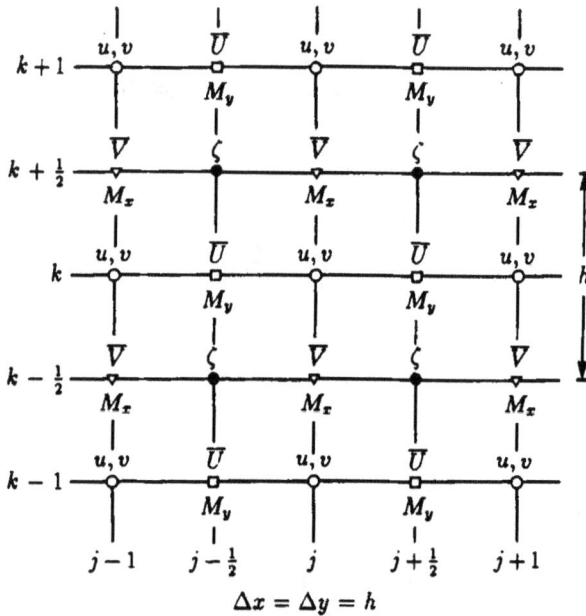

Fig. 3.30
Grid distribution, Sielecki and Wurtele (1970).

Scheme II is formulated using Arakawa's (1966) approach. The staggered grid for this scheme is shown in Fig.3.30 of Sielecki and Wurtele (1970). Also, define

$$\Delta_j \phi_{j,k}^m = \phi_{j+1/2,k}^m - \phi_{j-1/2,k}^m$$
$$\Delta_k \phi_{j,k}^m = \phi_{j,k+1/2}^m - \phi_{j,k-1/2}^m$$
$$\delta_j \phi_{j,k}^m = \tfrac{1}{2}(\phi_{j+1/2,k}^m + \phi_{j-1/2,k}^m)$$
$$\delta_k \phi_{j,k}^m = \tfrac{1}{2}(\phi_{j,k+1/2}^m + \phi_{j,k-1/2}^m)$$
$$M_{x,j,k+1/2} = (\delta_j D_{j,k+1/2})\delta_k(u_{j,k+1/2}) \qquad (3.260)$$
$$M_{y,j+1/2,k} = (\delta_k D_{j+1/2,k})(\delta_j v_{j+1/2,k})$$

$$U_{j+1/2,k} = \tfrac{1}{2}(\delta_j M_{x,j+1/2,k+1/2} + \delta_k M_{x,j+1/2,k-1/2}$$
$$V_{j,k+1/2} = \tfrac{1}{2}(\delta_k M_{y,j-1/2,k+1/2} + \delta_k M_{y,j+1/2,k+1/2}$$
$$\overline{D}_{j,k} = \tfrac{1}{2}(\delta_j D_{j,k+1/2} + \delta_j D_{j,k-1/2})$$

As with scheme I, this scheme has two parts in each time step. In the first part

$$D_{j+1/2,k+1/2}^{m+1} = D_{j+1/2,k+1/2}^{m} - 2\epsilon\Delta_j M_{x,j+1/2,k+1/2}^{m} + \Delta_k M_{y,j+1/2,k+1/2}^{m} \quad (3.261)$$

$$\overline{D}_{j,k}^{m+1}\tilde{u}_{j,k}^{m+1} = \overline{D}_{j,k}^{m} u_{j,k}^{m} - 2\epsilon\{\Delta_j \left[\overline{U}_{j,k}^{m}\delta_j u_{j,k}^{m} + G\delta_k(D_{j,k}^{m+1})\right]\}$$

$$-2\epsilon\Delta_k(\overline{V}_{j,k}^{m}\delta_k u_{j,k}^{m}) + \overline{D}_{j,k}^{m+1}(fTv_{j,k}^{m} + 2\epsilon g\Delta_j H_{j,k}) \quad (3.262)$$

$$\overline{D}_{j,k}^{m+1}\tilde{v}_{j,k}^{m+1} = \overline{D}_{j,k}^{m} v_{j,k}^{m} - 2\epsilon\Delta_j \left(\overline{U}_{j,k}^{m+1}\delta_j v_{j,k}^{m}\right)$$

$$-2\epsilon\{\Delta_k \left[\overline{V}_{j,k}^{m}\delta_k v_{j,k}^{m} + G\delta_j(D_{j,k}^{m+1})^2\right]\} + \overline{D}_{j,k}^{m+1}(-fT\tilde{u}_{j,k}^{m+1} + 2\epsilon g\Delta_k H_{j,k}) \quad (3.263)$$

In the second part,

$$\overline{D}_{j,k}^{m+1} u_{j,k}^{m+1} = \overline{D}_{j,k}^{m} u_{j,k}^{m} - 2\epsilon \left[\Delta_j\overline{U}_{j,k}^{m}\delta_j\tilde{u}_{j,k}^{m+1} + G\delta_k(D_{j,k}^{m+1})^2\right.$$

$$\left. + \Delta_k(\overline{V}_{j,k}^{m}\delta_k\tilde{u}_{j,k}^{m+1})\right] + \overline{D}_{j,k}^{m+1}(fT\tilde{v}_{j,k}^{m+1} + 2\epsilon g\Delta_j H_{j,k}) \quad (3.264)$$

and

$$\overline{D}_{j,k}^{m+1} v_{j,k}^{m+1} = \overline{D}_{j,k}^{m} v_{j,k}^{m} - 2\epsilon\Delta_j\{(\overline{U}_{j,k}^{m+1}\delta_j\tilde{u}_{j,k}^{m+1})$$

$$+\Delta_k \left[\overline{V}_{j,k}^{m}\delta_k\tilde{u}_{j,k}^{m+1} + G\delta_k(D_{j,k}^{m+1})\right]\} + \overline{D}_{j,k}^{m+1}(fTu_{j,k}^{m+1} + 2\epsilon g\Delta_k D_{j,k}) \quad (3.265)$$

where the symbol $\tilde{}$ over a variable denotes a temporary value. Note that in this scheme, a water level is calculated in the first half-step itself, whereas the final values of the currents are computed in the second half-step.

Scheme III also consists of two half-steps. In the first half-step,

$$\zeta_{j,k}^{m+1} = \zeta_{j,k}^{m} - \epsilon\{\Delta_j \left[(D_{j,k} + \zeta_{j,k}^{m})u_{j,k}^{m}\right] + \Delta_k \left[(D_{j,k} + \zeta_{j,k}^{m})v_{j,k}^{m}\right]\} \quad (3.266)$$

$$\tilde{u}_{j,k}^{m+1} = u_{j,k}^{m}(1 - \epsilon\Delta_j u_{j,k}^{m}) + v_{j,k}^{m}(fT - \epsilon\Delta_k u_{j,k}^{m}) - g\epsilon\Delta_j\zeta_{j,k}^{m+1} \quad (3.267)$$

$$\tilde{v}_{j,k}^{m+1} = v_{j,k}^{m}(1 - \epsilon\Delta_k v_{j,k}^{m}) + u_{j,k}^{m+1}(-fT - \epsilon\Delta_j v_{j,k}^{m}) - g\epsilon\Delta_k\zeta_{j,k}^{m+1} \quad (3.268)$$

In the second half-step, only the final values of the velocity components are computed:

$$u_{j,k}^{m+1} = u_{j,k}^{m}(1 - \epsilon\Delta_j\tilde{u}_{j,k}^{m+1}) - \epsilon v_{j,k}^{m}\Delta_k\tilde{u}_{j,k}^{m+1} + fT\tilde{v}_{j,k}^{m+1} - g\epsilon\Delta_j\zeta_{j,k}^{m+1} \quad (3.269)$$

$$v_{j,k}^{m+1} = v_{j,k}^m(1 - \epsilon\Delta_k\tilde{v}_{j,k}^{m+1}) - \epsilon\tilde{u}_{j,k}^{m+1}\Delta_j\tilde{v}_{j,k}^{m+1} - fTu_{j,k}^{m+1} - g\epsilon\Delta_k h_{j,k}^{m+1} \qquad (3.270)$$

In eqs.(3.266)–(3.270) the symbols used for the finite-differencing follow eqs.(3.252) and (3.253).

Sielecki and Wurtele (1970) faced no difficulty in the calculation for the case when water level is falling at the shore. On the other hand, when the water level is rising, a new underground point may have to be included. To determine the position of the new shoreline, one has to know the slope of the free surface at the present shoreline. To find the sea level at the first dry point they have used extrapolation formulae based on the continuity equation (see eq.(3.155)).

Reid and Whitaker (1976) and Reid et al. (1977a,b) allowed for vast stretches of vegetation and marsh grass (such as in Lake Okeechobee in Florida) in storm surge models. They showed that when the marsh grass extends above the water surface, a single canopy flow regime results whereas when the vegetation does not extend above the water surface, a two-layer regime exists. Flooded marsh areas are treated as an ensemble of subgrid scale obstacles.

For submerged vegetation, the model is similar to a two-layer system. The interfacial stress is formulated in terms of a coupling coefficient and the flow differential. The friction due to individual canopies is parameterized through a drag coefficient and the dimensions of the elements. When the canopy elements are not submerged, a sheltering factor is introduced.

Reid and Whitaker (1976) assumed that the obstacles are rigid elements of width w and height b and are oriented normal to the flow. It was also assumed that they are distributed evenly over the bottom. For water depth $D > b$, the vertically integrated equation of motion for the lower layer is

$$\frac{\partial M_1}{\partial t} + gb\nabla\zeta = (\tau_C - \tau_B - F_C)/\rho \qquad (3.271)$$

For the area above the obstacles,

$$\frac{\partial M_2}{\partial t} + g(D - b)\nabla\zeta = (\tau_a - \tau_C)/\rho \qquad (3.272)$$

where subscripts 1 and 2 refer, respectively, to the lower and upper layers. Here, D is the total water depth, τ_C is the interfacial stress, τ_B is the bottom stress, and F_C is any external force. M_1 and M_2, respectively, denote the transports in the lower and upper layers.

Let

$$Q = M_1 + M_2 \qquad (3.273)$$

From eqs.(3.271)–(3.273)

$$\frac{\partial Q}{\partial t} + gD\nabla\zeta = (\tau_a - \tau_B - F_C)/\rho \qquad (3.274)$$

For the total water column, the continuity equation is

$$\frac{\partial \zeta}{\partial t} + \nabla Q = 0 \tag{3.275}$$

Let u_1 and u_2 be the depth-averaged velocities for the two layers. Then, the stresses (in quadratic form) are

$$\tau_B = \rho f_1 |u_1| u_1 \tag{3.276}$$

$$\tau_C = \rho f_2 |u_2 - u_1| (u_2 - u_1) \tag{3.277}$$

Note that coefficients f_1 and f_2 are nondimensional, and f_1 can be found through comparison with formulae for the bottom stress given in Ch.I.

Let N be the number of obstacles per unit horizontal area and C_d be a dimensionless drag coefficient. The resistance per unit horizontal area to the flow in the lower layer is

$$F_C = \rho C_D w b N |u_1| u_1 \tag{3.278}$$

If $D \leq b$, the interfacial stress τ_C vanishes and in eq.(3.278) b has to be replaced with D.

When $D < b$, a sheltering coefficient S has to be introduced to model the modification of the wind stress due to the canopy. Reid and Whitaker (1976) wrote

$$S = 1 \Big/ \left(1 + \frac{\rho C_D N w D(C)}{K_1}\right) \tag{3.279}$$

where

$$\begin{aligned} D(C) &= b - D \quad &\text{if} \quad D < b \\ D(C) &= 0 \quad &\text{if} \quad D \geq b \end{aligned} \tag{3.280}$$

and K_1 is a wind stress coefficient similar to C_{10} given in Ch.I. The assumption here is that the wind stress is continuous at elevation b and that a quadratic form holds for the resistance to the wind provided by the individual elements. Note that eq.(3.279) for the dimensional coefficient S is valid for $D < b$ and $S \to 1$ as $N w D(C) \to 0$. Application of the parameters for the cases when canopy is completely covered by water and when the canopy extends above water is elucidated in Fig.3.31 and Fig.3.32.

The scheme for the numerical integration of these equations is similar to that of Platzman (1963) and Reid and Bodine (1968). This model was applied to calculate the storm surge in Lake Okeechobee, Florida due to the 1950 October storm. Earlier, Whitaker et al. (1973) showed that the dense marsh in the southwest quadrant of the lake has significant influence on the circulation. Their attempts to modify the bottom friction factor for the vegetation area proved to be useless. The water level computed with the inclusion of the canopy agreed well with observed values.

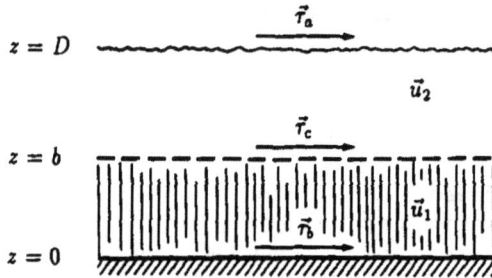

Fig. 3.31
Schematic diagram of stresses and water velocities in the case
where water completely covers the canopy.

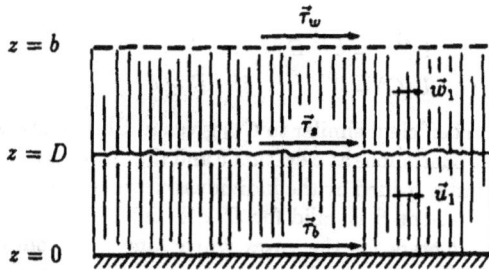

Fig. 3.32
Schematic diagram of stresses in the case where the canopy
extends above the water surface.

Danard and Murty (1992) extended this investigation by introducing a new method
for calculation wind stress for the case when vegetation extends above the water

surface. Observation indicated that in the Bay of Bengal storm surge damage was considerably less in the areas of extended vegetation.

Walton and Christensen (1980) developed a model for storm surges propagating onto a shore and applied this study to a hurricane storm surge on the west coast of Florida. Their theory includes a friction factor which varies spatially and depends not only on the local depth but also on the roughness elements and their spacings. The bottom stress components $\tau_b^{(x)}$ and $\tau_b^{(y)}$ are expressed as follows:

$$\tau_b^{(x)} = \frac{F\rho|M|M_x}{8D^2}$$
$$\tau_b^{(y)} = \frac{F\rho|M|M_y}{8D^2} \tag{3.281}$$

where $D(x, y)$ is the undisturbed water depth, ρ is the density of water, F is a Darcy–Weisbach friction factor, M_x and M_y are the transports in the x and y directions, and $|M| = (M_x^2 + M_y^2)^{1/2}$. The value of F in eq.(3.281) can be estimated through comparison to the bottom stress formulas given in Ch.I.

The friction factor F can be determined for the offshore regions knowing the sand roughness of the bed, the Reynolds number, the Froude number, and the Strouhal number. However, for the shore region, F cannot be easily determined. The Darcy–Weisbach formula for the energy loss ΔE in a water column of depth D is

$$\Delta E = \frac{FV_m^2}{2g}\frac{L}{4D} \tag{3.282}$$

where V_m is the spatially averaged velocity and L is the length over which the energy loss is determined. The commonly used expression for the shear (or friction) velocity V_* is

$$\frac{V_m}{V_*} = \sqrt{\frac{8}{F}} \tag{3.283}$$

For the travel of the storm surge over land, the following logarithmic distribution is assumed:

$$\frac{V}{V_*} = 8.48 + 2.5 \ln\left(\frac{z}{K} + 0.0338\right) \tag{3.284}$$

where V is the local velocity, z is the distance from the bottom, and K is the Nikuradse's equivalent sand roughness. The mean velocity V_m theoretically occurs at $z = 0.368D$ from the bed. Then, from eqs.(3.283) and (3.284)

$$F = \frac{1.28}{[\ln(10.94D/K + 1)]^2} \tag{3.285}$$

Equation (3.285) for F can be used for the overland travel of the storm surge. However, if vegetation and (man-made) structures are present, additional considerations are necessary.

Consider the travel of a storm surge over a vegetated area. Let the density of distribution of the trees per unit area be m and let δ be the average diameter of these trees. The energy loss ΔE, can be written as

$$\Delta E = \frac{FV_m^2}{2g}\frac{L}{4D}(1 - \epsilon) + m\delta DC_D\frac{V_m^2}{2g} \tag{3.286}$$

where ϵ is the fraction of the land occupied by the obstructions and C_D is the drag coefficient. An equivalent friction factor F_e can be introduced as follows:

$$F_e = F(1 - \epsilon) + 4m\delta DC_D \tag{3.287}$$

and

$$\epsilon = \frac{\pi}{4}\delta^2 m \tag{3.288}$$

$$F_e = \frac{1.28\left[1 - m(\pi/4)\delta^2\right]}{[\ln(10.94D/K + 1)]^2} + 4m\delta DC_D \tag{3.289}$$

For larger values of the Reynolds number (i.e., greater than 5×10^5), $C_D \sim 0.4$. Let S be the average spacing between the trees (obstacles). Then

$$m = \frac{1}{S^2} \tag{3.290}$$

Walton and Christensen (1980) wrote for a regular hexagonal pattern,

$$m = \frac{2.31}{S^2} \tag{3.291}$$

Averaging eqs.(3.290) and (3.291) gives

$$m = \frac{1.65}{S^2} \tag{3.292}$$

Equation (3.289) then becomes

$$F_e = \frac{1.28\left[-1.3(\delta/S)^2\right]}{[\ln(10.94D/K + 1)]^2} + 2.65\left(\frac{\delta}{S}\right)\left(\frac{D}{S}\right) \tag{3.293}$$

Here, the first term represents the roughness and the second term denotes the effect of the vegetation and structures. As mentioned earlier, this study was applied to a hurricane-generated surge on the west coast of Florida. As expected, inclusion of the friction factor reduced the peak surge and delayed it. The calculated results compare well with observations.

13. Nested grids and multiple grids

In this section, the use of multiple grids, such as combinations of coarse and fine grids to model storm surges in a water body, will be considered. The philosophy behind using multiple grids is to be able to reduce the total computational effort by placing a coarse grid in the deep (and offshore) region and couple this with a finer grid in the shallow coastal area. In Sec.6 of this chapter a simple approach to the grid interaction is delineated.

In connection with storm surge studies in the Beaufort Sea, Henry (1975) and Henry and Heaps (1976) used a combination of coarse and fine grids but the grids were not coupled dynamically. Examples of studies in which the grids are dynamically coupled are those of Abbott et al. (1973), Ramming (1976), Simons (1978), Johns and Ali (1980), and Ramming and Kowalik (1980).

Greenberg (1975, 1976, 1977, 1979) used a combination of grids in his numerical model for tides in the Bay of Fundy. Following Greenberg, a technique will be considered merging different mesh sizes for the simplified case of a rectangular basin of uniform depth D using the linearized version of the relevant equations. Rather than the traditional manner of using volume transports, Greenberg used depth-mean currents u and v in the x and y directions. Then the equation of continuity is

$$\frac{\partial \zeta}{\partial t} + \frac{\partial}{\partial x}(Du) + \frac{\partial}{\partial y}(Dv) = 0 \tag{3.294}$$

The equations of motion are

$$\frac{\partial u}{\partial t} + \frac{ku}{D} - fv = -g\frac{\partial \zeta}{\partial x} + \frac{\tau_x^s}{\rho_o D} \tag{3.295}$$

$$\frac{\partial v}{\partial t} + \frac{kv}{D} + fu = -g\frac{\partial \zeta}{\partial y} + \frac{\tau_y^s}{\rho_o D} \tag{3.296}$$

Note that these are eqs.(3.1)-(3.3) with the nonlinear, atmospheric pressure and horizontal friction terms neglected. The bottom friction term is taken in a linear form with coefficient k. The finite-difference forms of eqs.(3.294)-(3.296) are as follows:

$$\zeta_{j,k}^{m+1} = \zeta_{j,k}^{m} - DT\left[\frac{(u_{j,k}^{m} - u_{j-1,k}^{m})}{\Delta x} + \frac{(v_{j,k-1}^{m} - v_{j,k}^{m})}{\Delta y}\right] \tag{3.297}$$

$$u_{j,k}^{m+1} = \frac{1}{(1 + (kT/2D))}\left[u_{j,k}^{m}(1 - (kT/2D)) + fT\hat{v}_{j,k}^{m}\right.$$
$$\left. - \frac{gT}{\Delta x}(\zeta_{j+1,k}^{m+1} - \zeta_{j,k}^{m+1}) + \frac{T}{\rho_o D}\tau_x^s\right] \tag{3.298}$$

$$v_{j,k}^{m+1} = \frac{1}{(1 + (kT/2D))}\left[v_{j,k}^{m}(1 - (kT/2D)) - fT\hat{u}_{j,k}^{m+1}\right.$$

$$- \frac{gT}{\Delta y}(\zeta_{j,k}^{m+1} - \zeta_{j,k+1}^{m+1}) + \frac{T}{\rho_o D}\tau_y^* \Big] \tag{3.299}$$

where

$$\hat{u}_{j,k}^m \equiv \frac{1}{4}(u_{j-1,k}^m + u_{j,k}^m + u_{j-1,k+1}^m + u_{j,k+1}^m) \tag{3.300}$$

$$\hat{v}_{j,k}^m \equiv \frac{1}{4}(v_{j,k-1}^m + v_{j+1,k-1}^m + v_{j,k}^m + v_{j+1,k}^m) \tag{3.301}$$

Initially, u, v, and ζ were prescribed to be zero. The boundary conditions are $u = v = 0$ on all the boundaries. In addition to the usual CFL stability criterion, Greenberg (1977) gave another condition, namely

$$T < \frac{2}{f} \tag{3.302}$$

where f is the Coriolis parameter, which is satisfied easily. Also notice that the bottom friction term has been written in a semi–implicit form as

$$k\frac{u^m + u^{m+1}}{2D} \quad \text{and} \quad k\frac{v^m + v^{m+1}}{2D} \tag{3.303}$$

The grid distribution given in Fig.3.33 is similar to the one used by Flather and Heaps (1975). For the purpose of the grid enumeration the left–hand system is taken, so that the y axis is pointing downward, but actual derivative along y direction is taken for the y axis pointing upward, see. eq.(3.297).

The scheme for coupling grids is shown in Fig.3.34. It is based on the linear interpolation of the coarse grid variables and the fine grid variables to construct the boundary for the fine grid. A very similar approach is elucidated in Sec.6 of this chapter. Here the method is presented in a different enumeration similar to the one presented by Flather and Heaps (1975). The fine-grid calculations of the mixed grid commence at row $(q+1)$ of the coarse grid and row r of the fine grid. To be able to calculate ζ and u on the rth row, one must know v on the $(r-1)$th row. However, eq.(3.299) does not give these values directly. Hence, the qth row of the coarse grid and the rth row of the fine grid for the v values must be interpolated (linearly).

Two different approaches are needed here, one for the left boundary points and another one for the upper boundary points. To discriminate between the two grids, two different grid notations are applied i.e., for the coarse grid l and m indices are used and for the fine grid i and r indices. Fig.3.35 elucidates the construction of the left boundary point $v_{2,r-1}$. It is given as an average of $v_{2,r}$ and v_*. The velocity in a star point is calculated by extrapolation of $v_{2,q}$ and $v_{3,q}$, thus

$$v_{2,r-1} = \frac{1}{2}\left[v_{2,r} + v_{2,q} + \frac{1}{3}(v_{2,q} - v_{3,q})\right] \tag{3.304}$$

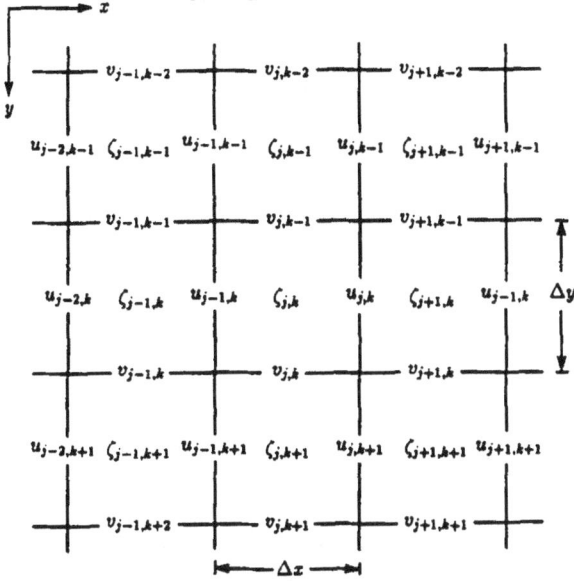

Fig. 3.33
Indexing scheme. (Greenberg, 1977)

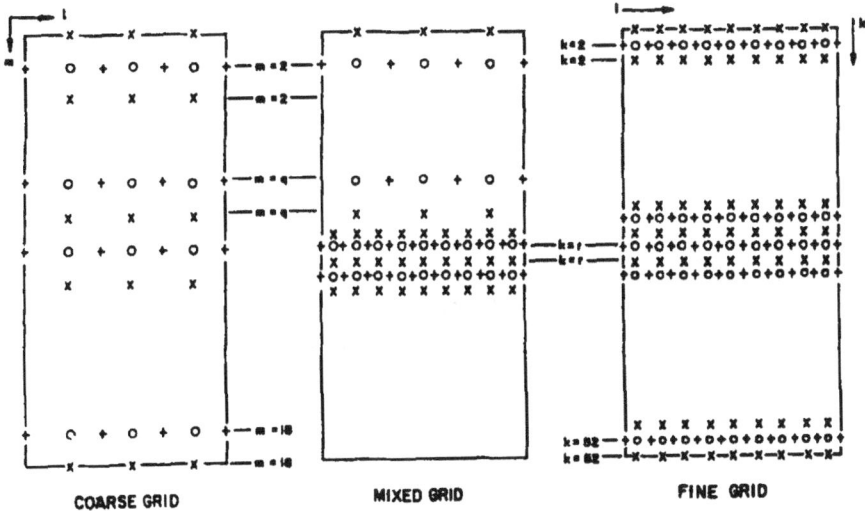

Fig. 3.34
Scheme for coupling grids in a multiple-grid computation.
o $-\zeta$ point, $+$ $-$ u point, $\times$ $-v$ point (Greenberg, 1977)

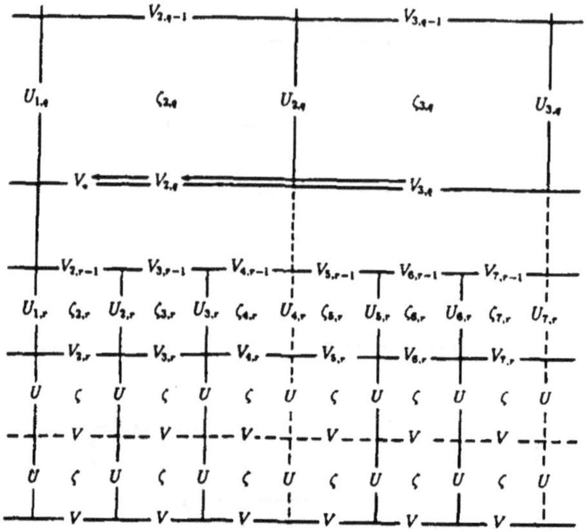

Fig. 3.35
Scheme for computation fine grid boundary
points at the left boundary. (Greenberg, 1977)

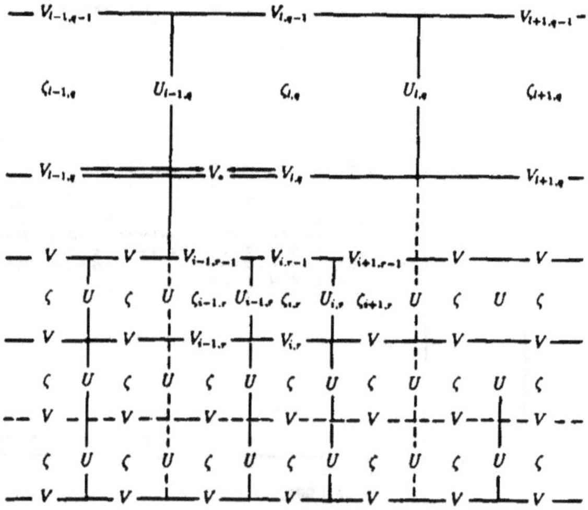

Fig. 3.36
Scheme for computation fine grid boundary
points at the upper boundary. (Greenberg, 1977)

The method to calculate the fine grid velocity at the upper boundary points is given in Fig.3.36. Consider point $v_{i-1,r-1}$, it is calculated as an average of $v_{i-1,r}$ and $(2/3)v_{l,q} + (1/3)v_{l-1,q}$, thus

$$v_{i-1,r-1} = \frac{1}{2}\left(v_{i-1,r} + \frac{2}{3}v_{l,q} + \frac{1}{3}v_{l-1,q}\right) \qquad (3.305)$$

Notice that the grid distance from the fine grid to the coarse grid is increased three times. As we have shown in Sec.6 this increase can be also done by any odd whole number.

Greenberg made further calculations with the inclusion of the nonlinear terms because these terms might be important in understanding the interaction between tide and storm surge. He found that when the nonlinear terms are present, the damping due to frictional terms is less. For details on the computational scheme when nonlinear (advective) terms and quadratic bottom friction terms are included, see Greenberg (1977). This report also includes details of the finite-difference forms, for the calculation of energy, and the stability criterion.

14. Stretched coordinates and transformed grid systems

Birchfield and Murty (1974) used a stretched coordinate system to study wind-generated circulation in the combined system of Lake Michigan, Straits of Mackinac, and Lake Huron. Although this study did not examine storm surges, the technique is applicable to simulation of storm surges in two water bodies connected by narrow straits.

The system studied here is shown in Fig.3.37a and the curvilinear grid used is shown in Fig.3.37b. The curvilinear grid is mapped onto a plane in which the irregular basin is transformed into a series of connected rectangles (Fig.3.38). An equispaced grid was used in the connected rectangle system, and all the calculations are performed conveniently in this system. However, for easy interpretation, the results of the output are printed in the geographical format.

Birchfield and Murty (1974) began with the formulation of Platzman (1963). The main difference is that, whereas Platzman used a no-slip condition at the bottom, these authors permitted slip. Their model has certain similarities to the model of Jelesnianski (1967).

The nonorthogonal curvilinear grid shown in Fig.3.37b is constructed freehand using hydrographic charts of the water bodies. The advantage of such a grid is that the boundary of the water body always falls on a coordinate line and the grid points are closely spaced in narrow and shallow areas. A rectangular area is designated for each major part of the basin (Fig.3.38) such that a one-to-one correspondence exists between each point on the curvilinear grid and a point on a square grid covering the rectangle.

The length-to-width ratio of the rectangle is then determined by the number of rows of points to the number of columns required to resolve the subregions and the

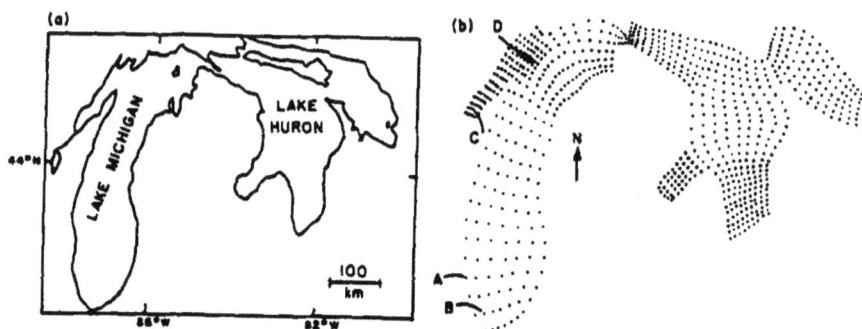

Fig. 3.37

(a) Basins of lakes Michigan and Huron joined at the Straits of Mackinac. (b) Curvilinear mesh for the basins of lakes Michigan and Huron. Note that the North Channel has been excluded. Points A,B,C, and D are special locations where observations were available for comparison with theory. (Birchfield and Murty, 1974)

entrances to the other regions. The finite-difference scheme used by Birchfield and Murty (1974) necessiates that the number of rows and columns in the rectangle be odd. When this procedure is used for all the major regions of the basin, the total basin consists of a series of interconnected rectangles.

Then, a table consisting of Cartesian coordinates (x, y) of the curvilinear grid can be developed on the hydrographic chart and each corresponding (ξ, η) on the rectangle. Thus, in essence, an empirical, nonconformal mapping of the (x, y) plane has been developed onto the (ξ, η) plane. For a one-to-one mapping the following can be written:

$$\xi = f_1(x, y)$$
$$\eta = f_2(x, y)$$

Then eqs.(2.145)–(2.147) in Murty (1984) can be written as follows, after defining some new coefficients (see Birchfield and Murty (1974) for details). We took the mass transport equations in the form originally derived by Platzman (1963). However, it is not essential for the stretching, and the usual set of equations (3.1)–(3.3) can be used as well.

$$\frac{\partial M_x}{\partial t} = gD\left(A_2\frac{\partial f_1}{\partial y} - A_1\frac{\partial f_1}{\partial x}\right)\frac{\partial \zeta}{\partial \xi} + gD\left(A_2\frac{\partial f_2}{\partial y} - A_2\frac{\partial f_2}{\partial x}\right)\frac{\partial \zeta}{\partial \eta}$$
$$- C_1M_x + C_2M_y + B_1\tau_{s(x)} - B_2\tau_{s(y)} \qquad (3.306)$$

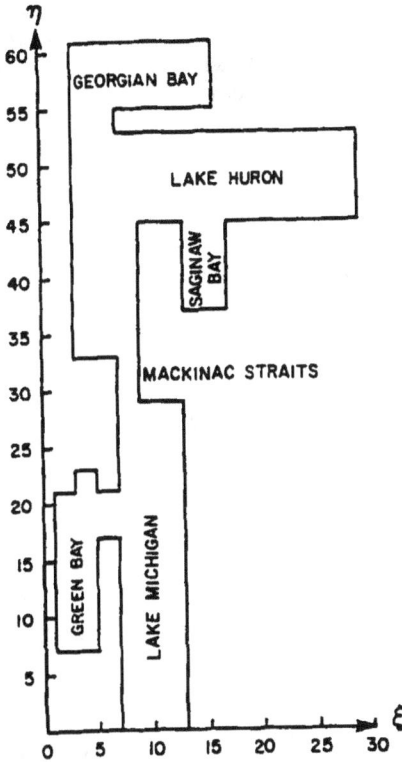

Fig. 3.38
Basins of lakes Michigan and Huron as arranged on the (ξ, η) plane.

$$\frac{\partial M_y}{\partial t} = gD\left(A_2 \frac{\partial f_1}{\partial x} + A_1 \frac{\partial f_1}{\partial y}\right)\frac{\partial \zeta}{\partial \xi} - gD\left(A_2 \frac{\partial f_2}{\partial x} + A_1 \frac{\partial f_2}{\partial y}\right)\frac{\partial \zeta}{\partial \eta}$$
$$- C_2 M_x - C_1 M_y + B_1 \tau_{s(y)} + B_2 \tau_{s(x)} \qquad (3.307)$$

$$\frac{\partial \zeta}{\partial t} = -\frac{\partial f_1}{\partial x}\frac{\partial M_x}{\partial \xi} - \frac{\partial f_2}{\partial x}\frac{\partial M_x}{\partial \eta} - \frac{\partial f_1}{\partial y}\frac{\partial M_y}{\partial \xi} - \frac{\partial f_2}{\partial y}\frac{\partial My}{\partial \eta} \qquad (3.308)$$

Note that these equations are no more difficult than the original equations. Whereas in the original equations the coefficients depend only on the depth D, in these equations they depend also on the derivatives of the mapping function. Since the curvilinear coordinates are not orthogonal, there is no particular advantage in resolving the motion into components along the local (ξ, η) axis. Also note that here, the map scale variations have not been taken into consideration.

For purposes of numerical integration it is convenient to rotate the (ξ, η) axis by 45°. Making use of two Richardson lattices, central space-differences, and forward-time differences, the equations are integrated in time.

Reid *et al.* (1977a,b) developed a transformed stretched coordinate system to calculate storm surges on a continental shelf. The principle underlying this scheme is to find a transformation involving mapping relations to keep the orthogonality and to make sure that the new independent variables, ξ and η, are continuous monotonic functions of the original independent variables, x and y. Furthermore, the transformation must map the coastline and seaward boundaries as isolines of the curvilinear coordinate, η.

A point (x, y) on the z plane will be transformed to (ξ, η) in a rectangular region on the ψ plane and will satisfy the above conditions provided the mapping relation is conformal:

$$\psi = \xi + i\eta = F(x + iy) \tag{3.309}$$

or conversely

$$z = x + iy = G(\xi + i\eta) \tag{3.310}$$

where F and G are single-valued real functions. The function $G(\psi)$ may be represented by a truncated Fourier series as

$$G(\psi) = P_0 + Q_0\psi + \sum_{n=1}^{N}[P_n \cos(n\psi) + Q_n \sin(n\psi)] \tag{3.311}$$

Here, the coefficients P_n and Q_n are complex constants.

The real and imaginary parts of eq.(3.311) give x and y in terms of ξ and η, respectively. The coefficients are determined in such a way that for constant $\eta = \beta$, the corresponding x and y as a function of ξ will map the coastline under consideration. Another constraint will be the representation of the seaward boundary (for example, the 200 m depth contour) as $\eta = -\beta$.

The coefficients P_n and Q_n can be evaluated by iteration using a least-square principle (Reid and Vastano, 1966):

$$\begin{aligned} P_n &= A_n + iB_n \\ Q_n &= C_n + iD_n \end{aligned} \tag{3.312}$$

where A_n, B_n, C_n, and D_n with $n = 1, 2\ldots, N$ are real constants. Then eq.(3.311) gives

$$x = A_0 + C_0\xi - D_0\eta + \sum_{n=1}^{N}[A_n \cosh{(\eta n)} - D_n \sinh{(\eta n)}] \cos(\eta\xi)$$

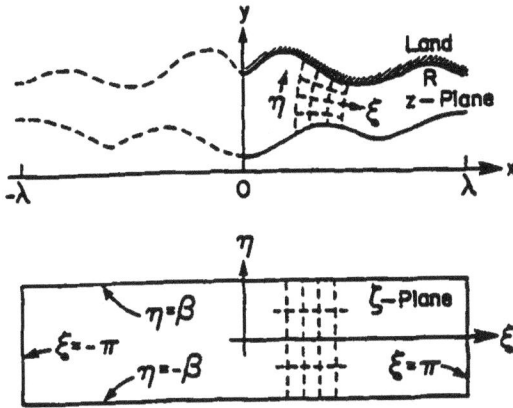

Fig. 3.39
Coast curves and seaward boundary to be mapped from (x, y) space
to straightline boundaries in (ξ, η) space. (Reid *et al.*, 1977a,b)

$$+ \sum_{n=1}^{N} [B_n \sinh(\eta n) + C_n \cosh(\eta n)] \sin(\eta \xi) \qquad (3.313)$$

and

$$y = B_0 + C_0 \eta + D_0 \xi + \sum_{n=1}^{N} [B_n \cosh(\eta n) + C_n \sinh(\eta n)] \cos(\eta \xi)$$

$$+ \sum_{n=1}^{N} [D_n \cosh(\eta n) - A_n \sinh(\eta n)] \sin(\eta \xi) \qquad (3.314)$$

Here, the range of ξ is $-\pi$ to $+\pi$.

The curves can be represented in the following parametric manner.

Seaward,

$$x = x_S(\xi, -\beta)$$
$$y = y_S(\xi, -\beta) \qquad (3.315)$$

Coastline,

$$x = x_C(\xi, +\beta)$$
$$y = y_C(\xi, +\beta) \qquad (3.316)$$

where

$$x_S(\xi, -\beta) = A_0 + C_0 \xi + D_0 \beta + \sum_{n=1}^{N} [C_n \cosh(n\beta) - B_n \sinh(n\beta)] \sin(n\xi)$$

$$+ \sum_{n=1}^{N} [A_n \cosh(n\beta) + D_n \sinh(n\beta)] \cos(n\xi) \qquad (3.317)$$

$$y_S(\xi, -\beta) = B_0 - C_0\beta + D_0\xi + \sum_{n=1}^{N} [B_n \cosh(n\beta) - C_n \sinh(n\beta)] \cos(n\xi)$$

$$+ \sum_{n=1}^{N} [A_n \sinh(n\beta) + D_n \cosh(n\beta)] \sin(n\xi) \qquad (3.318)$$

$$x_C(\xi, \beta) = A_0 + C_0\xi - D_0\beta + \sum_{n=1}^{N} [B_n \sinh(n\beta) + C_n \cosh(n\beta)] \sin(n\xi)$$

$$+ \sum_{n=1}^{N} [A_n \cosh(n\beta) - D_n \sinh(n\beta)] \cos(n\xi) \qquad (3.319)$$

$$y_C(\xi, \beta) = B_0 + C_0\beta + D_0\xi + \sum_{n=1}^{N} [B_n \cosh(n\beta) + C_n \sinh(n\beta)] \cos(n\xi)$$

$$+ \sum_{n=1}^{N} [D_n \cosh(n\beta) - A_n \sinh(n\beta)] \sin(n\xi) \qquad (3.320)$$

Note that $x_S(\xi, -\beta)$ and $x_C(\xi, +\beta)$ represent a periodic range of 2π and this corresponds to the distance 2λ in Fig.3.39. This implies that $C_0 = \pi/\lambda$ and $A_n = D_n = 0$; $n = 0, 1, 2, \ldots, N$ and B_0 is determined as the mean distance between the coast and seaward boundary curves. Note that β must be determined along with the coefficients in eqs.(3.317–3.320) using a curve-fitting scheme. However, the range of ξ and the scale factor C_0 are free parameters.

The coefficients and β must be determined for a given N such that eqs.(3.317)–(3.320) give an accurate approximation to the curves X_S, Y_S, X_C, and Y_C in a least-squares sense. Note that these curves represent the coastline and seaward boundary as a function of the arc length. One must begin with an initial approximation of arc length in terms of ξ for each curve and use an iteration technique to minimize the error function E:

$$E = \frac{1}{2\pi} \int_{-\pi}^{\pi} \left[(Y_S - y_S)^2 + (X_S - x_S)^2 + (Y_C - y_C)^2 + (X_C - x_C)^2 \right] d\xi \qquad (3.321)$$

For economy of computation, one would like to have more grid points in regions of specific interest and fewer grid points elsewhere. To be able to do this, one can stretch the orthogonal curvilinear grid system in both the shoreward and longshore directions. A curvilinear grid in the (x, y) space, which follows from the transformation, is shown in Fig.3.40a. The corresponding grid in the (ξ, η) space is shown in Fig.3.40b. From this grid, we would like to transform to another grid

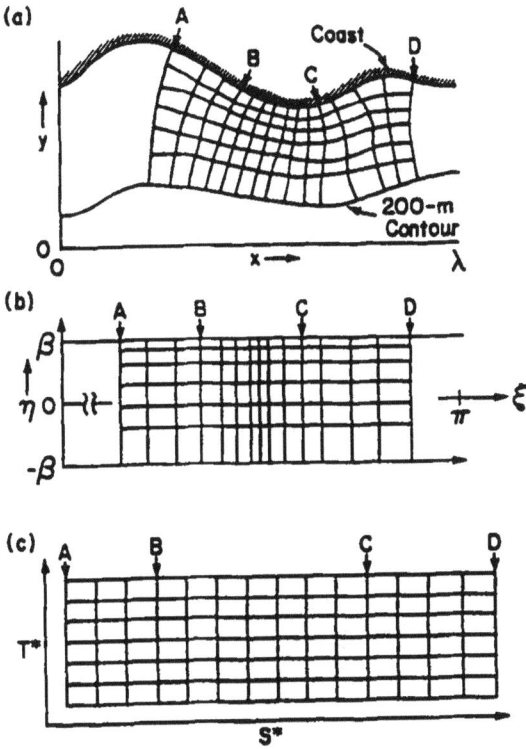

Fig. 3.40

(a) (x, y) space to be mapped by the transformation equations given in the text; (b) transformed system represented in the (ξ, η) space; (c) (ξ, η) space transformed to (S^*, T^*) space. (Reid *et al.*, 1977a)

with uniform grid increments. To be able to do this, a transformation is made to an (S^*, T^*) space in two steps.

The relation

$$S^* = S^* [S_p(\xi)] \qquad (3.322)$$

is used to generate a uniform ΔS^* spacing. Here, S_p is the arc length distance along the transform-generated coast. The relation in eq.(3.322) is generated by a choice of ΔS^* that gives the necessary longshore resolution. In the shoreward direction, the travel time, T^*, for a long wave to cover the distance between the seaward to shoreward boundary along a ξ line can be divided into a desired (from the point of view of resolution) number of increments. This can be written as

$$T^* = Y^* [S_\eta(\eta)] \tag{3.323}$$

where S_η is the distance along the ξ line and

$$T^* = \int_{S_\eta(\eta)} \frac{ds}{\sqrt{gD}}. \tag{3.324}$$

Here, D is the local water depth. Thus, choices of ΔS^* and ΔT^* generate the (S^*, T^*) grid shown in Fig.3.40. The shoreline and seaward boundaries are defined by constant values of T^* whereas constant values of S^* identify the lateral boundaries.

Let Q be the volume transport per unit width, τ the wind stress, ζ the perturbation of the water level relative to its undisturbed position, and ζ_B the hydrostatic elevation corresponding to the atmospheric pressure anomaly. Let F be a scale factor of the curvilinear coordinate system given by

$$F = \left[\left(\frac{\partial x}{\partial \xi} \right)^2 + \left(\frac{\partial y}{\partial \eta} \right)^2 \right]^{1/2} \tag{3.325}$$

Let μ and ν be scale factors representing the transformation to the (S^*, T^*) system:

$$\begin{aligned} \mu &= \frac{\partial \xi}{\partial S_p} \frac{\partial S_p}{\partial S^*} \\ \nu &= \frac{\partial \eta}{\partial S_\eta} \frac{\partial S_\eta}{\partial T^*}. \end{aligned} \tag{3.326}$$

Then, the vertically integrated equations in the (S^*, T^*) system are

$$\frac{\partial Q_{S^*}}{\partial t} - fQ_{T^*} + \frac{gD}{F\mu} \frac{\partial}{\partial S^*}(\zeta - \zeta_B) = \tau_{S^*} - \sigma_{S^*} \tag{3.327}$$

$$\frac{\partial Q_T^*}{\partial t} + fQ_{S^*} + \frac{gD}{F\nu} \frac{\partial}{\partial T^*}(\zeta - \zeta_B) = \tau_{T^*} - \sigma_{T^*} \tag{3.328}$$

$$\frac{\partial \zeta}{\partial t} + \frac{1}{F^2} \left[\frac{1}{\mu} \frac{\partial}{\partial S^*}(FQ_{S^*}) + \frac{1}{\nu} \frac{\partial}{\partial T^*}(FQ_{T^*}) \right] = 0 \tag{3.329}$$

The wind stress τ is determined from

$$\tau = \frac{\rho_a C_{10} W_{10}^2}{\rho_w} \tag{3.330}$$

where ρ_a and ρ_w are air and water densities, respectively, and C_{10} is a nondimensional drag coefficient such that

$$\frac{\rho_a C_D}{\rho_w} = K = \begin{cases} K_1 \text{ for } W_{10} < 7 \text{ m/s} \\ K_1 + \left(1 - \frac{7}{W_{10}}\right)^2 K_2 \text{ for } W \geq 7 \text{ m/s} \end{cases} \tag{3.331}$$

where $K_1 = 1.1 \times 10^{-6}$ and $K_2 = 2.5 \times 10^{-6}$ and W_{10} is the wind speed at a height of 10 m above the ground. The bottom stress terms are written as

$$\sigma_{S*} = \frac{K_0 Q Q_{S*}}{D^2}$$
$$\sigma_{T*} = \frac{K_0 Q Q_{T*}}{D^2} \tag{3.332}$$

where K_0 is a dimensionless drag coefficient and is taken as 2.5×10^{-3}. The surface stress (wind stress) components are written as

$$\tau_{S*} = \tau^{(x)} \cos \theta + \tau^{(y)} \sin \theta$$
$$\tau_{T*} = -\tau^{(x)} \sin \theta + \tau^{(y)} \cos \theta \tag{3.333}$$

where

$$\theta = \tan^{-1}\left(\frac{\partial y/\partial \xi}{\partial x/\partial \xi}\right) \tag{3.334}$$

The wind stress components $\tau^{(x)}$ and $\tau^{(y)}$ are assumed to be known.

At the lateral boundaries, the gradient of the volume transport in the $S*$ direction must vanish:

$$\frac{\partial Q_{S*}}{\partial S*} = 0 \tag{3.335}$$

On the seaward boundary, the water level is taken as the hydrostatic equivalent of the atmospheric pressure anomaly:

$$\zeta = \zeta_B \tag{3.336}$$

Note that Jelesnianski (1965) used the same condition. At the shoreline boundary the assumption is that (infinitely high wall) there is no transport in the shoreward direction:

$$Q_{T*} = 0 \tag{3.337}$$

For the numerical integration of eqs.(3.327)–(3.329), an explicit, central-difference, leapfrog scheme (Alvarez, 1973) was used. In this scheme Q_{S*} and Q_{T*} are specified at the same location (and time) but ζ is staggered in space and time from these.

This scheme was tested by Reid *et al.* (1977a,b) by simulating the surge on the coast of the Gulf of Mexico due to Hurricane Carla.

Jelesnianski (1976) used a sheared coordinate system for application to gently curved coastlines. This method is applicable for coastlines without bays, inlets, capes, etc. He stated:

> ... a mildly curved coastline is shifted or sheared onto a straight base line. A surface plane, truncated from the ocean shelf and containing the curved coast as a boundary, is fitted with a curved, nonorthogonal grid. The plane with curved boundaries is then transformed via a sheared coordinate system onto an image rectangle. In the transformed system, the computational grid is Cartesian, orthogonal, equally spaced and the coast lies exactly on and not across a grid line.

Butler (1979) used a stretched coordinate system for studying hurricane surges in Galveston Bay. The scheme is somewhat similar to that of Wanstrath (1976). Hamilton (1978), in his study of storm surges in the Thames Estuary, used a somewhat different approach, i.e., he wrote the equations in conformal coordinates.

The equations of motion in conformal coordinates for long waves (nondispersive) are

$$\frac{\partial u}{\partial t} - fv + \frac{\partial J}{\partial \xi}\left(\frac{1}{2}u^2 + \frac{1}{2}v^2\right) + Ju\frac{\partial u}{\partial \xi} + Jv\frac{\partial u}{\partial \psi} + g\frac{\partial \zeta}{\partial \xi} = \frac{\tau^\xi}{\rho J^{-1/2}} \quad (3.338)$$

$$\frac{\partial v}{\partial t} + fu + \frac{\partial J}{\partial \zeta}\left(\frac{1}{2}u^2 + \frac{1}{2}v^2\right) + Ju\frac{\partial v}{\partial \xi} + Jv\frac{\partial v}{\partial \psi} + g\frac{\partial \zeta}{\partial \psi} = \frac{\tau^\psi}{\rho J^{-1/2}} \quad (3.339)$$

$$\frac{\partial \zeta}{\partial t} + J\frac{\partial}{\partial \xi}(D+\zeta)u + J\frac{\partial}{\partial \psi}(D+\zeta)v = S(\xi, \psi, t) \quad (3.340)$$

where an arbitrary conformal mapping,

$$\xi + i\psi = w(\phi + i\Psi) \quad (3.341)$$

will produce the coordinates ξ and ψ and the metric

$$J(\xi, \psi) = \frac{d\xi\, d\psi}{dA} = R^2\text{sech}^2\mu\left[\left(\frac{\partial \mu}{\partial \xi}\right)^2 + \left(\frac{\partial \mu}{\partial \psi}\right)^2\right]^{-1} \quad (3.342)$$

Here, R is the radius of the earth, ϕ is the east longitude, θ is the north latitude, and μ is the Mercator coordinate given by

$$\tanh \mu = \sin \theta \quad (3.343)$$

and J is the ratio of an elemental area $d\xi\,d\psi$ to the actual area dA. Note that

$$u = \frac{1}{J}\frac{\partial \xi}{\partial t}$$
$$v = \frac{1}{J}\frac{\partial \psi}{\partial t} \qquad (3.344)$$

and ζ is the free surface height, D is the water depth, τ^{ξ} and τ^{ψ} are the stress components, and S is a source term. The velocities u and v in the ξ, ψ system are related to the actual velocities U and V through

$$u = UJ^{-1/2}$$
$$v = VJ^{-1/2} \qquad (3.345)$$

CHAPTER IV

THREE-DIMENSIONAL TIME-DEPENDENT MOTION

1. Introduction

Numerical models to study three-dimensional and time-dependent motion in the oceans are based on various numerical methods delineated in the previous chapters. One additional space coordinate brings about new computational problems because solution requires more CPU time and computer memory. Presently it is not feasible to search a numerical solution for any mesoscale process in the whole World Ocean. The economy consideration is probably the number one problem associated with construction of the numerical algorithms. In the process of devising a new algorithm one has to take into account stability and accuracy of the finite difference scheme. The variety of the available schemes to perform time integration is intricately related to economy, this is usually done by changing explicit time integration into implicit or semi-implicit schemes, which are free of stability requirements. Not always an economical solution exists in the mathematical domain, the physics of the process may as well bring about certain economical solutions. For example, the short time step which follows from the Courant–Friedrisch–Lewy (CFL) condition,

$$T \leq \frac{h}{\sqrt{2}\sqrt{gH}}$$

is needed to describe the surface gravity waves, otherwise the wave motion includes rather slow internal gravity waves. Surface gravity waves do not need to be described by the full three-dimensional equations because these are long waves and their velocity is constant over the whole depth. Thus the surface oscillations can be studied by the vertically integrated two-dimensional set of equations. This difference in physics of the surface and internal waves leads to a possible economical computational algorithm. Three-dimensional velocity distribution is calculated by two interconnected time steps; a short time step is applied to the vertically integrated equations and much longer time step is used for the three-dimensional equations from which the sea surface has been excluded.

Along with the CFL condition in the three-dimensional problem the condition due to vertical friction, often is considered,

$$T \leq \frac{h_z^2}{2N_z}$$

When the details of motion in the boundary layer has to be resolved the space step in the vertical direction (h_z) is diminished, and this requires a short time step of integration. The usual approach is to construct an implicit numerical scheme so that a long time step can be applied.

The construction of the numerical algorithms is so tedious that achieving economy and stability is usually the main purpose of the whole exercise. One can not forget to test the accuracy of the constructed numerical scheme. This may be especially important for an implicit scheme which allows long time steps and therefore the approximation and round off errors are plausible sources of reduced accuracy. A knowledge of the accuracy of a method is essential for the user. Unfortunately there is no simple answer to this problem. Here we show three possible approaches. The simplest approach is to compare the numerical solution against the exact solution of the partial differential equation. The main obstacle to this approach is that we know only a few analytical solutions. The second approach is to solve the same problem by two numerical models with different approximation and compare the derived solutions. In this approach we can find out about the error but we will not be able to tell how close the numerical solution is to the analytical solution. The third method of error estimation is to compare two (or more) numerical solutions derived from the grid nets of different space step (h). The error is calculated as the difference of two solutions.

2. Numerical modeling of the fjord circulation

2.1 Introduction

The general equation of motion relates the three components of velocity, pressure and density at a given time–space point to the external and internal forces taking into account constraints due to the initial and boundary conditions. Analytic solution of the full set of equations is not known, a numerical solution of the governing equation is also a complicated and time-consuming undertaking. Hence we shall start by analyzing a two-dimensional time-dependent problem, a density and tide driven flow in the deep fjords or channels. Even this problem is quite general since the time scale considered ranges from a few hours to several years. The problem can be again simplified by focusing on the narrow band of time scale or considering the range of variables specifically related to the problem. Let us say we intend to study the density and tide driven seasonal flow (circulation). The scale of motion to be resolved has 60 km in the horizontal direction and about 400 m in the vertical direction. In such a geometry the vertical component of motion ought to be much weaker than the horizontal velocity. (This assumption is somewhat doubtful around a sill area, where the vertical velocity may change significantly over a short horizontal distance). Typical vertical velocities are of the order 10^{-2} cm/s, whereas the horizontal velocities are 1000 times greater. If the flow is predominantly horizontal and the vertical acceleration is small compared to gravity acceleration, the equation of vertical motion can be reduced to the simple hydrostatic law.

Fjords can be idealized as elongated channels with dominant (horizontal) component of velocity along the axis of the channel. Thus, there arises a second possibility for simplifying the equations. The cross-channel motion can be neglected by averaging the equations of motion and continuity along a cross-channel coordinate.

A two-dimensional (horizontal and vertical) time-dependent models have been successful in reproducing the essential features of circulation in shallow estuaries (Wang and Kravitz, 1980; Nihoul et al., 1978; Perrels and Karelse, 1978; Niebauer, 1980). Numerical modeling problems have been dealt with extensively by Hamilton (1975), Elliot (1976) and Blumberg (1975). Their models were applied to shallow estuaries, where the process of salt and momentum exchange along the vertical are known and parametrized as function of flow characteristics (e.g., Bowden and Hamilton, 1975). Due to the large depth, complicated vertical and horizontal density structure, the situation is different in deep fjords. To study deep fjord dynamics a simplified numerical model of two-layered structure is often considered; e.g., Klinck et al., (1981) and Farmer (1976). In the theoretical approach initialized by Stommel and Farmer (1952, 1953), a layered model is developed to describe the mixing of fresh water river discharge with the salt water inflowing from outside the fjord. A short review of this line of fjord circulation theories is given by Pedersen (1978).

To reproduce adequately the complicated vertical structure of the density and current we shall describe a multilayered model along the vertical direction. The numerical algorithm should reproduce the main features of the fjord circulation but it must be also economical to allow for a long integration time. The selection we made, led to the development of an explicit–implicit, economical procedure related to fractional time steps (Yanenko, 1971; Ramming and Kowalik, 1980). To assure proper simulation of the fjord dynamics, appropriate boundary conditions have to be imposed at the mouth of the fjord. Both the seasonal signal and the tidal variations were considered in the measured salinity and currents. Because a uniqueness theorem is not specified yet for the general equation of motion, it is not clear how many and which variables should be specified at the boundary to derive a unique solution to the fjord circulation problem.

2.2 Governing equations for the fjord circulation and boundary conditions

The governing equations were introduced in Ch.I.Sec.2.2 setting the x coordinate along the fjord axis and directing it towards the head of the fjord. The following equations will be used in ensuing considerations; equation of continuity,

$$\frac{\partial bu}{\partial x} + \frac{\partial bw}{\partial z} = 0 \qquad (4.1)$$

equation of motion,

$$\frac{\partial u}{\partial t} + \frac{u}{b}\frac{\partial}{\partial x}ub + w\frac{\partial u}{\partial z} = -\frac{1}{\rho_o}\frac{\partial p_a}{\partial x} - g\frac{\partial \zeta}{\partial x} - \frac{g}{\rho_o}\frac{\partial}{\partial x}\int_z^\zeta \rho'dz + \frac{\partial}{\partial z}N_z\frac{\partial u}{\partial z} + \frac{N_h}{b}\frac{\partial}{\partial x}(b\frac{\partial u}{\partial x})$$
$$(4.2)$$

and equation of salt transport,

$$\frac{\partial S}{\partial t} + \frac{u}{b}\frac{\partial}{\partial x}Sb + w\frac{\partial S}{\partial z} = \frac{\partial}{\partial z}D_z\frac{\partial S}{\partial z} + \frac{D_h}{b}\frac{\partial}{\partial x}(b\frac{\partial S}{\partial x}) \qquad (4.3)$$

Integration of eq.(4.1) over the depth yields an additional equation for the unknown sea level,

$$\frac{\partial \zeta}{\partial t} + \frac{1}{b}\frac{\partial}{\partial x}(b\int_{-H}^\zeta udz) = 0 \qquad (4.4)$$

The water density (ρ) is assumed to be a linear function of salinity (Blumberg, 1978);

$$\rho = \rho_o(1 + \alpha S) \qquad (4.5)$$

where, $\alpha = 7.5 \cdot 10^{-4}, \rho_o = 0.99891 \text{gcm}^{-3}$. The governing equation defined above with a suitable boundary condition should provide a unique solution to the problem. We do not have a general theorem which can specify the necessary boundary condition. Additionally, as it is often the case with the numerical solutions some of the boundary conditions can be easily specified from the measurements, the other come from the simplified equation of motion or continuity. To illustrate some of these problems let us consider a channel open at one end. We would like to calculate a tide distribution starting from the known boundary condition. It is easy to measure sea level at the mouth of the channel, but it is more difficult to measure the amplitude and phase of the velocity distribution from surface to bottom at the open end. Let us assume that both the measurements are available and we carry out two experiments: in the first experiment the sea level is given at the open boundary and in the second one the velocity is specified. Even if the question about uniqueness is discarded, one still would like to know whether the same tides were generated in both experiments.

With this word of caution, we proceed to define the set of boundary condition for the motion in the fjord driven by tides and salt exchange.

At the mouth of fjord ($x = 0$) the following conditions are specified:
Sea level variation is known,

$$\zeta = \zeta_o(x = 0, t) \qquad (4.6)$$

Salinity is prescribed as a function of time and depth,

$$S = S(x = 0, z, t) \tag{4.7a}$$

or salinity is calculated from the advective term in equation (4.3),

$$\frac{\partial S}{\partial t} + u \frac{\partial S}{\partial x} = 0 \tag{4.7b}$$

Velocity is given as a function of time and depth,

$$u = u(x = 0, z, t) \tag{4.8a}$$

or velocity is computed from the simplified equation of motion,

$$\frac{\partial u}{\partial t} = -g \frac{\partial \zeta}{\partial x} - \alpha g \frac{\partial}{\partial x} \int_z^\zeta S dz + \frac{\partial}{\partial z}(N_z \frac{\partial u}{\partial z}) \tag{4.8b}$$

At the head of the fjord ($x = L$) the conditions are as follows:
Velocity due to the river inflow in the surface layer is given,

$$u = u(x = L, z = 0, t) \tag{4.9}$$

Salinity is specified in close proximity to the river,

$$S = S(x = L, z = 0, t) \tag{4.10}$$

At the water surface the salt flux vanishes along the normal to the surface,

$$D_z \frac{\partial S}{\partial z} = 0, \quad at \quad z = 0 \tag{4.11}$$

and exchange of momentum at the surface is due to the wind stress (τ_w),

$$\rho N_z \frac{\partial u}{\partial z} = \tau_w \tag{4.12}$$

Salt flux along the normal to the bottom contour should vanish. In the numerical approach, the bottom is approximated by straight-line segments parallel to the horizontal and vertical axes. Therefore, along a horizontal section,

$$D_z \frac{\partial S}{\partial z} = 0 \tag{4.13a}$$

and along a vertical section,

$$D_h \frac{\partial S}{\partial x} = 0 \tag{4.13b}$$

Velocity of the flow also vanishes at the bottom;

$$u = w = 0, \quad \text{at} \quad z = H(x, y) \tag{4.14}$$

Some of the above specified boundary conditions, such as (4.7a) and (4.7b), are redundant, and only one will be applied during the computation. If the velocity distribution, at the open boundary, is derived from the measurements, then condition (4.8a) is applied. On the other hand, when salinity and sea level are known at the mouth of the fjord, condition (4.8b) allows to compute the velocity.

The advective condition (4.7b) is applied to calculate the vertical salinity distribution at the open boundary. We shall assume that the vertical profile of salinity just outside the boundary is characterized by a prescribed oceanic distribution $S_o(z)$. Denoting the vertical profile of salinity at one grid-distance toward the head from the open boundary by $S_i(z)$, salinity distribution at the boundary (S_b) will be computed from a discretized form of (4.7b):

$$\frac{S_b^{t+\Delta t} - S_b^t}{\Delta t} = -(u + |u|)\frac{S_b^t - S_o^t}{\Delta x} - (u - |u|)\frac{S_i^t - S_b^t}{\Delta x} \tag{4.15}$$

where u is velocity of the flow at the boundary. According to (4.15), at a given depth, if the flow is directed into the fjord ($u > 0$), salinity is defined by the oceanic salinity; if an outflowing current ($u < 0$) advects salt from the fjord, the boundary value is defined by S_i.

2.3 Construction of the numerical algorithm

Before going into numerical formulation we pose to see what are the limits on the time step. Assuming that the maximum depth of the fjord is 400 m, and the horizontal space step $h = 1$ km, the CFL condition defines time step as

$$T \le \frac{h}{\sqrt{2gH}} = \frac{10^5}{\sqrt{2 \cdot 981 \cdot 4 \cdot 10^4}} = 11.3s \tag{4.16}$$

The limitation on the time step due to vertical friction are caused by the smallest vertical space step (h_z) and the largest value of vertical eddy viscosity (N_z),

$$T \le \frac{h_z^2}{2N_z} = \frac{(5 \cdot 10^2)^2}{2 \cdot 10^2} = 12.5 \cdot 10^2 s \tag{4.17}$$

In the above $h_z = 5$ m and $N_z = 10^2$ cm^2/s. Only when the vertical resolution is about 50 cm the restriction on the time step is similar to the CFL condition.

The system of equations (4.1)–(4.5) along with the boundary conditions will be solved by the finite difference approach. The space-staggered grid is used (Fig.4.1), with u, w and S set at different points. At the free surface, vertical velocity and sea level are computed at the common grid points. A variable density structure, with especially large variations in the proximity of the free surface, requires that a variable grid net be used to resolve the vertical distribution of the current and density.

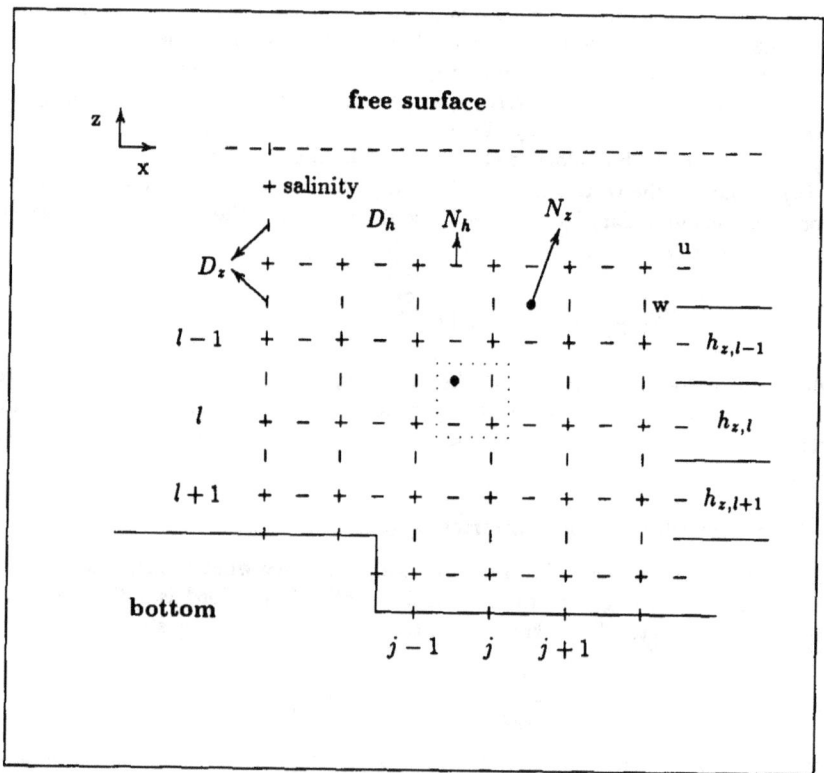

Fig. 4.1
Numerical grid distribution in the x–z plane.
Dotted rectangle contains variables with the same j, l indices.

The vertical step is denoted as h_z. The distance from the free surface is $z = l \cdot h_z$, where $l = 1, 2, \ldots, L$, and h_z is a function of l. The index enumeration proceeds from the free surface towards the bottom, but when constructing numerical formulas along the vertical direction we have to remember that z is directed upward. The horizontal grid step is h and distance is $x = h \cdot j$, where $j = 1, 2, \ldots, J$, and h is constant.

The grid is staggered, with the horizontal and vertical component of velocity and salinity being associated with different space locations. We can introduce indexing which will enumerate each variable separately but to save computer memory we have chosen a three-point arrangement shown in Fig.1. In this arrangement every triple considered in the computation contains salinity, one vertical, and one horizontal velocity point. Location of each triple is described by j and l indices. The time stepping will be done as usual with $t = m \cdot T$.

At this point we would like to start from a very simple numerical approach based on an explicit scheme so that only two time levels will be involved. First the equation of motion is considered, and term by term it will be changed from the differential form to the finite difference form.

The change of acceleration term into the difference form is straightforward,

$$\frac{\partial u}{\partial t} \simeq \frac{u_{j,l}^{m+1} - u_{j,l}^m}{T} \tag{4.18}$$

To the advective term the expressions derived in Ch II,4, can be applied, but we have to decide in which locations the width (b) is defined. In the ensuing computation we shall assume that b_j is defined in the sea level points (or vertical velocity points). Thus,

$$\frac{u}{b}\frac{\partial ub}{\partial x} \simeq \frac{u - |u|}{2bO}\frac{bFu_{j+1,l}^m - bOu_j^m}{h} + \frac{u + |u|}{2bO}\frac{bOu_{j,l}^m - bBu_{j-1,l}^m}{h} = UAD \tag{4.19}$$

where $u = u_{j,l}^m$; $bO = 0.5(b_j + b_{j-1})$; $bF = 0.5(b_j + b_{j+1})$; $bB = 0.5(b_{j-1} + b_{j-2})$. In a similar way the convective term is approximated:

$$w\frac{\partial u}{\partial z} \simeq \frac{wau - |wau|}{2}\frac{u_{j,l-1}^m - u_{j,l}^m}{hzn} + \frac{wau + |wau|}{2}\frac{u_{j,l}^m - u_{j,l+1}^m}{hzp} = UCON \tag{4.20}$$

Here,

$$hzp = 0.5(h_{z,l} + h_{z,l+1}); hzn = 0.5(h_{z,l} + h_{z,l-1})$$

and

$$wau = 0.25(w_{j,l}^m + w_{j-1,l}^m + w_{j,l+1}^m + w_{j-1,l+1}^m)$$

Nonsymmetry exhibited by (4.19) and (4.20) is due to the different enumeration method along the z axis.

The sea level term is easily approximated as

$$-g\frac{\partial \zeta}{\partial x} \simeq -g\frac{\zeta_j^m - \zeta_{j-1}^m}{h} = USL \tag{4.21}$$

It should be noted that the sea level points are located at the vertical velocity points and the sea level derivative is written for the horizontal velocity point. In the same way the atmospheric pressure is approximated:

$$-\frac{1}{\rho_0}\frac{\partial p_a}{\partial x} \simeq -\frac{1}{\rho_0}\frac{p_{a,j} - p_{a,j-1}}{h} = UAP \tag{4.22}$$

The vertical friction term is written along z axis as

$$\frac{\partial}{\partial z}\left(N_z\frac{\partial u}{\partial z}\right) \simeq \frac{1}{h_{z,l}}\left(N_{z,j,l}\frac{u_{j,l-1}^m - u_{j,l}^m}{hzn} - N_{z,j,l+1}\frac{u_{j,l}^m - u_{j,l+1}^m}{hzp}\right) = UVFR \tag{4.23}$$

The horizontal friction term is approximated as

$$\frac{N_h}{b}\frac{\partial}{\partial x}\left(b\frac{\partial u}{\partial x}\right) \simeq \frac{N_h}{bO}\frac{1}{h}\left(b_j\frac{u_{j+1,l}^m - u_{j,l}^m}{h} - b_{j-1}\frac{u_{j,l}^m - u_{j-1,l}^m}{h}\right) = UHFR \tag{4.24}$$

The horizontal derivative of the internal pressure term requires the vertical distribution of the density. Pressure in the first layer at the horizontal velocity location is

$$p_{j,2} = g\rho_{j,2}\frac{h_{z,2}}{2}$$

and in the second layer,

$$p_{j,3} = p_{j,2} + g\left(\rho_{j,2}\frac{h_{z,2}}{2} + \rho_{j,3}\frac{h_{z,3}}{2}\right)$$

therefore in the layer $l+1$,

$$p_{j,l} = p_{j,l-1} + g\left(\rho_{j,l-1}\frac{h_{z,l-1}}{2} + \rho_{j,l}\frac{h_{z,l}}{2}\right) \tag{4.25}$$

Notice that we start the vertical layer enumeration from $l = 2$, since $l = 1$ is located above the water body. The derivative of pressure is written in the following approximation:

$$-\frac{g}{\rho_0}\frac{\partial}{\partial x}\int_z^\zeta \rho' dz = -\frac{g}{\rho_0}\frac{\partial}{\partial x}\int_z^\zeta \rho dz = -\frac{1}{\rho_0}\frac{p_{j,l} - p_{j-1,l}}{h} = UDST \tag{4.26}$$

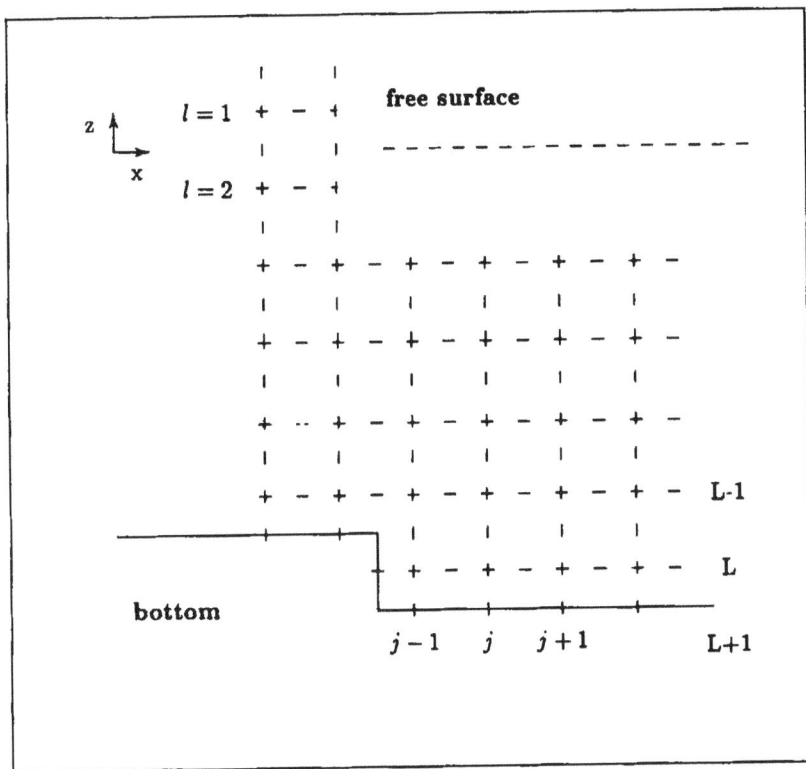

Fig. 4.2
Grid arrangement for the free surface and bottom boundary conditions.

Taking into account the above developed expressions, the general numerical form for the equation of motion can be written as

$$\frac{u_{j,l}^{m+1} - u_{j,l}^{m}}{T} + UAD + UCON = UAP + USL + UDST + UVFR + UHFR \quad (4.27)$$

All the terms in (4.27), except $u_{j,l}^{m+1}$, are known from the previous calculations, therefore (4.27) serves as an explicit algorithm to advance the calculations in time.

The unknown variables, i.e., vertical velocity, sea level and density are calculated from the set (4.1)–(4.5). A close relation exists between the equation of continuity (4.1) and the sea level (4.4). To show this let us integrate (4.1) from the bottom to the depth z,

$$w_z - w_{-H} = -\frac{1}{b}\int_{-H}^{z}\frac{\partial bu}{\partial x}dz \tag{4.28}$$

In (4.28) $w_{-H} = 0$, because the grid net has been chosen in such a way that the vertical velocity is equal to zero at the bottom. The vertical velocity at the surface follows from (4.28),

$$w_\zeta = -\frac{1}{b}\int_{-H}^{\zeta}\frac{\partial bu}{\partial x}dz \tag{4.29}$$

Rewriting (4.4) as

$$\frac{\partial\zeta}{\partial t} + \frac{\partial\zeta}{\partial x}u_\zeta + \frac{1}{b}\int_{-H}^{\zeta}\frac{\partial bu}{\partial x}dz = 0 \tag{4.30}$$

it is easy to see that the vertical velocity at the free surface is the necessary element of the sea level computing formula. For the computation of sea level, eq.(4.30) will be used, but first the distribution of the vertical velocity ought to be known. For this purpose the equation of continuity is vertically integrated from layer $l+1$ to l, the central point being the density point,

$$\int_{l+1}^{l}\frac{\partial bu}{\partial x}dz + b_j(w_{j,l}^m - w_{j,l+1}^m) = 0$$

Approximating the integral in the above formula,

$$h_{z,l}\frac{bFu_{j+1,l}^m - bOu_{j,l}^m}{h} + b_j(w_{j,l}^m - w_{j,l+1}^m) = 0$$

the vertical velocity at the layer l becomes

$$w_{j,l}^m = w_{j,l+1}^m - \frac{h_{z,l}}{b_j}\frac{bFu_{j+1,l}^m - bOu_{j,l}^m}{h} \tag{4.31}$$

Proceeding from the bottom, where the velocity is zero, we come to the free surface, and here,

$$w_{j,2}^m = w_{j,3}^m - \frac{h_{z,2}}{b_j}\frac{bFu_{j+1,2}^m - bOu_{j,2}^m}{h} \tag{4.32}$$

The above expression is introduced into (4.30) to derive the sea level variations,

$$\frac{\zeta_j^{m+1} - \zeta_j^m}{T} + \frac{u+|u|}{2}\frac{\zeta_j^m - \zeta_{j-1}^m}{h} + \frac{u-|u|}{2}\frac{\zeta_{j+1}^m - \zeta_j^m}{h} = w_{j,2}^m \tag{4.33}$$

Here, $u = u_\zeta = 0.5(u_{j,2}^m + u_{j+1,2}^m)$ The change from the differential to the difference form for the equation of salt transport is straightforward, and the above developed formulas can be easily applied to this equation. One has to remember that a staggered grid is used and that the equation has to be written at the salinity point. Approximating term by term, we have:

Change of salinity in time,

$$\frac{\partial S}{\partial t} \simeq \frac{S_{j,l}^{m+1} - S_{j,l}^m}{T} \qquad (4.34)$$

Advective transport,

$$\frac{u}{b}\frac{\partial Sb}{\partial x} \simeq \frac{u + |u|}{2b_j}\frac{b_j S_{j,l}^m - b_{j-1}S_{j-1,l}^m}{h} + \frac{u - |u|}{2b_j}\frac{b_{j+1}S_{j+1,l}^m - b_j S_j^m}{h} = SADV \quad (4.35)$$

Here, $u = 0.5(u_{j,l}^m + u_{j+1,l}^m)$

Convective transport,

$$w\frac{\partial S}{\partial z} \simeq \frac{w + |w|}{2}\frac{S_{j,l}^m - S_{j,l+1}^m}{hzp} + \frac{w - |w|}{2}\frac{S_{j,l-1}^m - S_{j,l}^m}{hzn} = SCON \qquad (4.36)$$

Here, $w = 0.5(w_{j,l}^m + w_{j,l+1}^m)$

Vertical diffusion term,

$$\frac{\partial}{\partial z}D_z\frac{\partial S}{\partial z} \simeq \frac{1}{h_{z,l}}\left(D_{z,j,l}\frac{S_{j,l-1}^m - S_{j,l}^m}{hzn} - D_{z,j,l+1}\frac{S_{j,l}^m - S_{j,l+1}^m}{hzp}\right) = SDFV \qquad (4.37)$$

Horizontal diffusion,

$$\frac{D_h}{b}\frac{\partial}{\partial x}(b\frac{\partial S}{\partial x}) = \frac{D_h}{b_j}\frac{1}{h}\left(bF\frac{S_{j+1,l}^m - S_{j,l}^m}{h} - bO\frac{S_{j,l}^m - S_{j-1,l}^m}{h}\right) = SDFH \qquad (4.38)$$

General computational formula follows from the above partial expressions as

$$\frac{S_{j,l}^{m+1} - S_{j,l}^m}{T} + SADV + SCON = SDIFV + SDIFH \qquad (4.39)$$

Finally, the salinity distribution is used to calculate density by (4.5). Here we may as well discuss a few problems which relate to the density calculations and an integral derived from the density. It is not clear how to compute the density perturbation (ρ') at each time step. In (4.2) only the density derivative is required, and it does not matter whether density or density perturbation is used. Therefore, we have substituted ρ for ρ' in (4.26). Integrating from the bottom to the surface may result in large values of $\int \rho dz$, but the difference of these values is usually a small value. Thus, here is a possible source of the error in computing the density driven current. To diminish this error, instead of density, the smaller quantity such as σ_{Θ} may be used. For example, to estimate the above integral one can use only the variable part of (4.5) – $\rho_o \alpha S$, because $\frac{\partial \rho_o}{\partial x} = 0$.

2.4 Numerical formulas for the boundary conditions

Several types of boundary conditions occur in fjord circulation problems. Conditions at the wall (closed boundary) are quite easy to rewrite in a numerical form. Free surface open boundary conditions are also readily available. The portion of the domain connected to the open ocean requires a special treatment since measurements are not always available.

At the closed boundary the conditions defined by (4.13) and (4.14) require that the diffusive flux of salt and velocity be equal to zero at the wall. The staggered grid employed in the calculation sets the normal velocity to zero at the wall. Equating the tangential velocity to zero can be done in various ways. Assuming (Fig.4.2),

$$u^m_{j,L+1} = 0 \qquad (4.40a)$$

or

$$u^m_{j,L} = -u^m_{j,L+1} \qquad (4.40b)$$

The latter is the velocity at the bottom constructed as an average of the two points situated above and below the bottom. It is of interest to notice that both formulations result in a different value of the bottom stress.

Boundary condition implementation for the salt transport again can be carried out through various approaches. Assuming the horizontal eddy diffusivity to be a function of x and z and rewriting (4.38) in flux form as (4.37), no flux condition is achieved by setting $D_h(x, z) = 0$ outside of the computational domain. The same result is achieved by introducing a variable coefficient which equals to one inside, and to zero, outside the computational domain.

Boundary condition at the free surface (4.11) and (4.12) require an additional grid point located above the free surface –(see Fig.4.2). This is the so-called false point. No-flux condition (4.11) follows by setting $D_{z,j,2} = 0$ (see Fig.4.1 and Fig.4.2).

Condition (4.12) becomes,

$$0.5(\rho_{j,2} + \rho_{j-1,2})N_{z,j,2} \frac{u^m_{j,1} - u^m_{j,2}}{0.5(h_{z,1} + h_{z,2})} = \tau_w \qquad (4.41)$$

Actually there is no need to know $u^m_{j,1}$ because condition (4.41) does not work alone, it will be introduced into the equation of motion. To elucidate this point, let us consider only the part of the equation of motion where the above condition will be introduced:

$$\frac{\partial u}{\partial t} = \frac{\partial}{\partial z} N_z \frac{\partial u}{\partial z}$$

By combining (4.18) and (4.23) the numerical form for the above equation is obtained and at the free surface it becomes

$$\frac{u_{j,2}^{m+1} - u_{j,2}^m}{T} = \frac{1}{h_{z,2}} \left(N_{z,j,2} \frac{u_{j,1}^m - u_{j,2}^m}{hzn} - N_{z,j,3} \frac{u_{j,2}^m - u_{j,3}^m}{hzp} \right) \qquad (4.42)$$

The first term in the parenthesis at the right side of (4.42) is exactly (4.41) and it equals to the wind stress; therefore the calculation can be carried out and the variable $u_{j,1}^m$ in (4.41) does not need to be specified.

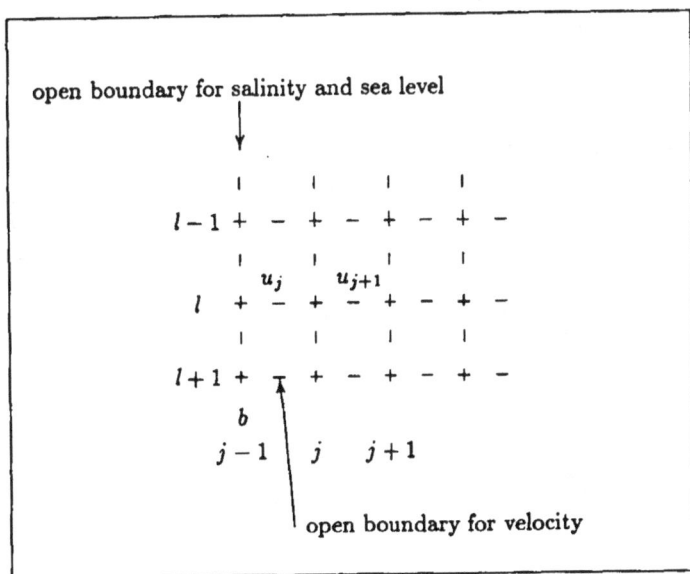

Fig. 4.3
Boundary condition at the mouth of a fjord.

Open boundary conditions (4.7b) and (4.8b) specify the flux of salt and momentum through the vertical boundary - (see Fig.4.3). These are the equation of motion and diffusion tailored to the boundary needs. In the equation of motion the horizontal advection and diffusion are neglected. The geometry taken to construct the boundary condition is depicted in Fig.4.3. The differential–difference form of the momentum boundary condition is

$$\frac{u_{j,l}^{m+1} - u_{j,l}^m}{T} = -g\frac{\zeta_j^m - \zeta_{j-1}^m}{h} - \alpha g\frac{\partial}{\partial x}\int_z^{\zeta} S^m dz + \frac{\partial}{\partial z}(N_z\frac{\partial u^m}{\partial z}) \qquad (4.43)$$

To obtain a vertical distribution of the horizontal velocity one has to specify the salinity and sea level distribution. If the sea level is unknown a few approaches can be considered, such as: a) definition of the sea level from the radiating boundary condition, and, b) assuming zero sea level slope.

The advective condition for the salinity transport (4.7b) or (4.15), using the geometry from Fig.4.3, becomes

$$\frac{S_{j,l}^{m+1} - S_{j,l}^m}{T} + \frac{u - |u|}{2}\frac{S_{j+1,l}^m - S_{j,l}^m}{h} + \frac{u + |u|}{2}\frac{S_{j,l}^m - S_{b,l}^m}{h} = 0 \qquad (4.44)$$

Here, $u = 0.5(u_{j,l}^m + u_{j+1,l}^m)$.

2.5 Methods for improving accuracy

The above constructed algorithm was based on two time-level explicit formulas. It has a first order of approximation in time and generally a second order of approximation in space (excluding the advective terms). This is a relatively simple model and therefore all the elements can be easily checked and rechecked. It may be very useful as a basic model to compare against the complicated algorithms. Below we delineate some more complicated approaches leading to improved accuracy. To derive the second-order of approximation in time a leapfrog time integration is introduced. The simpler and shorter notation is achieved through the difference–differential form of the equation. Applying a three time-level approach to the equation of motion (4.2), it becomes

$$\frac{u_{j,l}^{m+1} - u_{j,l}^{m-1}}{2T} + UADLF + UCONLF$$

$$= -\frac{1}{\rho_o}\frac{\partial p_a^m}{\partial x} - g\frac{\partial \zeta^m}{\partial x} - g\alpha\int_z^{\zeta} S^m dz + \frac{\partial}{\partial z}N_z\frac{\partial u}{\partial z}^{m-1} + \frac{N_h}{b}\frac{\partial}{\partial x}(b\frac{\partial u}{\partial x}^{m-1}) \qquad (4.45)$$

In (4.45) the pressure terms are taken at the middle time point but due to stability requirements the frictional terms have to be computed at $m - 1$ time step. Finite-differencing in the space remain the same as in the two time-level approach. The advective terms will assume a new form which is shortly described in the ensuing considerations. First, the advective terms

$$\frac{u}{b}\frac{\partial}{\partial x}ub + w\frac{\partial u}{\partial z}$$

with the help of the equation of continuity, are transformed to

$$\frac{1}{b^2}\frac{\partial}{\partial x}(ub)^2 + \frac{1}{b^2}\frac{\partial}{\partial z}(wbub) = \frac{1}{b^2}\frac{\partial}{\partial x}(ub)^2 + \frac{\partial}{\partial z}(uw) \qquad (4.46)$$

This equation will serve to construct a finite-difference form in space at the time level m. In Fig.1 at the horizontal velocity point, term

$$\frac{1}{b^2}\frac{\partial}{\partial x}(ub)^2$$

becomes

$$\frac{1}{(bO)^2}\frac{1}{h}[(\frac{u_{j+1,l}^m + u_{j,l}^m}{2}b_j)^2 - (\frac{u_{j,l}^m + u_{j-1,l}^m}{2}b_{j-1})^2] = UADLF \qquad (4.47)$$

Term

$$\frac{\partial}{\partial z}(wu)$$

becomes

$$\frac{1}{h_{z,l}}[uan\frac{w_{j,l}^m + w_{j-1,l}^m}{2} - uap\frac{w_{j,l+1}^m + w_{j-1,l+1}^m}{2}] = UCONLF \qquad (4.48)$$

Here,

$$uap = \frac{u_{j,l}^m h_{z,l+1} + u_{j,l+1}^m h_{z,l}}{h_{z,l} + h_{z,l+1}} \qquad (4.49a)$$

$$uan = \frac{u_{j,l-1}^m h_{z,l} + u_{j,l}^m h_{z,l-1}}{h_{z,l} + h_{z,l-1}} \qquad (4.49b)$$

To derive the vertical velocity distribution the equation of continuity (4.1) can be solved by (4.32).

The sea level variations (4.4), previously expressed by the two time-level formula (4.33), now become

$$\frac{\zeta_j^{m+1} - \zeta_j^{m-1}}{2T} + \frac{1}{h}[u_{j+1,2}^m\frac{\zeta_{j+1}^m + \zeta_j^m}{2} - u_{j,2}^m\frac{\zeta_j^m + \zeta_{j-1}^m}{2}] = w_{j,2}^m \qquad (4.50)$$

The diffusion equation is integrated in time as

$$\frac{S_{j,l}^{m+1} - S_{j,l}^{m-1}}{2T} + SADVLF + SCONLF = \frac{\partial}{\partial z}D_z\frac{\partial S^{m-1}}{\partial z} + \frac{D_h}{b}\frac{\partial}{\partial x}(b\frac{\partial S^{m-1}}{\partial x}) \qquad (4.51)$$

Again, first we apply the equation of continuity to transform the transport terms. The advective term

$$\frac{1}{b^2}\frac{\partial}{\partial x}(Sbub)$$

written at the salinity point in Fig.4. 1, becomes

$$\frac{1}{b_j^2}\frac{1}{h}\left[u_{j+1,l}^m\frac{S_{j+1,l}^m + S_{j,l}^m}{2}(\frac{b_j + b_{j+1}}{2})^2 - u_{j,l}^m\frac{S_{j,l}^m + S_{j-1,l}^m}{2}(\frac{b_j + b_{j-1}}{2})^2\right] = SADVLF$$

(4.52)

The convective term

$$\frac{\partial}{\partial z}(wS)$$

becomes

$$\frac{SANw_{j,l}^m - SAPw_{j,l+1}^m}{h_{z,l}} = SCONLF$$

(4.53)

Here

$$SAN = \frac{S_{j,l-1}^m h_{z,l} + S_{j,l}^m h_{z,l-1}}{h_{z,l} + h_{z,l-1}}$$

(4.54a)

$$SAP = \frac{S_{j,l}^m h_{z,l+1} + S_{j,l+1}^m h_{z,l}}{h_{z,l+1} + h_{z,l}}$$

(4.54b)

Thus we may conclude that the leapfrog method has a second-order approximation in time (frictional and diffusion terms excluded), and the same accuracy in space. Comparison with the two time-level method shows that the leapfrog method will require more computer memory, because of an additional time level. One more problem is due to the false (numerical) solutions which tend to be generated by the three time-level finite-difference schemes (see Ch.II. Sec.7). Several remedies can be prescribed. The simplest one is to use the leapfrog and every 10–30 time steps alternate it with the two time-level method. The numerical solution can be also deleted through a filtering technique; this is possible due to different behavior of the physical and numerical solutions in time (see Ch.III).

The name of the method is related to the way this three time-level formula is applied. For the sake of simplicity, let the calculation in time be carried out by a simple formula

$$u^{m+1} = u^{m-1} + 2TLu^{m,m-1}$$

(4.55)

Here L is a space operator.

From (4.55) it is clear that we are stepping with double time step, from $m - 1$ time point to $m + 1$ time point.

After the first step of computation is finished a simple change of variables from a new one to an old one is done. Which can go like,

$$u^{m-1} = u^m, \quad u^m = u^{m+1}$$

Here the new variables are on the left-hand side. The second time step of numerical calculation starts from the m point and not $m + 1$. Thus we compute a new value

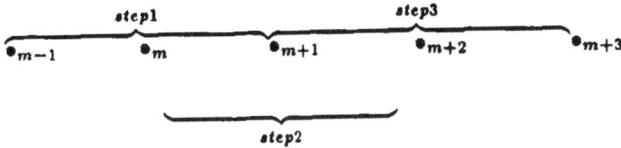

Fig. 4.4
Time stepping in the leapfrog method.

applying $2T$ time step but effective time step is T. The process resembles a frog leaping (Fig.4.4).

2.6 Improving the accuracy and stability of the frictional–diffusive terms

Because of stability requirements the frictional and diffusive terms are taken at the old time step. A better approximation and stability are achieved through an implicit and semi-implicit schemes. The new problem is an inversion of the implicit operator. We shall apply an implicit approach only in connection with the line inversion methods (see Appendix 1). In the ensuing consideration an approach for the two time-level schemes is delineated; application to the three time-level formulas is straightforward. An example with the vertical friction is considered.

For the vertical friction in equation

$$\frac{u_{j,l}^{m+1} - u_{j,l}^m}{T} = Lu^m + UVFR \tag{4.56}$$

a variety of time–differencing schemes can be obtained by introducing a parameter q,

$$UVFR = q\frac{\partial}{\partial z}N_z\frac{\partial u}{\partial z}^m + (1-q)\frac{\partial}{\partial z}N_z\frac{\partial u}{\partial z}^{m+1} \tag{4.57}$$

Various explicit–implicit schemes can be constructed for $0 \leq q \leq 1$ ranging from the completely explicit scheme (q=1) to the fully implicit (q=0). We shall set $q = 1/2$, which leads to the stable scheme of the second-order approximation in time and space,

$$\frac{u_{j,l}^{m+1} - u_{j,l}^m}{T} = Lu^m + \frac{1}{2}\frac{\partial}{\partial z}N_z\frac{\partial u}{\partial z}^m + \frac{1}{2}\frac{\partial}{\partial z}N_z\frac{\partial u}{\partial z}^{m+1} \tag{4.58}$$

In the practical realization one can proceed with the above formulation, or to make computations simpler a split-up method can be used by introducing an intermediate step $m + 1/2$ as

$$\frac{u_{j,l}^{m+1/2} - u_{j,l}^m}{T} = Lu^m + \frac{1}{2}\frac{\partial}{\partial z}N_z\frac{\partial u}{\partial z}^m \tag{4.59a}$$

and

$$\frac{u_{j,l}^{m+1} - u_{j,l}^{m+1/2}}{T} = \frac{1}{2}\frac{\partial}{\partial z}N_z\frac{\partial u}{\partial z}^{m+1} \tag{4.59b}$$

The sum of these equations is equal to (4.58). Anyway we proceed; the end product is an implicit equation:

$$u_{j,l}^{m+1} - \frac{T}{2}\frac{\partial}{\partial z}N_z\frac{\partial u}{\partial z}^{m+1} = H^m \tag{4.60}$$

Here in H^m all the terms known from the previous computations have been included. Applying the space differencing constructed for (4.23), the above equation becomes

$$u_{j,l}^{m+1} - \frac{T}{2}\frac{1}{h_{z,l}}(N_{z,j,l}\frac{u_{j,l-1}^{m+1} - u_{j,l}^{m+1}}{hzn} - N_{z,j,l+1}\frac{u_{j,l}^{m+1} - u_{j,l+1}^{m+1}}{hzp}) = H^m \tag{4.61}$$

We transform the above implicit equation into a three-point formula;

$$-a_l u_{j,l-1}^{m+1} + b_l u_{j,l}^{m+1} - c_l u_{j,l+1}^{m+1} = d_l \tag{4.62}$$

Coefficients in (4.62) are

$$a_l = \frac{TN_{z,j,l}}{2h_{z,l}hzn} \qquad c_l = \frac{TN_{z,j,l+1}}{2h_{z,l}hzp}$$

$$b_l = 1 + a_l + c_l \qquad d_l = H^m \tag{4.63}$$

Conditions required to apply the line inversion method, i.e., $a_l > 0, b_l > 0, c_l > 0$ and $a_l + c_l \le b_l$, are fulfilled, and we may proceed with further computations. The method (Appendix 1) is based on the btransformation of the three-point space formula (4.62) into a two-point formula

$$u_{j,l}^{m+1} = s_l u_{j,l+1}^{m+1} + e_l \tag{4.64}$$

New coefficients s_l and e_l are defined through the recursion formulas by the old coefficients a_l, b_l, c_l and d_l. Since the two point expression (4.64) requires only one boundary condition to find the unknown function inside the domain, the second boundary is used to begin recursion which defines the coefficients.

    Let us assume that the free surface boundary condition is given by the wind stress (Fig.4.2),

$$N_{z,j,2}\frac{u_{j,1}^{m+1} - u_{j,2}^{m+1}}{0.5(h_{z,1} + h_{z,2})} = \tau_w/\rho \tag{4.65}$$

This condition serves to find e_2 and s_2. (We start the calculation from $l = 2$). Rewriting (4.61) at $l = 2$

$$u_{j,2}^{m+1} - \frac{T}{2h_{z,2}}(N_{z,j,2}\frac{u_{j,1}^{m+1} - u_{j,2}^{m+1}}{hzn} - N_{z,j,3}\frac{u_{j,2}^{m+1} - u_{j,3}^{m+1}}{hzp}) = H^m \qquad (4.66)$$

and introducing (4.65)

$$u_{j,2}^{m+1} - \frac{T}{2\rho}\frac{\tau_w}{h_{z,2}} + c_2(u_{j,2}^{m+1} - u_{j,3}^{m+1}) = H^m \qquad (4.67)$$

From above

$$u_{j,2}^{m+1} = u_{j,3}^{m+1}c_2/(1 + c_2) + (H^m + \frac{T\tau_w}{2\rho h_{z,2}})\frac{1}{1 + c_2} \qquad (4.68)$$

Comparing (4.68) and (4.64) we obtain

$$s_2 = c_2/(1 + c_2) \qquad e_2 = (H^m + \frac{T\tau_w}{2\rho h_{z,2}})\frac{1}{1 + c_2} \qquad (4.69)$$

to start the recursive process for the coefficients. As soon as these coefficients are defined we apply (4.64) to find the velocity distribution starting from the bottom. Let us assume that the velocity vanishes at the bottom (4.40b),

$$u_{j,L}^{m+1} + u_{j,L+1}^{m+1} = 0 \qquad (4.70)$$

(Here index L is for the first point above the bottom) and from (4.64)

$$u_{j,L}^{m+1} = s_L u_{j,L+1}^{m+1} + e_L = -s_L u_{j,L}^{m+1} + e_L$$

thus,

$$u_{j,L}^{m+1} = e_L/(1 + s_L) \qquad (4.71)$$

With the boundary value given, the process will proceed into the domain through the formula (4.64).

2.7 Physical processes versus numerical processes

The transport equation describes vertical mixing in a stable stratified water column. Unstable stratification is resolved by mechanical mixing. It is achieved by averaging salinity or temperature over the depth in such way that stability is restored. The average is defined as

$$S = \frac{\sum_l S_{j,l} h_{z,l}}{\sum_l h_{z,l}} \qquad (4.72)$$

Here summation is over the number of layers with unstable stratification. The mechanical mixing can be achieved with the help of the subroutine below, written in FORTRAN.

```
          ms=0
C Index ms counts number of unstable layers
          do 42 j=2,jx
          do 42 l=2,le(j)
C Index j serves to integrate along x direction
C Index l serves to integrate along vertical direction
C le(j) denotes number of layers in the j column
          if(ms.eq.0.0)then
          p7=s(j,l)*hz(l)
          p8=hz(l)
C Notice that the average salinity in l layer is p7/p8
          end if
          st=s(j,l)-s(j,l+1)
C st tests stability between layer l+1 and l
          if(st.gt.0.0)then
C st greater than 0, denotes instability
          ms=ms+1
C start to count the number of unstable layers by ms
          p7=p7+s(j,l+1)*hz(l+1)
          p8=p8+hz(l)
          s(j,l+1)=p7/p8
C This salinity in layer l+1, and also average salinity. It will
C be ascribed to the layers above l+1, which were previously unstable.
          do 43 n=1,ms
          k=l+1-n
C This loop ascribes salinity from layer l+1 to above (previously unstable)
C layers.
43        s(j,k)=s(j,k+1)
          if(l.eq.le(j))then
C When the bottom is reached, start again computation in the
C next column
          ms=0
          p7=0
          p8=0
          go to 42
          end if
          go to 42
C Unstable layer was found go to 42 with ms not equal to zero.
```

```
          end if
C Unstable layer was not found, continue same column with ms=0
          ms=0
          p7=0
          p8=0
42        continue
```

The mixing routine starts to work when instability occurs and generates stable stratification. How it interacts with the diffusion equation and whether the vertical processes are changed by the mixing is an important question.

Consider vertical convection (upwelling) of salinity with a sharp gradient — an initial profile depicted in Fig.4.5. The time changes are calculated by the first-order directional derivative method and by the second-order three time-level method – (see Fig.4.5).

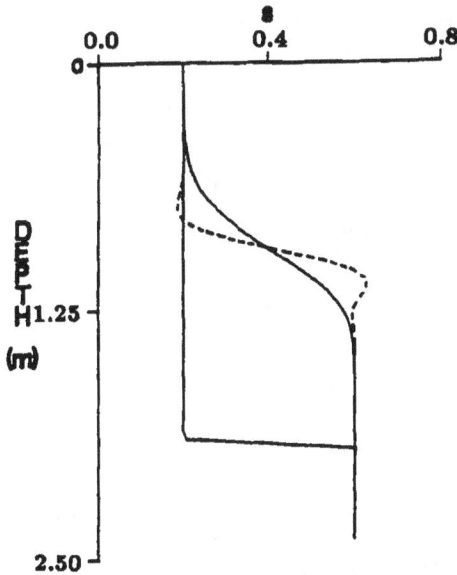

Fig. 4.5
Salinity front movement calculated by the first-order scheme (continuous line) and second-order scheme (broken line).

The distribution calculated by the directional derivative display quite a smooth gradient since the method has excessive numerical diffusion. The vertical profile possesses a positive definite property because this method does not generate any dispersive waves. The second-order method reproduces well the discontinuity but it also generates oscillations leading to local instabilities. These instabilities create a potential danger that the physics of the vertical mixing can be distorted by mechanical mixing. The mixing routine applied to the dispersive vertical distribution removes the numerical dispersive effects as well. Thus we actually do have a numerical–mechanical scheme without dispersion! It is difficult to conclude whether this process conserves heat and salinity. Removal of the dispersive effects in the numerical scheme is also possible to accomplish through a diffusion coefficient. Its value can be in excess of the coefficient required for the proper diffusive processes. It is possible to apply the numerical methods with very small dispersive–diffusive properties, such as flux-corrected method, but they require larger computational resources. This situation calls for a careful approach and additional means to check the results of computations. In the wind and density driven model the differences in salinity profile due to various numerical methods can be very small and thus difficult to trace. One possible approach is to introduce an additional equation for the nonbuoyant passive admixture. The source of this suspended matter can be set anywhere in the water body. If, for example, suspended matter is located at the bottom layer and excessive numerical diffusion–dispersion will transport this matter into the surface layer on the time scale which is incompatible with the vertical velocity or diffusion coefficient, we have definitely a case of a poor numerical approximation.

3. Three-dimensional motion in the shallow seas

Three-dimensional models for shallow seas have been originated by many modelers. A set of models were fully developed and tested in a series of works by Leendertse and Liu. Starting from Leendertse's (1967) two-dimensional model for the long wave propagation, he and subsequently Liu built, step by step, a 3D model (Leendertse *et al.*, 1973; Leendertse and Liu, 1975) and afterwards realistic turbulent exchange processes were included (Liu and Leendertse, 1978). The model was tested and applied to various phenomena including tides, wind and density driven motion and transport of pollutants.

We shall continue to construct finite-difference schemes for relatively shallow water bodies. A straightforward approach based on two and three time-level formulation will be presented. Basic difference between 2D (vertical) and 3D motion is due to Coriolis force. Before starting on the general problem we shall develop a numerical scheme for the Coriolis term.

3.1 Numerical scheme for the Coriolis force term

Consider stability of the finite-difference, two time-level numerical scheme

$$\frac{u^{m+1} - u^m}{T} - fv^m = 0 \qquad (4.73a)$$

$$\frac{v^{m+1} - v^m}{T} + fu^m = 0 \qquad (4.73b)$$

In the ensuing considerations for the sake of simplicity the space indices are omitted. In order to determine the stability of (4.73) we shall assume $u, v \sim (u^*, v^*)\lambda^m$, and from (4.73) the determinant to define λ is deduced:

$$\det \begin{vmatrix} \lambda - 1 & -fT \\ fT & \lambda - 1 \end{vmatrix} = 0 \qquad (4.74)$$

Thus equation

$$(\lambda - 1)^2 + (fT)^2 = 0$$

defines

$$|\lambda_{1,2}| = \sqrt{1 + fT} \qquad (4.75)$$

which is always greater than 1, and we can conclude that (4.73) is unstable.

One simple change from u explicit to implicit in the second equation of (4.73) will stabilize the time integration.

$$\frac{u^{m+1} - u^m}{T} - fv^m = 0 \qquad (4.76a)$$

$$\frac{v^{m+1} - v^m}{T} + fu^{m+1} = 0 \qquad (4.76b)$$

Although the second equation is implicit in time, it does not require any new effort because the new value of u was just computed. The determinant provides again a second-order equation for λ,

$$(\lambda - 1)^2 + (fT)^2\lambda = 0 \qquad (4.77)$$

The two roots of (4.77)

$$\lambda_{1,2} = \frac{2 - (fT)^2 \pm fT\sqrt{(fT)^2 - 4}}{2}$$

lead to the stability coefficient $|\lambda_{1,2}| = 1$, subject to an inequality

$$fT < 2 \qquad (4.78)$$

The time step of the numerical integration chosen according to (4.78) assures the stability of (4.76).

The above scheme, although stable, is only of first-order approximation in time. Let us consider three time-level numerical schemes which are of second-order approximation.

$$\frac{u^{m+1} - u^{m-1}}{2T} - fv^m = 0 \qquad (4.79a)$$

$$\frac{v^{m+1} - v^{m-1}}{2T} + fu^m = 0 \tag{4.79b}$$

Searching for stability we arrive at the determinant

$$\det \begin{vmatrix} \lambda^2 - 1 & -f2T\lambda \\ f2T\lambda & \lambda^2 - 1 \end{vmatrix} = 0 \tag{4.80}$$

which leads to a biquadratic equation. The stability parameter is located on the unit circle $|\lambda_{1,2,3,4}| = 1$, subject to an inequality

$$fT < \sqrt{2} \tag{4.81}$$

This result is similar to (4.78). In many applications the long time stepping is quite important, but such a time step can cause according to (4.81) numerical instability of (4.79). To obtain an unconditional stability, the following scheme of second-order approximation, was proposed by Fischer (1965b),

$$\frac{u^{m+1} - u^{m-1}}{2T} - f\frac{(v^{m+1} + v^{m-1})}{2} = 0 \tag{4.82a}$$

$$\frac{v^{m+1} - v^{m-1}}{2T} + f\frac{(u^{m+1} + u^{m-1})}{2} = 0 \tag{4.82b}$$

Stability analysis leads to the biquadratic equations with all roots on the unit circle $|\lambda| = 1$. These equations are always stable and no condition is imposed on the time step.

As we know, an implicit formulation is sometimes difficult to put into effect with the other terms being implicit as well. For the shallow water model we may like to have the vertical friction implicit as well. To accomplish only one implicit operation in time, we shall use splitting in time with the fractional time step $m+1/2$.

At the first substep the vertical friction is implicit

$$\frac{u^{m+1/2} - u^m}{T} - fv^m = \frac{1}{2}\frac{\partial}{\partial z}N_z\frac{\partial u}{\partial z}^{m+1/2} \tag{4.83a}$$

$$\frac{v^{m+1/2} - v^m}{T} + fu^m = \frac{1}{2}\frac{\partial}{\partial z}N_z\frac{\partial v}{\partial z}^{m+1/2} \tag{4.83b}$$

and at the second substep, the Coriolis term is implicit

$$\frac{u^{m+1} - u^{m+1/2}}{T} - fv^{m+1} = \frac{1}{2}\frac{\partial}{\partial z}N_z\frac{\partial u}{\partial z}^{m+1/2} \tag{4.84a}$$

$$\frac{v^{m+1} - v^{m+1/2}}{T} + fu^{m+1} = \frac{1}{2}\frac{\partial}{\partial z}N_z\frac{\partial v}{\partial z}^{m+1/2} \tag{4.84b}$$

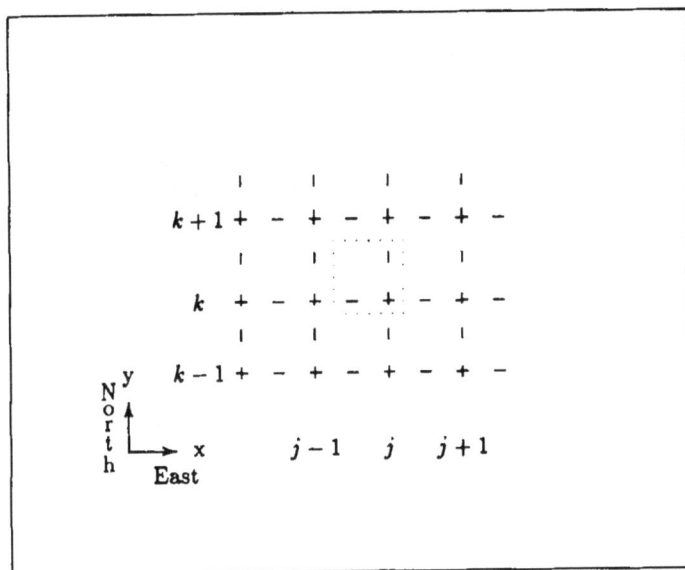

Fig. 4.6
Location of the grid points in the horizontal plane.
Horizontal bar denotes u component, vertical bar denotes v component.
Dotted rectangle contains variables with the same j, k indices.

Now we can tackle a practical implementation of these schemes for the given space grid. The problems, which require further study, are due to the staggered grid. Let us consider a staggered grid on the horizontal plane with u (East) velocity, v (North) velocity, and pressure located at different points as indicated in the Fig.4.6. As before, equation for the East component is considered at the u grid point, it means that the v component which occurs in the Coriolis term has to be taken as an average for the four points located around the u point. A similar approach is taken for the North component of the momentum equation. Thus in the u point

$$\bar{v}^u = 0.25(v_{j,k-1,l} + v_{j,k,l} + v_{j-1,k,l} + v_{j-1,k-1,l}) \tag{4.85}$$

and in the v point

$$\bar{u}^v = 0.25(u_{j+1,k,l} + u_{j+1,k+1,l} + u_{j,k,l} + u_{j,k+1,l}) \tag{4.86}$$

In the above notation an explicit scheme (4.76) becomes

$$\frac{u_{j,k}^{m+1} - u_{j,k}^m}{T} - f\bar{v}^{u,m} = 0 \tag{4.87a}$$

$$\frac{v_{j,k}^{m+1} - v_{j,k}^{m}}{T} + f\bar{u}^{v,m+1} = 0 \tag{4.87b}$$

The three time-level equation (4.79) can be written in a similar way.

For convenience the implicit numerical scheme (4.82), to carry out the computations, ought to be changed into an explicit one (or it can be solved by iteration method on every time step). To elucidate this approach, we first perform the operation in the nonstaggered grid so that the all variables are defined at the same grid point.

From (4.82a)

$$u^{m+1} = u^{m-1} + fT(v^{m+1} + v^{m-1}) \tag{4.88a}$$

and from (4.82b)

$$v^{m+1} = v^{m-1} - fT(u^{m+1} + u^{m-1}) \tag{4.88b}$$

Inserting (4.88b) into the right-hand side of (4.88a), the u component of velocity at the new time step becomes

$$u^{m+1} = \frac{u^{m-1}[1 - (fT)^2] + 2fTv^{m-1}}{1 + (fT)^2} \tag{4.89a}$$

Introducing (4.88a) into (4.88b) we arrive at

$$v^{m+1} = \frac{v^{m-1}[1 - (fT)^2] - 2fTu^{m-1}}{1 + (fT)^2} \tag{4.89b}$$

Both (4.89a) and (4.89b) are functions of the known variables and are quite easy to program. The approach delineated above has to be modified for the staggered grid. To elucidate this point let us write (4.82) on the staggered grid depicted in Fig.4.6.

$$\frac{u^{m+1} - u^{m-1}}{2T} - f\frac{(\bar{v}^{u,m+1} + \bar{v}^{u,m-1})}{2} = 0 \tag{4.90a}$$

$$\frac{v^{m+1} - v^{m-1}}{2T} + f\frac{(\bar{u}^{v,m+1} + \bar{u}^{v,m-1})}{2} = 0 \tag{4.90b}$$

From the above equation we can deduce that the average and point values are non-compatible, or simply speaking we have two equations and four unknown. We shall solve this problem by writing both equations of momentum first at the u point and afterwards at the v point. At the u point

$$\frac{u^{m+1} - u^{m-1}}{2T} - f\frac{(\bar{v}^{u,m+1} + \bar{v}^{u,m-1})}{2} = 0 \tag{4.91a}$$

$$\frac{\bar{v}^{u,m+1} - \bar{v}^{u,m-1}}{2T} + f\frac{(u^{m+1} + u^{m-1})}{2} = 0 \tag{4.91b}$$

At the point v

$$\frac{\bar{u}^{v,m+1} - \bar{u}^{v,m-1}}{2T} - f\frac{(v^{m+1} + v^{m-1})}{2} = 0 \qquad (4.92a)$$

$$\frac{v^{m+1} - v^{m-1}}{2T} + f\frac{(\bar{u}^{v,m+1} + \bar{u}^{v,m-1})}{2} = 0 \qquad (4.92b)$$

Equations (4.91) can be solved to find u^{m+1} and equations (4.92) will determine v^{m+1}

$$u^{m+1} = \frac{u^{m-1}[1 - (fT)^2] + 2fT\bar{v}^{u,m-1}}{1 + (fT)^2} \qquad (4.93a)$$

$$v^{m+1} = \frac{v^{m-1}[1 - (fT)^2] - 2fT\bar{u}^{v,m-1}}{1 + (fT)^2} \qquad (4.93b)$$

3.2 Two time–level numerical schemes

Three-dimensional motion in a shallow stratified sea or in the shelf area of the oceans will be described with the help of the equation of continuity (1.39), equations of motion (1.62) and (1.63), equation of state (1.40), and equations of transport for the heat (1.43) and salt (1.44). The approach which was delineated for the two-dimensional motion in the channels (Ch.IV.Sec.2.3) will be now applied for the three-dimensions. The formulation of the open and closed boundary conditions is omitted since it is well elucidated in Ch.IV.Sec. 2.2.

To show the arrangement for the time stepping we shall write first the equations of motion

$$\frac{u^{m+1} - u^m}{T} + (u\frac{\partial u}{\partial x})^m + (v\frac{\partial u}{\partial y})^m + (w\frac{\partial u}{\partial z})^m - fv^m$$

$$= -\frac{1}{\rho_o}\frac{\partial p_a}{\partial x}^m - g\frac{\partial \zeta}{\partial x}^m - \frac{g}{\rho_o}(\frac{\partial}{\partial x}\int_z^\zeta \rho' dz)^m + \frac{\partial}{\partial z}N_z\frac{\partial u}{\partial z}^m + N_h\Delta u^m \qquad (4.94)$$

and

$$\frac{v^{m+1} - v^m}{T} + (u\frac{\partial v}{\partial x})^m + (v\frac{\partial v}{\partial y})^m + (w\frac{\partial v}{\partial z})^m + fu^{m+1}$$

$$= -\frac{1}{\rho_o}\frac{\partial p_a}{\partial y}^m - g\frac{\partial \zeta}{\partial y}^m - \frac{g}{\rho_o}(\frac{\partial}{\partial y}\int_z^\zeta \rho' dz)^m + \frac{\partial}{\partial z}N_z\frac{\partial v}{\partial z}^m + N_h\Delta v^m \qquad (4.95)$$

Construction of the finite-difference equations will be carried out with the help of Fig.4.7, in which three-dimensional grid distribution is shown. Grid distribution in the horizontal plane is given in Fig.4.6 and in the vertical plane is depicted in Fig.4.1. A large portion of the algorithm has already been constructed in Ch.IV.Sec.2.3; these formulas need to be rewritten for three dimensions. Consider first the equation of motion in the East direction.

The first nonlinear term in (4.94) is

$$(u\frac{\partial u}{\partial x})^m \simeq \frac{u_{j,k,l}^m + |u_{j,k,l}^m|}{2}\frac{u_{j,k,l}^m - u_{j-1,k,l}^m}{h}$$

$$+ \frac{u_{j,k,l}^m - |u_{j,k,l}^m|}{2}\frac{u_{j+1,k,l}^m - u_{j,k,l}^m}{h} = UADU \qquad (4.96)$$

The second nonlinear term is

$$(v\frac{\partial u}{\partial y})^m \simeq \frac{\bar{v}^{u,m} + |\bar{v}^{u,m}|}{2}\frac{u_{j,k,l}^m - v_{j,k-1,l}^m}{h} + \frac{\bar{v}^{u,m} - |\bar{v}^{u,m}|}{2}\frac{u_{j,k+1,l}^m - u_{j,k,l}^m}{h} = UADV$$

$$(4.97)$$

The third nonlinear term is equal to (4.20),

$$w\frac{\partial u}{\partial z} \simeq \frac{wau - |wau|}{2}\frac{u_{j,k,l-1}^m - u_{j,k,l}^m}{hzn} + \frac{wau + |wau|}{2}\frac{u_{j,k,l}^m - u_{j,k,l+1}^m}{hzp} = UCON$$

$$(4.98)$$

Here

$$hzp = 0.5(h_{z,l} + h_{z,l+1}); \qquad hzn = 0.5(h_{z,l} + h_{z,l-1})$$

and

$$wau = 0.25(w_{j,k,l}^m + w_{j-1,k,l}^m + w_{j,k,l+1}^m + w_{j-1,k,l+1}^m)$$

Continuing with the construction we define the Coriolis term by (4.85). In the ensuing consideration we shall repeat the construction from Ch.IV. Sec.2.3.

The sea level term is easily approximated as

$$-g\frac{\partial \zeta}{\partial x} \simeq -g\frac{\zeta_{j,k}^m - \zeta_{j-1,k}^m}{h} = USL \qquad (4.99)$$

In the same way the atmospheric pressure is approximated:

$$-\frac{1}{\rho_o}\frac{\partial p_a}{\partial x} \simeq -\frac{1}{\rho_o}\frac{p_{a,j,k} - p_{a,j-1,k}}{h} = UAP \qquad (4.100)$$

The vertical friction term is written along the z axis as

$$\frac{\partial}{\partial z}(N_z\frac{\partial u}{\partial z}) \simeq \frac{1}{h_{z,l}}(\tilde{N}_{z,j,k,l}\frac{u_{j,k,l-1}^m - u_{j,k,l}^m}{hzn} - \tilde{N}_{z,j,k,l+1}\frac{u_{j,k,l}^m - u_{j,k,l+1}^m}{hzp}) = UVFR$$

$$(4.101)$$

Here $\tilde{N}_{z,j,k} = 0.5(N_{z,j,k} + N_{z,j,k-1})$ is an average along the y direction at the depth l or $l+1$. The horizontal friction term is approximated as

$$N_h\Delta u \simeq \frac{N_h}{h^2}(u_{j+1,k,l}^m + u_{j-1,k,l}^m + u_{j,k+1,l}^m + u_{j,k-1,l}^m - 4u_{j,k,l}^m) = UHFR \qquad (4.102)$$

The horizontal derivative of the internal pressure term requires the vertical distribution of density. Pressure in the first layer at the horizontal velocity location is (Fig.4.1),

$$p_{j,k,2} = g\rho_{j,k,2}\frac{h_{z,2}}{2}$$

and in the second layer,

$$p_{j,k,3} = p_{j,k,2} + g(\rho_{j,k,2}\frac{h_{z,2}}{2} + \rho_{j,k,3}\frac{h_{z,3}}{2})$$

therefore in the layer $l + 1$,

$$p_{j,k,l} = p_{j,k,l-1} + g(\rho_{j,k,l-1}\frac{h_{z,l-1}}{2} + \rho_{j,k,l}\frac{h_{z,l}}{2}) \tag{4.103}$$

The derivative of pressure is written to the following approximation:

$$-\frac{g}{\rho_o}\frac{\partial}{\partial x}\int_z^\zeta \rho' dz = -\frac{g}{\rho_o}\frac{\partial}{\partial x}\int_z^\zeta \rho dz = -\frac{1}{\rho_o}\frac{p_{j,k,l} - p_{j-1,k,l}}{h} = UDST \tag{4.104}$$

Summarizing the above expressions, the equation of motion along the x direction becomes

$$\frac{u_{j,k,l}^{m+1} - u_{j,k,l}^m}{T} + UADU + UADV + UCON - f\bar{v}^{u,m}$$
$$= UAP + USL + UDST + UVFR + UHFR \tag{4.105}$$

We can now tackle eq.(4.95); to facilitate construction we shall use Figs.4.7 and 4.8. The first nonlinear term becomes

$$(u\frac{\partial v}{\partial x})^m \simeq \frac{\bar{u}^{v,m} + |\bar{u}^{v,m}|}{2}\frac{v_{j,k,l}^m - v_{j-1,k,l}^m}{h} + \frac{\bar{u}^{v,m} - |\bar{u}^{v,m}|}{2}\frac{v_{j+1,k,l}^m - v_{j,k,l}^m}{h} = VADU \tag{4.106}$$

Here $\bar{u}^{v,m}$ is calculated by (4.86).
  The second nonlinear term is

$$(v\frac{\partial v}{\partial y})^m \simeq \frac{v_{j,k,l}^m + |v_{j,k,l}^m|}{2}\frac{v_{j,k,l}^m - v_{j,k-1,l}^m}{h} + \frac{v_{j,k,l}^m - |v_{j,k,l}^m|}{2}\frac{v_{j,k+1,l}^m - v_{j,k,l}^m}{h} = VADV \tag{4.107}$$

The third nonlinear term in (4.95) becomes (see Fig.4.8),

$$(w\frac{\partial v}{\partial z})^m \simeq \frac{wav + |wav|}{2}\frac{v_{j,k,l}^m - v_{j,k,l+1}^m}{hzp} + \frac{wav - |wav|}{2}\frac{v_{j,k,l-1}^m - v_{j,k,l}^m}{hzn} = VCON \tag{4.108}$$

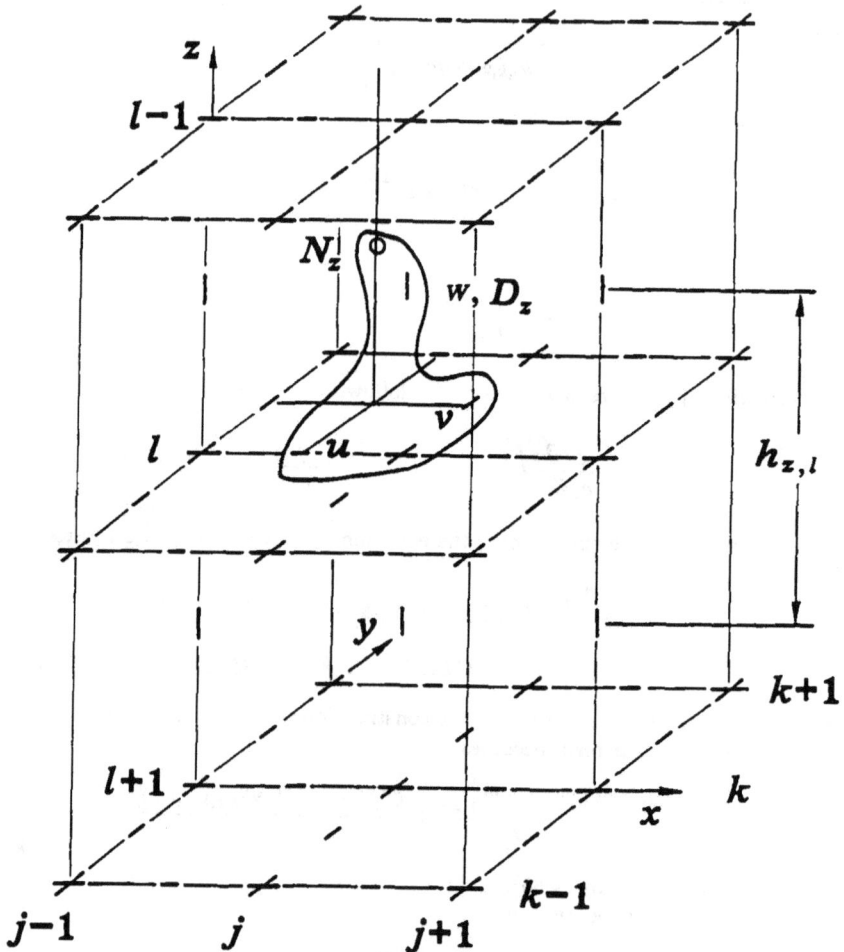

Fig. 4.7
Grid distribution in space.

Here

$$wav = 0.25(w_{j,k,l}^m + w_{j,k+1,l}^m + w_{j,k,l+1}^m + w_{j,k+1,l+1}^m)$$

The sea level change is approximated as

$$-g\frac{\partial \zeta}{\partial y}^m \simeq -g\frac{\zeta_{j,k+1}^m - \zeta_{j,k}^m}{h} = VSL \qquad (4.109)$$

The atmospheric pressure derivative is

$$-\frac{1}{\rho_o}\frac{\partial p_a}{\partial y} \simeq \frac{1}{\rho_o}\frac{p_{a,j,k+1} - p_{a,j,k}}{h} = VAP \qquad (4.110)$$

The finite-difference form for the vertical friction is

$$\frac{\partial}{\partial z}(N_z\frac{\partial v}{\partial z}) \simeq \frac{1}{h_{z,l}}(\hat{N}_{z,j,k,l}\frac{v_{j,k,l-1}^m - v_{j,k,l}^m}{hzn} - \hat{N}_{z,j,k,l+1}\frac{v_{j,k,l}^m - v_{j,k,l+1}^m}{hzp}) = VVFR$$
$$(4.111)$$

Here $\hat{N}_{z,j,k} = 0.5(N_{z,j,k} + N_{z,j+1,k})$ is an average along the x direction taken at the layer l and $l+1$.

Horizontal friction is rendered to the following numerical form:

$$N_h\Delta v \simeq \frac{N_h}{h^2}(v_{j+1,k,l}^m + v_{j-1,k,l}^m + v_{j,k+1,l}^m + v_{j,k-1,l}^m - 4v_{j,k,l}^m) = VHFR \quad (4.112)$$

The internal pressure calculated by expression (4.104) can again serve to construct the derivative, this time along the y direction:

$$-\frac{g}{\rho_o}\frac{\partial}{\partial y}\int_z^\zeta \rho' dz = -\frac{g}{\rho_o}\frac{\partial}{\partial y}\int_z^\zeta \rho dz = -\frac{1}{\rho_o}\frac{p_{j,k+1,l} - p_{j,k,l}}{h} = VDST \qquad (4.113)$$

Summarizing the above results, the finite-difference equation of motion along the y direction becomes

$$\frac{v_{j,k,l}^{m+1} - v_{j,k,l}^m}{T} + VADU + VADV + VCON + f\bar{u}^{v,m+1}$$

$$= VAP + VSL + VDST + VVFR + VHFR \qquad (4.114)$$

The unknown sea level and vertical velocity will be searched through the equation of continuity. Integrating (1.39) from the bottom to the depth z,

$$w_z - w_{-H} = -\int_{-H}^z (\frac{\partial u}{\partial x} + \frac{\partial v}{\partial y})dz \qquad (4.115)$$

the vertical velocity at the free surface is obtained as

$$w_\zeta = -\int_{-H}^\zeta (\frac{\partial u}{\partial x} + \frac{\partial v}{\partial y})dz \qquad (4.116)$$

Thus from (1.64) and (4.116) the relation between sea level and vertical velocity at the free surface follows:

$$\frac{\partial \zeta}{\partial t} + \frac{\partial \zeta}{\partial x} u_\zeta + \frac{\partial \zeta}{\partial y} v_\zeta = w_\zeta \tag{4.117}$$

Therefore, as soon as the vertical and horizontal velocities are defined at the free surface, the sea level distribution can be calculated from (4.117). To derive the vertical velocity at any layer l, the equation of continuity is vertically integrated from $l+1$ to l, with the density point being the central point.

$$\int_{l+1}^{l} \left(\frac{\partial u}{\partial x} + \frac{\partial v}{\partial y} \right) dz + w_{j,k,l}^m - w_{j,k,l+1}^m = 0$$

Approximating the above integral as

$$h_{z,l} \left(\frac{u_{j+1,k,l}^m - u_{j,k,l}^m}{h} + \frac{v_{j,k,l}^m - v_{j,k-1,l}^m}{h} \right) + w_{j,k,l}^m - w_{j,k,l+1}^m = 0$$

the vertical velocity at the l layer becomes

$$w_{j,k,l}^m = w_{j,k,l+1}^m - h_{z,l} \left(\frac{u_{j+1,k,l}^m - u_{j,k,l}^m}{h} + \frac{v_{j,k,l}^m - v_{j,k-1,l}^m}{h} \right) \tag{4.118}$$

This expression will serve to calculate the vertical velocity in any layer. The vertical velocity at the free surface can be introduced into (4.117), and an equation for the sea level follows:

$$\frac{\zeta_{j,k}^{m+1} - \zeta_{j,k}^m}{T} + \frac{u + |u|}{2} \frac{\zeta_{j,k}^m - \zeta_{j-1,k}^m}{h} + \frac{u - |u|}{2} \frac{\zeta_{j+1,k}^m - \zeta_{j,k}^m}{h}$$

$$+ \frac{v + |v|}{2} \frac{\zeta_{j,k}^m - \zeta_{j,k-1}^m}{h} + \frac{v - |v|}{2} \frac{\zeta_{j,k+1}^m - \zeta_{j,k}^m}{h} = w_{j,k,2}^m \tag{4.119}$$

Here, $u = u_\zeta = 0.5(u_{j,k,2}^m + u_{j+1,k,2}^m)$, and $v = v_\zeta = 0.5(v_{j,k,2}^m + v_{j,k-1,2}^m)$. To close the calculations of the velocity a new distribution of the density is required. It is given by the equation of state (1.40). Because density is calculated as a function of salinity and temperature, eqs.(1.43) and (1.44) need to be written in the finite-difference form. For this purpose we use the approach from Ch.IV.Sec.2.3. Construction will be carried out for the salinity equation only; the formulas for the heat equation are identical.

The change of salinity in time is approximated as

$$\frac{\partial S}{\partial t} \simeq \frac{S_{j,k,l}^{m+1} - S_{j,k,l}^m}{T} \tag{4.120}$$

Advective transport along x direction becomes

$$u\frac{\partial S}{\partial x} \simeq \frac{u+|u|}{2}\frac{S_{j,k,l}^m - S_{j-1,k,l}^m}{h} + \frac{u-|u|}{2}\frac{S_{j+1,k,l}^m - S_{j,k,l}^m}{h} = SADVU \qquad (4.121)$$

Here, $u = 0.5(u_{j,k,l}^m + u_{j+1,k,l}^m)$.

Advective transport along y direction is

$$v\frac{\partial S}{\partial y} \simeq \frac{v+|v|}{2}\frac{S_{j,k,l}^m - S_{j,k-1,l}^m}{h} + \frac{v-|v|}{2}\frac{S_{j,k+1,l}^m - S_{j,k,l}^m}{h} = SADVV \qquad (4.122)$$

Here, $v = 0.5(v_{j,k,l}^m + v_{j,k-1,l}^m)$.

Convective transport is rendered as

$$w\frac{\partial S}{\partial z} \simeq \frac{w+|w|}{2}\frac{S_{j,k,l}^m - S_{j,k,l+1}^m}{hzp} + \frac{w-|w|}{2}\frac{S_{j,k,l-1}^m - S_{j,k,l}^m}{hzn} = SCON \qquad (4.123)$$

Here, $w = 0.5(w_{j,k,l}^m + w_{j,k,l+1}^m)$.

Vertical diffusion term becomes

$$\frac{\partial}{\partial z}D_z\frac{\partial S}{\partial z} \simeq \frac{1}{h_{z,l}}\left(D_{z,j,k,l}\frac{S_{j,k,l-1}^m - S_{j,k,l}^m}{hzn} - D_{z,j,k,l+1}\frac{S_{j,k,l}^m - S_{j,k,l+1}^m}{hzp}\right) = SDFV$$
$$(4.124)$$

Horizontal diffusion term is similar to the horizontal friction,

$$D_h\Delta S \simeq \frac{D_h}{h^2}\left(S_{j+1,k,l}^m + S_{j-1,k,l}^m + S_{j,k-1,l}^m + S_{j,k+1,l}^m - 4S_{j,k,l}^m\right) = SDFH \qquad (4.125)$$

General computational formula follows from the above partial expressions as

$$\frac{S_{j,k,l}^{m+1} - S_{j,k,l}^m}{T} + SADVU + SADVV + SCON = SDIFV + SDIFH \qquad (4.126)$$

In summary, the solution proceeds as follows:

Step 1   Set spatial and temporal grid, and use basic stability criteria (CFL).

Step 2   Set initial and boundary conditions.

Step 3   Set $UN = u^{m+1}; UO = u^m, VN = v^{m+1}, VO = V^m$, and solve equations of motion (4.105) and (4.114).

Step 4   Solve equation for the vertical velocity (4.118).

Step 5   Set $ZN = \zeta^{m+1}, ZO = \zeta^m$ and solve equation for the sea level changes (4.119).

Step 6   Set $SN = s^{m+1}, SO = s^m$ and find density distribution from equation of state by solving transport equation (4.126).

Do not use density to calculate integrals (4.104) and (4.113) but rather σ_Θ.

Step 7   Change variables to start new time loop by setting
$$UO = UN, \quad VO = VN, \quad ZO = ZN, \quad SO = SN.$$

3.3 Three time-level finite-difference approach

In constructing a three-dimensional model we shall follow Ch.IV.Sec.2.5. The second-order approximation in space and time will be achieved for all terms but the horizontal and vertical friction/diffusion. Whenever better accuracy or stronger stability is required for these terms the relevant formulas are given in Ch.IV.Sec.2.6. Since the leapfrog method is applied one may expect both physical and numerical solutions. The numerical solutions can be deleted by periodic application of the two time-level algorithm given above. Let us start construction in the usual fashion by writing the difference–differential form for equations of motion

$$\frac{u^{m+1} - u^{m-1}}{T} + (\frac{\partial uu}{\partial x})^m + (\frac{\partial vu}{\partial y})^m + (\frac{\partial wu}{\partial z})^m - f\bar{v}^{u,m}$$

$$= -\frac{1}{\rho_o}\frac{\partial p_a}{\partial x}^m - g\frac{\partial \zeta}{\partial x}^m - \frac{g}{\rho_o}(\frac{\partial}{\partial x}\int_z^\zeta \rho' dz)^m + \frac{\partial}{\partial z}N_z\frac{\partial u}{\partial z}^{m-1} + N_h\Delta u^{m-1} \quad (4.127)$$

and

$$\frac{v^{m+1} - v^{m-1}}{T} + (\frac{\partial uv}{\partial x})^m + (\frac{\partial vv}{\partial y})^m + (\frac{\partial wv}{\partial z})^m + f\bar{u}^{v,m} +$$

$$= -\frac{1}{\rho_o}\frac{\partial p_a}{\partial y}^m - g\frac{\partial \zeta}{\partial y}^m - \frac{g}{\rho_o}(\frac{\partial}{\partial y}\int_z^\zeta \rho' dz)^m + \frac{\partial}{\partial z}N_z\frac{\partial v}{\partial z}^{m-1} + N_h\Delta v^{m-1} \quad (4.128)$$

We shall write here only the finite-difference forms for the nonlinear terms, the remaining terms in the above set will have the same numerical forms as the two time-level equations. Atmospheric pressure is calculated by (4.100) and and (4.110), sea level by (4.99) and (4.109), pressure due to the density stratification by (4.104) and (4.113), vertical friction by (4.101) and (4.111), horizontal friction by (4.102) and (4.112). Using the grid distribution given in Fig.4.1, the term

$$\frac{\partial}{\partial x}(u)^2$$

at the u velocity grid point becomes

$$\frac{1}{h}[(\frac{u_{j+1,k,l}^m + u_{j,k,l}^m}{2})^2 - (\frac{u_{j,k,l}^m + u_{j-1,k,l}^m}{2})^2] = UADLF \quad (4.129)$$

Using the grid distribution from Fig.4.7, the term

$$\frac{\partial}{\partial y}(vu)$$

at the u grid point becomes

$$\frac{1}{h}\left[0.5(v^m_{j,k,l} + v^m_{j-1,k,l})0.5(u^m_{j,k+1,l} + u^m_{j,k,l})\right.$$

$$\left. -0.5(v^m_{j,k-1,l} + v^m_{j-1,k-1,l})0.5(u^m_{j,k,l} + u^m_{j,k-1,l})\right] \tag{4.130}$$

The term

$$\frac{\partial}{\partial z}(wu)$$

becomes (see Fig.4.1)

$$\frac{1}{h_{z,l}}\left[uan\frac{w^m_{j,k,l} + w^m_{j-1,k,l}}{2} - uap\frac{w^m_{j,k,l+1} + w^m_{j-1,k,l+1}}{2}\right] = UCONLF \tag{4.131}$$

Here,

$$uap = \frac{u^m_{j,k,l}h_{z,l+1} + u^m_{j,k,l+1}h_{z,l}}{h_{z,l} + h_{z,l+1}} \tag{4.132}$$

$$uan = \frac{u^m_{j,k,l-1}h_{z,l} + u^m_{j,k,l}h_{z,l-1}}{h_{z,l} + h_{z,l-1}} \tag{4.133}$$

The finite-difference forms for the nonlinear terms in (4.128) are given below. Using the grid distribution from Fig.4.7 the term

$$\frac{\partial}{\partial x}(uv)$$

at the v velocity grid point is

$$\frac{1}{h}\left[0.5(u^m_{j+1,k,l} + u^m_{j+1,k+1,l})0.5(v^m_{j,k,l} + v^m_{j+1,k,l})\right.$$

$$\left. -0.5(u^m_{j,k,l} + u^m_{j,k+1,l}) + 0.5(v^m_{j,k,l} + v^m_{j-1,k,l})\right] \tag{4.134}$$

Using grid distribution depicted in Fig.4.8 or Fig.4.7 the term

$$\frac{\partial}{\partial y}(v^2)$$

becomes

$$\frac{1}{h}\left[\left(\frac{v^m_{j,k,l} + v^m_{j,k+1,l}}{2}\right)^2 - \left(\frac{v^m_{j,k,l} + v^m_{j,k-1,l}}{2}\right)^2\right] \tag{4.135}$$

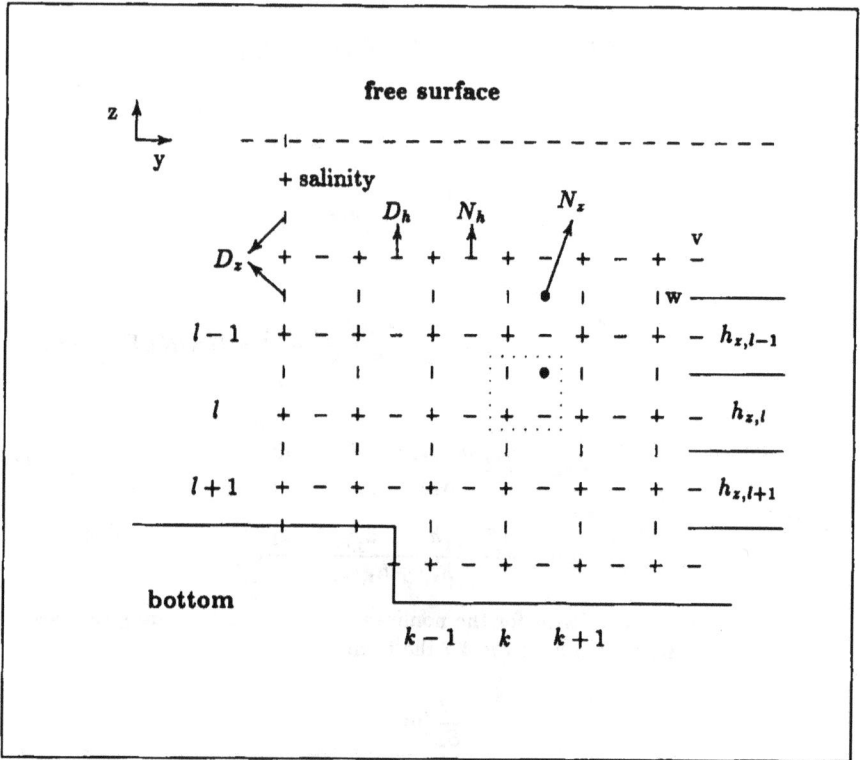

Fig. 4.8
Numerical grid distribution in the y–z plane.
Dotted rectangle contains variables with the same k, l indices.

Fig.4.8 will be of help in constructing the finite-difference form for the term

$$\frac{\partial}{\partial z}(wv)$$

as

$$\frac{1}{h_{z,l}}\left[van\frac{w^m_{j,k,l}+w^m_{j,k+1,l}}{2} - vap\frac{w^m_{j,k,l+1}+w^m_{j,k+1,l+1}}{2}\right] \qquad (4.136)$$

Here,

$$van = \frac{v^m_{j,k,l-1}h_{z,l} + v^m_{j,k,l}h_{z,l-1}}{h_{z,l} + h_{z,l-1}} \tag{4.137}$$

$$vap = \frac{v^m_{j,k,l}h_{z,l+1} + v^m_{j,k,l+1}h_{z,l}}{h_{z,l} + h_{z,l+1}} \tag{4.138}$$

The vertical velocity distribution for the leapfrog method can be derived from eq.(4.118). The sea level change, previously expressed by the two time-level formula (4.119) now is (Fig.4.7),

$$\frac{\zeta^{m+1}_{j,k} - \zeta^{m-1}_{j,k}}{2T} + \frac{1}{h}\left[u^m_{j+1,k,2}\frac{\zeta^m_{j+1,k} + \zeta^m_{j,k}}{2} - u^m_{j,k,2}\frac{\zeta^m_{j,k} + \zeta^m_{j-1,k}}{2}\right]$$

$$+ \frac{1}{h}\left[v^m_{j,k,2}\frac{\zeta^m_{j,k+1} + \zeta^m_{j,k}}{2} - v^m_{j,k-1,2}\frac{\zeta^m_{j,k} + \zeta^m_{j,k-1}}{2}\right] = w^m_{j,k,2} \tag{4.139}$$

The diffusion equation is advanced in time through the following formula:

$$\frac{S^{m+1}_{j,k,l} - S^{m-1}_{j,k,l}}{2T} + SADVULF + SADVVLF + SCONLF$$

$$= \frac{\partial}{\partial z}D_z\frac{\partial S}{\partial z}^{m-1} + D_h\Delta S^{m-1}) \tag{4.140}$$

Again, we consider only the advective–convective terms. The advective term along the x direction

$$\frac{\partial}{\partial x}(Su)$$

written at the salinity point in Fig.4.1 or Fig.4.7 becomes

$$\frac{1}{h}\left[u^m_{j+1,k,l}\frac{S^m_{j+1,k,l} + S^m_{j,k,l}}{2} - u^m_{j,k,l}\frac{S^m_{j,k,l} + S^m_{j-1,k,l}}{2}\right] = SADVULF \tag{4.141}$$

The advective term along y direction

$$\frac{\partial}{\partial y}(Sv)$$

is

$$\frac{1}{h}\left[v^m_{j,k,l}\frac{S^m_{j,k+1,l} + S^m_{j,k,l}}{2} - v^m_{j,k-1,l}\frac{S^m_{j,k,l} + S^m_{j,k-1,l}}{2}\right] \tag{4.142}$$

The convective term

$$\frac{\partial}{\partial z}(wS)$$

becomes

$$\frac{SANw_{j,k,l}^m - SAPw_{j,k,l+1}^m}{h_{z,l}} = SCONLF \qquad (4.143)$$

Here

$$SAN = \frac{S_{j,k,l-1}^m h_{z,l} + S_{j,k,l}^m h_{z,l-1}}{h_{z,l} + h_{z,l-1}} \qquad (4.144a)$$

$$SAP = \frac{S_{j,k,l}^m h_{z,l+1} + S_{j,k,l+1}^m h_{z,l}}{h_{z,l+1} + h_{z,l}} \qquad (4.144b)$$

We can briefly summarize the computational procedure for the leap–frog method in the following steps:

Step 1   Set spatial and temporal grid, and use basic stability criteria (CFL).

Step 2   Set initial and boundary conditions.

Step 3   Set $UN = u^{m+1}; U = u^m; UO = u^{m-1}; VN = v^{m+1}; V = V^m$; $VO = V^{m-1}$, and solve equations of motion (4.127) and (4.128).

Step 4   Solve equation for the vertical velocity (4.118).

Step 5   Set $ZN = \zeta^{m+1}; Z = \zeta^m; ZO = \zeta^{m-1}$ and solve equation for the sea level changes (4.139).

Step 6   Set $SN = S^{m+1}; S = S^m; SO = S^{m-1}$, solve (4.140), and apply the same procedure for the temperature. Find density distribution from equation of state. Do not use density to calculate integrals (4.104) and (4.113) but rather σ_θ.

Step 7   Change variables to start the new time loop by setting $UO = U, U = UN, VO = V, V = VN, ZO = Z, Z = ZN$, $SO = S, S = SN$.

4. Three-dimensional modeling utilizing the mode splitting and sigma coordinate

One of the major problems related to three-dimensional modeling has been too much time consumption in the computer simulations. To study 3D† time dependent dynamical processes and to reduce the large amount of computational work, a splitting algorithm for the fast (external) and slow (internal) modes has been used. For the integration of the equations of ocean dynamics, the method was utilized by Simons (1974, 1980), Madala and Piacsek (1977), Berntsen et al. (1981), Blumberg and Mellor (1983) and Hess (1985). In Berntsen et al. (1981) in addition to splitting into physical modes, a time splitting into fractional time steps was applied as well (Yanenko, 1971; Ramming and Kowalik, 1980). The stability conditions

---

† abbreviation for three-dimensional

of an external or barotropic mode of motion in the numerical computation is governed by the CFL condition, which requires quite a short computational time step (Ramming and Kowalik, 1980). This mode does not need to be described by the full three-dimensional equations, it can be studied by a two-dimensional, vertically-averaged set of equations. The internal mode allows a much longer time step since it is related to the internal wave propagation. The computations are performed in two time steps, a short time step is applied to the barotropic mode described by the vertically integrated equations and much longer time step is applied to the three-dimensional problem. A rectangular system of coordinates is used because the model has been usually applied to mesoscale problems. In further considerations the vertical coordinate (z) is changed to a dimensionless vertical coordinate (σ).

4.1 Rudiments of the mode splitting approach

All the equations were derived in Ch.I, and here we rewrite the required set. These are:

the equation of continuity (1.39)

$$\frac{\partial u}{\partial x} + \frac{\partial v}{\partial y} + \frac{\partial w}{\partial z} = 0 \tag{4.145}$$

the equations of motion along the x and y directions (1.62) and (1.63)

$$\frac{Du}{Dt} - fv = -\frac{1}{\rho_o}\frac{\partial p_a}{\partial x} - g\frac{\partial \zeta}{\partial x} - \frac{g}{\rho_o}\frac{\partial}{\partial x}\int_z^\zeta \rho' dz + \frac{\partial}{\partial z}N_z\frac{\partial u}{\partial z} + N_h\Delta u \tag{4.146}$$

$$\frac{Dv}{Dt} + fu = -\frac{1}{\rho_o}\frac{\partial p_a}{\partial y} - g\frac{\partial \zeta}{\partial y} - \frac{g}{\rho_o}\frac{\partial}{\partial y}\int_z^\zeta \rho' dz + \frac{\partial}{\partial z}N_z\frac{\partial v}{\partial z} + N_h\Delta v \tag{4.147}$$

Equation of state (1.40);

$$\rho = \rho(S,\Theta) \tag{4.148}$$

defines density from salinity and temperature. Temperature and salinity are calculated from equations of conservation (1.43), (1.44);

$$\frac{\partial c}{\partial t} + u\frac{\partial c}{\partial x} + v\frac{\partial c}{\partial y} + w\frac{\partial c}{\partial z} = \frac{\partial}{\partial z}D_z\frac{\partial c}{\partial z} + D_h\Delta c \tag{4.149}$$

Here c stands for either temperature Θ or salinity S. It is also useful at this point to write the pertinent boundary condition because of the splitting procedure. The equation of continuity will be integrated from the bottom to the surface with the bottom velocity equated to zero. Equations of motion (4.146) and (4.147) are subject to the vertical exchange of momentum – at the free surface due to the wind stress and at the bottom due to the bottom stress. Thus at the surface,

$$\tau_x^s = \tau_x^w = \rho N_z\frac{\partial u}{\partial z} \qquad \tau_y^s = \tau_y^w = \rho N_z\frac{\partial v}{\partial z} \tag{4.150}$$

and at the bottom

$$\tau_z^b = \rho N_z \frac{\partial u}{\partial z} \qquad \tau_y^b = \rho N_z \frac{\partial v}{\partial z} \tag{4.151}$$

A few simple but important questions may be asked. Why the bottom condition is defined by stress and not by the velocity? It is important to understand that interaction between the vertically averaged and nonaveraged equation takes place by stresses and not velocities. How does one actually define the bottom stress? There are a few ways to approach this problem. One method defines the bottom stress as a quadratic function of the velocity. This velocity may be taken at the grid location adjacent to the bottom. Another way is to apply the definition given above; along the x direction

$$\tau_z^b \simeq \rho_{j,k,L} 0.5 (N_{z,j,k,L} + N_{z,j,k-1,L}) \frac{u_{j,k,L} - u_{j,k,L+1}}{0.5(h_{z,L} + h_{z,L+1})}$$

$$= \rho_{j,k,L}(N_{z,j,k,L} + N_{z,j,k-1,L}) \frac{u_{j,k,L}}{0.5(h_{z,L} + h_{z,L+1})} \tag{4.152}$$

and a similar expression for the y coordinate

$$\tau_y^b \simeq \rho_{j,k,L}(N_{z,j,k,L} + N_{z,j+1,k,L}) \frac{v_{j,k,L}}{0.5(h_{z,L} + h_{z,L+1})} \tag{4.153}$$

To derive the above result, we have assumed that the tangential velocity at the bottom is equal to zero (see 4.40b). Capital index L is adjacent to the bottom, and $L + 1$ is actually below the bottom. For the grid point location see Figs. 4.2, 4.7 and 4.8.

Continuing with the boundary conditions we shall assume that the horizontal velocity vanishes at the vertical boundary (closed boundary).

Boundary conditions for the salt transport equation were discussed in Ch.IV. Sec.2.2 and 2.4. For the heat exchange processes very similar conditions can be devised, the difference is at the free surface; where the heat flux occurs as

$$F^q = c\rho D_z \frac{\partial \Theta}{\partial z} \simeq c\rho D_{z,j,k,2} \frac{\Theta_{z,j,k,1} - \Theta_{z,j,k,2}}{0.5(h_{z,1} + h_{z,2})} \tag{4.154}$$

Here c is the specific heat of sea water. At this point we assume the heat flux F^q to be known.

The next step is to apply the mode splitting procedure. As we mentioned above the barotropic component of the flow is described by the vertically-averaged equations. We integrate the above equations in the same way as we did in Ch.I. Sec.2.1 (here bar denotes vertical averaging):

$$\frac{\partial \bar{u}}{\partial t} + A_z - f\bar{v} = -\frac{1}{\rho_o}\frac{\partial p_a}{\partial x} - g\frac{\partial \zeta}{\partial x} - B_z + C_z + N_h \Delta \bar{u} \tag{4.155}$$

$$\frac{\partial \bar{v}}{\partial t} + A_y + f\bar{u} = -\frac{1}{\rho_o}\frac{\partial p_a}{\partial y} - g\frac{\partial \zeta}{\partial y} - B_y + C_y + N_h \Delta \bar{v} \qquad (4.156)$$

In the above equations only the higher-order terms arising from the integration of the horizontal friction have been neglected. A few terms are singled out because these are the very same terms which will appear in the 3D equations. The nonlinear terms,

$$A_x = \frac{1}{H}\left[\frac{\partial}{\partial x}\int_{-H}^{\zeta} u^2 dz + \frac{\partial}{\partial y}\int_{-H}^{\zeta} uv\, dz\right] \qquad (4.157a)$$

$$A_y = \frac{1}{H}\left[\frac{\partial}{\partial x}\int_{-H}^{\zeta} uv\, dz + \frac{\partial}{\partial y}\int_{-H}^{\zeta} v^2 dz\right] \qquad (4.157b)$$

Integrals of the density gradients,

$$B_x = \frac{g}{H\rho_o}\int_{-H}^{\zeta}\int_{z}^{\zeta}\frac{\partial \rho'}{\partial x}dz\,dz \qquad (4.158a)$$

$$B_y = \frac{g}{H\rho_o}\int_{-H}^{\zeta}\int_{z}^{\zeta}\frac{\partial \rho'}{\partial y}dz\,dz \qquad (4.158b)$$

Surface and bottom stress,

$$C_x = \tau_x^s/(H\rho_o) - \tau_x^b/(H\rho_o) \qquad (4.159a)$$

$$C_y = \tau_y^s/(H\rho_o) - \tau_y^b/(H\rho_o) \qquad (4.159b)$$

The equation of continuity for the vertically-averaged flow (1.80) will serve to calculate a sea level change

$$\frac{\partial \bar{u}D}{\partial x} + \frac{\partial \bar{v}D}{\partial y} + \frac{\partial \zeta}{\partial t} = 0 \qquad (4.160)$$

Here $D = H + \zeta$ is the total depth.

To derive the internal mode equations we define the velocity components as a sum of the average and variations around this average,

$$u = \bar{u} + u' \quad \text{and} \quad v = \bar{v} + v' \qquad (4.161)$$

Subtracting eq.(4.155) from eq.(4.146) and eq.(4.156) from eq.(4.147) we arrive at

$$\frac{\partial u'}{\partial t} + u\frac{\partial u}{\partial x} + v\frac{\partial u}{\partial y} + w\frac{\partial u}{\partial z} - A_x - fv'$$

$$= B_x - \frac{g}{\rho_o} \frac{\partial}{\partial x} \int_z^\zeta \rho' dz + \frac{\partial}{\partial z} N_z \frac{\partial u'}{\partial z} - C_x + N_h \Delta u' \qquad (4.162)$$

and

$$\frac{\partial v'}{\partial t} + u \frac{\partial v}{\partial x} + v \frac{\partial v}{\partial y} + w \frac{\partial v}{\partial z} - A_y + f u'$$

$$= B_y - \frac{g}{\rho_o} \frac{\partial}{\partial y} \int_z^\zeta \rho' dz + \frac{\partial}{\partial z} N_z \frac{\partial v'}{\partial z} - C_y + N_h \Delta v' \qquad (4.163)$$

By this means we have constructed a set of equations which does not contain explicitly barotropic oscillations since the sea level variations were deleted in the subtraction process. The numerical solution to the equations of motion may therefore be searched in two stages. First the two-dimensional problem defined by eqs.(4.155), (4.156) and (4.160) can be solved by advancing in time with the short time step (T_e) as defined by the CFL condition. The three-dimensional problem will be solved with a much longer time step, let say $10T_e$, and afterwards the velocity distribution follows from (4.161).

The temperature and salinity fields given by eq.(4.149) do not require splitting and they can be calculated by an explicit method with quite a long time step. This is actually an unexpected bonus for every explicit numerical model. The fields of salinity and temperature change in time rather slowly, therefore there is no need to calculate these variables at each time step, we can do calculations with a much longer time step. This conclusion is applied, as well, to the shallow water models discussed previously.

Before proceeding to write a numerical form of the above set of equations let us scrutinize the terms A, B and C. Some of them, like B_x and B_y, change slowly in time and do not undergo modifications when average velocities are calculated. Other terms, like A_x and A_y, alter value both at the short and the long time steps.

The nonlinear terms can be rewritten as

$$A_x = \frac{1}{H} \frac{\partial}{\partial x} \left[(\bar{u}^2 H) + \int_{-H}^\zeta u'^2 dz \right] + \frac{1}{H} \frac{\partial}{\partial y} \left[(\bar{u}\bar{v}H) + \int_{-H}^\zeta u'v' dz \right] \qquad (4.164)$$

$$A_y = \frac{1}{H} \frac{\partial}{\partial x} \left[(\bar{u}\bar{v}H) + \int_{-H}^\zeta u'v' dz \right] + \frac{1}{H} \frac{\partial}{\partial y} \left[(\bar{v}^2 H) + \int_{-H}^\zeta v'^2 dz \right] \qquad (4.165)$$

thus it becomes clear which terms are modified by the average velocities and by the variations around the average. In the C terms, bottom stress is especially important. It is expressed by (4.152) and (4.153), and it can be rewritten as

$$\tau_x^b = c_z^b u_{j,k,L} = c_z^b (\bar{u}_{j,k,L} + u'_{j,k,L}) \qquad (4.166)$$

$$\tau_y^b = c_y^b v_{j,k,L} = c_y^b(\bar{v}_{j,k,L} + v'_{j,k,L}) \tag{4.167}$$

Here c_z^b and c_y^b are bulk coefficients in (4.152) and (4.153).

Finally one additional simplification – all vertical derivatives of the velocity may as well be altered to the derivative of the velocity variations, because the average component of velocity does not change along the vertical direction.

4.2 Numerical solution and implementation of the split method

The time interaction of the 2D and 3D models depicts Fig.4.9.

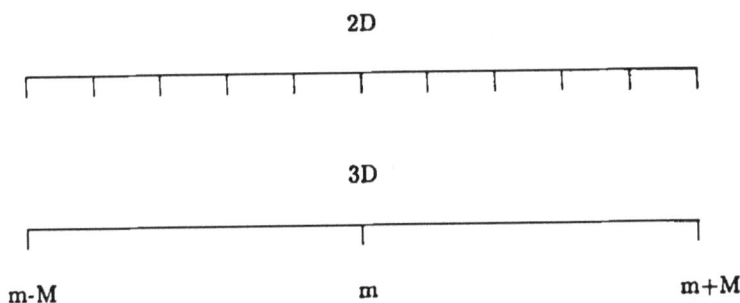

2D

3D

m-M   m   m+M

Fig. 4.9

Time-stepping of the split equations.

The 3D model at the slow mode of computations operates with a time step $T_{3D} = t_{m+M} - t_m$ while the 2D model of fast mode computations is advanced with a time-step $T_{2D} = T = t_{m+1} - t_m$, i.e., $T_{3D} = MT_{2D}$. A typical value of M ranges from 10 to 50.

In the ensuing considerations a three time-level numerical forms will be implemented. Starting with the depth-integrated equations (4.155), (4.156) and (4.160), one can notice that these are the same equations as derived by integration in Ch.I.Sec.2.1, augmented by the source-like terms A, B and C. Let us imagine that we have just advanced in time the 3D model to the time $m + M$, and the terms A, B and C have been transferred to the 2D model. The aim is right now to advance the 2D computations from time m to $m + M$. The variable terms, i.e., A and C are composed of two parts, one which is a function of the average velocity and it will change at each time step T. The other part stays constant during the 2D computations because it depends on the velocity variations only. Term B is a function of density and it is updated when the 3D model is advanced in time.

The depth-integrated 2D model is advanced in time by the leapfrog method,

$$\frac{\zeta^{m+1} - \zeta^{m-1}}{2T} = -\frac{\partial}{\partial x}(\bar{u}D)^m - \frac{\partial}{\partial y}(\bar{v}D)^m \qquad (4.168)$$

$$\frac{\bar{u}^{m+1} - \bar{u}^{m-1}}{2T} + A_x^m - f\bar{v}^m = -\frac{1}{\rho_o}\frac{\partial p_a^m}{\partial x} - g\frac{\partial \zeta^m}{\partial x} - B_x^m + C_x^m + N_h\Delta\bar{u}^{m-1} \quad (4.169)$$

$$\frac{\bar{v}^{m+1} - \bar{v}^{m-1}}{2T} + A_y^m + f\bar{u}^m = -\frac{1}{\rho_o}\frac{\partial p_a^m}{\partial y} - g\frac{\partial \zeta^m}{\partial y} - B_y^m + C_y^m + N_h\Delta\bar{v}^{m-1} \quad (4.170)$$

After M time steps the computation is advanced to the time level $m + M$. Now the new velocities can be calculated by (4.161) and a new density distribution is obtained by solving (4.149) for the temperature and salinity as

$$\frac{c^* - c^{m-1}}{2T} + (u\frac{\partial c}{\partial x})^m + (v\frac{\partial c}{\partial y})^m + (w\frac{\partial c}{\partial z})^m = D_h\Delta c^{m-1} \qquad (4.171a)$$

$$\frac{c^{m+1} - c^*}{2T} = \frac{\partial}{\partial z}D_z\frac{\partial c^{m+1}}{\partial z} \qquad (4.171b)$$

Anticipating possible instability due to vertical diffusion we have split the equation into two equations, and a substep variable is denoted as c^*. The form of the space finite-differencing has been discussed previously in Ch.IV.Sec.3.1 and 3.3. The implicit equation (4.171b) can be solved by the line inversion method delineated in Ch.IV.Sec.2.6 (eq.(4.83) gives second-order approximation).

The new fields of temperature and salinity through eq.(4.148) define a new density, which is required for an update of the various terms dependent on the density stratification.

Now, construction of the 3D model can be approached. At this juncture the A, B and C terms ought to be transferred from the 2D model and as soon as the set of new (primed) velocities is calculated these terms will be updated and used in the depth-averaged model. A very handy solution of the transfer process was developed by Berntsen, (see Berntsen et al, 1981).

The 3D algorithm starts by computing the vertical velocities from the equation of continuity by expression (4.118).

Next eqs.(4.162) and (4.163) are written in the differential–difference form. The space derivatives will be replaced by the finite-difference form through the results from Ch.IV.Sec.3.3. By this means

$$\frac{u^* - u'^{m-M}}{2TM} + (u\frac{\partial u}{\partial x})^m + (v\frac{\partial u}{\partial y})^m + (w\frac{\partial u'}{\partial z})^m - A_x^m - f(v')^m$$

$$= B_x^m - \frac{g}{\rho_o}(\frac{\partial}{\partial x}\int_z^\zeta \rho' dz)^m + N_h \Delta u'^{m-M} \tag{4.172a}$$

$$\frac{u'^{m+M} - u^*}{2TM} = \frac{\partial}{\partial z} N_z \frac{\partial u'^{m+M}}{\partial z} - C_x^m \tag{4.172b}$$

and

$$\frac{v^* - v'^{m-M}}{2T} + (u\frac{\partial v}{\partial x})^m + (v\frac{\partial v}{\partial y})^m + (w\frac{\partial v'}{\partial z})^m - A_y^m + fu'^m$$

$$= B_y^m - \frac{g}{\rho_o}(\frac{\partial}{\partial y}\int_z^\zeta \rho' dz)^m + N_h \Delta v'^{m-M} \tag{4.173a}$$

$$\frac{v'^{m+M} - u^*}{2T} = \frac{\partial}{\partial z} N_z \frac{\partial v'^{m+M}}{\partial z} - C_y^m \tag{4.173b}$$

In the above equations the split method has been applied again to derive a stable numerical solution to the vertical friction term. It is obvious that splitting of the equations should go together with splitting of the boundary condition. Thus in the above set the first equation will be solved subject to the condition on the vertical wall, and the second equation will be solved with the boundary conditions at the bottom and the surface. Let us assume for the moment that only half of the vertical term is taken to the second substep and the other half is retained in the first equation (see Ch.IV.Sec.2.6). How shall we proceed to split the boundary condition, and what shall happen to the solution? Although one can devise very intricate, beautiful and stable schemes through the split method (Yanenko, 1971), it is very difficult to construct the appropriate boundary conditions. Too many substeps will often generate numerical solutions quite similar to the leapfrog numerical solutions.

4.3 Sigma coordinate transformation

The resolution of the vertical processes by a constant layer thickness often fails to reproduce the processes in the region of abrupt topography variations or in the surface or bottom boundary layers. To overcome these obstacles a dimensionless coordinate is introduced through the transformation

$$\sigma = \frac{z - \zeta}{H + \zeta} \tag{4.174}$$

The total depth $H + \zeta$ will be denoted as D. The new coordinate transforms the column of water from the surface ($z = \zeta$) to the bottom ($z = -H$) into a uniform depth ranging from 0 to -1. The vertical discretization in σ coordinate can be chosen in such manner as to provide an optimal vertical resolution. In Fig.4.10 an underwater mountain is resolved by two numerical grids. A constant layer thickness with $h_z=100$ m, and $\Delta\sigma=0.25$ are used. Comparing both discretizations one can see that σ coordinate works better. It follows smoothly the contours of the variable topography with the number of the vertical layers being conserved over the whole computational domain.

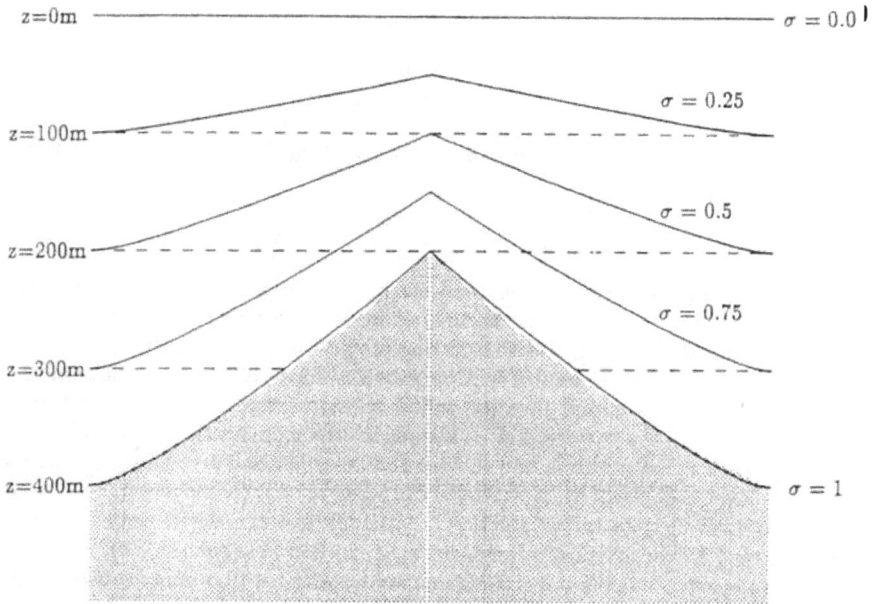

Fig. 4.10
Bottom topography and depth distribution.
Resolved by z coordinate (dash line), and by σ coordinate (continuous line).

From (4.174) it is obvious that the sigma coordinate not only transforms the vertical direction but it also depends on the horizontal coordinates. Thus all partial derivatives must be restated in the new system of coordinates. Derivative with respect to z is

$$\frac{\partial\Phi}{\partial z} = \frac{1}{D}\frac{\partial\Phi}{\partial\sigma} \qquad (4.175)$$

Here Φ denotes a dependent variable. Transformation from the old variables to the new ones defines the following derivative with respect to the horizontal coordinate:

$$\left.\frac{\partial \Phi}{\partial x}\right|_z = \left.\frac{\partial \Phi}{\partial x}\right|_\sigma - \frac{1}{D}(\sigma\frac{\partial D}{\partial x} + \frac{\partial \zeta}{\partial x})\frac{\partial \Phi}{\partial \sigma} \tag{4.176}$$

To simplify the above equation we set

$$\sigma\frac{\partial D}{\partial x} + \frac{\partial \zeta}{\partial x} = Q_x$$

A similar notation is applied to the y coordinate. The time derivative is constructed as

$$\left.\frac{\partial \Phi}{\partial t}\right|_z = \left.\frac{\partial \Phi}{\partial t}\right|_\sigma - \frac{1}{D}(1+\sigma)\frac{\partial \zeta}{\partial t}\frac{\partial \Phi}{\partial \sigma} \tag{4.177}$$

To simplify the derivation we shall not use the above notation to discern between the derivatives in z and σ coordinates, it will be obvious from the text which system is actually applied. The vertical velocity changes in the new system into a somewhat complicated expression:

$$w = \frac{dz}{dt} = \frac{\partial}{\partial t}(D\sigma + \zeta) + u\frac{\partial(D\sigma + \zeta)}{\partial x} + v\frac{\partial(D\sigma + \zeta)}{\partial y}$$

$$= D\frac{\partial \sigma}{\partial t} + (\sigma + 1)\frac{\partial \zeta}{\partial t} + uQ_x + vQ_y \tag{4.178}$$

To transform the continuity equation we shall use expressions for the first derivatives (4.175) and (4.176), thus,

$$\frac{\partial u}{\partial x} + \frac{\partial v}{\partial y} + \frac{\partial w}{\partial z} = \frac{\partial u}{\partial x} - \frac{Q_x}{D}\frac{\partial u}{\partial \sigma} + \frac{\partial v}{\partial y} - \frac{Q_y}{D}\frac{\partial v}{\partial \sigma} + \frac{1}{D}\frac{\partial w}{\partial \sigma} = 0 \tag{4.179}$$

The vertical velocity derivative in (4.179) is

$$\frac{\partial w}{\partial \sigma} = D\frac{\partial}{\partial \sigma}\frac{\partial \sigma}{\partial t} + \frac{\partial \zeta}{\partial t} + Q_x\frac{\partial u}{\partial \sigma} + Q_y\frac{\partial v}{\partial \sigma} + u\frac{\partial D}{\partial x} + v\frac{\partial D}{\partial y} \tag{4.180}$$

Introducing the expression (4.180) into (4.179) yields

$$\frac{\partial Du}{\partial x} + \frac{\partial Dv}{\partial y} + D\frac{\partial}{\partial \sigma}\frac{\partial \sigma}{\partial t} + \frac{\partial \zeta}{\partial t} = 0 \tag{4.181}$$

Since velocity is a superposition of the two components (depth-average and variations around the average) therefore eq.(4.181) can be split into

$$\frac{\partial Du'}{\partial x} + \frac{\partial Dv'}{\partial y} + D\frac{\partial}{\partial \sigma}\frac{\partial \sigma}{\partial t} = 0 \tag{4.182a}$$

and

$$\frac{\partial D\bar{u}}{\partial x} + \frac{\partial D\bar{v}}{\partial y} + \frac{\partial \zeta}{\partial t} = 0 \qquad (4.182b)$$

Equation (4.182b) can also be constructed straight from the equation of continuity for the vertically-averaged motion (4.160).

Vertically integrated equations of motion (4.155) and (4.156) are easily transformed into the new system because averaged variables do not depend on z or on σ coordinate (obviously (4.182b) written in σ coordinate is (4.160), which is written in z coordinate). Thus space and time transformation are derived accordingly to the rules,

$$\frac{\partial}{\partial x} \Rightarrow \frac{\partial}{\partial x} - \frac{Q_x}{D}\frac{\partial}{\partial \sigma} \Rightarrow \frac{\partial}{\partial x} \qquad (4.183)$$

$$\frac{\partial}{\partial t} \Rightarrow \frac{\partial}{\partial t} - \frac{1}{D}(1+\sigma)\frac{\partial \zeta}{\partial t}\frac{\partial}{\partial \sigma} \Rightarrow \frac{\partial}{\partial t} \qquad (4.184)$$

The terms which require special attention in the vertically-averaged equations are integrals related to the horizontal density stratification. The integral along z axis,

$$-B_z = -\frac{g}{H\rho_0}\int_{-H}^{\zeta}\int_{z}^{\zeta}\frac{\partial \rho'}{\partial x}dzdz$$

when transformation $dz = d\sigma D \approx d\sigma H$ is applied, becomes

$$-\frac{gD}{\rho_0}\int_{-1}^{0}\int_{\sigma}^{0}\frac{\partial \rho'}{\partial x}d\sigma d\sigma = -\frac{gD}{\rho_0}\int_{-1}^{0}\int_{\sigma}^{0}(\frac{\partial \rho'}{\partial x} - \frac{Q_x}{D}\frac{\partial \rho'}{\partial \sigma})d\sigma d\sigma \qquad (4.185)$$

In a similar way the integrals in eqs.(4.146) and (4.147) can also be transformed,

$$-\frac{g}{\rho_0}\int_{z}^{\zeta}\frac{\partial \rho'}{\partial x}dz = -\frac{gD}{\rho_0}\int_{\sigma}^{0}(\frac{\partial \rho'}{\partial x} - \frac{Q_x}{D}\frac{\partial \rho'}{\partial \sigma})d\sigma$$

$$= -\frac{gD}{\rho_0}\int_{\sigma}^{0}\frac{\partial \rho'}{\partial x}d\sigma + \frac{g}{\rho_0}\int_{\sigma}^{0}Q_x\frac{\partial \rho'}{\partial \sigma}d\sigma \qquad (4.186)$$

Now we can carry over transformation of the internal mode equations (4.162) and (4.163). We shall perform detailed analysis for the eq.(4.162) only. First the acceleration terms are considered,

$$\frac{\partial u'}{\partial t} \Rightarrow \frac{\partial u'}{\partial t} - \frac{1}{D}(1+\sigma)\frac{\partial \zeta}{\partial t}\frac{\partial u'}{\partial t} \qquad (4.187a)$$

$$u\frac{\partial u'}{\partial x} \Rightarrow u(\frac{\partial u'}{\partial x} - \frac{Q_x}{D}\frac{\partial u'}{\partial \sigma}) \qquad (4.187b)$$

$$u\frac{\partial \bar{u}}{\partial x} \Rightarrow u(\frac{\partial \bar{u}}{\partial x} - \frac{Q_x}{D}\frac{\partial \bar{u}}{\partial \sigma}) \Rightarrow u\frac{\partial \bar{u}}{\partial x} \tag{4.187c}$$

$$v\frac{\partial u'}{\partial y} \Rightarrow v(\frac{\partial u'}{\partial y} - \frac{Q_y}{D}\frac{\partial u'}{\partial \sigma}) \tag{4.187d}$$

$$v\frac{\partial \bar{u}}{\partial y} \Rightarrow v(\frac{\partial \bar{u}}{\partial y} - \frac{Q_y}{D}\frac{\partial \bar{u}}{\partial \sigma}) \Rightarrow v\frac{\partial \bar{u}}{\partial y} \tag{4.187e}$$

$$w\frac{\partial u'}{\partial z} \Rightarrow \frac{w}{D}\frac{\partial u'}{\partial \sigma} \Rightarrow \frac{1}{D}[D\frac{\partial \sigma}{\partial t} + (\sigma+1)\frac{\partial \zeta}{\partial t} + Q_x u + Q_y v]\frac{\partial u'}{\partial \sigma} \tag{4.188}$$

$$= \frac{\partial \sigma}{\partial t}\frac{\partial u'}{\partial \sigma} \quad \text{this term will remain in the equation of motion}$$

$$+ \frac{(\sigma+1)}{D}\frac{\partial \zeta}{\partial t}\frac{\partial u'}{\partial \sigma} \quad \text{this terms cancels against the term in eq.(4.187a)}$$

$$+ \frac{Q_x}{D}u\frac{\partial u'}{\partial \sigma} \quad \text{this terms cancels against the term in eq.(4.187b)}$$

$$+ \frac{Q_y}{D}v\frac{\partial u'}{\partial \sigma} \quad \text{this term cancels against the term in eq.(4.187d)}$$

The remaining terms in eq.(4.162) are easy to derive; here we shall consider the horizontal friction term only:

$$N_h \Delta u' \Rightarrow N_h \Delta u' + HOT \tag{4.189}$$

The higher order terms (HOT) express a dependence of u' on σ and will be neglected in the ensuing computations. Taking into account eqs.(4.185), (4.186), (4.187) and (4.188) the equation of motion for the internal mode in the new system of coordinate becomes

$$\frac{\partial u'}{\partial t} + u\frac{\partial u'}{\partial x} + u\frac{\partial \bar{u}}{\partial x} + v\frac{\partial u'}{\partial y} + v\frac{\partial \bar{u}}{\partial y} + \frac{\partial \sigma}{\partial t}\frac{\partial u'}{\partial \sigma} - A_x - fv'$$

$$= \frac{gD}{\rho_o}\int_{-1}^{0}\int_{\sigma}^{0}(\frac{\partial \rho'}{\partial x} - \frac{Q_x}{D}\frac{\partial \rho'}{\partial \sigma})d\sigma d\sigma - \frac{gD}{\rho_o}\int_{\sigma}^{0}(\frac{\partial \rho'}{\partial x} - \frac{Q_x}{D}\frac{\partial \rho'}{\partial \sigma})d\sigma$$

$$+ D^{-2}\frac{\partial}{\partial \sigma}(N_\sigma\frac{\partial u'}{\partial \sigma}) - [\tau_x^s/(H\rho_o) - \tau_x^b/(H\rho_o)] + N_h \Delta u' \tag{4.190}$$

and

$$\frac{\partial v'}{\partial t} + u\frac{\partial v'}{\partial x} + u\frac{\partial \bar{v}}{\partial x} + v\frac{\partial v'}{\partial y} + v\frac{\partial \bar{v}}{\partial y} + \frac{\partial \sigma}{\partial t}\frac{\partial v'}{\partial \sigma} - A_y + fu'$$

$$= \frac{gD}{\rho_o}\int_{-1}^{0}\int_{\sigma}^{0}(\frac{\partial \rho'}{\partial y} - \frac{Q_y}{D}\frac{\partial \rho'}{\partial \sigma})d\sigma d\sigma - \frac{gD}{\rho_o}\int_{\sigma}^{0}(\frac{\partial \rho'}{\partial y} - \frac{Q_y}{D}\frac{\partial \rho'}{\partial \sigma})d\sigma$$

$$+D^{-2}\frac{\partial}{\partial\sigma}(N_\sigma\frac{\partial v'}{\partial\sigma}) - [\tau_y^s/(H\rho_o) - \tau_y^b/(H\rho_o)] + N_h\Delta v' \qquad (4.191)$$

In the above set A_x and A_y are expressed by (4.164) and (4.165), thus before starting computation the integral have to be altered to the σ coordinate.

Equation of salt (or heat) conservation in the new system is

$$\frac{\partial DS}{\partial t} + \frac{\partial uDS}{\partial x} + \frac{\partial vDS}{\partial y} + \frac{\partial}{\partial\sigma}(\frac{\partial\sigma}{\partial t}DS) = D^{-1}\frac{\partial}{\partial\sigma}(D_\sigma\frac{\partial S}{\partial\sigma}) - DD_h\Delta S \qquad (4.192)$$

or assuming small depth variations,

$$\frac{\partial S}{\partial t} + u\frac{\partial S}{\partial x} + v\frac{\partial S}{\partial y} + D^{-1}D\frac{\partial\sigma}{\partial t}\frac{\partial S}{\partial\sigma} = D^{-2}\frac{\partial}{\partial\sigma}(D_\sigma\frac{\partial S}{\partial\sigma} - D_h\Delta S \qquad (4.193)$$

The system of equations is complete and ready for the next step, that is construction of the numerical algorithm. Such approach has been already considered in the previous section and can be applied successfully to the present set.

Application of the σ coordinate in meteorology to mountain flow resulted in the large errors especially in a reproduction of the pressure term. Based on (4.176), consider the pressure gradient along the x direction:

$$\frac{\partial p}{\partial x}\Big|_z = \frac{\partial p}{\partial x}\Big|_\sigma - \frac{1}{D}(\sigma\frac{\partial D}{\partial x} + \frac{\partial\zeta}{\partial x})\frac{\partial p}{\partial\sigma} \qquad (4.194)$$

As the changes of the sea level are usually very small when compared to the bottom variations, (4.194) becomes

$$\frac{\partial p}{\partial x}\Big|_z = \frac{\partial p}{\partial x}\Big|_\sigma - \frac{\sigma}{D}\frac{\partial D}{\partial x}\frac{\partial p}{\partial\sigma} \qquad (4.195)$$

The influence of topography is obvious when the second term on the right-hand side is rewritten as $\frac{1}{D}\frac{\partial p}{\partial ln\sigma}\frac{\partial D}{\partial x}$. We are dealing here with the logarithmic stretching of the coordinate; this technique is very prone to any approximation error (Haney, 1991). To eliminate the errors in the gradient of pressure, Gary (1973) suggested decomposition of the pressure into a hydrostatic portion and the superposed changes, or into any other average and deviations from this average. Therefore, in the process of calculation of the pressure gradient the large (hydrostatic or average) components of pressure cancel out and only pressure variations, which generate horizontal motion, are left in the pressure gradient. This problem is not quite new to us, and we have recognized the importance of pressure reduction even in the z coordinate. To refresh this problem see Ch.IV.Sec.2.3, also notice how to avoid the messy negative values of z or σ when integral for the density is estimated.

5. General circulation model – rigid lid condition

This 3D model is based on the splitting of the flow into the depth-average component and the changes around the average. The same approach has been applied in the mode splitting approach in the previous section but an additional assumption is introduced here, namely the vertical velocity of the free surface is equal to zero. This actually eliminates the surface long gravity waves from the equations and the depth-average motion in the numerical formulation is not governed by the CFL stability condition. A numerical model was constructed by Bryan (1969), but the application of this model to study general oceanic motion are numerous. The best known numerical implementations were constructed by Semtner (1974) and Cox (1984). The model has second order of approximation in time and space (but not for the friction/diffusion terms), it also conserves mass, momentum and energy. The time stepping is based on the leapfrog approach, space staggered grid is used in which both horizontal velocities are located at the same point. The model has one flaw which makes it difficult to compare model results with the satellite observations, the change of the sea surface in time is not calculated. Actually, the sea level slope can be retrieved from the model, thus we only need to know the sea level at one grid point of the computational domain and at the remaining points the seal level will be calculated from the full differential as

$$d\zeta = \frac{\partial \zeta}{\partial x}dx + \frac{\partial \zeta}{\partial y}dy \qquad (4.196)$$

5.1 Basic equations and boundary conditions

We shall start from the equation of motion in the spherical system of coordinates (1.45), (1.46), (1.47), and by taking into account the Boussinesq and hydrostatic approximations the set of equations similar to (1.62) and (1.63) is obtained:

$$\rho_o \frac{Du}{Dt} - \rho_o (2\Omega + \frac{u}{R\cos\Phi})v\sin\Phi$$

$$= -\frac{1}{R\cos\Phi}\frac{\partial p_a}{\partial \lambda} - \frac{\rho_o g}{R\cos\Phi}\frac{\partial \zeta}{\partial \lambda} - \frac{g}{R\cos\Phi}\int_z^\zeta \frac{\partial \rho'}{\partial \lambda}dz + \rho_o A'_\lambda \qquad (4.197)$$

$$\rho_o \frac{Dv}{Dt} + \rho_o (2\Omega + \frac{u}{R\cos\Phi})u\sin\Phi$$

$$= -\frac{1}{R}\frac{\partial p_a}{\partial \lambda} - \frac{\rho_o g}{R}\frac{\partial \zeta}{\partial \lambda} - \frac{g}{R}\int_z^\zeta \frac{\partial \rho'}{\partial \Phi}dz + \rho_o A'_\Phi \qquad (4.198)$$

The time operator is

$$\frac{D}{Dt} = \frac{\partial}{\partial t} + \frac{u}{R\cos\Phi}\frac{\partial}{\partial\lambda} + \frac{v}{R}\frac{\partial}{\partial\Phi} + w\frac{\partial}{\partial z} \tag{4.199}$$

and the frictional forces are written in somewhat complicated form,

$$A'_\lambda = A_1 u - \frac{N_h}{R^2\cos^2\Phi}(u + 2\frac{\partial}{\partial\lambda}(v\sin\Phi)) \tag{4.200}$$

$$A'_\Phi = A_1 v + N_h(-\frac{v}{R^2\cos^2\Phi} + \frac{2\sin\Phi}{R^2\cos^2\Phi}\frac{\partial u}{\partial\lambda}) \tag{4.201}$$

where operator A_1 has the following form:

$$A_1 = N_h(\frac{1}{R^2\cos^2\Phi}\frac{\partial^2}{\partial\lambda^2} + \frac{1}{R^2\cos\Phi}\frac{\partial}{\partial\Phi}(\cos\Phi\frac{\partial}{\partial\Phi})) + \frac{\partial}{\partial z}(N_z\frac{\partial}{\partial z}) \tag{4.202}$$

In the above equations based on scale considerations, we have additionally neglected the vertical velocity (w) in the horizontal friction terms and in the Coriolis terms.

Equation of continuity in the spherical system is

$$\frac{1}{R\cos\Phi}\frac{\partial u}{\partial\lambda} + \frac{1}{R\cos\Phi}\frac{\partial}{\partial\Phi}(v\cos\Phi) + \frac{\partial w}{\partial z} = 0 \tag{4.203}$$

Finally equation of diffusion can be expressed in the following way:

$$\frac{Dc}{Dt} = A_2 c \tag{4.204}$$

here operator A_2 is

$$A_2 = D_h(\frac{1}{R^2\cos^2\Phi}\frac{\partial^2}{\partial\lambda^2} + \frac{1}{R^2\cos\Phi}\frac{\partial}{\partial\Phi}(\cos\Phi\frac{\partial}{\partial\Phi})) + \frac{\partial}{\partial z}(D_z\frac{\partial}{\partial z}) \tag{4.205}$$

Variable c stands for concentration, and it can denote salinity, temperature or any passive admixture in the sea water.

Equation of state (1.40), as usual, serves to calculate density from salinity and temperature. The approximation of equation of state in the numerical form with the high order of accuracy was given by Bryan and Cox (1972), and Friedrich and Levitus (1972). Less accurate formulations by Mamayev (1975) and Eckart (1960) are still used. Density instabilities in the model are dealt with by the mixing routine described in Ch.IV.Sec.2.7.

The boundary conditions at the horizontal surfaces (free surface and bottom) for the momentum equations are expressed by the stresses. This is again caused by the approach taken – the basic equations will be split into a vertical average and changes around the average. It is easy to see that for the depth-averaged equations only the stress boundary condition is relevant. Thus to keep all boundary conditions consistent we also set here the stress at the free surface and at the bottom:

$$\rho N_z \frac{\partial u}{\partial z} = \tau_z^s = \tau_z^w \quad \text{and} \quad \rho N_z \frac{\partial v}{\partial z} = \tau_y^s = \tau_z^w \quad \text{at} \quad z = \zeta(x,y,t) \qquad (4.206)$$

$$\tau_z^b = 0 \quad \text{and} \quad \tau_y^b = 0 \quad \text{at} \quad z = -H(x,y) \qquad (4.207)$$

The flux of heat and the salinity flux are set to zero at the bottom, but the values at the free surface are prescribed:

$$F^q = cD_z\rho\frac{\partial \theta}{\partial z} = Q_2(\theta_1 - \theta_A) \qquad (4.208)$$

$$D_z\frac{\partial S}{\partial z} = S_1(P - E) \qquad (4.209)$$

Here E and P res-ect5ve3y denote evaporation and precipitation change in time (cm/s), θ_1 and S_1 denote salinity and temperature in the first layer, θ_A is the air temperature at 10 m, c is specific heat of sea water and Q_2 is a coupling coefficient defined by Haney (1971, 1974).

The boundary conditions at the vertical wall prescribe zero velocity and no flux for the salt or heat

$$\frac{\partial \theta}{\partial n} = \frac{\partial S}{\partial n} = 0 \qquad (4.210)$$

Here n denote normal to the wall.

The vertical velocity at the free surface is

$$w = 0 \quad \text{at} \quad z = 0, \qquad (4.211)$$

and at the bottom a kinematic condition (1.65a) is implemented in the spherical coordinates

$$w = -\frac{DH}{Dt} = -\frac{u}{R\cos\Phi}\frac{\partial H}{\partial \lambda} - \frac{v}{R}\frac{\partial H}{\partial \Phi} \quad \text{at} \quad z = -H(\lambda, \Phi) \qquad (4.212)$$

5.2 Rudiments of the splitting method

The equations to study general oceanic circulation incorporate the so-called rigid lid assumption (Bryan, 1969). It allows a more efficient calculation since the time step of the numerical calculation can be substantially increased. The rigid lid condition (4.211)

$$w \simeq \frac{\partial \zeta}{\partial t} = 0 \quad \text{at} \quad z = 0 \qquad (4.213)$$

requires the vertical velocity at the free surface to be zero. Integrating the equation of continuity (4.203) from the free surface to the bottom we arrive at (see Ch.I.Sec.2.2, 3.1)

$$\frac{1}{R\cos\Phi}(\frac{\partial\bar{u}H}{\partial\lambda} + \frac{\partial}{\partial\Phi}(\bar{v}H\cos\Phi)) + \frac{\partial\zeta}{\partial t} = 0 \qquad (4.214)$$

Introducing the rigid lid condition into the above equation we derive an equation of continuity for the steady state

$$\frac{\partial\bar{u}H}{\partial\lambda} + \frac{\partial}{\partial\Phi}(\bar{v}H\cos\Phi)) = 0 \qquad (4.215)$$

which permits introduction of a stream function as

$$\bar{v}HR\cos\Phi = \frac{\partial\Psi}{\partial\lambda} \qquad \text{and} \qquad \bar{u}HR = -\frac{\partial\Psi}{\partial\Phi} \qquad (4.216)$$

Here $\bar{u}$ and $\bar{v}$ are the depth-average velocities.

Depth-averaged equation of motion is obtained in the usual way by integrating eqs.(4.197) and (4.198) from the surface to the bottom, but now we can add one additional step since velocities are expressed by the stream function (4.216), thus

$$-\frac{1}{HR}\frac{\partial}{\partial\Phi}\frac{\partial\Psi}{\partial t} - \frac{f}{HR\cos\Phi}\frac{\partial\Psi}{\partial\lambda}$$

$$= -\frac{1}{\rho_o R\cos\Phi}\frac{\partial p_a}{\partial\lambda} - \frac{g}{R\cos\Phi}\frac{\partial\zeta}{\partial\lambda} - \frac{g}{\rho_o HR\cos\Phi}\int_{-H}^{0}\int_z^0\frac{\partial\rho'}{\partial\lambda}dzdz + \frac{1}{H}\int_{-H}^0 G^\lambda dz \qquad (4.217)$$

$$\frac{1}{HR\cos\Phi}\frac{\partial}{\partial\lambda}\frac{\partial\Psi}{\partial t} - \frac{f}{HR}\frac{\partial\Psi}{\partial\Phi}$$

$$= -\frac{1}{R\rho_o}\frac{\partial p_a}{\partial\Phi} - \frac{g}{R}\frac{\partial\zeta}{\partial\Phi} - \frac{g}{HR\rho_o}\int_{-H}^{0}\int_z^0\frac{\partial\rho'}{\partial\Phi}dzdz + \frac{1}{H}\int_{-H}^0 G^\Phi dz \qquad (4.218)$$

By cross-differentiation of the above equations and subtracting on either side, the equation for the stream function is obtained:

$$\frac{\partial}{\partial\lambda}(\frac{1}{H\cos\Phi}\frac{\partial^2\Psi}{\partial\lambda\partial t}) + \frac{\partial}{\partial\Phi}(\frac{\cos\Phi}{H}\frac{\partial^2\Psi}{\partial\Phi\partial t}) + \frac{\partial}{\partial\Phi}(\frac{f}{H}\frac{\partial\Psi}{\partial\lambda}) - \frac{\partial}{\partial\lambda}(\frac{f}{H}\frac{\partial\Psi}{\partial\Phi})$$

$$= \frac{\partial}{\partial\Phi}(\frac{g}{H\rho_o}\int_{-H}^0\int_z^0\frac{\partial\rho'}{\partial\lambda}dzdz) - \frac{\partial}{\partial\lambda}(\frac{g}{H\rho_o}\int_{-H}^0\int_z^0\frac{\partial\rho'}{\partial\Phi}dzdz)$$

$$+\frac{\partial}{\partial\lambda}(\frac{R}{H}\int_{-H}^0 G^\Phi dz) - \frac{\partial}{\partial\Phi}(\frac{R\cos\Phi}{H}\int_{-H}^0 G^\lambda dz) \qquad (4.219)$$

The stream function approach simplifies this problem, now instead of the two equations for the $\bar{u}$ and $\bar{v}$ the one equation of higher order is obtained. It remains to explain that G^λ and G^Φ represent nonlinear and viscous terms in equations (4.217) and (4.218), these terms are analogous to the $A, B,$ and C terms from section 4.1 of this chapter.

Let us shortly delineate a possible way to the construction of a general solution. We know from the previous section that along with the depth-average equations the complementary set of equations for the depth-dependent velocity component is needed. To obtain this complementary set the previous method can be applied as well, i.e., eq.(4.217) is subtracted from (4.197), and eq.(4.218) from (4.198), and we arrive at the equations for u' and v', similar to (4.162) and (4.163). Such equations were actually obtained by Haney (1974). Here we demonstrate an approach presented by Bryan (1969) and Semtner (1974). First of all it is useful to notice that the depth-dependent velocity can be derived from a general solution of (4.197) and (4.198) as

$$u' = u - \frac{1}{H}\int_{-H}^{0} u\,dz \quad \text{and} \quad v' = v - \frac{1}{H}\int_{-H}^{0} v\,dz \qquad (4.220)$$

If general set of equations have been linear then by adding or subtracting any body forces (such as sea level slope) the resulting depth-dependent velocity will always stay unchanged. We shall take this idea and apply it to the numerical solution by the leapfrog method. Let us consider only the equation along the x coordinate

$$\frac{u^{*m+1} - u^{m-1}}{T} + (\text{Nonlinear terms})^m = (\text{All other terms})^{m,m-1} \qquad (4.221)$$

Now we can scrutinize the above equations in two aspects: a) nonlinear terms, and b) application (4.220) to obtain a depth-dependent part of the velocity. If in (4.221) full velocity (i.e. $\bar{u} + u', \bar{v} + v'$) is used in the nonlinear term calculation the nonlinear interactions are well reproduced and the presence of the sea level in (All other terms) influence the value u^{*m+1} in a linear manner. We may conclude that the presence or absence of the sea level slope in (4.221) will not influence the velocities derived from (4.220). On the other hand the presence or absence of the sea level does change the full velocity, this is why we have introduced the notation u^{*m+1} instead u^{m+1} in (4.221). In summary, (4.221) will serve to calculate u' and v' when the sea level slope is unknown.

5.3 Finite differencing

Bryan (1969) constructed the finite difference equations by considering boxes from the surface to the bottom with the horizontal dimension along λ axis equal to $\Delta\lambda R\cos\Phi$, and along Φ axis equal to $R\Delta\Phi$. The vertical dimension of a box will be denoted in the usual way as h_z. (note that our notation differs from Bryan (1969) and Semtner (1974)). The finite difference is based on a staggered numerical grid

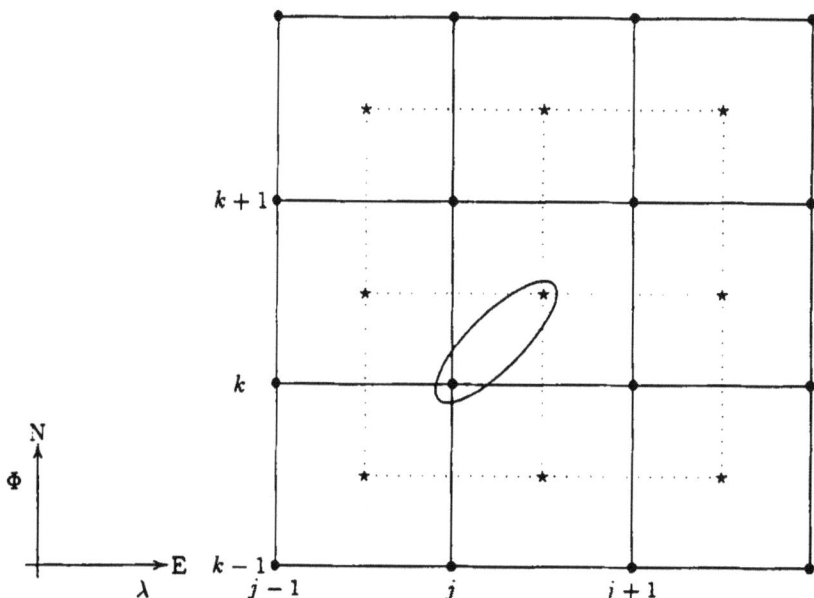

Fig. 4.11
Grid point distribution in the horizontal plane.
Bullets denote T, S and Ψ grid points, stars are for the u and v.
Bullet and star inside an ellipse has the same j, k indices.

with u and v located at the same grid point (star), and temperature (θ), salinity (S) and stream function (Ψ) located in another grid point (bullet) – see Fig.4.11.

Along the vertical direction two columns are considered: for the θ, S point and for the u, v points; in both the columns the vertical velocity (w) is given as well – Fig.4.12. The u, v point and θ, S point are located at the center of the boxes depicted in Fig.4.11. Bryan (1969) considered a momentum flux (for u, v box) or a mass flux (for θ, S box) taking into account the volume and the flux through all (six) surfaces of a box. When all the expressions containing a quantity inside a given box and fluxes through the surfaces are written they do look quite impressive and confusing. But all this is not nearly as confusing as it seems to be, one may forget about the volumes and surfaces and apply the rules of the finite differencing.

In every case it is quite important to use the equation of motion in the flux form.
The nonlinear terms in the time operator (4.199)

$$L(\sigma) = \frac{u}{R\cos\Phi}\frac{\partial\sigma}{\partial\lambda} + \frac{v}{R}\frac{\partial\sigma}{\partial\Phi} + w\frac{\partial\sigma}{\partial z}$$

ought to be rewritten in the flux form

$$= \frac{1}{R\cos\Phi}\frac{\partial(\sigma u)}{\partial\lambda} + \frac{1}{R}\frac{\partial(\sigma v)}{\partial\Phi} + \frac{\partial(\sigma w)}{\partial z} \qquad (4.222)$$

Here σ stands for the dependent variables (u, v, S, θ).

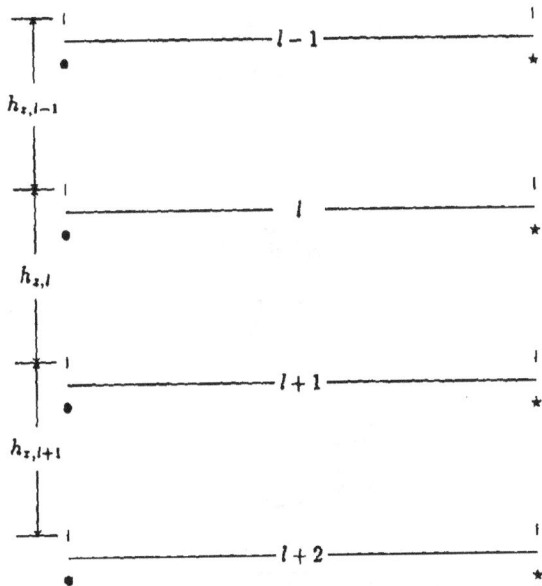

Fig. 4.12

Grid point distribution in the vertical plane. Left-hand side describes T,
S column (bullets). Righ-hand side describes u, v column (stars). Vertical
bar stands for the vertical velocity (w) grid point.

As an example of the numerical grid application from Fig.4.11 and Fig.4.12 let us render the nonlinear portion of equation of motion along λ axis into the finite difference form

$$LN(u) = \frac{1}{\Delta\lambda R \cos\Phi_{k+1/2}}\left[\left(\frac{u_{j,k,l} + u_{j+1,k,l}}{2}\right)^2 - \left(\frac{u_{j,k,l} + u_{j-1,k,l}}{2}\right)^2\right]$$

$$+\frac{1}{R\Delta\Phi}\left[\left(\frac{v_{j,k,l} + v_{j,k+1,l}}{2}\right)\left(\frac{u_{j,k,l} + u_{j,k+1,l}}{2}\right) - \left(\frac{v_{j,k,l} + v_{j,k-1,l}}{2}\right)\left(\frac{u_{j,k,l} + u_{j,k-1,l}}{2}\right)\right]$$

$$+\frac{u^*_{j,k,l}w_{j,k,l} - u^*_{j,k,l+1}w_{j,k,l+1}}{h_{z,l}} \tag{4.223}$$

Here LN is a numerical rendition of the operator L, $\sigma = u$, and

$$u^*_{j,k,l} = \frac{u_{j,k,l}h_{z,l-1} + u_{j,k,l-1}h_{z,l}}{h_{z,l} + h_{z,l-1}}$$

$$u^*_{j,k,l+1} = \frac{u_{j,k,l}h_{z,l+1} + u_{j,k,l+1}h_{z,l}}{h_{z,l} + h_{z,l+1}}.$$

To stay with the nonlinear operator let us consider $L(\sigma)$ for the transport equation ($\sigma = S$)

$$L(S) = \frac{1}{R\cos\Phi}\frac{\partial(Su)}{\partial\lambda} + \frac{1}{R}\frac{\partial(Sv)}{\partial\Phi} + \frac{\partial(Sw)}{\partial z} \tag{4.224}$$

To render (4.224) into the finite difference form we consider for the horizontal advection a dotted line box in Fig.4.11 with the j, k grid point for the salinity S located at the center of this box, and for the convection we refer to Fig.4.12.

$$LN(S) = \frac{1}{\Delta\lambda R\cos\Phi_k}\left[\left(\frac{S_{j,k,l} + S_{j+1,k,l}}{2}\right)\left(\frac{u_{j,k,l} + u_{j,k-1,l}}{2}\right)\right.$$

$$-\left(\frac{S_{j,k,l} + S_{j-1,k,l}}{2}\right)\left(\frac{u_{j-1,k,l} + u_{j-1,k-1,l}}{2}\right)\right]$$

$$+\frac{1}{R\Delta\Phi}\left[\left(\frac{S_{j,k+1,l} + S_{j,k,l}}{2}\right)\left(\frac{v_{j,k,l} + v_{j-1,k,l}}{2}\right)\right.$$

$$-\left(\frac{S_{j,k,l} + S_{j,k-1,l}}{2}\right)\left(\frac{v_{j,k-1,l} + v_{j-1,k-1,l}}{2}\right)\right]$$

$$+\frac{S^*_{j,k,l}w_{j,k,l} - S^*_{j,k,l+1}w_{j,k,l+1}}{h_{z,l}} \tag{4.225}$$

Here

$$S^*_{j,k,l} = \frac{S_{j,k,l}h_{z,l-1} + S_{j,k,l-1}h_{z,l}}{h_{z,l} + h_{z,l-1}}$$

$$S_{j,k,l+1}^* = \frac{S_{j,k,l}h_{z,l+1} + S_{j,k,l+1}h_{z,l}}{h_{z,l} + h_{z,l+1}}$$

As we know, velocity components u and v are expressed as a sum of $\bar{u} + u'$ and $\bar{v} + v'$. The depth-average components are actually calculated by (4.216) as derivatives of the stream function (Ψ) and they are located in the middle point between the stream function points. This observation can simplify the expressions derived above at the u, v boxes. For example the average of two velocity points in the first term on the right-hand side of (4.223) is

$$\frac{u_{j,k,l} + u_{j+1,k,l}}{2} = \frac{u'_{j,k,l} + u'_{j+1,k,l}}{2} + \frac{\bar{u}_{j,k} + \bar{u}_{j+1,k}}{2}$$

$$= \frac{u'_{j,k,l} + u'_{j+1,k,l}}{2} - \frac{(\Psi_{j+1,k+1} - \Psi_{j+1,k})}{HR\Delta\Phi} \qquad (4.226)$$

Now that we have the knowledge of the grid and the basic numerical approach, let us consider two important steps of the method, i.e., calculation of the depth-dependent and depth–average velocities.

Prediction of the depth-dependent velocities. This task was shortly discussed above and the main problem here is a calculation of the u^* and v^* velocities. In the next step they will be introduced into (4.220) and the internal mode velocities u' and v' are obtained. Now, the idea depicted in (4.221) will be fully explored. Let us rewrite equation of motion (4.197) and (4.198) in the difference–differential form

$$\frac{u^{*,m+1} - u^{m-1}}{2T} - f\frac{(v^{*,m+1} + v^{m-1})}{2} = F^\lambda \qquad (4.227)$$

$$\frac{v^{*,m+1} - v^{m-1}}{2T} + f\frac{(u^{*,m+1} + u^{m-1})}{2} = F^\Phi \qquad (4.228)$$

Here

$$F^\lambda = -LN(u^m) + \frac{(uv)^m}{R}\tan\Phi - \frac{1}{R\rho_o\cos\Phi}\frac{\partial p_a^m}{\partial\lambda} - \frac{g}{\rho_o R\cos\Phi}\int_z^\zeta \frac{\partial\rho'^m}{\partial\lambda}dz + A_\lambda'^{m-1}$$

and

$$F^\Phi = -LN(v^m) - \frac{(uv)^m}{R}\tan\Phi - \frac{1}{R\rho_o}\frac{\partial p_a^m}{\partial\Phi} - \frac{g}{R\rho_o}\int_z^\zeta \frac{\partial\rho'^m}{\partial\Phi}dz + A_\Phi'^{m-1}$$

Above we have denoted LN as the numerical rendition of the $L(\sigma)$ given by (4.222); A_λ' and A_Φ' are given by (4.200) and (4.201) respectively.

An implicit set of equations (4.227) and (4.228) is well known to us from section 3.1 of this chapter. Solving above set for the unknown $u^{*,m+1}$ and $v^{*,m+1}$ we arrive at

$$u^{*,m+1} = \left\{ 2TF^\lambda + 2fT^2F^\Phi + [1 - (fT)^2]u^{m-1} + 2fTv^{m-1} \right\}/[1 + (fT)^2] \quad (4.229)$$

$$v^{*,m+1} = \left\{ 2TF^\Phi - 2fT^2F^\lambda + [1 - (fT)^2]v^{m-1} - 2fTu^{m-1} \right\}/[1 + (fT)^2] \quad (4.230)$$

Now the depth-dependent velocities are easily obtained through (4.220) by inserting $u^{*,m+1}$ for u and $v^{*,m+1}$ for v.

Prediction of the depth-average velocities. This task is straightforward because equation for the stream function is already available (4.219), and the depth-average velocity then follows from (4.216). A glance at (4.219) may raise a question whether the stream function or time derivative of the stream function is an unknown variable here. This is an equation for the time derivative, and the stream function can be obtained from the time derivative. Let us rewrite (4.219) as

$$\frac{\partial}{\partial \lambda}\left(\frac{1}{H\cos\Phi}\frac{\partial^2\Psi}{\partial\lambda\partial t}\right) + \frac{\partial}{\partial\Phi}\left(\frac{\cos\Phi}{H}\frac{\partial^2\Psi}{\partial\Phi\partial t}\right) - \frac{\partial}{\partial\Phi}\left(\frac{f}{H}\frac{\partial\Psi}{\partial\lambda}\right) + \frac{\partial}{\partial\lambda}\left(\frac{f}{H}\frac{\partial\Psi}{\partial\Phi}\right)$$

$$= \frac{\partial}{\partial\Phi}\left(\frac{g}{H\rho_o}\int_{-H}^0\int_z^0\frac{\partial\rho'}{\partial\lambda}dzdz\right) - \frac{\partial}{\partial\lambda}\left(\frac{g}{H\rho_o}\int_{-H}^0\int_z^0\frac{\partial\rho'}{\partial\Phi}dzdz\right)$$

$$+ \frac{\partial}{\partial\lambda}\left(\frac{R}{H}\int_{-H}^0 G^\Phi dz\right) - \frac{\partial}{\partial\Phi}\left(\frac{R\cos\Phi}{H}\int_{-H}^0 G^\lambda dz\right), \quad (4.231)$$

or denoting the space operator on the left-hand side as $\mathcal{L}$ and the expressions on the right-hand side of (4.231) as $\mathcal{F}$, above equation becomes

$$\mathcal{L}\frac{\partial\Psi}{\partial t} = \mathcal{F} \quad (4.231a)$$

This equations can be solved for the unknown $\dfrac{\partial\Psi}{\partial t} = D$ by any iteration method. The new value of the stream function then follow from $\dfrac{\Psi^{m+1} - \Psi^{m-1}}{2T} = D$. To start this computational process we can assume that the initial value $\Psi^0 = 0$, and at the first time step an expression $\dfrac{\Psi^1 - \Psi^0}{T} = D$ is used.

Instead of the above delineated approach one may resort to the primary numerical equations (similar to (4.227) and (4.228)) and construct a stream function equation. This new stream function equation may differ somewhat from (4.231) due to the numerical scheme employed. We now explore this route starting from eqs.(4.227) and (4.228) but written for the full velocity

$$\frac{u^{m+1} - u^{m-1}}{2T} - f\frac{(v^{m+1} + v^{m-1})}{2} = F^\lambda - \frac{g}{R\cos\Phi}\frac{\partial\zeta}{\partial\lambda} \qquad (4.232)$$

$$\frac{v^{m+1} - v^{m-1}}{2T} + f\frac{(u^{m+1} + u^{m-1})}{2} = F^\phi - \frac{g}{R}\frac{\partial\zeta}{\partial\lambda} \qquad (4.233)$$

For the ensuing calculations we rearrange the Coriolis term and the right-hand sides of this set.

$$\frac{u^{m+1} - u^{m-1}}{2T} - f\frac{(v^{m+1} - v^{m-1})}{2} = F^{\prime\lambda} - \frac{g}{R\cos\Phi}\frac{\partial\zeta}{\partial\lambda} \qquad (4.234)$$

$$\frac{v^{m+1} - v^{m-1}}{2T} + f\frac{(u^{m+1} - u^{m-1})}{2} = F^{\prime\phi} - \frac{g}{R}\frac{\partial\zeta}{\partial\lambda} \qquad (4.235)$$

Here $F^{\prime\lambda} = F^\lambda + fv^{m-1}$, and $F^{\prime\phi} = F^\phi - fu^{m-1}$.

Now we can construct the stream function equation in a systematic way. First we introduce an average velocity by the vertical integration of the above set

$$\frac{\bar{u}^{m+1} - \bar{u}^{m-1}}{2T} - f\frac{(\bar{v}^{m+1} - \bar{v}^{m-1})}{2} = \frac{1}{H}\int_{-H}^{0} F^{\prime\lambda}dz - \frac{g}{R\cos\Phi}\frac{\partial\zeta}{\partial\lambda} \qquad (4.236)$$

$$\frac{\bar{v}^{m+1} - \bar{v}^{m-1}}{2T} + f\frac{(\bar{u}^{m+1} - \bar{u}^{m-1})}{2} = \frac{1}{H}\int_{-H}^{0} F^{\prime\phi}dz - \frac{g}{R}\frac{\partial\zeta}{\partial\lambda} \qquad (4.237)$$

The average velocity can be expressed by the stream function through the definition (4.216), and above set becomes

$$-\frac{1}{HR}\frac{\partial}{\partial\Phi}\left(\frac{\Psi^{m+1} - \Psi^{m-1}}{2T}\right) - \frac{f}{HR\cos\Phi}\frac{\partial}{\partial\lambda}\left(\frac{\Psi^{m+1} - \Psi^{m-1}}{2}\right)$$

$$= \frac{1}{H}\int_{-H}^{0} F^{\prime\lambda}dz - \frac{g}{R\cos\Phi}\frac{\partial\zeta}{\partial\lambda} \qquad (4.238)$$

$$\frac{1}{HR\cos\Phi}\frac{\partial}{\partial\lambda}\left(\frac{\Psi^{m+1} - \Psi^{m-1}}{2T}\right) - \frac{f}{HR}\frac{\partial}{\partial\Phi}\left(\frac{\Psi^{m+1} - \Psi^{m-1}}{2}\right)$$

$$= \frac{1}{H}\int_{-H}^{0} F^{\prime\phi}dz - \frac{g}{R}\frac{\partial\zeta}{\partial\lambda} \qquad (4.239)$$

From these equations one equation for the stream function will be obtained and in the process the sea level terms will be deleted. For this purpose eq.(4.238) is multiplied by $R \cos \Phi$ and then differentiated by $\frac{\partial}{\partial \Phi}$; eq.(4.239) is multiplied by R and afterwards differentiated by $\frac{\partial}{\partial \lambda}$. When the resulting equations are deducted on either side, we arrive at

$$\frac{\partial}{\partial \lambda}\left(\frac{1}{H \cos \Phi}\frac{\partial D}{\partial \lambda}\right) + \frac{\partial}{\partial \Phi}\left(\frac{\cos \Phi}{H}\frac{\partial D}{\partial \Phi}\right) + \frac{\partial}{\partial \Phi}\left(\frac{fT}{H}\frac{\partial D}{\partial \lambda}\right) - \frac{\partial}{\partial \lambda}\left(\frac{fT}{H}\frac{\partial D}{\partial \Phi}\right)$$

$$= \frac{\partial}{\partial \lambda}\left(\frac{R}{H}\int_{-H}^{0} F'^{\Phi}\,dz\right) - \frac{\partial}{\partial \Phi}\left(\frac{R \cos \Phi}{H}\int_{-H}^{0} F'^{\lambda}\,dz\right) \qquad (4.240)$$

Unknown D is the time derivative of the stream function

$$D = \frac{\Psi^{m+1} - \Psi^{m-1}}{2T} \qquad (4.241)$$

Equation (4.240) is of elliptical type, its solution can be searched by any iteration method.

5.4 Solution of the stream function equation around islands

We are left with one additional problem to be addressed, i.e., integration of an elliptical equation in the multiply connected domains. For the depth-averaged flow a multiply connected domain is any water body with islands. If we solve a stream function equation in such a domain applying the standard boundary equation ($\Psi = 0$) everywhere, i.e., at the mainland and at the islands the derived solution will behave in a strange manner. To show this consider a line integral around any island $\oint \frac{\partial \Psi}{\partial s}\,ds$ and travel around the island to the point we have started from; the resulting integral does not vanish. From above we may as well conclude that the solution which we have obtained does not possess the property of uniqueness. This problem is well known in dynamical oceanography since in the 50's the main tool to study the wind and density driven motion was the stream function equation. The problem was posed and solved by Kamenkovitch (1961). In the ensuing considerations we shall use the results derived by Kowalik (1969) and Semtner (1973, 1974).

Let us start from the elliptical equation (4.231a)

$$\mathcal{L}D = \mathcal{F} \qquad (4.242)$$

for the dependent variable $D = \frac{\partial \Psi}{\partial t}$. Notice that this is a steady state equation (!), as such it is solved at every time step. The boundary conditions for (4.242) can be found by assuming that the coast of the ocean is described by an analytical curve, $\Gamma(x, y)$. We choose two directions: normal ($\vec{n}$) and tangential ($\vec{s}$) to this curve, as shown in Fig.4.13.

With this assumption the boundary condition at the mainland (Γ_0) becomes

$$\bar{u}_n = \frac{\partial \Psi}{\partial s} \quad \text{or} \quad \Psi(s)|_{\Gamma_0} = C_0',$$

and similarly for the variable D,

$$\frac{\partial \bar{u}_n}{\partial t} = \frac{\partial D}{\partial s} = 0 \quad \text{or} \quad D(s)|_{\Gamma_0} = C_0 \qquad (4.243)$$

Here C_o is a constant to be chosen later on.

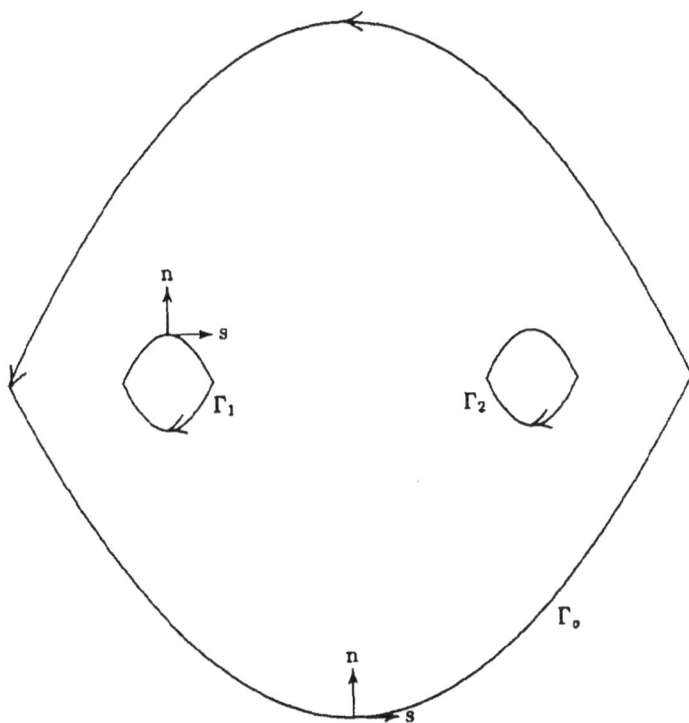

Fig. 4.13
Contour integration around the mainland coast Γ_o and around the islands Γ_1 and Γ_2.
The movement around the main boundary is counterclockwise,
while around the islands is clockwise.

Let $\Gamma_1, \Gamma_2, \ldots, \Gamma_n$ be the contours around the islands located inside the main contour Γ_0 as shown in Fig.4.13. Then the conditions (4.243) apply to every contour as

$$D|_{\Gamma_1} = C_1; \quad D|_{\Gamma_2} = C_2; \ldots, \quad D|_{\Gamma_n} = C_n \qquad (4.244)$$

At this point the problem is unspecified because the values of the integration constants are not known. When there are no islands (simply connected domain) one can take $\Psi|_{\Gamma_0} = C_0$ as zero and obtain a solution of (4.242). This is not the case for the multiply connected domain. It is possible to assign a definite value to one of the arbitrary constants of integration $(C_0, C_1, \ldots, C_n)$ and for our later use we take the value C_0 on the contour Γ_0 as zero $(C_0 = 0)$. Therefore we shall search a solution of (4.242) subject to the following boundary conditions,

$$D|_{\Gamma_0} = 0; \quad D|_{\Gamma_1} = C_1; \quad D|_{\Gamma_2} = C_2; \ldots, \quad D|_{\Gamma_n} = C_n \qquad (4.244a)$$

Let us assume for the moment that the constants $C_1, C_2, \ldots, C_n$ are known. Equation (4.242) is linear and so are the boundary conditions (4.244a). This gives us an opportunity to use the principle of superposition in constructing a solution of (4.242) as a sum of solutions of the nonhomogeneous equation (4.242) with homogeneous boundary conditions $(C_0 = C_1 =, \ldots, C_n = 0)$, and homogeneous equation (i.e. $\mathcal{F} = 0$) with nonzero boundary conditions $(C_0 = 0, C_1 \neq 0, \text{or } C_2 \neq 0, \ldots, \text{or } C_n \neq 0)$. In accordance with the above, the solution of (4.242) is

$$D = D_0 + \sum_{q=1}^{n} C_q D_q \qquad (4.245)$$

Here, D_0 is a solution of (4.242) with the homogeneous boundary conditions:

$$D|_{\Gamma_0} = D|_{\Gamma_1} = D|_{\Gamma_2} = \ldots = D|_{\Gamma_n} = 0 \qquad (4.246)$$

and D_q is solution of the homogeneous part of (4.242) with the boundary conditions:

$$D_q|_{\Gamma_q} = 1, \quad \text{and} \quad D_q|_{\Gamma_0} = D_q|_{\Gamma_1} = \ldots = D_q|_{\Gamma_n} = 0 \qquad (4.247)$$

Searching a solution of the elliptical equation in the form of the sum (4.245), the boundary values on the contours $\Gamma_1, \Gamma_2, \ldots, \Gamma_n$ are defined as 1; however, the values of the constants $C_1, C_2, \ldots, C_n$ are still unknown. Here we involve some auxiliary conditions relevant to integration along a contour in a multiply connected domain. Suppose we move along contour Γ_q, after arriving at the same point we have started from, we observe that the values of the field variables do not change, since this is the equation for the steady state. Say for the function D, $dD = 0$ along Γ_q, and for the sea level variations, $d\zeta = 0$. The last condition in the integral form is

$$\oint_{\Gamma_q} \frac{\partial \zeta}{\partial s} ds = \oint_{\Gamma_q} (\nabla \zeta \cdot d\vec{s}) = \oint_{\Gamma_q} \frac{\partial \zeta}{\partial \lambda} d\lambda + \frac{\partial \zeta}{\partial \Phi} d\Phi = 0 \qquad (4.248)$$

The value of such an integral depends on the direction of motion along the contour Γ_q, as shown in Fig.4.13 .

If one performs integration along every contour $\Gamma_1, \Gamma_2, \ldots, \Gamma_n$ it results in n conditions for determining the n constants $C_1, C_2, \ldots, C_n$. We shall use the equations of motion to construct the above contour integral. Let us multiply (4.217) by $d\lambda R \cos \Phi$ and (4.218) by $R d\Phi$ add the resulting equations on either side and applying the contour integral, we arrive at:

$$\oint [-\frac{\cos \Phi}{H} \frac{\partial D}{\partial \Phi} d\lambda + \frac{1}{H \cos \Phi} \frac{\partial D}{\partial \lambda} d\Phi] - \oint (\frac{f}{H} \frac{\partial \Psi}{\partial \lambda} d\lambda + \frac{f}{H} \frac{\partial \Psi}{\partial \Phi} d\Phi)$$

$$= -\oint (\frac{\partial p_a}{\partial \lambda} d\lambda + \frac{\partial p_a}{\partial \Phi} d\Phi) - \oint (\frac{\partial \zeta}{\partial \lambda} d\lambda + \frac{\partial \zeta}{\partial \Phi} d\Phi)$$

$$+ \oint [(\frac{1}{H} \int_{-H}^{0} G^\lambda dz) d\lambda + (\frac{1}{H} \int_{-H}^{0} G^\Phi dz) d\Phi]$$

$$- \oint \frac{g}{\rho_o H} [(\int_{-H}^{0} \int_{z}^{0} \frac{\partial \rho'}{\partial \lambda} dz dz) d\lambda + (\int_{-H}^{0} \int_{z}^{0} \frac{\partial \rho'}{\partial \Phi} dz dz) d\Phi] \tag{4.249}$$

In (4.249) the terms due to the Coriolis force, atmospheric pressure and sea level are equal to zero according to the condition (4.248).

We now apply (4.249) to define the arbitrary constants in (4.245)

$$\oint_{\Gamma_q} [-\frac{\cos \Phi}{H} (\frac{\partial D_0}{\partial \Phi} + \sum_{p=1}^{n} C_p \frac{\partial D_p}{\partial \Phi}) d\lambda + \frac{1}{H \cos \Phi} (\frac{\partial D_0}{\partial \lambda} + \sum_{p=1}^{n} C_p \frac{\partial D_p}{\partial \lambda}) d\Phi]$$

$$+ \oint_{\Gamma_q} [(\frac{1}{H} \int_{-H}^{0} G^\lambda dz) d\lambda + (\frac{1}{H} \int_{-H}^{0} G^\Phi dz) d\Phi]$$

$$- \oint_{\Gamma_q} \frac{g}{\rho_o H} [(\int_{-H}^{0} \int_{z}^{0} \frac{\partial \rho'}{\partial \lambda} dz dz) d\lambda + (\int_{-H}^{0} \int_{z}^{0} \frac{\partial \rho'}{\partial \Phi} dz dz) d\Phi] \tag{4.250}$$

Here both p and q change from 1 to n.

We may stop here, estimate the contour integrals, find the constants, and close the whole solution. But since the numerical computation is an art rather than a science we shall proceed to simplify the above result. We change the contour integral to the surface integral by Stokes theorem,

$$\oint_\Gamma (\vec{a} \cdot d\vec{s}) = \iint_S \frac{1}{R \cos \Phi} (\frac{\partial a_\lambda}{\partial \Phi} - \frac{\partial a_\Phi}{\partial \lambda}) R^2 \cos \Phi d\Phi d\lambda \tag{4.251}$$

Here, Γ is the boundary of the surface S; $\vec{a}$ is vector with the components a_λ and a_Φ; ds is a directed element of the contour Γ.

Let us apply this result to the first integral in (4.249), it can be rewritten as

$$\oint [-\frac{\cos\Phi}{H}\frac{\partial D}{\partial\Phi}d\lambda + \frac{1}{H\cos\Phi}\frac{\partial D}{\partial\lambda}d\Phi] = \oint(a_\lambda d\lambda + a_\Phi d\Phi \qquad (4.252)$$

With the values of a_λ and a_Φ defined,

$$(\frac{\partial a_\lambda}{\partial\Phi} - \frac{\partial a_\Phi}{\partial\lambda})$$

is equal to:

$$\frac{\partial}{\partial\lambda}(\frac{1}{H\cos\Phi}\frac{\partial D}{\partial\lambda}) + \frac{\partial}{\partial\Phi}(\frac{\cos\Phi}{H}\frac{\partial D}{\partial\Phi})$$

which is exactly the leading expression in (4.231) and (4.240). Thus these equations may serve to estimate the surface integral instead of the line integral. The surface integral is much simpler than the line integral because the sense of direction is not attached to it. What is the meaning of the surface integral taken over an island? We do not know, and we do not need to know since we are only interested in the coastal points, therefore let us set all dependent variables over an island equal to zero and then only the grid points which are at the shore will generate nonzero values of the surface integral.

6. A three-dimensional semi-implicit model

A three-dimensional model was developed by J. Backhaus (1985) for the North Sea. The model is implicit for the barotropic mode so that relatively large computational time steps can be employed. To facilitate further consideration we recal the basic equations and boundary conditions from Chapter 1. The continuous partial differential equations of the model are as follows:
Mass Conservation:

$$\frac{\partial u}{\partial x} + \frac{\partial v}{\partial y} + \frac{\partial w}{\partial z} = 0 \qquad (4.253)$$

x-directed Momentum Conservation:

$$\frac{\partial u}{\partial y} - fv + \frac{1}{\rho}\frac{\partial p}{\partial x} + u\frac{\partial u}{\partial x} + v\frac{\partial u}{\partial y} + w\frac{\partial u}{\partial z}$$
$$- \frac{\partial}{\partial x}\left(N_h\frac{\partial u}{\partial x}\right) - \frac{\partial}{\partial y}\left(N_h\frac{\partial u}{\partial y}\right) - \frac{\partial}{\partial z}\left(N_z\frac{\partial u}{\partial z}\right) = 0 \qquad (4.254)$$

y-directed Momentum Conservation:

$$\frac{\partial v}{\partial y} + fu + \frac{1}{\rho}\frac{\partial p}{\partial y} + u\frac{\partial v}{\partial x} + v\frac{\partial v}{\partial y} + w\frac{\partial v}{\partial z}$$
$$- \frac{\partial}{\partial x}\left(N_h\frac{\partial v}{\partial x}\right) - \frac{\partial}{\partial y}\left(N_h\frac{\partial v}{\partial y}\right) - \frac{\partial}{\partial z}\left(N_z\frac{\partial v}{\partial z}\right) = 0 \qquad (4.255)$$

Density Conservation:

$$\frac{\partial \sigma_t}{\partial t} + u\frac{\partial \sigma_t}{\partial x} + v\frac{\partial \sigma_t}{\partial y} + w\frac{\partial \sigma_t}{\partial z} = 0 \tag{4.256}$$

Hydrostatic Equation:

$$\frac{\partial p}{\partial z} = -\rho g \tag{4.257}$$

with

$$\rho = \rho_0 + \sigma_t \tag{4.258}$$

Here, $\sigma_t(x, y, z, t) = 1000(\rho - 1)$. A few comments are in order with regard to the above equations. Most notable is that the conservation equations are not written for the separate fields of salt and temperature. Instead, there is a conservation equation for the density field alone. This simplification is due to the strong control that salinity exerts on density in the estuarine waters to be modelled: the three variables would need to be retained in models of other regions. In fact, the original North Sea model code incorporated separate equations for each of temperature and salinity, and computed the density field using the complete equation of state, incorporating compressibility effects for deep ocean modelling. Here, the compressibility effects were eliminated because the waters under consideration are relatively shallow.

In the momentum equations the stress terms are expressed in terms of eddy coefficients, as that is how these terms are evaluated in the model. The horizontal eddy coefficient N_h is taken as constant, except that it is set to zero when evaluating the flows parallel to boundaries in order to implement a free slip condition. This lateral boundary condition is used throughout this section, although in principal more complex side-wall conditions could be imposed. The vertical eddy viscosity N_z is assumed to depend on vertical shear and on the Richardson number in the case of baroclinic flows.

The following boundary conditions will be used:
At the surface:

$$\frac{\partial \zeta}{\partial t} = w_{z=\zeta} - u\frac{\partial \zeta}{\partial x} - v\frac{\partial \zeta}{\partial y} \qquad \text{at } z = \zeta \tag{4.259}$$

At the bottom:

$$w + u\frac{\partial D}{\partial x} + v\frac{\partial D}{\partial y} = 0 \qquad \text{at } z = -D \tag{4.260}$$

A condition of no flow normal to the solid walls of the modelled water body is enforced, analogous to the seafloor boundary condition:

$$\vec{u} \cdot \vec{n} \equiv 0 \tag{4.261}$$

where $\vec{n}$ is a unit vector in the direction normal to the wall. Moreover, conditions for the bottom and surface stresses will be formulated in the following way (see Ch. I):

$$N_z \frac{\partial u}{\partial z}\bigg|_{\text{bottom}} = \tau_z^b = r u_{\text{bottom}} \sqrt{u_{\text{bottom}}^2 + v_{\text{bottom}}^2} \qquad (4.262a)$$

$$N_z \frac{\partial v}{\partial z}\bigg|_{\text{bottom}} = \tau_y^b = r u_{\text{bottom}} \sqrt{u_{\text{bottom}}^2 + v_{\text{bottom}}^2} \qquad (4.262b)$$

$$N_z \frac{\partial u}{\partial z}\bigg|_{\text{surface}} = \tau_z^s = \rho_{\text{air}} C_{10} U_{\text{wind}} \frac{\sqrt{U_{\text{wind}}^2 + V_{\text{wind}}^2}}{\rho_{\text{water}}} \qquad (4.262c)$$

$$N_z \frac{\partial v}{\partial z}\bigg|_{\text{surface}} = \tau_y^s = \rho_{\text{air}} C_{10} U_{\text{wind}} \frac{\sqrt{U_{\text{wind}}^2 + V_{\text{wind}}^2}}{\rho_{\text{water}}} \qquad (4.262d)$$

It is also assumed that no stress is transferred laterally from the side walls, which is implemented by requiring that at wall boundaries:

$$N_{h_{\text{wall}}} = 0 \qquad (4.263)$$

Although the model is written in terms of velocities rather than transports, it is still necessary to integrate the relevant equations over each layer to arrive at the proper vertical averages to use in the finite difference equations. The model described here assumes that there is no vertical variation of density or horizontal velocity within each layer. The vertical velocity must consequently vary linearly within each layer, in order to satisfy the differential form of the continuity equation. For the derivations in this section, subscript 1 refers to the bottom of a particular layer, and subscript 2 refers to the top. These notations are replaced by explicit labels denoting the sea bed and the sea surface, where relevant.

6.1 Continuity equation

Integrating the equation of continuity (4.253) over the entire water column gives:

$$\frac{\partial U}{\partial x} + \frac{\partial V}{\partial y} = \frac{\partial \zeta}{\partial t} \qquad (4.264)$$

where the kinematic boundary conditions have been applied at the sea bed and the free surface, and where

$$U = \int_{-D}^{\zeta} u \, dz \qquad (4.265)$$

$$V = \int_{-D}^{\zeta} v \, dz \qquad (4.266)$$

The continuity eq.(4.253) is next integrated over a discrete layer of thickness dz. Three cases are considered: the bottom layer, the interior layers, and the surface layer.

Bottom layer:

$$\int_{-D}^{h_2} \frac{\partial u}{\partial x}\, dz + \int_{-D}^{h_2} \frac{\partial v}{\partial y}\, dz = w_{-D} - w_{h_2} \qquad (4.267)$$

Interior layers:

$$\int_{h_1}^{h_2} \frac{\partial u}{\partial x}\, dz + \int_{h_1}^{h_2} \frac{\partial v}{\partial y}\, dz = w_{h_1} - w_{h_2} \qquad (4.268)$$

Top layer:

$$\int_{h_1}^{\zeta} \frac{\partial u}{\partial x}\, dz + \int_{h_1}^{\zeta} \frac{\partial v}{\partial y}\, dz = w_{h_1} - w_{\zeta} \qquad (4.269)$$

Using the boundary conditions (4.259) and (4.260), and Leibnitz's rule to bring the derivative in front of the integrals, these equations simplify to:

Bottom layer:

$$\frac{\partial}{\partial x} \int_{-D}^{h_2} u\, dz + \frac{\partial}{\partial y} \int_{-D}^{h_2} v\, dz = w_{h_2} \qquad (4.270)$$

Interior layers:

$$\frac{\partial}{\partial x} \int_{h_1}^{h_2} u\, dz + \frac{\partial}{\partial y} \int_{h_1}^{h_2} v\, dz = w_{h_1} - w_{h_2} \qquad (4.271)$$

Top layer:

$$\frac{\partial}{\partial x} \int_{h_1}^{\zeta} u\, dz + \frac{\partial}{\partial y} \int_{h_1}^{\zeta} v\, dz = w_{h_1} - \frac{\partial \zeta}{\partial t} \qquad (4.272)$$

The assumption of uniform horizontal velocity and density within each layer will be invoked later where the integrals are evaluated to arrive at the layer-averaged equations. Note that if the individual layer continuity equations above are summed over the entire water column, an equivalent equation to (4.264) is obtained: the internal vertical velocities cancel out.

6.2 Equation of motion

For a layer lying between h_1 and h_2, the pressure term in eq.(4.254), integrated over the layer, is given by

$$\int_{h_1}^{h_2} \frac{1}{\rho} \frac{\partial p}{\partial x} \, dz \qquad (4.273)$$

As stated in eq.(4.258), the density field has two components, a uniform one and one which varies both horizontally and vertically. The term due to the uniform part of the density field gives rise to the surface pressure gradient term. The term which varies in the vertical gives rise to baroclinic motions. Writing these two contributions explicitly:

$$p = p_{\text{barotropic}} + p_{\text{baroclinic}} = \int_z^\zeta \rho_0 g \, dz + \int_z^\zeta \sigma_t g \, dz \qquad (4.274)$$

Evaluating the barotropic term, the barotropic pressure and its horizontal gradient are given respectively by:

$$p_{\text{barotropic}} = \rho_0 g(\zeta - z) \qquad (4.275)$$

$$\frac{\partial p}{\partial x}\bigg|_{\text{barotropic}} = g \frac{\partial \zeta}{\partial x} \qquad (4.276)$$

The term under the integral in expression (4.273) arising from the barotropic pressure is independent of z, and the integral is readily evaluated, giving for the x and y components of the barotropic pressure force:

$$g(h_2 - h_1)\frac{\partial \zeta}{\partial x}, \qquad g(h_2 - h_1)\frac{\partial \zeta}{\partial y} \qquad (4.277)$$

For the remainder of this discussion, the symbol p will be used to refer to the baroclinic part of the pressure field. The Boussinesq approximations will be used to evaluate the baroclinic part of the expression (4.273). Using Leibnitz's rule to bring the derivative out of the integral, one obtains:

$$\int_{h_1}^{h_2} \frac{\partial p}{\partial x} \, dz = \frac{\partial}{\partial x} \int_{h_1}^{h_2} p \, dz - p_{h_2} \frac{\partial h_2}{\partial x} + p_{h_1} \frac{\partial h_1}{\partial x} \qquad (4.278)$$

Considering again the three classes of layers, interior, bottom and top:
Interior layers:

In this case, spatial derivatives of the upper and lower interfaces h_1 and h_2 vanish, resulting in

$$\int_{h_1}^{h_2} \frac{\partial p}{\partial x}\, dz = \frac{\partial}{\partial x} \int_{h_1}^{h_2} p\, dz \qquad (4.279)$$

which is unambiguously evaluated, once the vertical variations of density are specified.

Top layer:

The baroclinic pressure is zero at the sea surface:

$$p_\zeta \equiv 0 \qquad (4.280)$$

so that the interior layer expression applies.

Bottom layer:

In this case, the surface describing the top of the layer has no horizontal gradient. Thus,

$$\int_{-D}^{h_2} \frac{\partial p}{\partial x}\, dz = \frac{\partial}{\partial x} \int_{-D}^{h_2} p\, dz + p_{-D}\frac{\partial(-D)}{\partial x} \qquad (4.281)$$

The integral in eqs.(4.279) and (4.281) is evaluated as follows. If one assumes that ρ is uniform within each layer, then the baroclinic part of the pressure field varies linearly within the layer:

$$p = a - bz \qquad (4.282)$$

Thus, the integral term in these equations becomes

$$\begin{aligned}
\frac{\partial}{\partial x} \int_{h_1}^{h_2} p\, dz &= \frac{\partial}{\partial x} \int_{h_1}^{h_2} (a - bz)\, dz \\
&= \frac{\partial}{\partial x}\left[(h_2 - h_1)\left(a - \frac{(h_2 + h_1)}{2}b\right)\right] \\
&= \frac{\partial}{\partial x}[h p_{z=\text{mid-layer}}] \qquad (4.283)
\end{aligned}$$

Equation (4.283) indicates that the baroclinic pressure term is given by the gradient of the layer thickness $(h_2 - h_1)$, multiplied by the baroclinic pressure at mid-layer depth. Equation (4.283) is applicable to the bottom layer, in which h_1 is replaced by $-D$. Adding the second term of eq.(4.281) gives, for the bottom layer:

$$\int_{-D}^{h_2} \frac{\partial p}{\partial x}\, dz = \frac{\partial}{\partial x} \int_{-D}^{h_2} (a - bz)\, dz + (a - b(-D))\frac{\partial(-D)}{\partial x}$$

$$= (h_2 + D)\left(\frac{\partial a}{\partial x} - \frac{(h_2 - D)}{2}\frac{\partial b}{\partial x}\right)$$

$$= h\frac{\partial p}{\partial x}\Big|_{z=\text{mid-layer}} \tag{4.284}$$

For the interior layers, the layer thickness is independent of the horizontal position, so eq.(4.284) also applies, although the finite-difference equations for the interior layer are based on eq.(4.283). For the top layer, h_2 is replaced by ζ, and the small horizontal variation in layer thickness caused by the spatial variation of the water surface enters into both the baroclinic and barotropic components of pressure.

By incorporating the continuity equation, the complete acceleration terms in the x-momentum equation can be written:

$$\frac{\partial u}{\partial t} + \frac{\partial uu}{\partial x} + \frac{\partial uv}{\partial y} + \frac{\partial uw}{\partial z} \tag{4.285}$$

These terms, integrated over a generalized depth range extending from h_1 to h_2 become

$$\int_{h_1}^{h_2}\frac{\partial u}{\partial t}\,dz + \int_{h_1}^{h_2}\frac{\partial uu}{\partial x} + \int_{h_1}^{h_2}\frac{\partial uv}{\partial y}\,dz + (uw)_{z=h_2} - (uw)_{z=h_1} \tag{4.286}$$

Using Leibnitz's rule the advective terms become

$$\frac{\partial}{\partial t}\int_{h_1}^{h_2}u\,dz + \frac{\partial}{\partial x}\int_{h_1}^{h_2}uu\,dz + \frac{\partial}{\partial y}\int_{h_1}^{h_2}uv\,dz + (uw)_{z=h_2} - (uw)_{z=h_1}$$

$$- u_{h_2}\frac{\partial h_2}{\partial t} + u_{h_1}\frac{\partial h_1}{\partial t} - u_{h_2}^2\frac{\partial h_2}{\partial x} + u_{h_1}^2\frac{\partial h_1}{\partial x} - (uv)_{h_2}\frac{\partial h_2}{\partial y} + (uv)_{h_1}\frac{\partial h_1}{\partial y} \tag{4.287}$$

Once again, it is necessary to consider the three types of layers. The interior layer is simplest, because the two bounding surfaces have no temporal or spatial variation. Interior layers:

$$\frac{\partial}{\partial t}\int_{h_1}^{h_2}u\,dz + \frac{\partial}{\partial x}\int_{h_1}^{h_2}uu\,dz + \frac{\partial}{\partial y}\int_{h_1}^{h_2}uv\,dz + (uw)_{h_2} - (uw)_{h_1} \tag{4.288}$$

Bottom layer:

$$\frac{\partial}{\partial t}\int_{-D}^{h_2}u\,dz + \frac{\partial}{\partial x}\int_{-D}^{h_2}uu\,dz + \frac{\partial}{\partial y}\int_{-D}^{h_2}uv\,dz + (uw)_{h_2} \tag{4.289}$$

Top layer:

In this case, the surface boundary condition eliminates considerable complexity which arises in the intermediate steps, giving:

$$\frac{\partial}{\partial t}\int_{h_1}^{h_2} u\,dz + \frac{\partial}{\partial x}\int_{h_1}^{h_2} uu\,dz + \frac{\partial}{\partial y}\int_{h_1}^{h_2} uv\,dz + (uw)_{h_1} \qquad (4.290)$$

The above equations require further manipulation before they can be written in finite difference form. First, it is assumed that u and v are vertically uniform within each layer, allowing the integrals in the previous equations to be readily evaluated, as shown in eqs.(4.291).

$$\int_{h_1}^{h_2} uu\,dz = uuh, \qquad \int_{h_1}^{h_2} uv\,dz = uvh, \qquad \int_{h_1}^{h_2} vv\,dz = vvh \qquad (4.291)$$

The assumption that the velocity profile is uniform in the vertical within each layer is not strictly required, and is made merely for clarity of presentation. As discussed by Kuipers and Vreugdenhil (1973), one can consider that the velocity varies with vertical position within the layer, the vertically-averaged part giving rise to the nonlinear terms as discussed here, and the nonuniform part contributing to the horizontal eddy viscosity term. The present model makes this latter interpretation. The effects arising from the velocity shear within a layer then become part of the horizontal eddy viscosity, which is controlled by an adjustable parameter of the model.

Interior layers:

The momentum terms for the interior layers simplify considerably, as the spatial and temporal derivatives of h_1 and h_2 vanish. Defining $h_z = h_2 - h_1$, the nonlinear terms for the interior layers may be written:

$$\frac{\partial u}{\partial t} + \frac{\partial uu}{\partial x} + \frac{\partial uv}{\partial y} + \frac{1}{h_z}(uw)_{h_2} - \frac{1}{h_z}(uw)_{h_1} \qquad (4.292)$$

Bottom layer:

For the bottom layer, the layer thickness, which depends on the local depth of water, must be retained within the differentiated expressions:

$$\frac{\partial u}{\partial t} + \frac{1}{h_z}\frac{\partial uuh_z}{\partial x} + \frac{1}{h_z}\frac{\partial uvh_z}{\partial y} + \frac{1}{h_z}(uw)_{h_2} \qquad (4.293)$$

Top layer:

For the top layer, expression (4.290) simplifies to:

$$\frac{\partial u}{\partial t} + \frac{1}{h_z}\frac{\partial h_z}{\partial t} + \frac{\partial uu}{\partial x} + \frac{\partial uv}{\partial y} + \frac{1}{h_z}(uw)_{h_1} + \frac{uu}{h_z}\frac{\partial h_z}{\partial x} + \frac{uv}{h_z}\frac{\partial h_z}{\partial y} \tag{4.294}$$

The last two terms are generally quite small, and were ignored in the simulations presented here.

The vertical eddy viscosity term is readily integrated over the layer depth, as it involves a complete derivative with respect to z:

$$\rho\int_{h_1}^{h_2} \frac{\partial}{\partial z}\left(N_z\frac{\partial u}{\partial z}\right)\,dz = \rho\left(N_z\frac{\partial u}{\partial z}\right)_{h_2} - \rho\left(N_z\frac{\partial u}{\partial z}\right)_{h_1}$$

$$= \tau_{z,\text{upper}} - \tau_{z,\text{lower}} \tag{4.295}$$

The evaluation of the vertical derivatives of u and v is complicated by the nonuniform grid spacing in the vertical, and is discussed in more detail when finite-difference equations are presented.

6.3 Density advection

Integrating eq.(4.256) between layers h_1 and h_2, one obtains

$$\int_{h_1}^{h_2} \frac{\partial \sigma_t}{\partial t}\,dz + \int_{h_1}^{h_2} \frac{\partial u\sigma_t}{\partial x}\,dz + \int_{h_1}^{h_2} \frac{\partial v\sigma_t}{\partial y}\,dz + \int_{h_1}^{h_2} \frac{\partial w\sigma_t}{\partial z}\,dz = 0 \tag{4.296}$$

Considering three classes of layers, and dividing by the layer thickness h_z, the density advection equations may be written as follows.
Interior layers:

$$\frac{\partial \sigma_t}{\partial t} + \frac{\partial}{\partial x}(u\sigma_t) + \frac{\partial}{\partial y}(v\sigma_t) + \left(\frac{w\sigma_t}{h_z}\right)_{h_2} - \left(\frac{w\sigma_t}{h_z}\right)_{h_1} = 0 \tag{4.297}$$

Multiplying the layer continuity eq.(4.301), presented in the next section, by the layer density, and subtracting from eq. (4.296) gives:

$$\frac{\partial \sigma_t}{\partial t} + u\frac{\partial \sigma_t}{\partial x} + v\frac{\partial \sigma_t}{\partial y} + \frac{w_{h_2}}{h_z}(\sigma_{t_{h_2}} - \sigma_t) - \frac{w_{h_1}}{h_z}(\sigma_{t_{h_1}} - \sigma_t) = 0 \tag{4.298}$$

where $\sigma_{t_{h_1}}$ and $\sigma_{t_{h_2}}$ indicate the value of density at the interface between layers.
Bottom layer:

The density flux through the bottom of the layer is zero. Hence

$$\frac{\partial \sigma_t}{\partial t} + u \frac{\partial \sigma_t}{\partial x} + v \frac{\partial \sigma_t}{\partial y} + \frac{w_{h_2}}{h_z}(\sigma_{t_{h_2}} - \sigma_t) = 0 \qquad (4.299)$$

Top layer:

Since the flux through the top of this layer is zero,

$$\frac{\partial \sigma_t}{\partial t} + u \frac{\partial \sigma_t}{\partial x} + v \frac{\partial \sigma_t}{\partial y} - \frac{w_{h_1}}{h_z}(\sigma_{t_{h_1}} - \sigma_t) = 0 \qquad (4.300)$$

6.4 The layer-averaged equations

The results of the previous section may be combined to give the following layer-averaged equations. In these equations, it is assumed that u and v represent layer-averaged values. Each layer is described by a set of identical equations. The kinematic boundary conditions are satisfied in these equations if the vertical velocity w is set to zero at the bottom of the water column, and is set to $\partial \zeta / \partial t$ at the free surface.

Continuity:

$$w_{\text{upper}} = \frac{\partial u h_z}{\partial x} + \frac{\partial u h_z}{\partial y} + w_{\text{lower}} \qquad (4.301)$$

x-momentum:

$$\frac{\partial u}{\partial t} + \frac{1}{h_z}\left(\frac{\partial uu h_z}{\partial x} + \frac{\partial uv h_z}{\partial y} + (wu)_{\text{lower}} - (wu)_{\text{upper}}\right) - fv + g\frac{\partial \zeta}{\partial x} + \frac{\partial p_{\text{baroclinic}}}{\partial x}$$
$$- N_h \frac{\partial^2 u}{\partial x^2} - N_h \frac{\partial^2 u}{\partial y^2} - \frac{1}{h_z}\tau_{x,\text{upper}} + \frac{1}{h_z}\tau_{x,\text{lower}} = 0 \quad (4.302)$$

y-momentum:

$$\frac{\partial v}{\partial t} + \frac{1}{h_z}\left(\frac{\partial uv h_z}{\partial x} + \frac{\partial vv h_z}{\partial y} + (wv)_{\text{lower}} - (wv)_{\text{upper}}\right) + fu + g\frac{\partial \zeta}{\partial y} + \frac{\partial p_{\text{baroclinic}}}{\partial y}$$
$$- N_h \frac{\partial^2 v}{\partial x^2} - N_h \frac{\partial^2 v}{\partial y^2} - \frac{1}{h_z}\tau_{y,\text{upper}} + \frac{1}{h_z}\tau_{y,\text{lower}} = 0 \quad (4.303)$$

Density:

$$\frac{\partial \sigma}{\partial t} + \frac{\partial u\sigma}{\partial x} + \frac{\partial v\sigma_t}{\partial y} + \frac{1}{h_z}(w\sigma)_{\text{lower}} - \frac{1}{h_z}(w\sigma)_{\text{upper}} = 0 \qquad (4.304)$$

where the t-subscript on σ_t has been dropped for clarity in later discussion.

Overall continuity:

$$\frac{\partial \zeta}{\partial t} + \frac{\partial U}{\partial x} + \frac{\partial V}{\partial y} = 0 \qquad (4.305)$$

where U and V represent the vertical integrals of the velocity components.

The layer-averaged equations derived above are solved on a three-dimensional grid (Fig. 4.14), where the x variable is associated with the index j, the y variable is associated with the index variable k, and the z variable associated with the index variable l. Index variable j increases with increasing x, index variable k increases with decreasing y, and index variable l increases with decreasing z. The horizontal grid spacing is uniform, but the vertical grid spacing varies in the vertical, while remaining the same with respect to horizontal location within the grid. Thus, the grid is defined by the horizontal grid spacing, and by a set of layer depths which define the vertical resolution. This set of layer depths is readily modified in the present model and is chosen to maximize the resolution of vertical density and velocity gradients, usually by choosing the layers to be thinner at the top of the water column.

Variables are distributed over the grid according to the Arakawa C grid (Arakawa and Lamb,1977). The fundamental computational cell can be visualized as a cube, with velocity values defined at the centres of each face of the cube, and density and pressure defined at the centre (S) of the cube itself (Fig.4.14). Notice that index used in this section differs from the notation used throughout this chapter. The velocity and density variables are staggered in space, which will allow many of the spatial gradients to be space-centred. Over the bulk of the modelled basin, lateral boundaries of the grid are adjusted to match the basin geometry in full grid cell increments: all cells have the same horizontal dimensions, and the grid is chosen to conform to the coastline within one half a grid spacing by requiring each cell to be either a land or a water cell. Narrow passes are accommodated by allowing the plan area and vertical cross-sections of the cells to be different from their nominal values. The requirement of no flow through the solid boundaries (eq.(4.261)) can be readily satisfied using the Arakawa C grid, since the velocity component normal to the model grid at a solid boundary is a computational variable and can be held zero throughout the simulation.

Two time-levels are used in the model, the present time level and the advanced time. In order to avoid confusion in describing relative locations in the grid, the following convention is used:

east and west: respectively increasing and decreasing values of x;
north and south: respectively increasing and decreasing values of y;
above and below: respectively increasing and decreasing values of z.

In the finite-difference operators and finite-difference equations, time step is denoted by T, and h is the horizontal grid spacing between like points. Variables with superscript (0) are evaluated at the present time, and a superscript (1) indicates a variable at the advanced time.

To simplify the appearance of the finite-difference equations, operator notation will be used for first-order differences and averages, as defined in the following

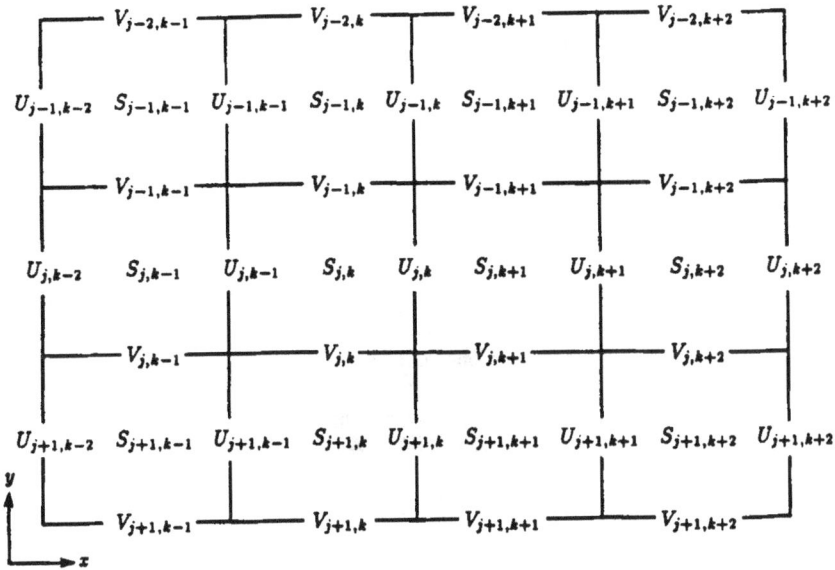

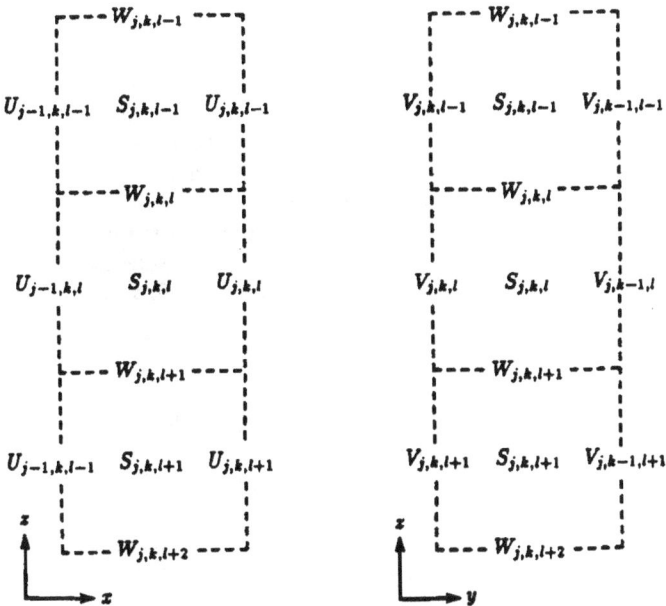

Fig. 4.14
Grid distribution

equations.

The horizontal differencing operators are defined in eqs.(4.306) and (4.307).

$$\delta_x \phi = \phi(x + h/2) - \phi(x - h/2) \tag{4.306}$$
$$\delta_y \phi = \phi(y + h/2) - \phi(y - h/2) \tag{4.307}$$

The u and w locations are staggered horizontally in the Arakawa C grid, as shown in Fig. 4.14. When δ_x is applied to the u variable in the equation for w, it determines the difference in value of its argument u between values one half grid to the east, and one half grid to the west of the location of w. The v and w locations are also staggered horizontally in the Arakawa C grid. When δ_y is applied to the v variable in the equation for w, it determines the difference in value of its argument v between values one half grid to the north, and one half grid to the south of the location of w.

The overbar notation is used to denote averages: the horizontal averaging operators are defined in eqs.(4.308) and (4.309) by

$$\overline{\phi}^x = \frac{\phi(x + h/2) + \phi(x - h/2)}{2} \tag{4.308}$$
$$\overline{\phi}^y = \frac{\phi(y + h/2) + \phi(y - h/2)}{2} \tag{4.309}$$

The variable h_z denotes layer thickness, and is staggered with respect to the u field. When $\overline{h_z}^x$ is evaluated at the location of the variable u, it causes the average of h_z one half grid to the west and one half grid to the east of the u variable to be calculated; h_z is also staggered with respect to the v field. When $\overline{h_z}^y$ is evaluated at the location of the variable v, it causes the average of h_z one half grid to the north and one half grid to the south of the v variable to be calculated.

The corresponding operators in the z-direction are more complex because of the variable grid spacing. Consider first the difference operator applied to a variable located midway between the top and bottom of a cell, such as u or v.

$$\delta_z \phi_l = \phi(z + \Delta z_{l-1}/2) - \phi(z - \Delta z_l/2) \tag{4.310}$$

The layer index is included because layer thickness varies from one layer to another. The u and N_z locations are staggered vertically in the Arakawa C grid. When δ_z is applied to the u variable to determine the vertical shear in the equation for N_z, it determines the difference in value of its argument one half grid above and one half grid below the location of N_z.

When the difference operator is applied to a variable located at the top or bottom of a cell, the above operator definition continues to apply, except that the subscript is l for all terms.

$$\delta_z \phi_l = \phi(z + \Delta z_l/2) - \phi(z - \Delta z_l/2) \tag{4.311}$$

Such is the case in evaluating the uw terms in the eqution for u, eq.(4.302).

The vertical average terms must be weighed by layer thickness, so that they represent the linear interpolations at the required location. When determining a value midway between the top and bottom of a cell, such as at a u point, the vertical averaging operation is given by

$$\overline{\phi}^z = \frac{\phi(z + \Delta z/2) + \phi(z - \Delta z/2)}{2} \tag{4.312}$$

When determining a value at the lower or upper interface of a cell, the expression becomes

$$\overline{\phi_l}^z = \frac{\phi(z + \Delta z_{l-1}/2) + \phi(z - \Delta z_l/2)}{2} \tag{4.313}$$

The spatially averaged layer thickness, for two cells connected by either a u-point or a v-point, frequently appears in the finite difference equations. For layers which are neither the top nor the bottom layer in both cells, the spatially averaged layer thickness is the nominal fixed layer thickness. For the top layer the instantaneous water level, ζ, must be included in calculating the average. For the bottom layer, the situation is more complex, depending on whether the two cells to be averaged have the same number of layers or not. If they have the same number of layers, the average of the two bottom layer thicknesses is used, shown in Fig.4.15a. If not, the spatially averaged layer depth is taken as the average of the bottom layer thickness in the shallower cell, and the nominal layer thickness in the deeper cell, as shown in Fig.4.15b.

In the following finite-difference equations, the grid indices (j, k, l) are suppressed, except in the layer continuity equation, where they are required in order to explicitly indicate the sweep through the water column.

Layer continuity equation (eq.(4.301))

$$w^{(0)}_{j,k,l} = w^{(0)}_{j,k,l+1} + \frac{T}{h}\left(\delta_x(u\overline{h_z}^x)^{(0)} + \delta_y(u\overline{h_z}^z)^{(0)}\right) \tag{4.314}$$

x-momentum equation (eq.(4.302)):

$$u^{(1)} = u^{(0)} - \frac{T}{h}\frac{1}{\overline{h_z}^z}\left(\delta_x(uu\overline{h_z}^x) + \delta_y(\overline{u}^y\overline{(\overline{vh_z}^y)}^x)\right)^{(0)} - \frac{T}{\Delta z}\delta_z\left(\overline{w}^x\overline{(\overline{u}^z)}^x\right)^{(0)}$$

$$+ T(f\overline{v}^y)^{(0)} + \frac{T}{h^2}\left(\delta_x(N_h\delta_x u) + \delta_y(N_h\delta_y u)\right)^{(0)}$$

$$- \frac{T}{h}(\delta_z p_{\text{baroclinic}} + (1-\alpha)g\delta_z\zeta)^{(0)} - T\left(\frac{\alpha g}{h}\delta_z\zeta + \delta_x\tau_x\right)^{(1)} \tag{4.315}$$

where α indicates the relative weighing to be applied to the implicit terms.

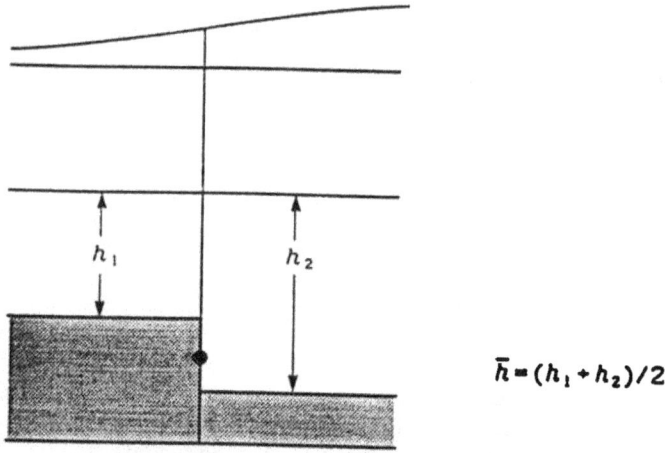

$$\bar{h} = (h_1 + h_2)/2$$

Fig. 4.15a

The definition of horizontally-averaged bottom layer thickness when the two adjacent cells have the same number of layers. The solid circle is drawn at the location of $\bar{h}$

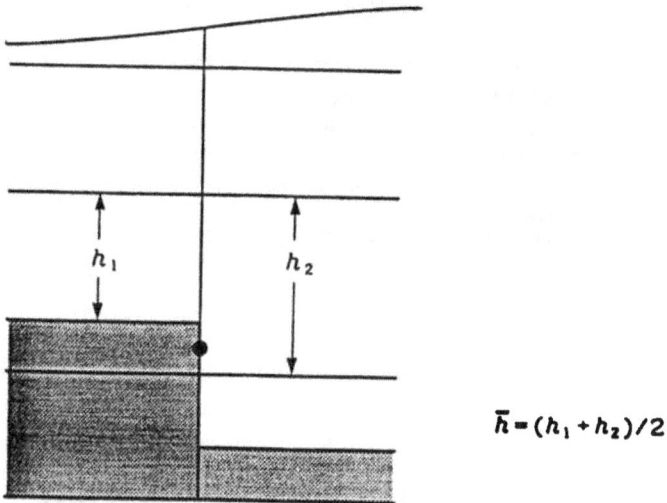

$$\bar{h} = (h_1 + h_2)/2$$

Fig. 4.15b

The definition of horizontally-averaged bottom layer thickness when the two adjacent cells have differing number of layers. The solid circle is drawn at the location of $\bar{h}$

y-momentum equation (eq.(4.303)):

$$v^{(1)} = v^{(0)} - \frac{T}{h}\frac{1}{\overline{h}_z^y}\left(\delta_y(vv\overline{h}_z^y) + \delta_z\left(\overline{v}^z\overline{(\overline{uh}_z^z)}^y\right)\right)^{(0)} - \frac{T}{\Delta z}\delta_z\left(\overline{w}^y\overline{(\overline{v}^y)}^z\right)^{(0)}$$

$$- T(f\overline{u}^z)^{(0)} + \frac{T}{h^2}\left(\delta_z(N_h\delta_z v) + \delta_y(N_h\delta_y v)\right)^{(0)}$$

$$- \frac{T}{h}(\delta_y p_{\text{baroclinic}} + (1-\alpha)g\delta_y\zeta)^{(0)} - T\left(\frac{\alpha g}{h}\delta_y\zeta + \delta_z\tau_y\right)^{(1)} \qquad (4.316)$$

where α indicates the relative weighting to be applied to the implicit terms. Overall continuity equation (eq.(4.305)):

$$\zeta^{(1)} = \zeta^{(0)} - \alpha\frac{T}{h}(\delta_z U^{(1)} + \delta_y V^{(1)}) - (1-\alpha)\frac{T}{h}(\delta_z U^{(0)} + \delta_y V^{(0)}) \qquad (4.317)$$

where

$$U = \sum_l u_l h_{z,l} \qquad (4.318)$$

$$V = \sum_l v_l h_{z,l} \qquad (4.319)$$

Density equation (eq.(4.304))

$$\sigma^{(1)} = \sigma^{(0)} - \frac{T}{h}(\beta\delta_z^*\sigma + \gamma\delta_y^*\sigma) - \frac{T}{h_z}\overline{w}^z\delta_z(\overline{\sigma}^z) + \frac{T^2}{h_z}\overline{w}^z\delta_z\left(\frac{w}{h_z}\delta_z\sigma\right) \qquad (4.320)$$

The density equation requires further explanation. The variables β and γ are the x and y distances travelled over one time step, normalized by the grid size. The horizontal advection is computed by the method of characteristics. The superscript * on the δ operators indicate that they represent the gradient operators acting on a density surface defined by three points: the grid point in question, the grid point in the upstream direction of the faster of u or v, and the grid point which is diagonal to the grid point in question, and closest to the direction of the velocity vector. Figure 4.16 illustrates the grids involved for the case where u and v are both positive, and v is larger than u.

Equation (4.320) uses a second-order scheme for the density advection in the vertical. The second-order vertical scheme is similar to the one-dimensional Lax-Wendroff scheme (Roache, 1972), modified for nonuniform grid sizes.

The method of solution for the complete set of finite difference equations involves five steps. First, the layer continuity eq.(4.314) is solved for the vertical velocity field. Second, the density field is updated, using the updated horizontal and vertical velocities in eq.(4.320). Third, the explicit parts of the momentum

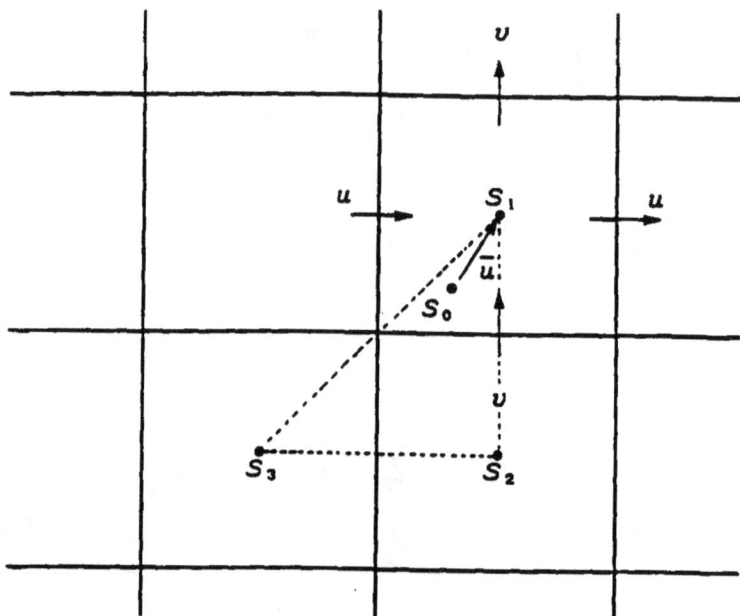

Fig. 4.16

Schematic representation of the horizontal advection term in the density equation. The characteristic from S_1 projected backwards one time step intersects the planar surface defined by S_1, S_2 and S_3, where an interpolated density S_0 is calculated, and projected to the location of S_1 at the advanced time. The characteristic velocity is defined by the two u–components and two v–components surrounding S_1.

equations, that is, all the terms at time level (0) in eqs.(4.315) and (4.316), are evaluated, and incorporated into intermediate values of the u and v fields. At this point, the u and v fields are out of balance with the barotropic pressure gradients, and so are not valid velocity fields. Fourth, the vertically integrated continuity eq.(4.317) is converted to an elliptic equation and solved using successive over relaxation. Fifth, the velocity fields and water levels are updated by the implicit parts of the relevant equations, the time level (1) terms of eqs.(4.315), (4.316), and (4.317). Two of the above five steps require further discussion: the solution of the overall continuity equation by successive over-relaxation and the solution of the implicit vertical eddy viscosity part of the momentum equations.

Next, consider the solution of the implicit form of the continuity equation. The partial differential forms of the equations are manipulated to form an elliptical equation for the water level at the advanced time. This equation is then cast into finite difference form for solution by successive over-relaxation. Following Backhaus (1985), the two momentum conservation eqs.(4.302) and (4.303) are schematized as

$$u_l^{(1)} = u_l^{(0)} - g\alpha T\zeta_x^{(1)} - g(1-\alpha)T\zeta_x^{(0)} + TX_l^{(0)} + \frac{T}{h_{z,l}}(\tau_l^{(1)} - \tau_{l+1}^{(1)}) \quad (4.321)$$

$$v_l^{(1)} = v_l^{(0)} - g\alpha T\zeta_y^{(1)} - g(1-\alpha)T\zeta_y^{(0)} + TY_l^{(0)} + \frac{T}{h_{z,l}}(\tau_l^{(1)} - \tau_{l+1}^{(1)}) \quad (4.322)$$

where superscript (1) refers to the advanced time level, superscript (0) to the earlier time level. Subscripts x and y refer to derivatives, not differential operators. The term α represents the degree of implicitness of the scheme: a value of 1 represents a fully explicit scheme, a value of 0 represents a fully implicit scheme. Subscript l refers to the layer. The terms $X_l^{(0)}$ and $Y_l^{(0)}$ incorporate internal pressure gradients, Coriolis forcing, advective terms and horizontal eddy viscosity at time level 0, and are evaluated in the previous phase of the solution.

For the lowest layer, L, the bottom stress, τ_{L+1} is given by

$$\tau_{L+1} = ru_L^{(1)}S, \qquad S = \sqrt{(u_L^2 + v_L^2)^{(0)}} \quad (4.323)$$

The bottom friction term is treated semi-implicitly, to eliminate any numerical stability problems arising from the large time steps employed in the model. The lowest layer momentum equation for the x-component becomes

$$u_L^{(1)}(1 + rST) = u_L^{(0)} - g\alpha T\zeta_x^{(1)} - g(1-\alpha)T\zeta_x^{(0)} + TX_L^{(0)} + \frac{T}{h_{z,L}}\tau_L^{(1)} \quad (4.324)$$

With

$$F = (1 + rST)^{-1} \quad (4.325)$$

it may be written as

$$u_L^{(1)} = F\left(u_L^{(0)} - g\alpha T\zeta_x^{(1)} - g(1-\alpha)T\zeta_x^{(0)} + TX_L^{(0)} + \frac{T}{h_{z,L}}\tau_L^{(1)}\right) \quad (4.326)$$

To solve the implicit equations, the vertically integrated continuity equation in the form

$$\zeta^{(1)} = \zeta^{(0)} - \alpha T(U_x^{(1)} + V_y^{(1)}) - T(1-\alpha)(U_x^{(0)} + V_y^{(0)}) \quad (4.327)$$

where

$$U = \sum_l u_l h_{z,l}$$
$$V = \sum_l v_l h_{z,l} \tag{4.328}$$

is recast by substituting the expressions (4.321) and (4.322) for u_l and v_l to give:

$$\zeta^{(1)} = \zeta^{(0)} - (1 - \alpha)T \sum_L^1 h_{z,l} u_l^{(0)}$$

$$- \alpha T \sum_{L-1}^l \left[h_{z,l} \left(u_l^{(0)} - g\alpha T \zeta_z^{(1)} - g(1-\alpha)T\zeta_z^{(0)} + TX_l^{(0)} + \frac{T}{h_{z,l}}(\tau_l^{(1)} - \tau_{l+1}^{(1)}) \right) \right]_z$$

$$- \alpha T \left[h_{z,L} F \left(u_L^{(0)} - g\alpha T\zeta_z^{(1)} - g(1-\alpha)T\zeta_z^{(0)} + TX_L^{(0)} + \frac{T}{h_{z,L}}\tau_L^{(1)} \right) \right]_z$$

$$+ V \text{ terms.} \tag{4.329}$$

All of the internal stresses in the water column cancel when vertically integrated, except for a term involving the stress at the top of the bottom layer. In order to evaluate this term using time-level (0) quantities, it will be assumed in eq. (4.329) that the change in the stress at the top of the bottom layer over one time step is negligible:

$$\tau_L^{(1)} = \tau_L^{(0)} \tag{4.330}$$

This is the only inconsistent approximation required to solve the SOR set of equations. It would not be required if an explicit form for the bottom stress were used, but the large time steps allowable in the present model require animplicit form. With this approximation, the internal stresses do not appear in the final form of the overall continuity equation, except for the stress at the top of the bottom layer, which may be evaluated from the time-level (0) variables. Thus, all quantities in eq.(4.329) are either time level (0) quantities, or the elevation field at time level (1).

For notational convenience, the friction factor F_l is defined such that:

$$F_L = F \qquad F_L = 1 \qquad \text{for } l \neq L \tag{4.331}$$

Thus, the continuity eq.(4.329) becomes

$$\zeta^{(1)} = \zeta^{(0)} - T(1 - \alpha) \sum [h_{z,l} u_l^{(0)}]_z - \alpha T \sum [h_{z,l} F_l u_l^{(0)}]_z - \alpha T \sum [h_l F_{z,l} X_l^{(0)}]_z$$

$$- \alpha T[\tau_l^{(1)}(1 - F_l)\tau_l^{(0)}]_z + \alpha(1-\alpha)gT \left[\zeta_z^{(0)} \sum F_l h_{z,l} \right]_z + \alpha^2 gT \left[\zeta_z^{(1)} \sum F_l h_{z,l} \right]_z$$

$$+ V \text{ terms} \tag{4.332}$$

In eq.(4.332), the first five terms on the right-hand side are obtained from the time-level (0) field exclusively, except for the surface stress, which is assumed known from the boundary condition for the top layer. Terms two to five collectively are referred as

$$U_x^* = \frac{\partial U^*}{\partial x} \tag{4.333}$$

the divergence of an intermediate vertically-integrated velocity. Two new symbols are also defined:

$$\sum F_l h_{x,l} = H \tag{4.334}$$

$$\delta = \zeta^{(1)} - \zeta^{(0)} \tag{4.335}$$

where δ is the change in water level over one time step. Thus,

$$\delta = -T(U_x^* + V_y^*) + \alpha^2 g T[(H\delta_x)_x + (H\delta_y)_y] + \alpha g T\left[(H\zeta_x^{(0)})_x + (H\zeta_y^{(0)})_y\right] \tag{4.336}$$

This is an elliptic equation in δ. A further variable ST^{**} is defined:

$$ST^{**} = -T(U_x^* + V_y^*) \tag{4.337}$$

Introducing the subscript notation appropriate to the Arakawa C grid, and defining:

$$g^* = gT/h^2 \tag{4.338}$$

$$ce^* = g^* \overline{H}_{j,k}^x \tag{4.339}$$

$$cw^* = g^* \overline{H}_{j-1,k}^x \tag{4.340}$$

$$cn^* = g^* \overline{H}_{j,k-1}^y \tag{4.341}$$

$$cs^* = g^* \overline{H}_{j,k}^y \tag{4.342}$$

$$ccc = ce^* + cw^* + cn^* + cs^* \tag{4.343}$$

eq.(4.336) becomes

$$\delta_{j,k}(1 + \alpha^2 ccc) - \alpha^2(ce^* \delta_{j+1,k} + cw^* \delta_{j-1,k} + cn^* \delta_{j,k-1} + cs^* \delta_{j,k+1})$$
$$= ST_{j,k}^{**} - \alpha ccc \zeta_{j,k}^{(0)} + \alpha(ce^* \zeta_{j+1,k}^{(0)} + cw^* \zeta_{j-1,k}^{(0)} + cn^* \zeta_{j,k-1}^{(0)} + cs^* \zeta_{j,k+1}^{(0)})$$
$$= ST_{j,k}^*. \tag{4.344}$$

Redefining:

$$cd = ce^* \frac{\alpha^2}{1 + \alpha^2} \tag{4.345}$$

and similarly for cw, cs and cn, the continuity equation may be simplified to

$$\delta_{j,k} - (ce_{j,k}\delta_{j+1,k} + cw_{j,k}\delta_{j-1,k} + cn_{j,k}\delta_{j,k-1} + cs_{j,k}\delta_{j,k+1}) = ST_{j,k} \tag{4.346}$$

Straighforward iteration of equations of the type (4.346) has been shown to lead to a very slow rate of convergence. The convergence may be accelerated by the method of successive over-relaxation (Varga, 1962) in which the contributions of the off-grid elements are increased relative to the solution point (see Ch. VI). Also, as each value is updated, it is used in the remaining part of the solution as one completes the sweep through the grid.

In the successive over-relation technique presented here, a superscript m represents the mth iteration to the solution. Time levels need not be indicated, as all solution fields are at the advanced time level. The fundamental iteration equation in the SOR technique is then:

$$\delta_{j,k}^{m+1} = (1 - \omega)\delta_{j,k}^m$$
$$+ \omega \left[ce_{j,k}\delta_{j+1,k}^m + cw_{j,k}\delta_{j-1,k}^{m+1} + cn_{j,k}\delta_{j,k-1}^{m+1} + s_{j,k}\delta_{j,k+1}^m + ST_{j,k} \right] \qquad (4.347)$$

where ω is a weighting factor which may be adjusted to maximize the rate of convergence. Proof of convergence is difficult, and in fact is generally considered impossible in a non-rectangular solution domain. In the present model the method has always proven to converge, provided the other aspects of the model were operating correctly. A controlling factor in the rate of convergence is the diagonal dominance of the solution matrix, i.e. the amount by which the following expression exceeds unity:

$$\frac{1 + T\sqrt{gH}/h}{T\sqrt{gH}/h} \qquad (4.348)$$

The closer the expression (4.348) is to unity, the slower is the rate of convergence. If the time step is relatively small, this expression will be considerably larger than unity; if the time step is relatively large, it will be close to unity. This is the manner in which the Courant–Friedrichs–Levy criterion enters the solution: the more the model exceeds the value set by the CFL criterion, the less diagonally dominant is the solution and hence the larger the number of iterations required to achieve convergence, that is, by doubling the extent of violation of the CFL criterion by halving grid size, the number of iterations was approximately doubled.

6.5 Vertical eddy viscosity term

The fifth phase of the solution involves evaluating the implicit part of the pressure gradient term and the vertical eddy viscosity term. The pressure gradient term is straightforward, and can be calculated from the elevation field determined by the SOR solution. The vertical eddy viscosity term is more complex, and discussed below.

The distribution of dependent variables amongst the layers, relevant to the vertical eddy viscosity term, are shown in Fig. 4.17. In the equations for the x-component of momentum, to follow, the index notation for the y direction will be omitted. The index l labels the stress and eddy viscosity at the top of the lth layer. A general expression for the stress at the top of the lth layer is

$$\tau_l = \frac{N_{j,l} + N_{j+1,l}}{h_{z,l-1} + h_{z,l}} [u_{l-1} - u_l] \qquad (4.349)$$

The momentum eq.(4.321) takes the following form:

$$u_l^{(1)} = \frac{T(\tau_l^{(1)} - \tau_{l+1}^{(1)})}{h_{z,l}} + R_l \qquad (4.350)$$

where the superscripts (1) and (0) refer to the advanced and present times, respectively. R_l incorporates both the value of velocity at time level zero, all the explicit acceleration terms and the accelerations due to the implicit pressure gradient.

Equation (4.350) becomes, on expanding the eddy terms:

$$u_l^{(1)} - \frac{T}{h_{z,l}} \frac{N_{j,l} + N_{j+1,l}}{h_{z,l-1} + h_l} \left[u_{l-1}^{(1)} - u_l^{(1)} \right]$$
$$+ \frac{T}{h_{z,l}} \frac{N_{j,l+1} + N_{j+1,l+1}}{h_{z,l} + h_{z,l+1}} \left[u_l^{(1)} - u_{l+1}^{(1)} \right] \frac{1}{h_{z,l}} = R_l \qquad (4.351)$$

This equation may be rewritten in the form

$$-a_l u_{l-1}^{(1)} + b_l u_l^{(1)} - c_l u_{l+1}^{(1)} = d_l \qquad (4.352)$$

where

$$a_l = \frac{T}{h_{z,l}} \frac{N_{k,l} + N_{j+1,l}}{h_{z,l-1} + h_{z,l}} \qquad (4.353)$$

$$c_l = \frac{T}{h_{z,l}} \frac{N_{j,l+1} + N_{j+1,l+1}}{h_{z,l} + h_{z,l+1}} \qquad (4.354)$$

$$b_l = 1 + \frac{T}{h_{z,l}} \frac{N_{j,l} + N_{j+1,l}}{h_{z,l-1} + h_{z,l}} + \frac{T}{h_{z,l}} \frac{N_{j,l+1} + N_{j+1,l+1}}{h_{z,l} + h_{z,l+1}}$$
$$= 1 + (a_l + c_l) \qquad (4.355)$$

$$d_l = R_l \qquad (4.356)$$

Equation (4.352) will be solved by the line inversion method (Appendix 1). Before proceeding with the solution, it is necessary to determine the boundary conditions which must be satisfied at the top and bottom of the water column. These boundary conditions are discussed below.

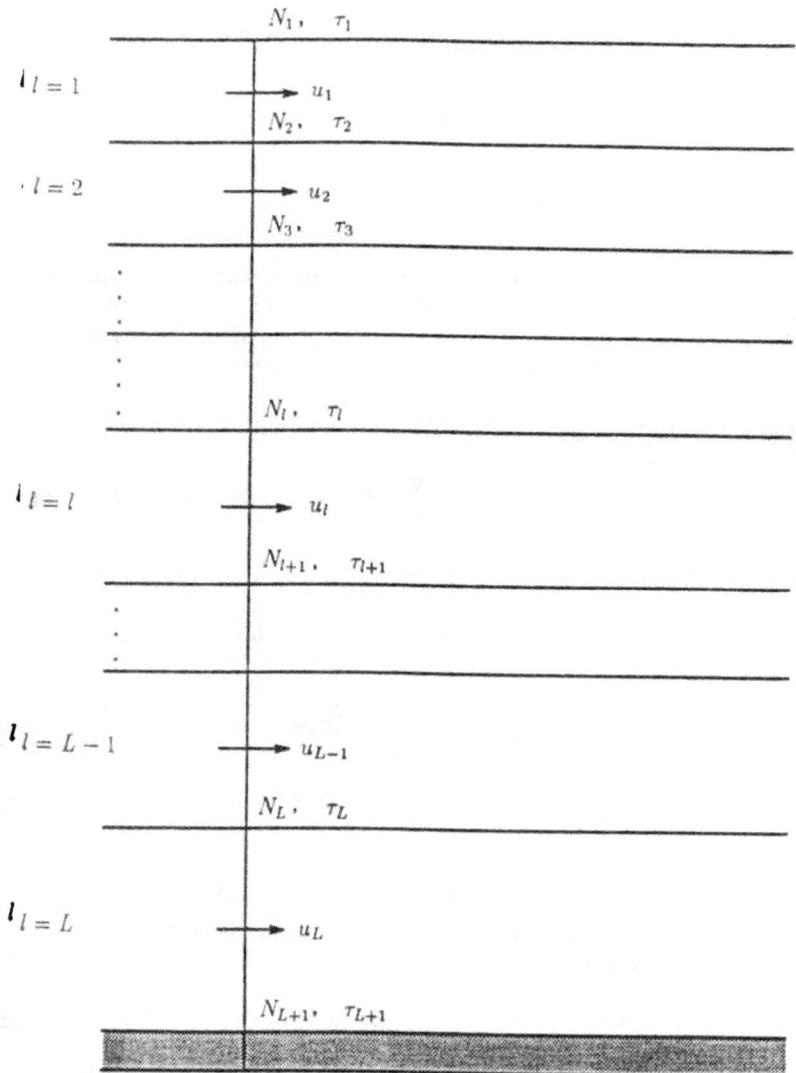

Fig. 4.17
The grid used to define the eddy viscosity terms.

Wind stress on the top of the uppermost layer is added to the acceleration terms for that layer during the explicit part of the solution. Hence, for the implicit part of the solution, the stress at the top of the uppermost layer is set to zero, and the stress which was applied during the explicit part, contained in the term d_1, is redistributed through the water column. Equation (4.351) then reduces to

$$u_1^{(1)} = R_1 - \frac{T}{h_{z,1}}\tau_2^{(1)} \tag{4.357}$$

This expands to

$$u_1^{(1)} + \frac{T}{h_{z,1}}\frac{N_{j,2}+N_{j+1,2}}{h_{z,1}+h_{z,2}}\left[u_1^{(1)} - u_2^{(1)}\right] = R_1 \tag{4.358}$$

or

$$u_1^{(1)}\left(1 + \frac{T}{h_{z,1}}\frac{N_{j,2}+N_{j+1,2}}{h_{z,1}+h_{z,2}}\right) - \left(\frac{T}{h_{z,1}}\frac{N_{j,2}+N_{j+1,2}}{h_{z,1}+h_{z,2}}\right)u_2^{(1)} = R_1 \tag{4.359}$$

which retains the general form

$$-a_1 u_0^{(1)} + b_1 u_1^{(1)} - c_1 u_2^{(1)} = d_1 \tag{4.360}$$

with

$$a_1 = 0 \tag{4.361}$$

$$c_1 = \frac{T}{h_{z,1}}\frac{N_{j,2}+N_{j+1,2}}{h_{z,1}+h_{z,2}} \tag{4.362}$$

$$b_1 = 1 + (a_1 + c_1) \tag{4.363}$$

$$d_1 = R_1 \tag{4.364}$$

Thus, the surface layer equation has the same form as (4.352) where a_1 is equal to zero.

Bottom boundary condition:

The general equation of motion for the bottom layer is

$$u_L^{(1)} = u_L^{(0)} + T\left(X_L + \frac{(\tau_L - \tau_{L+1})}{h_{z,L}}\right) \tag{4.365}$$

The bottom stress is evaluated as

$$\tau_{L+1} = ru_L^{(1)}S, \qquad S = \frac{\sqrt{(u_L^2+v_L^2)^{(0)}}}{h_{z,L}} \tag{4.366}$$

that is, the bottom friction term is treated implicitly although the coefficient S depends on the velocity field at the earlier time. Setting

$$R_L = u_L^{(0)} + TX_L \tag{4.367}$$

eq.(4.365) becomes

$$u_L^{(1)} = R_L + \frac{T}{h_{z,L}}\tau_L^{(1)} - rTSu_L^{(1)} \tag{4.368}$$

or

$$u_L^{(1)}(1 + rST) = R_L + \frac{T}{h_{z,L}}\frac{N_{j,L} + N_{j+1,L}}{h_{z,L-1} + h_{z,L}}\left[u_{L-1}^{(1)} - u_L^{(1)}\right] \tag{4.369}$$

By defining

$$F = \frac{1}{1 + rST} \tag{4.370}$$

then

$$u_L^{(1)} = FR_L + F\frac{T}{h_{z,L}}\frac{N_{j,L} + N_{j+1,L}}{h_{z,L-1} + h_{z,L}}\left[u_{L-1}^{(1)} - u_L^{(1)}\right] \tag{4.371}$$

or

$$-F\frac{T}{h_{z,L}}\frac{N_{j,L} + N_{j+1,L}}{h_{z,L-1} + h_{z,L}}u_{L-1}^{(1)} + u_L^{(1)}\left(1 + F\frac{T}{h_{z,L}}\frac{N_{j,L} + N_{j+1,L}}{h_{z,L-1} + h_{z,L}}\right) = FR_L \tag{4.372}$$

Thus, the momentum equation for the bottom layer reduces to

$$-a_L u_{L-1} + u_L(1 + a_L) = d_L \tag{4.373}$$

where

$$a_L = F\frac{T}{h_{z,L}}\frac{N_{j,L} + N_{j+1,L}}{h_{z,L-1} + h_{z,L}} \tag{4.374}$$

$$c_L = 0 \tag{4.375}$$

$$b_L = 1 + (a_L + c_L) \tag{4.376}$$

$$d_L = FR_L \tag{4.377}$$

Thus, the equation for the bottom layer also has the general form (4.352) although the individual matrix coefficients are multiplied by the frictional attenuation factor. Now, the algorithm given in Appendix 1 can be applied.

7. A two-layer model

We will discuss a hierarchy of three-dimensional numerical models, starting with the simplest, namely, a two-layer model. The particular model we discuss here was originally developed by O'Brien and Hurlburt (1972). The momentum and continuity equations for a two-layer model can be written as (for convenience we retain the authors' notation):

$$\frac{\partial u_1}{\partial t} + u_1 \frac{\partial u_1}{\partial x} + g(h_1 + h_2 + D)_x = f v_1 + (\tau_x^S - \tau_x^I)/(\rho h_1) + A \frac{\partial^2 u_1}{\partial x^2} \qquad (4.378)$$

$$\frac{\partial v_1}{\partial t} + u_1 \frac{\partial v_1}{\partial x} = -f u_1 + (\tau_y^S - \tau_y^I)/(\rho h_1) + A \frac{\partial^2 v_1}{\partial x^2} \qquad (4.379)$$

$$\frac{\partial h_1}{\partial t} + \frac{\partial h_1 u_1}{\partial x} = 0 \qquad (4.380)$$

$$\frac{\partial u_2}{\partial t} + u_2 \frac{\partial u_2}{\partial x} + g(h_1 + h_2 + D)_x - g' \frac{\partial h_1}{\partial x} = f v_2 + (\tau_x^I - \tau_x^B)/(\rho h_2) + A \frac{\partial^2 u_2}{\partial x^2} \quad (4.381)$$

$$\frac{\partial v_2}{\partial t} + u_2 \frac{\partial v_2}{\partial x} = -f u_2 + (\tau_y^I - \tau_y^B)/(\rho h_2) + A \frac{\partial^2 v_2}{\partial x^2} \qquad (4.382)$$

$$\frac{\partial h_2}{\partial t} + \frac{\partial h_2 u_2}{\partial x} = 0 \qquad (4.383)$$

where

$$\left.\begin{aligned}
\tau_x^I &= \rho c \bar{q}(u_1 - u_2) \\
\tau_y^I &= \rho c \bar{q}(v_1 - v_2) \\
\tau_x^B &= \rho c q_2 u_2 \\
\tau_y^B &= \rho c q_2 v_2 \\
q_i &= (u_i^2 + v_i^2)^{1/2} \\
\bar{q} &= (q_1 + q_2)/2 \\
g' &= g(\rho_2 - \rho_1)/\rho_2
\end{aligned}\right\} \qquad (4.384)$$

and where subscripts 1 and 2, respectively, denote the top and bottom layers, τ^S and τ^B are the stresses at the top of the upper layer and the bottom of the lower layer, w and v are the vertically averaged horizontal components of velocity along the x and y directions (x points towards east, y points towards north, and it is assumed that $\partial/\partial y = 0$, i.e. no variations in the north-south direction), h_1 and h_2 are the deviations of the free surface and interface from their respective undisturbed positions, g is gravity, ρ is water density, t is time, A is the horizontal eddy viscosity coefficient, D is indicated in Fig.4.18.

$$q = w + iv$$

where $i = \sqrt{-1}$. The bar denotes an average and c is a coefficient to be prescribed. For the values chosen by O'Brien and Hurlburt (1972), the interior stresses play no role in the model dynamics. O'Brien and Hurlburt (1972) developed an efficient semi-implicit method, following Kwizak and Robert (1971). It was shown by

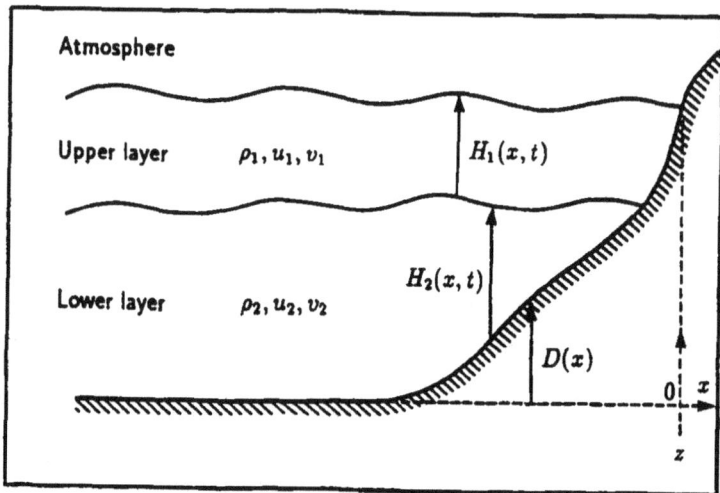

Fig. 4.18

Typical geometry for a two-layer model with a free surface and one interface. The bottom top topography is $D(x)$, and h_1 and h_2 are the thicknesses of each layer of density ρ_1 and ρ_2. The velocities are depth-averaged.
From O'Brien and Hurlburt (1972).

Kwizak and Robert that if an implicit scheme were chosen for the terms representing the fastest travelling waves, a larger time step can be used without causing
computational instability. O'Brien and Hurlburt treated in an implicit manner,
the pressure gradient terms in the momentum equations and the divergence term
in the continuity equation, thereby enabling them to use a time step of about one
hour. For convenience, we will use the same notation as employed by the authors.
Consider the model equations in the form:

$$\frac{\partial u_1}{\partial t} + g\frac{\partial}{\partial x}(h_1 + h_2 + D) = U_1 \tag{4.385}$$

$$\frac{\partial h_1}{\partial t} + H_1\frac{\partial u_1}{\partial x} = -h_1'\frac{\partial u_1}{\partial x} - u_1\frac{\partial h_1}{\partial x} \equiv D_1 \tag{4.386}$$

$$\frac{\partial u_2}{\partial t} + g\frac{\partial}{\partial x}(h_1 + h_2 + D) - g'\frac{\partial h_1}{\partial x} = U_2 \tag{4.387}$$

$$\frac{\partial h_2}{\partial t} + (H_2 + D)\frac{\partial u_2}{\partial x} = -h_2'\frac{\partial u_2}{\partial x} - u_2\frac{\partial h_2}{\partial x} + D\frac{\partial u_2}{\partial x} \equiv D_2 \tag{4.388}$$

where H_1 is the mean depth of the upper layer and $H_2 + D$ the mean height of the interface between the two layers. No approximations have been made; U_1 and U_2 denote all the remaining terms in the x-directed momentum equations. The linear part of the divergence term in the continuity equation is retained on the left-hand side. The primed quantities are defined by

$$h_1 = h_1' + H_1$$
$$h_2 = h_2' + H_2 \qquad\qquad (4.389)$$

It is important to realize that H_1 and $H_2 + D$ are constants. We will treat the left-hand side of (4.385–4.388) implicitly and the right-hand side explicitly. (We have also made the horizontal diffusive terms implicit using the Crank–Nicholson method. The tendencies are evaluated over $2\Delta t$ and the pressure gradient terms and the divergence terms are averaged between time levels $(n-1)\Delta t$ and $(n+1)\Delta t$.

The time-differenced equations are

$$u_{1j}^{n+1} + \left[\Delta tg\frac{\partial}{\partial x}(h_1 + h_2 + D)\right]_j^{n+1}$$
$$= 2\Delta t U_{1j}^n + u_{1j}^{n-1} - \left[\Delta tg\frac{\partial}{\partial x}(h_1 + h_2 + D)\right]_j^{n-1} \equiv L_1 \qquad (4.390)$$

$$h_{1j}^{n+1} + \left[\Delta tH_1\frac{\partial u_1}{\partial x}\right]_j^{n+1} = 2\Delta t D_{1j}^n + h_{1j}^{n-1} - \left[\Delta tH_1\frac{\partial u_1}{\partial x}\right]_j^{n-1} \equiv L_2 \qquad (4.391)$$

$$u_{2j}^{n+1} + \left[\Delta tg\frac{\partial}{\partial x}(h_1 + h_2 + D) - \Delta tg'\frac{\partial h_1}{\partial x}\right]_j^{n+1}$$
$$= 2\Delta t U_{2j}^n + u_{2j}^{n-1} - \left[\Delta tg\frac{\partial}{\partial x}(h_1 + h_2 + D) - \Delta tg'\frac{\partial h_1}{\partial x}\right]_j^{n-1} \equiv L_3 \qquad (4.392)$$

$$h_{2j}^{n+1} + \left[\Delta tg(H_2 + D)\frac{\partial u_2}{\partial x}\right]_j^{n+1}$$
$$= 2\Delta t D_{2j}^n + h_{2j}^{n-1} - \left[\Delta t(H_2 + D)\frac{\partial u_2}{\partial x}\right]_j^{n-1} \equiv L_4 \qquad (4.393)$$

All spatial derivatives are replaced by second-order finite differences. The subscripts j and n imply for any scalar q that

$$q_j^n = q(-j\Delta x, n\Delta t) \qquad\qquad (4.394)$$

All of the right-hand side terms, defined now as L_i $(i = 1, 2, 3, 4)$, are "known" at any time level in the calculations. We wish to find $(u_1, u_2, h_1, h_2)_j^{n+1}$ for all j at time level $(n + 1)\Delta t$. Kwizak and Robert recommend elimination of the velocities and the solution of coupled Helmholtz equations for h_1 and h_2. Since we have homogeneous boundary conditions for u_1 and u_2, we elect to eliminate h_1 and h_2 from the above and obtain

$$u_{2j}^{n+1} - b\left[\frac{\partial^1 u_2}{\partial x^2}\right]_j^{n+1} + (c - a)\left[\frac{\partial^2 u_1}{\partial x^2}\right]_j^{n+1} = d \tag{4.395}$$

$$-u_{1j}^{n+1} + u_{2j}^{n+1} + c\left[\frac{\partial^1 u_1}{\partial x^2}\right]_j^{n+1} = e \tag{4.396}$$

where $a = gH_1\Delta t^2$, $b = g(H_2 + D)\Delta t^2$, $c = g'H_1\Delta t^2$, and d and e are a linear combination of the L_i. These equations are coupled one-dimensional Helmholtz equations in the unknowns u_1^{n+1} and u_2^{n+1} for all space points j. When the spatial derivatives are replaced by standard second-order finite differences, the resulting algebraic equations are tridiagonal and are easily solved by the special "up-down" variant of Gaussian elimination. In practice, we solve eqs.(4.395) and (4.396) iteratively by solving (4.396) for u_{1j}^{n+1} using "up-down" and then eq.(4.395) for u_{2j}^{n+1}. Only a few scans are needed for convergence.

The y-directed momentum equations are solved using well-known techniques — leapfrog for time differences, a quadratic-averaging method known as Scheme F from Grammelvedt (1969) for the advective terms, Crank–Nicholson for the diffusive terms, and centered-in-time Coriolis terms.

8. Mode splitting and reduced gravity model

In Section 7 we discussed a two-layer model as a simple representation of a three-dimensional model. Here we will discuss another simpler version of a three-dimensional model, namely the so-called reduced gravity model. However, to be able to derive a reduced gravity model, certain assumptions and approximations have to be made. These will be discussed at first, following Moore and Philander (1977).

In any three-dimensional representation, in the linearized shallow water model, the vertical structure can be represented in terms of barotropic and baroclinic modes. It can be shown that the phase speed for the barotropic modes is given by $\sqrt{gD}$, and an equivalent depth can be defined for each baroclinic mode (Moore and Philander, 1977). For the oceanic scale, according to these authors, the time scale for the barotropic mode is about four hours, whereas the time scale for the first five baroclinic modes are 1.5, 2.0, 2.6, 3.1, and 3.6 days, respectively. This clear separation in the time scales between the barotropic and baroclinic modes helps in representing a three-dimensional circulation problem through a reduced gravity model. Following Camerlengo and O'Brien (1980), we assume that the lower layer is infinitely deep ($D_2 \gg D_1$). This allows us to assume that the lower layer is of uniform density and is at rest for all x, y, and t. These assumptions automatically filter out the barotropic mode of the motion (Camerlengo and O'Brien, 1980). If we do not consider turbulent mixing between the ocean layers, i.e. ρ_1 being constant, we can define the reduced gravity g' as

$$g' = \frac{\rho_2 - \rho_1}{\rho_2} g \qquad (4.397)$$

With these assumptions, the forced linear shallow-water wave equations for the first baroclinic mode can be written as follows:

$$\frac{\partial u}{\partial t} = \beta y v - g' \frac{\partial h}{\partial x} + A\nabla^2 u + \frac{\tau^x}{\rho H}$$

$$\frac{\partial v}{\partial t} = -\beta y u - g' \frac{\partial h}{\partial y} + A\nabla^2 v + \frac{\tau^y}{\rho H} \qquad (4.398)$$

$$\frac{\partial h}{\partial t} = -H\left(\frac{\partial u}{\partial x} + \frac{\partial v}{\partial y}\right)$$

Here A is the horizontal eddy viscosity and β is the variation of the Coriolis parameter with latitude.

Next following Camerlengo and O'Brien (1980) we will discuss the open boundary conditions. The shallow-water wave equations are hyperbolic in nature. The most accurate way to prescribe the outflow at a certain boundary for the hyperbolic system is to use a Sommerfeld radiation condition:

$$\frac{\partial \psi}{\partial t} + F \frac{\partial \psi}{\partial x} = 0 \qquad (4.399)$$

where ψ is any variable, and F, the phase speed. There are many methods for implementing eq.(4.399). The experiments here are conducted using the open boundary conditions from Orlanski (1976). In addition, a modified version of Orlanski's boundary condition is implemented.

In the Orlanski method, the phase speed is numerically evaluated for the interior points close to the open boundary. By using a leapfrog representation of eq.(4.399) for each variable, the evaluated phase speed, F_ψ, yields (see Ch. II):

$$\left(\frac{\psi_{B-1}^n - \psi_{B-1}^{n-2}}{2\Delta t}\right) + F_\psi\left[\left(\frac{\psi_{B-1}^n + \psi_{B-1}^{n-2}}{2}\right) - \psi_{B-1}^{n-2}\right]\frac{1}{\Delta x} = 0 \qquad (4.400)$$

where the index, B, denotes a point at the boundary, $B-1$, the first interior point, etc. In other words, values of ψ near the boundary and previous in time are used to estimate F_ψ for each variable at each time step. Equation (4.399) is used again to evaluate the extrapolated value of ψ_B^{n+1} as a function of F. The final formulation for the phase speed is then:

$$
\begin{aligned}
F_\psi &= 0 && \text{if } F_\psi \leq 0 \\
F_\psi &= \frac{\Delta x}{\Delta t} && \text{if } F_\psi > \Delta x/\Delta t \\
F_\psi &= -\psi_t/\psi_x && \text{if } 0 < F_\psi < \Delta x/\Delta t
\end{aligned}
\qquad (4.401)
$$

where the approximation (4.400) is implied. However, because a staggered grid in space and time is used, the specific numerical formulation of F_ψ is slightly different at different boundaries. Camerlengo and O'Brien (1980) also used a modified Orlanski method.

Next, following Greatbatch and Otterson (1991) we will discuss open boundary conditions at the mouth of a bay. As pointed out by these authors, the formulation of the open boundary conditions problem is complicated by the fact that coastal Kelvin waves propagate only one way around a bay. The implication of this is that an open boundary from which such waves can propagate into the model domain (which is referred to as the upstream boundary) must be treated in a special way from other open boundaries, since otherwise, spurious effects can propagate in from the boundary and contaminate the solution.

Greatbatch and Otterson (1991) identified two problems with using a radiation condition on an upstream boundary, perpendicular to the coast. First, in the case with Ekman transport away from the coast immediately downstream from the boundary, upwelling occurs indefinitely and propagates into the solution domain by means of Kelvin waves and completely dominates the model response. Second, a radiation condition was found to generate Kelvin waves of near-inertial period which contaminated the solution.

To suppress these problems, Greatbatch and Otterson (1991) extended the coastline out to sea on the upstream side of the bay mouth and used a condition of zero normal gradient in interfacial displacement on the artificial part of the boundary. This ensures that Kelvin waves cannot propagate along this stretch of the model boundary, thus eliminating the problems without being unduly reflective to Poincaré waves generated in the bay. It is necessary that the other open boundaries are placed sufficiently far out from the bay.

Greatbatch and Otterson (1991) essentially made use of the Orlanski radiation condition. However, they applied it only to the interface displacement (i.e. the pressure field) and not to the velocities. They also experimented with the use of sponge layers, rather than radiation conditions.

9. Quasi-geostrophic models

We will discuss briefly the so-called quasi-geostrophic version of shallow-water three-dimensional models and we follow the paper by Holland (1986). For a historical development discussion, the relevant papers are those of Charney (1949) and Philipps (1951). Pedlosky (1982) has given an excellent presentation of the problem. From the set of shallow water equations given in Ch. I, let us write the following simplified version:

$$\frac{\partial u}{\partial t} + u\frac{\partial u}{\partial x} + v\frac{\partial u}{\partial y} - fv = -g\frac{\partial \zeta}{\partial x}$$

$$\frac{\partial v}{\partial t} + u\frac{\partial v}{\partial x} + v\frac{\partial v}{\partial y} + fu = -g\frac{\partial \zeta}{\partial y} \qquad (4.402)$$

$$\frac{\partial H}{\partial t} + \frac{\partial}{\partial x}(uH) + \frac{\partial}{\partial y}(vH) = 0$$

Here u and v are the depth-averaged velocities in the x and y directions, g is gravity, t is time, f is the Coriolis parameter (assumed to be uniform), ζ is the surface height perturbation, H is the net thickness of the fluid layer which is related to the constant average thickness of the fluid layer through the relation

$$H = D + \zeta - H_B \qquad (4.403)$$

where H_B is the variable bottom topography perturbation.

We will restrict our discussion to the time scales greater than f^{-1}. Let L, T, U be typical scales of length, time and velocity and primes denote dimensionless variables.

$$x, y = L(x', y')$$
$$t = T \cdot t'$$
$$u, v = U(u', v') \qquad (4.404)$$
$$\zeta = N_0 \cdot \zeta'$$

Substitute (4.403) and (4.404) into (4.402) and also assume that

$$\epsilon = \frac{U}{fL} \ll 1$$

Then, it can be seen that the nonlinear advective terms in eq.(4.402) are much smaller compared to the Coriolis terms. It follows that

$$N_0 = \frac{fUL}{g} = \epsilon\frac{f^2L^2}{g}$$

and

$$H = D\left[H\epsilon\frac{L^2}{R^2}\zeta' - \frac{H_B}{D}\right] \qquad (4.405)$$

where $R = \sqrt{gD}/f$ is the Rossby radius of deformation.

For the case when L does not exceed R, the layer thickness departs from its value in the absence of motion by, at most, $\theta(\epsilon)$. Then eqs.(4.402) and (4.403) can be written as

$$\epsilon\frac{\partial u}{\partial t} + \epsilon\left(u\frac{\partial u}{\partial x} + v\frac{\partial u}{\partial y}\right) - v = -\frac{\partial \zeta}{\partial x}$$

$$\epsilon\frac{\partial v}{\partial t} + \epsilon\left(u\frac{\partial v}{\partial x} + v\frac{\partial v}{\partial y}\right) + u = -\frac{\partial \zeta}{\partial y} \qquad (4.406)$$

$$\epsilon F\frac{\partial \zeta}{\partial t} + \epsilon F\left(u\frac{\partial \zeta}{\partial x} + v\frac{\partial \eta}{\partial y}\right) - u\frac{\partial}{\partial x}\left(\frac{H_B}{D}\right) - v\frac{\partial}{\partial y}\left(\frac{H_B}{D}\right)$$

$$+\left(1 + \epsilon F\zeta - \frac{H_B}{d}\right)\left(\frac{\partial u}{\partial x} + \frac{\partial v}{\partial y}\right) = 0$$

where

$$F = \frac{f^2 L^2}{gD} = \left(\frac{L}{R}\right)^2 = \theta(1)$$

Since $\epsilon \ll 1$, one can expand u, v, ζ in an asymptotic series:

$$u = u_0 + \epsilon u_1 + \epsilon^2 u_2 + \dots$$

and collect terms with the same power of ϵ to give:

$$v_0 = \frac{\partial \zeta_0}{\partial x}$$

$$u_0 = -\frac{\partial \zeta_0}{\partial y} \qquad (4.407)$$

$$\frac{\partial u_0}{\partial x} + \frac{\partial v_0}{\partial y} = 0$$

This means, to the lowest order, the fields are geostrophic and nondivergent.

Next assume that

$$\frac{H_B}{D} \sim \epsilon \quad \text{i.e.} \quad \frac{H_B}{D} = \epsilon\zeta_B \quad \text{where} \quad \zeta_B \sim \theta(1)$$

Then the equations to order ϵ are

$$\frac{\partial u_0}{\partial t} + u_0\frac{\partial u_0}{\partial x} + v_0\frac{\partial u_0}{\partial y} - v_1 = -\frac{\partial \zeta_1}{\partial x}$$

$$\frac{\partial v_0}{\partial t} + u_0\frac{\partial v_0}{\partial x} + v_0\frac{\partial v_0}{\partial y} + u_1 = -\frac{\partial \zeta_1}{\partial y} \qquad (4.408)$$

$$F\left(\frac{\partial \zeta_0}{\partial t} + u_0 \frac{\partial \zeta_0}{\partial x} + v_0 \frac{\partial \zeta_0}{\partial y}\right) - u_0 \frac{\partial \zeta_B}{\partial x} - v_0 \frac{\partial \zeta_B}{\partial y} + \left(\frac{\partial u_1}{\partial x} + \frac{\partial v_1}{\partial y}\right) = 0$$

By forming the vorticity equation, one can solve for the zero order fields:

$$\frac{d\xi_0}{dt} = \frac{\partial \xi_0}{\partial t} + u_0 \frac{\partial \xi_0}{\partial x} + v_0 \frac{\partial \xi_0}{\partial y} = -\left(\frac{\partial u_1}{\partial x} + \frac{\partial v_1}{\partial y}\right)$$

where the vorticity ξ_0 is given by

$$\xi_0 = \frac{\partial v_0}{\partial x} - \frac{\partial u_0}{\partial y} = \nabla^2 \zeta_0 \qquad (4.409)$$

then

$$\frac{d}{dt}(\nabla^2 \zeta_0 - F\zeta_0 + \zeta_B) = 0$$

This is the statement of the conservation of potential vorticity. To summarize, derivation of the quasi-geostrophic model involved the following assumptions:

$$\text{small Rossby number:} \quad \epsilon \ll 1$$

$$\text{small bottom slope:} \quad \frac{H_B}{D} \ll 1$$

$$\text{slope of perturbed density surfaces must be small:} \quad \frac{\zeta}{D} \ll 1$$

The third assumption is extremely important for three-dimensional models. The quasi-geostrophic (QG) model is computationally efficient because of the filtering out of gravity waves. It is dynamically simple and can be used to study adiabatic (no heat diffusion) problems. However, it has several drawbacks. Gravity wave problems cannot be studied, thermally forced situations cannot be easily treated, since $f \to 0$, problems close to the equator cannot be studied, Rossby number must be small and isopycnal surfaces cannot deviate much from level surfaces.

Pedlosky (1982) extended the shallow water QG model to the full three-dimensional equations in spherical geometry. In this system, the statement of the conservation of potential vorticity becomes:

$$\frac{D}{Dt}\left[\nabla^2 \psi + \frac{\partial}{\partial z}\left(\frac{1}{S}\frac{\partial \psi}{\partial z}\right) + \beta y\right] = W \qquad (4.410)$$

where ψ is three-dimensional quasi-geostrophic streamfunction, W represents the nonconservative effects of forcing and friction, $S = S(z)$ is a static stability parameter determined by the mean background stratification, and β is the local rate of change of the Coriolis parameter f. This equation forms the basis for the QG model developed by Holland (1978) to study eddy-resolved ocean circulation problems of the kind begun earlier with the Holland–Lin primitive equation model.

The quasi-geostrophic model formulation with N arbitrary layers is a straighforward extension of the two-layer case described by Holland (1978). Here we shall present the form of the equations in which the vertical discretization has already been done. The horizontal discretization and the form of the finite-difference equations will not be discussed here.

The governing equations are the vorticity and interface height perturbation equations and the thermal wind relation:

$$\frac{\partial}{\partial t}\nabla^2\psi_k = J(f + \nabla^2\psi_k, \psi_k) + \frac{f_0}{H_k}(w_{k-\frac{1}{2}} - w_{k+\frac{1}{2}}) + F_k + T_k \qquad : k = 1 \text{ to } N$$

$$\frac{\partial}{\partial t}h_{k+\frac{1}{2}} = J(h_{k+\frac{1}{2}}, \psi_{k+\frac{1}{2}}) + w_{k+\frac{1}{2}} \qquad : k = 1 \text{ to } N - 1 \qquad (4.411)$$

$$h_{k+\frac{1}{2}} = \frac{f_0}{g'_{k+\frac{1}{2}}}(\psi_{k+1} - \psi_k)$$

Whole number subscripts (k) denote the vertical layers (k increasing downward) in which the quasi-geostrophic streamfunction is defined (nominally at the center of each of the layers) while fractional subscripts $(k + \frac{1}{2})$ denote the interfaces between layers where vertical velocity and interface height perturbations are defined. The variables are the quasi-geostrophic streamfunction (ψ_k) with horizontal velocity components $(u = -\psi_y, v = \psi_x)$, the interface height perturbation $(h_{k+\frac{1}{2}})$, positive upward, and the vertical velocity $(w_{k+\frac{1}{2}})$, also positive upward. The horizontal coordinates are x (eastward) and y (northward), the Coriolis parameter is $f = f_0 + \beta y$, and the mean layer thicknesses are H_k. The values of f_0 and β are chosen to represent typical mid-latitude gyre values. The basic background vertical stratification is written in terms of the reduced gravity $g' = g\Delta\rho_{k+\frac{1}{2}}/\rho_0$, where $\Delta\rho_{k+\frac{1}{2}}$ is the (positive) density difference between layers $k+1$ and k. Frictional effects, written symbolically in eq.(4.411) as F_k, have been parameterized in different ways in various calculations — as lateral friction of the Laplacian or biharmonic kind (Holland, 1978), in which $F_k = A_2\nabla^4\psi_k$ or $F_k = -A_4\nabla^6\psi_k$, respectively. In addition, in many experiments, F_k includes a bottom friction, $-\epsilon\nabla^2\psi_N$, when $k = N$ (the bottom layer). Note that the effect of the wind forcing T_1, equal to curl, τ/H_1, produces an Ekman pumping stretching tendency in the upper layer that is equivalent to a body force acting on the upper layer. The T_k for $k > 1$ are zero. Also note that a gentle bottom slope can be consistently included in the QG framework so that the influence of variable depth can be examined, at least within the limitations of quasi-geostrophy. This is taken care of by the bottom boundary conditions on w, i.e. $w_{N+\frac{1}{2}} = J(\psi_N, H_B)$, where $H_B(x, y)$ is the variable bottom topography (positive upward). At the sea surface, $w_{\frac{1}{2}} = 0$. The advective velocities at the interfaces, needed in eq.(4.411), are calculated from a weighted average of the velocities in the layers, i.e. $\psi_{k+\frac{1}{2}} = \alpha_{k+\frac{1}{2}}\psi_{k+1} + (1 - \alpha_{k+\frac{1}{2}})\psi_k$, where $\alpha_{k+\frac{1}{2}} = H_k/(H_{k+1})$.

These equations can be written in potential vorticity form

$$\frac{DQ_k}{Dt} = T_k + F_k \tag{4.412}$$

where

$$Q_k = \nabla^2\psi_k + f + \frac{f_0}{H_k}(h_{k+\frac{1}{2}} - h_{k-\frac{1}{2}})$$

For consistency in this equation, $h_{\frac{1}{2}} = 0$ and $h_{N+\frac{1}{2}} = H_B$, the variable bottom topography. Potential vorticity is conserved except for wind forcing and frictional effects.

Note that, although not included here, thermal forcing can be easily added (within the restrictions of QG physics) to the above model. Also note that auxillary momentum conditions are needed in multiply-connected geometry (McWilliams, 1977). The model numerics used in the following examples make use of finite differences in space and time, include conservation properties for energy, potential vorticity, and potential enstrophy (Q^2), and allow a time step of several hours (a factor of 10 greater than that used by Holland and Lin (1975a,b)).

Next we will consider eddy-resolving general circulation models (EGCM) following Haidvogel (1983). Although a few EGCM simulations have utilized primitve equation dynamics, periodic and regional modelling studies have typically been applied to the mid-ocean regions, for which the quasi-geostrophic approximation is appropriate (McWilliams, 1976). Prompted in part by the initial observational and theoretical interest in describing mesoscale eddy properties in mid-latitude regions, a local β-plane representation of the planetary vorticity gradient has also generally been used. Under these dynamical assumptions and, for definiteness, adopting a two-level vertical structure (Phillips, 1951),[1] the associated quasi-geostrophic equations can be written in nondimensional form as:

$$\frac{\partial q_1}{\partial t} + J(\psi_1, q_1) = -v\nabla^6\psi_1 + \mathcal{F}_1 \tag{4.413}$$

$$\frac{\partial q_2}{\partial t} + J(\psi_2, q_2) = -\kappa\nabla^6\psi_2 - v\nabla^6\psi_2 + \mathcal{F}_2 \tag{4.414}$$

$$\qquad\qquad\qquad \text{SD} \qquad \text{LM} \qquad \mathcal{F}$$

---

[1] (In general, the leveled models to be discussed are strictly equivalent to a layered model only in circumstances for which fluctuations in layer depth in the latter can be ignored. However, we will follow conventional practice and use the terms "multi-level" and "multi-layer" synonymously.)

where

$$q_1 = \nabla^2 \psi_1 + \beta y + \left(\frac{1}{1+\delta}\right)(\psi_2 - \psi_1) \tag{4.415}$$

and

$$q_2 = \nabla^2 \psi_2 + \beta y + \left(\frac{1}{1+\delta}\right)(\psi_2 - \psi_1) + \alpha h \tag{4.416}$$

$$\text{RV} \quad \text{BY} \qquad\qquad\qquad \text{VS} \qquad \text{T}$$

Equations (4.413) and (4.414) are statements of conservation of upper and lower level potential vorticity — q_1 and q_2, respectively — following the horizontal fluid motion at each level. As indicated, nonconservative effects, represented by the terms on the righthand sides of eqs.(4.413) and (4.414), are provided by surface drag (SD), lateral mixing (LM), and forcing ($\mathcal{F}$) of the system by external driving agents and/or larger-scale motions. Contributions to the potential vorticities are made by the relative vorticities [RV; $\nabla^2 \psi_i$ $(i = 1, 2)$], planetary vorticity (BY), and vortex stretching due to equivalent "interfacial" displacements between levels 1 and 2 (VS) and topographic roughness (T).

The nondimensional parameters which enter the physical description of the two-level quasi-geostrophic system are:

$$\beta = \left(\frac{\beta_{\text{dim}} \lambda^2}{U}\right)$$

$$\delta = \left(\frac{H_1}{H_2}\right)$$

$$\alpha = \left(\frac{f_0 h_0 \lambda}{HU}\right)$$

$$\kappa = \left(\frac{\kappa_{\text{dim}} \lambda}{U}\right)$$

$$v = \left(\frac{v_{\text{dim}}}{U \lambda^3}\right)$$

representing respectively the nondimensionalized effects of the planetary vorticity and bottom topographic gradients (β and α, respectively), and dissipation (κ, v), as well as the ratio of the thickness of the fluid layers surrounding levels 1 and 2. The subscript "dim" denotes a dimensional quantity. These two-level equations have been nondimensionalized by an RMS velocity scale U and the Rossby deformation radius, λ. By suppressing the vortex stretching term in eq.(4.416), eq.(4.414) describes the dynamics of a single-layer (or barotropic) system, although Miller *et al.* (1983) formulated an open ocean model. For other studies using QG models, see Cummins and Mysak (1988).

10. Streamfunction models

We have shown in Section 3.1 of Ch. I that in the steady state, for the vertically integrated case, the transport is completely determined by the streamfunction ψ. Here the transport components are given by

$$M_x = -\frac{\partial\psi}{\partial y}, \qquad M_y \frac{\partial\psi}{\partial x} \tag{4.417}$$

From equation (4.417) we can write the vorticity equation for the vertically averaged current, in the Cartesian coordinates in the follwoing form:

$$J\left(\psi, \frac{f}{H}\right) = \operatorname{curl}\left(\frac{\tau_s - \tau_b}{\rho H}\right) \tag{4.418}$$

where J is the Jacobian operator defined as

$$J(a,b) \equiv \nabla a(\nabla b \times k) = \frac{\partial a}{\partial x}\frac{\partial b}{\partial y} - \frac{\partial a}{\partial y}\frac{\partial b}{\partial x} \tag{4.419}$$

and H is the water depth, ρ is the density of water, f is the Coriolis parameter, and τ_s and τ_b are the wind and bottom stresses, respectively. It is not uncommon in the oceanographic literature to use a different form of the vorticity equation, namely one referring to the vertically integrated transport instead of the mean current. However, the basic concepts of wind-driven circulations in the prescence of topography are more easily visualized in terms of the vorticity balance (4.418).

Because the normal transport component at the shores must be zero, the lateral boundary conditions for a closed basin are simply:

$$\psi = 0 \quad \text{on the boundary} \tag{4.420}$$

The horizontal circulation problem is fully specified only if the bottom stress is given in terms of the stream function and wind stress. Because the vertical problem is assumed to have been solved, the transport as well as the bottom stress are known in terms of surface gradient and wind stress. The surface gradient can then be eliminated to find a relationship between the other three quantities. Alternatively, one can simply postulate any functional relationship between bottom stress and stream function.

Inspection of the vorticity balance (4.418) immediately reveals basically two nontrivial classes of closed basin circulations; the first associated with variation of the Coriolis parameter with latitude, the second due to depth variations. Historically, the first class of circulation has received most attention in the oceanographic literature. For lakes and shallow seas, however, it is the second class of circulation that is of primary interest, as demonstrated by early treatments of the North and

Baltic seas circulations. Recently, there has also been a renewed interest in topographic effects in the ocean, but this interest centers on the joint effect of topography and baroclinicity. The present discussion is restricted to homogeneous circulations and concentrates on basins of horizontal scales sufficiently small for the variation of Coriolis parameter to be negligible. However, in view of the close correspondence between the effects of spatial variations in rotation and depth, respectively, it is useful to briefly recall the basic concepts of large-scale ocean dynamics, and to contrast them with the dynamics of lakes and shallow seas.

Simons (1980) derived the vorticity equation corresponding to eq.(4.434) which does not explicitly contain the bottom stress term. One can ignore terms of $\theta(\Delta/H)$ compared to terms of order one, to obtain

$$J\left(\psi, \frac{f}{H}\right) = \text{curl}\left(\frac{\tau_s}{\rho H}\right) - \text{div}\left(\frac{K}{H^2}\nabla\psi\right) \qquad (4.421)$$

where $K \equiv f/2$. Equating this result with eq.(4.418) shows that the bottom stress for deep water may be approximated by

$$\frac{\tau_b}{\rho} = \frac{K}{H}V \qquad (4.422)$$

at least insofar as the vorticity equation is concerned. In other words, only the component of bottom stress in the direction of transport enters in the vorticity balance. The expression (4.422) may also be recognized as a linearized version of the familiar hydraulic formulation, relating bottom stress to the square of the vertical mean current. Thus, the coefficient, K, can be visualized as the product of a skin friction coefficient and a typical mean current speed.

First consider a deep ocean for which depth variations may be neglected by comparison to spatial variations of wind stress and rotation. If the x axis is taken to point eastward and the y axis northward, then (4.418) becomes

$$\rho\beta\frac{\partial\psi}{\partial x} = \text{curl } \tau_s - K'\nabla^2\psi \qquad (4.423)$$

where $K' \equiv \rho K/H$ and β is the latitudinal derivative of the Coriolis parameter, $\partial f/\partial y$. The term on the left is known as the "planetary vorticity tendency" and represents the contribution to the vorticity balance arising from a movement of water masses to regions of higher or lower rotation. It is often referred to as the "β-effect" and plays a major role in the large-scale dynamics of the oceans as well as the atmosphere.

Following Stommel and Leetmaa (1972) one may arrive at a qualitative interpretation of the vorticity balance (4.423) by assuming that the ocean is subjected to a large-scale anticyclonic (i.e. clockwise) wind field, such as that from mid-latitude westerlies and low latitude easterlies over the Atlantic. One would naturally expect the resulting ocean circulation to also adopt an anticyclonic structure, but it is immediately clear that a symmetric circulation does not satisfy eq.(4.423). As the curl of the wind stress is negative for a clockwise shear, it can be balanced by the planetary vorticity tendency only on the eastern side of the basin where water presumably flows to the south, thus the transport has a negative sign. On the western side, the contribution from the planetary vorticity tendency is of the same sign as the wind curl and this tends to speed up the local northward flow until the bottom stress curl, as well as the planetary vorticity term, become much larger than the wind curl. The result is that the northward current becomes concentrated in a narrow boundary layer because the net water transport across any longitudinal cross-section must vanish. This is known as the "westward intensification" of ocean circulations. For a formal solution of the problem and further refinements of the theory, see Stommel and Leetmaa (1972).

It follows from the above that the problem of large-scale ocean dynamics can be successfully approached with the aid of boundary-layer methods. The basin may be visualized as a large area without appreciable friction, with coastal boundary regions where friction dominates. In the first region an approximate balance exists between planetary vorticity tendency and wind stress curl. This is usually referred to as the "Sverdrup interior". Conversely, in the boundary layer, the planetary vorticity term is approximately balanced by the bottom stress term or similar effects. In the above example, the Sverdrup region extends all the way to the eastern boundary whereas a boundary layer is established along the western side.

To obtain a general idea of the effects of topography, the vorticity equation for the Sverdrup regime may be written as

$$\frac{d\psi}{ds} = \frac{\partial \psi}{\partial x}\frac{\partial x}{\partial s} + \frac{\partial \psi}{\partial y}\frac{\partial y}{\partial s} = -\text{curl}\left(\frac{\tau_s}{\rho H}\right) \tag{4.424}$$

where

$$\frac{\partial x}{\partial s} = -\frac{\partial}{\partial y}\left(\frac{f}{H}\right), \qquad \frac{\partial y}{\partial s} = \frac{\partial}{\partial x}\left(\frac{f}{H}\right)$$

This will be recognized as a partial differential equation with characteristics given by lines of constant (f/H). The solution is obtainable by integration along the characteristics, starting from any curve intersecting the latter in areas where frictional effects are negligible. This interior solution cannot satisfy all boundary conditions, so a boundary-layer correction is included where necessary. In this layer, gradients of the stream function normal to the boundary are taken to be much larger than any othe terms. Thus, according to eq.(4.421), if the $x$axis is placed normal to the boundary, the boundary-layer correction, say ψ^*, must satisfy the equation:

$$\frac{\partial}{\partial x}\left(\frac{K}{H^2}\frac{\partial \psi^*}{\partial x}\right) + \frac{\partial}{\partial y}\left(\frac{f}{H}\right)\frac{\partial \psi^*}{\partial x} = 0 \tag{4.425}$$

This generalized oceanic circulation problem, including topography, has been treated by Holland (1977) and Welander (1961). The special case of constant Coriolis parameter was considered earlier by Weenink (1956).

By the way of illustration, consider the simple case where the depth contours run from east to west and the wind is independent of longitude and has no latitudinal component. The vorticity balance for the Sverdrup region (4.424) becomes:

$$\frac{d}{dy}\left(\frac{f}{H}\right)\frac{\partial\psi}{\partial x} = -\frac{d}{dy}\left(\frac{\tau_\bullet}{\rho H}\right) \tag{4.426}$$

Ignore the northern and southern boundaries for a moment and concentrate on an east–west cross-section, then the boundary conditions are

$$\psi = 0 \quad \text{for} \quad x = 0, L \tag{4.427}$$

which expresses the fact that the net latitudinal transport must vanish for any longitudinal cross-section.

With reference to the above result for the case of constant depth, postulate that an interior inviscid solution extends all the way to either one of the two side boundaries. Then eq.(4.426) may be integrated with respect to x, starting from this boundary. If the resulting value of the stream function at the opposite wall is denoted by ψ_0, then

$$\psi_0(y) = \pm\frac{d}{dy}\left(\frac{\tau_\bullet}{\rho H}\right)\bigg/\frac{d}{dy}\left(\frac{f}{H}\right) \tag{4.428}$$

where the plus sign applies to the case where the Sverdrup regime extends to the eastern wall, and the minus sign refers to the case of the interior solution extending to the western wall. To satisfy the boundary condition (4.427) at the opposite boundary, the above result must be exactly cancelled out by the boundary layer correction, which must vanish for large distances from the wall. The appropriate solution to eq.(4.425) is

$$\psi^* = -\psi_0 e^{-x/w}, \qquad w \equiv \frac{K}{H^2}\bigg/\frac{d}{dy}\left(\frac{f}{H}\right) \tag{4.429}$$

where w is clearly a representative scale for the width of the boundary layer. Inspection of eq.(4.429) shows that the boundary layer solution decays in positive x-direction if the gradient of (f/H) is positive and vice versa. Consequently, a boundary layer is established on the western side for the northward increase of (f/H) independent of the direction of the wind. This resolves the question of the sign in the interior solution (4.428).

Next, following Simons (1980) we will consider circulation in shallow basins. We consider basins with horizontal dimensions large enough for rotation to play a significant role in the circulation, but not so large that variation of the Coriolis parameter has to be taken into consideration. Then the vorticity equation (4.419) becomes

$$\rho f J(H, \psi) = H \operatorname{curl}(\tau_s) + \tau_{sx} \frac{\partial H}{\partial y} - \tau_{sy} \frac{\partial H}{\partial x} - H^2 \operatorname{curl}\left(\frac{\tau_b}{H}\right) \qquad (4.430)$$

If this equation is contrasted with the deep ocean equation (4.423), important differences become apparent. The first is the previously discussed replacement of the β-effect by topographic vorticity terms on the left-hand side of the equation. The second distinction relates to wind forcing and, although implicitly present in the foregoing topographic solutions, it will now be analyzed more explicitly. Whereas nontrivial solutions of eq.(4.421) are found only for a nonzero wind stress curl, it follows from eq.(4.430) that, in shallow basins, water circulations can be excited by depth variations perpendicular to a spatially constant wind stress. In the North Sea literature this effect was referred to as the "bottom slope current" in contrast to the "wind curl current" (Weenink, 1958; Weenink and Groen, 1958; Groen and Groves, 1963). Weenink (1958) noted that this generation of vorticity is due to the fact that wind stress per unit mass is smaller in deep water than in shallow water, thus causing a differential acceleration. It is of particular importance in the study of lakes and shallow seas, because the horizontal scales are usually much smaller than those of typical weather systems and, consequently, spatial variations of the stress field are often negligble.

Next following Holland (1977), we will review the classical horizontal circulation problem in a basin of oceanic scale, in contrast to the somewhat smaller scale considered by Simons (1980). On an oceanic scale, the variation β of the Coriolis parameter f with latitude must be included.

One class of analytical, wind-driven ocean models is concerned with the horizontal transport in a homogeneous, rectangular ocean basin of constant depth. When lateral friction is included by a simple eddy viscosity hypothesis (Munk, 1950), and vertical friction is included by means of Ekman layers at the top and bottom, a single vorticity equation governing the mass transport streamfunction ψ includes virtually all the models and encompasses all the results of the early steady, wind-driven theories. That equation is

$$\frac{1}{\rho H} \frac{\partial(\psi, \nabla^2 \psi)}{\partial(x, y)} + \beta \psi_x = \operatorname{curl}_z \tau - \frac{1}{H}\left(\frac{K_m f}{2}\right)^{1/2} \nabla^2 \psi + A_m \nabla^4 \psi \qquad (4.431)$$

where x and y are the eastward and northward coordinates, β is the northward gradient of the Coriolis parameter f, H is the depth, and τ is the wind stress. The vertical and horizontal coefficients of eddy viscosity are K_m and A_m, respectively. This equation can be written, with some simplifications, in the nondimensional form:

$$Ro \frac{\partial(\psi', \nabla^2 \psi')}{\partial(x', y')} + \psi'_x = \operatorname{curl}_z \tau' - Ek_v \nabla^2 \psi' + Ek_h \nabla^4 \psi' \qquad (4.432)$$

where Ro is the Rossby number, and Ek_v and Ek_h are the vertical and horizontal Ekman numbers. In some studies the Reynolds number, $Re = Ro/Ek_h$, is introduced. This nondimensionalization is accomplished by the scaling $(x, y) = L(x', y')$ and $\psi = T_0 \psi'/\beta$ where L is the length of the basin and T_0 is the amplitude of the wind stress. Then

$$Ro = \frac{T_0}{\rho H \beta^2 L^3} \tag{4.433}$$

$$Ek_v = \left(\frac{K_m f}{2}\right)^{1/2} (\beta H L) \tag{4.434}$$

$$Ek_h = \frac{A_m}{\beta L^3} \tag{4.435}$$

The classical work of Ekman (1905) showed how the wind stress, acting at the sea surface, could put momentum into the ocean in a thin boundary layer at the sea surface and how bottom friction could remove it in a thin layer at the bottom. The theory is simple, and the vertically integrated effects of these Ekman layers are included in eq.(4.431) as the first and second terms on the right-hand side. A more elaborate treatment of the surface mixed layer is probably necessary to reproduce details of the vertical momentum flux; in the simple Ekman theory the influence of stratification is ignored, and a simple prescription of the vertical mixing process is chosen. At this time much observational and theoretical work is underway on this problem. Here, though, we stick to the simple Ekman model that has been used in nearly all of the numerical models, but it is clear that a proper parameterization of the vertical momentum (and heat) flux near the sea surface will have to be incorporated into future large-scale ocean models.

The influence of bottom topography has been examined in a number of homogeneous and baroclinic (both diagnostic and prognostic) studies. In a simple extension of the constant-depth model of Bryan (1963), Holland (1977) introduced variable depth into the homogeneous case. The appropriate nondimensional equations are

$$-\psi_y' \left(\frac{Ro\zeta' + f'}{h'}\right)_x + \psi_x' \left(\frac{Ro\zeta' + f'}{h}\right)_y = +\text{curl}_z \left(\frac{\tau'}{h'}\right) + Ek_h \nabla^2 \zeta' \tag{4.436}$$

where h' is the nondimensional depth and ζ' is the nondimensional vorticity equal to $\zeta' = \nabla \cdot (\nabla \psi'/h')$. In this study it was found that when a simple sloping region was included leading away from the western boundary, the western boundary current could be induced to leave the coast and follow topographic contours (actually f'/h' contours) seaward. As the Rossby number was increased, steady meanders in the free current began to be important. Thus, the results of Bryan (1963) were extended to incorporate the ideas suggested by Warren (1963) concerning the meandering nature of the path of the Gulf Stream after it had left Cape Hatteras and turned seaward.

It is of interest to ask how these rather frictional results behave in the highly inertial limit. As Veronis (1966) and Niiler (1966) showed, the Fofonoff (1954) solution gives a good representation of the inertial limit of Stommel's (1948) model with nonlinear terms added. Therefore it is useful to extend Fofonoff's model to the variable-depth case. The vorticity equation then would be

$$\frac{D}{Dt}\left(\frac{\zeta + f}{h}\right) = 0 \qquad (4.437)$$

where

$$\zeta = \nabla \cdot \frac{\nabla \psi}{h} \qquad (4.438)$$

Then, following Fofonoff, we look for special solutions such that

$$\frac{\zeta + f}{h} = A + B\psi \qquad (4.439)$$

The second-order, elliptic differential equation is easily solved by relaxation for any $h(x, y)$.

Next, following Haidvogel *et al.* (1980) let us consider the barotropic vorticity equation on a β-plane:

$$\left(\frac{\partial}{\partial t} + J(\psi, \ )\right)(\zeta + f) = F(x, y), \qquad 0 \leq x \leq L_x, \quad 0 \leq y \leq L_y \qquad (4.440)$$

where

$$f = f_0 + \beta y$$

and

$$J(\psi, \ ) = \frac{\partial \psi}{\partial x}\frac{\partial}{\partial y} - \frac{\partial \psi}{\partial y}\frac{\partial}{\partial x}$$

In addition:

$$\zeta = \nabla^2 \psi \qquad (4.441)$$

is the relationship between streamfunction (ψ) and (ζ), and $F(x, y)$ represents the effects of a body force, if any. Dissipation has been neglected for three reasons. First, the inviscid system poses a simpler physical and numerical problem, one for which many analytic and perturbation solutions are available. This is the basis of our testing of the limited-area models. Second, quadratic conservation laws are available for nondissipative phsyical and numerical systems. This property also contributes to the evaluation of model performance. Lastly, by ignoring explicit higher-order friction, we sidestep for the moment the question of the correct specification of vorticity boundary conditions on outflow, which are not formally needed for integration of the inviscid system. The assumption of inviscid dynamics does, however, require that greater care be taken to construct a numerical scheme which is stable in the absence of explicit dissipative (that is, smoothing) mechanisms. As we shall see, such filtering is in fact necessary to maintain stability in some cases.

If we now nondimensionalize x, y, t and ψ by $d, d, (\beta d)^{-1}$, and $(V_0 d)$, respectively, then eq.(4.440) becomes, in nondimensional form:

$$\left(\frac{\partial}{\partial t} + \epsilon J(\psi, \ )\right)\nabla^2\psi + \psi_x = F(x, y) \qquad 0 \leq x \leq x_B, \quad 0 \leq y \leq y_B \qquad (4.442)$$

where the Rossby number

$$\epsilon = V_0/\beta d^2$$

and

$$x_B = L_x/d, \qquad y_B = L_y/d$$

The parameters d and V_0 are taken to be the characteristic length and velocity scales of the anticipated field of motion. Note that the length scale d does not correspond to the basin dimensions L_x or L_y. Hence, x_B and y_B are, in general, greater than one.

Arakawa (1966) developed a method for the suppression of nonlinear instability. Haltiner and Williams (1980) illustrated this technique for the case of a nondivergent wind field:

$$\frac{\partial \zeta}{\partial t} = -\mathbf{V} \cdot \mathbf{V}\zeta = -\nabla \cdot (\zeta \mathbf{V}), \qquad \mathbf{V} = \mathbf{k} \times \nabla\psi, \qquad \zeta = \nabla^2\psi \qquad (4.443)$$

Write this as

$$\frac{\partial \zeta}{\partial t} = -J(\psi, \zeta) \qquad (4.444)$$

Write the Jacobian as

$$J(\psi, \zeta) = \frac{\partial \psi}{\partial x}\frac{\partial \zeta}{\partial y} - \frac{\partial \psi}{\partial y}\frac{\partial \zeta}{\partial x} = \frac{\partial}{\partial x}\left(\psi\frac{\partial \zeta}{\partial y}\right) - \frac{\partial}{\partial y}\left(\psi\frac{\partial \zeta}{\partial x}\right)$$

$$= \frac{\partial}{\partial y}\left(\zeta\frac{\partial \psi}{\partial x}\right) - \frac{\partial}{\partial x}\left(\zeta\frac{\partial \psi}{\partial y}\right) \qquad (4.445)$$

With reference to Fig. 4.19, the finite difference forms of the Jacobians are given by

$$J^{++} = [(\psi_1 - \psi_3)(\zeta_2 - \zeta_4) - (\psi_2 - \psi_4)(\zeta_1 - \zeta_3)]/4h^2$$
$$J^{+x} = [(\psi_1(\zeta_1 - \zeta_8) - \psi_3(\zeta_6 - \zeta_7) - \psi_2(\zeta_5 - \zeta_6) + \psi_4(\zeta_8 - \zeta_7)]/4h^2 \qquad (4.446)$$
$$J^{x+} = [\zeta_2(\psi_5 - \psi_6) - \zeta_4(\psi_8 - \psi_7) - \zeta_1(\psi_5 - \psi_8) + \zeta_3(\psi_6 - \psi_7)]/4h^2$$

$$
\begin{array}{ccc}
\overset{\displaystyle\bullet}{6} & \overset{\displaystyle\bullet}{2} & \overset{\displaystyle\bullet}{5} \\[2em]
\overset{\displaystyle\bullet}{3} & \overset{\displaystyle\bullet}{0} & \overset{\displaystyle\bullet}{1} \\[2em]
\overset{\displaystyle\bullet}{7} & \overset{\displaystyle\bullet}{4} & \overset{\displaystyle\bullet}{8}
\end{array}
$$

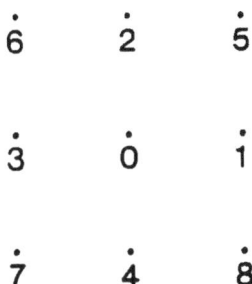

Fig. 4.19
Grid distribution.

where the symbols $+$ and x refer to the points in the grid in Fig. 4.19 used in the derivatives of ψ and ζ in that order.

Consider now the gain in the square vorticity at point zero due to the value at point 1, and vice versa. The results, including the factor $4h^2$ are

$$
\begin{aligned}
\zeta_0 J_0^{++} &\sim -\zeta_0\zeta_1(\psi_2 - \psi_4) + \text{ other terms} \\
\zeta_1 J_1^{++} &\sim \zeta_1\zeta_0(\psi_5 - \psi_8) + \text{ other terms}
\end{aligned}
\tag{4.447}
$$

Since the terms do not cancel and no other opportunity for the product $\zeta_0\zeta_1$ occurs over the grid, the sum $\overline{\zeta J^{++}}$ cannot vanish in general, and $\overline{\zeta^2}$ will not be conserved (i.e., $\overline{\partial\zeta^2/\partial t} \neq 0$). Similarly:

$$
\begin{aligned}
\zeta_0 J_0^{+x} &\sim \zeta_0\zeta_5(\psi_1 - \psi_2) + \cdots \\
\zeta_5 J_5^{+x} &\sim \zeta_5\zeta_0(\psi_2 - \psi_1) + \cdots \\
\zeta_0 J_0^{x+} &\sim \zeta_0\zeta_1(\psi_5 - \psi_8) + \cdots \\
\zeta_1 J_1^{x+} &\sim \zeta_1\zeta_0(\psi_2 - \psi_4) + \cdots
\end{aligned}
\tag{4.448}
$$

We can see from eqs.(4.447) and (4.448) that the Jacobians J^{+x} and $\frac{1}{2}(J^{++} + J^{x+})$ conserve $\overline{\zeta^2}$. One can examine the finite difference approximation of $\psi J(\psi,\zeta)$ to show that J^{x+} and $\frac{1}{2}(J^{++} + J^{+x})$ conserve $\overline{K} = \frac{1}{2}V^2$. A combination of these two Jacobians gives the Arakawa Jacobian $\frac{1}{3}(J^{++} + J^{+x} + J^{x+})$ which conserves both $\overline{\zeta^2}$ and $\overline{K}$. Hence, the average wave number is also conserved, and the nonlinear instability is suppressed. The above results hold for the time-continuous case. Haltiner and Williams (1980) showed that with the leapfrog scheme and making use of the Arakawa Jacobian gives rise to

$$
\overline{\zeta_{0,n}(\zeta_{0,n+1} - \zeta_{0,n-1})} = 2\Delta t\overline{\zeta_{0,n} J_{0,n}} = 0
$$

and hence

$$\overline{\zeta_{0,n}\zeta_{0,n+1}} = \overline{\zeta_{0,n}\zeta_{0,n-1}}$$

However, for strict conservation of mean square vorticity the result should be:

$$\overline{\zeta_{0,n+1}^2} = \overline{\zeta_{0,n}^2} = \overline{\zeta_{0,n-1}^2}, \text{ etc.}$$

Now if $\zeta_{0,n} = (\zeta_{0,n+1} + \zeta_{0,n-1})/2$ the above property holds, but it would take a rather complex scheme to achieve this property. The desired result is achieved in practice if ζ varies smoothly between successive time steps, but this is not always the case and occassional smoothing and restarting is necessary to avoid solution decoupling at adjacent time steps.

11. The Bidston models

Researchers at the Bidston Observatory in the U.K. developed sophisticated models with somewhat different features (e.g. Heaps, 1972; Davies, 1985a,b). Whereas most tidal storm surge and tsunami modellers used vertically integrated two-dimensional (in the horizontal) models, and some type of three-dimensional models for ocean circulation problems, the Bidston group is among the first to use three-dimensional models for tide and storm surge computations and in some sense, bridged the gap between two-dimensional coastal models and three-dimensional circulation models. Here we will discuss two particular models in some detail, starting with the model developed by Heaps for the three-dimensional computation of tides and storm surges. We will write the equation of continuity and motion in the following form:

$$\frac{\partial}{\partial x}\int_0^D u \, dz + \frac{\partial}{\partial y}\int_0^D v \, dz + \frac{\partial \zeta}{\partial t} = 0 \qquad (4.449)$$

$$\frac{\partial u}{\partial t} - fv = -g\frac{\partial}{\partial x}(\zeta - \zeta') + \frac{\partial}{\partial z}\left(N\frac{\partial u}{\partial z}\right) \qquad (4.450)$$

$$\frac{\partial v}{\partial t} + fu = -g\frac{\partial}{\partial y}(\zeta - \zeta') + \frac{\partial}{\partial z}\left(N\frac{\partial v}{\partial z}\right) \qquad (4.451)$$

where x, y, z are a left-handed Cartesian coordinate set, and D is the water depth in the undisturbed state. Here,

$$\frac{\partial \zeta'}{\partial x} = -\frac{1}{\rho g}\frac{\partial p_a}{\partial x}$$
$$\frac{\partial \zeta'}{\partial y} = -\frac{1}{\rho g}\frac{\partial p_a}{\partial y} \qquad (4.452)$$

where p_a is the atmospheric pressure at sea level and N is the coefficient of vertical eddy viscosity. Also note that

$$\tau_{s_x} = -\rho \left(N \frac{\partial u}{\partial z} \right)_0$$
$$\tau_{s_y} = -\rho \left(N \frac{\partial v}{\partial z} \right)_0$$

(4.453)

where subscript 0 denotes the undisturbed sea surface ($z = 0$) and τ_{s_x} and τ_{s_y} are the x and y components of wind stress. Assume that the bottom friction varies with the bottom current according to the following relation (τ_{B_x} and τ_{B_y} are x and y components of bottom friction):

$$\tau_{B_x} = K \rho u_D$$
$$\tau_{B_y} = K \rho v_D$$

also noting that

$$-\rho \left(N \frac{\partial u}{\partial z} \right)_D = \tau_{B_x}$$
$$-\rho \left(N \frac{\partial v}{\partial z} \right)_D = \tau_{B_y}$$

We get

$$\left(N \frac{\partial u}{\partial z} \right)_D + K u_D = 0 \quad \text{and} \quad \left(N \frac{\partial v}{\partial z} \right)_D + K v_D = 0 \qquad (4.454)$$

We will now transform eq.(4.449) to (4.451) to eliminate the depth coordinate z. Consider the differential equation:

$$\frac{d}{dz}[N f'(z)] = -\lambda f(z) \qquad (4.455)$$

where $\lambda = \lambda_r$ denotes the Eigenvalues in ascending order of magnitude ($r = 1, 2, 3, \ldots, \infty$) and $f = f_r(z)$ are the corresponding Eigenfunctions. Equation (4.455) should be solved for the range $0 \leq z \leq D$ under the following conditions:

$$f'(0) = 0 \qquad (4.456)$$

$$N_D f'(D) + K f D = 0 \qquad (4.457)$$

$$f(0) = 1 \qquad (4.458)$$

Here $f' = df/dz$. For each value of r, one can define a pair of integral transforms:

$$u_r = \frac{1}{D} \int_0^D u f_r(z) \, dz$$

$$v_r = \frac{1}{D} \int_0^D v f_r(z) \, dz \qquad (4.459)$$

Multiply eqs.(4.450) and (4.451) with $f_r(z)$ and integrate with respect to z from $z = 0$ to D to give

$$\frac{\partial u_r}{\partial t} - f v_r = -g a_r \frac{\partial}{\partial x}(\zeta - \zeta') + \frac{1}{D} \int_0^D f_r(z) \frac{\partial}{\partial z}\left(N \frac{\partial u}{\partial z}\right) dz \qquad (4.460)$$

$$\frac{\partial v_r}{\partial t} - f u_r = -g a_r \frac{\partial}{\partial y}(\zeta - \zeta') + \frac{1}{D} \int_0^D f_r(z) \frac{\partial}{\partial z}\left(N \frac{\partial v}{\partial z}\right) dz \qquad (4.461)$$

where

$$a_r = \frac{1}{D} \int_0^D f_r(z) \, dz \qquad (4.462)$$

In eqs.(4.460) and (4.461), integrating by parts twice and use of eqs.(4.453) and (4.454) as well as the following relationships which follow (4.456) to (4.458) leads to

$$\frac{d}{dz}[N f_r'(z)] = -\lambda_r f_r(z) \qquad (4.463)$$

$$f_r'(0) = 0 \qquad (4.464)$$

$$N_D f_r'(D) = -K f_r(D) \qquad (4.465)$$

$$f_r(0) = 1 \qquad (4.466)$$

We get

$$\frac{1}{D} \int_0^D f_r(z) \frac{\partial}{\partial z}\left(N \frac{\partial u}{\partial z}\right) dz = \frac{T_{s_x}}{\rho d} - \lambda_r u_r$$

$$\frac{1}{D} \int_0^D f_r(z) \frac{\partial}{\partial z}\left(N \frac{\partial v}{\partial z}\right) dz = \frac{T_{s_y}}{\rho d} - \lambda_r u_r$$

Thus from eqs.(4.460) and (4.461) we get the following integrated forms of the equations of motion involving u_r, v_r and ζ as the dependent variables:

$$\frac{\partial u_r}{\partial t} + \lambda_r u_r - f v_r = -g a_r \frac{\partial}{\partial x}(\zeta - \zeta') + \frac{\tau_{s_x}}{\rho D} \tag{4.467}$$

$$\frac{\partial v_r}{\partial t} + \lambda_r v_r + f u_r = -g a_r \frac{\partial}{\partial y}(\zeta - \zeta') + \frac{\tau_{s_y}}{\rho D} \tag{4.468}$$

Next we will express u, v in terms of the transforms u_r, v_r and substitute these into the continuity equation (4.449):

$$u = \sum_{r=1}^{\infty} A_r f_r(z) \qquad \text{and} \qquad v = \sum_{r=1}^{\infty} B_r f_r(z) \tag{4.469}$$

where A_r and B_r are independent of z. From eqs.(4.463) to (4.465) one can obtain the following orthogonality relationship:

$$\int_0^D f_r(z) f_s(z) \, dz = 0 \qquad r \neq s \tag{4.470}$$

and from eq.(4.455):

$$\int_0^D f_r^2(z) \, dz = I_r(D) - I_r(0) \tag{4.471}$$

where

$$I_r(z) = N \left[f' \frac{\partial f}{\partial \lambda} - f \frac{\partial f'}{\partial \lambda} \right]_{\lambda=\lambda_r} \tag{4.472}$$

one can determine the coefficients from eqs.(4.469) to (4.471). Using these in eq.(4.469) gives

$$u = \sum_{r=1}^{\infty} \phi_r u_r f_r(z) \qquad \text{and} \qquad v = \sum_{r=1}^{\infty} \phi_r v_r f_r(z) \tag{4.473}$$

where

$$\phi_r = \frac{d}{I_r(D) - I_r(0)} \tag{4.474}$$

Substitute eq.(4.473) into (4.449) to give

$$\frac{\partial \zeta}{\partial t} + \sum_{r=1}^{\infty} \left(\frac{\partial}{\partial x}(D a_r \phi_r w_r) + \frac{\partial}{\partial y}(D a_r \phi_r v_r) \right) = 0 \tag{4.475}$$

Equations (4.467), (4.468) and (4.475), valid for all positive integer values of r constitute a system of partial differential equations where u_r, v_r and ζ are the dependent variables. Once u_r and v_r are solved, one can recover u and v from eq.(4.473). Note that the variation with depth z enters through the Eigenfunctions $f_r(z)$.

Next, we will consider very briefly a three-dimensional model formulated by Davies (1985a,b), using sigma coordinates in the vertical direction. Starting essentially with the same equations as Heaps (1972), transformation from the interval $-\zeta \leq z \leq D$ into the constant interval $0 \leq \sigma \leq 1$ can be done through the σ transformation:

$$\sigma = \frac{z + \zeta}{D + \zeta}. \tag{4.476}$$

In the σ coordinates, the equations of motion and continuity can be written as

$$\frac{\partial u}{\partial t} + u\frac{\partial u}{\partial x} + v\frac{\partial u}{\partial y} + v\frac{\partial u}{\partial \sigma} - fv = -g\frac{\partial \zeta}{\partial x} + \frac{1}{(D+\zeta)^2}\frac{\partial}{\partial \sigma}\left(N\frac{\partial u}{\partial \sigma}\right) \tag{4.477}$$

$$\frac{\partial v}{\partial t} + u\frac{\partial v}{\partial x} + v\frac{\partial v}{\partial y} + v\frac{\partial v}{\partial \sigma} + fu = -g\frac{\partial \zeta}{\partial y} + \frac{1}{(D+\zeta)^2}\frac{\partial}{\partial \sigma}\left(N\frac{\partial v}{\partial \sigma}\right) \tag{4.478}$$

$$\frac{\partial \zeta}{\partial t} + \frac{\partial}{\partial x}\left[(D+\zeta)\int_0^1 u\,d\sigma\right] + \frac{\partial}{\partial y}\left[(D+\zeta)\int_0^1 v\,d\sigma\right] = 0 \tag{4.479}$$

where

$$v = \frac{1}{(D+\zeta)}\left(\frac{\partial \zeta}{\partial t}(1-\sigma)\right)$$

$$+ \frac{1}{(D+\zeta)}\frac{\partial}{\partial x}\left((D+\zeta)\int_\sigma^1 u\,d\sigma\right) + \frac{1}{(D+\zeta)}\frac{\partial}{\partial y}\left((D+\zeta)\int_\sigma^1 v\,d\sigma\right). \tag{4.480}$$

In the σ coordinates, the free surface and bottom boundary conditions can be written as follows.

At the free surface:

$$-\rho\left(N\frac{\partial u}{\partial \sigma}\right)_0 = (D+\zeta) - \tau_{s_x} \quad \text{and} \quad -\rho\left(N\frac{\partial v}{\partial \sigma}\right)_0 = (D+\zeta) - \tau_{s_y} \tag{4.481}$$

and at the bottom:

$$-\rho\left(N\frac{\partial u}{\partial \sigma}\right)_1 = (D+\zeta) - \tau_{B_x} \quad \text{and} \quad -\rho\left(N\frac{\partial v}{\partial \sigma}\right)_1 = (D+\zeta) - \tau_{B_y} \tag{4.482}$$

12. Haidvogel et al.'s model

Haidvogel *et al.* (1991) developed a primitve equation ocean circulation model using σ coordinates in the vertical and orthogonal curvilinear coordinates in the horizontal plane.

As is traditional in diabatic, large- and regional-scale ocean circulation modelling (see, e.g. Cox (1985)), we adopt the hydrostatic primitive equations, which can be written:

$$\frac{\partial u}{\partial t} + \mathbf{v} \cdot \nabla u - fv = -\frac{\partial}{\partial x}\phi + \mathcal{F}_u + D_u \qquad (4.483)$$

$$\frac{\partial v}{\partial t} + \mathbf{v} \cdot \nabla v + fu = -\frac{\partial}{\partial y}\phi + \mathcal{F}_v + D_v \qquad (4.484)$$

$$\frac{\partial \phi}{\partial z} = \frac{-\rho g}{\rho_0} \qquad (4.485)$$

$$\frac{\partial \rho}{\partial t} + \mathbf{v} \cdot \nabla \rho = \mathcal{F}_\rho + D_\rho \qquad (4.486)$$

and

$$\frac{\partial u}{\partial x} + \frac{\partial v}{\partial y} + \frac{\partial w}{\partial z} = 0, \qquad (4.487)$$

where, in standard notation, u, v, w equals x, y, z components of vector velocity $\mathbf{v}$, $\rho_0 + \rho(x, y, z, t)$ equals the total density, ϕ is the dynamic pressure p/ρ_0, f is the Coriolis parameter, and g is acceleration of gravity. Equations (4.483) and (4.484) express the momentum balance in the x and y directions, respectively. In the Boussinesq approximation, density variations are neglected in the momentum equations except in their contribution to the buoyancy force in the vertical momentum equation (4.485). Under the hydrostatic approximation, it is further assumed that the vertical pressure gradient balances the buoyancy force. The time evolution of the density field, $\rho(x, y, z, t)$ is governed by the advective–diffusive equation (4.486). Lastly, eq.(4.487) expresses continuity for an incompressible fluid. For the moment, the effects of forcing and dissipation will be represented by the schematic terms $\mathcal{F}$ and D, respectively. Formulations of the dissipation terms D presently in use in oceanographic modelling include smoothing based on harmonic and/or biharmonic operators, and Shapiro filters (Gent and Kane, 1989). The separation of eq.(4.486) into two independent prognostic equations for temperature and salinity and an equation of state is straightforward.

It is computationally convenient to introduce a stretched vertical coordinate system which conforms to the variable bottom at $z = -h(x, y)$. Such "sigma" coordinte systems have long been used in both meteorology and oceanography (e.g. Phillips, 1951; Freeman *et al.*, 1972). To proceed, we make the coordinate transformation:

$$\hat{x} = x$$

$$\hat{y} = y$$

$$\sigma = 1 + 2\left(\frac{z}{h}\right)$$

and

$$t = t$$

In the stretched system, the vertical coordinate σ spans the range $-1 \leq \sigma \leq 1$. We are therefore left with level upper ($\sigma = 1$) and lower ($\sigma = -1$) bounding surfaces. As a trade-off for this geometric simplification, the dynamic equations become somewhat more complicated. The resulting dynamic equations are, after dropping the carats:

$$\frac{\partial u}{\partial t} - fv + \mathbf{v} \cdot \nabla u = -\frac{\partial \phi}{\partial x} + (1 - \sigma)\left(\frac{g\rho}{2\rho_0}\right)\frac{\partial h}{\partial x} + \mathcal{F}_u + D_u \tag{4.488}$$

$$\frac{\partial v}{\partial t} + fu + \mathbf{v} \cdot \nabla v = -\frac{\partial \phi}{\partial y} + (1 - \sigma)\left(\frac{g\rho}{2\rho_0}\right)\frac{\partial h}{\partial y} \mathcal{F}_v + D_v \tag{4.489}$$

$$\frac{\partial \phi}{\partial \sigma} = \left(\frac{-gh\rho}{2\rho_0}\right) \tag{4.490}$$

$$\frac{\partial \rho}{\partial t} + \mathbf{v} \cdot \nabla \rho = \mathcal{F}_\rho + D_\rho \tag{4.491}$$

and

$$\frac{\partial}{\partial x}(hu) + \frac{\partial}{\partial y}(hv) + h\frac{\partial \Omega}{\partial \sigma} = 0, \tag{4.492}$$

where

$$\mathbf{v} = (u, v, \Omega)$$

$$\mathbf{v} \cdot \nabla = u\frac{\partial}{\partial x} + v\frac{\partial}{\partial y} + \Omega\frac{\partial}{\partial \sigma},$$

and the vertical velocity in sigma coordinates:

$$\Omega(x, y, \sigma, t) = \frac{1}{h}\left[(1 - \sigma)u\frac{\partial h}{\partial x}(1 - \sigma)v\frac{\partial h}{\partial y} + 2w\right]$$

In real water bodies, the coastlines are always more or less irregular and a boundary-fitted coordinate system is more appropriate. The requirement for a boundary-following coordinate system which allows laterally variable grid resolution can be met (for suitably smooth domains) by introducing an appropriate orthogonal coordinate transformation in the horizontal. Let the new coordinates by $\xi(x, y)$ and

$\eta(x, y)$ where the relationship of horizontal arc length to the differential distances is given by

$$(ds)_\xi = \left(\frac{1}{m}\right) d\xi \tag{4.493a}$$

and

$$(ds)_\eta = \left(\frac{1}{m}\right) d\eta \tag{4.493b}$$

Here, $m(\xi, \eta)$ and $n(\xi, \eta)$ are the metric factors of the coordinate transformation which relate the differential distances $(\Delta\xi, \Delta\eta)$ to the actual (physical) arc lengths.

Denoting the velocity components in the new coordinate system by

$$\mathbf{v} \cdot \hat{\xi} = u \tag{4.494a}$$

and

$$\mathbf{v} \cdot \hat{\eta} = v \tag{4.494b}$$

the equations of motion (4.488) to (4.492) can be rewritten (Arakawa and Lamb, 1977) as

$$\frac{\partial}{\partial t}\left(\frac{hu}{mn}\right) + \frac{\partial}{\partial\xi}\left(\frac{hu^2}{n}\right) + \frac{\partial}{\partial\eta}\left(\frac{huv^2}{m}\right) + \frac{\partial}{\partial\sigma}\left(\frac{hu\Omega}{mn}\right)$$
$$- \left[\left(\frac{f}{mn}\right)\right] + v\frac{\partial}{\partial\xi}\left(\frac{1}{n}\right) - u\frac{\partial}{\partial\eta}\left(\frac{1}{n}\right)\right] hv$$
$$= -\left(\frac{h}{n}\right)\frac{\partial\phi}{\partial\xi} + (1 - \sigma)\left(\frac{gh\rho}{2\rho_0 n}\right)\frac{\partial h}{\partial\xi} + \mathcal{F}_u + D_u \tag{4.495}$$

$$\frac{\partial}{\partial t}\left(\frac{hv}{mn}\right) + \frac{\partial}{\partial\xi}\left(\frac{huv}{n}\right) + \frac{\partial}{\partial\eta}\left(\frac{hv^2}{m}\right) + \frac{\partial}{\partial\sigma}\left(\frac{hv\Omega}{mn}\right)$$
$$+ \left[\left(\frac{f}{mn}\right)\right] + v\frac{\partial}{\partial\xi}\left(\frac{1}{n}\right) - u\frac{\partial}{\partial\eta}\left(\frac{1}{m}\right)\right] hu$$
$$= -\left(\frac{h}{m}\right)\frac{\partial\phi}{\partial\eta} + (1 - \sigma)\left(\frac{gh\rho}{2\rho_0 m}\right)\frac{\partial h}{\partial\eta} + \mathcal{F}_v + D_v, \tag{4.496}$$

$$\frac{\partial\phi}{\partial\sigma} = -\left(\frac{gh\rho}{2\rho_0}\right) \tag{4.497}$$

$$\frac{\partial}{\partial t}\left(\frac{h\rho}{mn}\right) + \frac{\partial}{\partial\xi}\left(\frac{hu\rho}{n}\right)\frac{\partial}{\partial\eta}\left(\frac{hv\rho}{m}\right) + \frac{\partial}{\partial\sigma}\left(\frac{h\Omega\rho}{mn}\right) = \mathcal{F}_\rho + D_\rho \tag{4.498}$$

and

$$\frac{\partial}{\partial \xi}\left(\frac{hu}{n}\right) + \frac{\partial}{\partial \eta}\left(\frac{hv}{m}\right) \frac{\partial}{\partial \sigma}\left(\frac{h\Omega}{mn}\right) = 0 \qquad (4.499)$$

Next, we will discuss the boundary and initial conditions. Consider a fluid lying within the three-dimensional domain $0 \leq \xi \leq L_x$, $0 \leq \eta \leq L_y$, and $-1 \leq \sigma \leq +1$, the latter corresponding to the depth range $0 \leq z \leq -h(x, y)$. Then, given an appropriate set of boundary and initial conditions, eqs.(4.495) to (4.499) can, in principle, be solved for the time evolution of the fluid motion. In all the test problems described below, the upper and lower bounding surfaces are taken to be rigid. Hence

$$(\Omega)_{\sigma = \pm 1} = 0 \qquad (4.500)$$

Applying the so-called "rigid lid" surface boundary condition eliminates from the solution high speed surface gravity waves which would otherwise severely constrain the model time step. The lateral boundaries at $\eta = 0$, L_y will likewise be impermeable:

$$(v)_{\eta = 0, L_y} = 0 \qquad (4.501)$$

Finally, the remaining lateral boundaries ($\xi = 0, L_x$) will be treated in one of three ways — either as periodic:

$$(u, v, \Omega, \rho, \phi)_{\xi = 0} = (u, v, \Omega, \rho, \phi)_{\xi = L_x} \qquad (4.502)$$

as closed:

$$(u)_{\xi = 0, L_x} = 0 \qquad (4.503)$$

or as open. In the latter case, an Orlanski radiation condition (Orlanski, 1976; Chapman, 1985) is used, as described below.

Next, the numerical solution techniques will be considered, starting with the vertical spectral method. The vertical (σ) dependence of the model variables is represented as an expansion in the finite polynomial basis set $P_k(\sigma)$; that is, we set

$$b(\xi, \eta, \sigma) = \sum_{k=0}^{N} P_k(\sigma) \hat{b}_k(\xi, \eta), \qquad (4.504)$$

where the arbitrary variable b, introduced here for the purposes of discussion, is any of $u, v, \rho, \phi,$ or Ω. The only restriction placed on the form of the basis polynomials is that $\frac{1}{2} \int_{-1}^{1} P_k(\sigma) \, d\sigma = \delta_{0k}$, where δ_{0k} is the Kronecker delta — i.e., that only the lowest order polynomial ($k = 0$) may have a non-zero vertical integral. This isolates the external mode (or depth-averaged component of the field) in the amplitude of the $k = 0$ polynomial. In practice, the spectral technique does not explicitly solve for the polynomial coefficients $\hat{b}_k$ but rather for the actual variable values at "collocation" points (or equivalent grid points) σ_n chosen to correspond to the location of the extrema of the highest order polynomial. The collocation point values b_n are thus defined as

$$b_n = b(\sigma_n) = \sum_{k=0}^{N} P_k(\sigma_n)\hat{b}_k, \qquad 0 \le n \le N, \qquad (4.505)$$

and will be functions of (ξ, η). The polynomial coefficients $\hat{b}_k$ can be recovered from the collocation point values b_n by a linear matrix transformation. Consider a matrix F whose elements are

$$F_{nk} = P_k(\sigma_n) \qquad (4.506)$$

and let b and $\hat{b}$ be column vectors whose elements are b_n and $\hat{b}_k$, respectively. Then,

$$b = F\hat{b} \qquad (4.507)$$

and, hence,

$$\hat{b} = F^{-1}b \qquad (4.508)$$

It is now straightforward to represent vertical differentiation and integration in terms of matrix operators. Consider a matrix R whose elements are

$$R_{nk} = \frac{\partial P_k}{\partial \sigma}(\sigma_n) \qquad (4.509)$$

Then the matrix $C_{DZ} = RF^{-1}$ will perform differentiation of the model field (b) in the vertical direction, denoted in the following subsections by the δ_σ operation since

$$\delta_\sigma b = \frac{\partial b}{\partial \sigma} = \sum_{k=0}^{N} \frac{\partial P_k}{\partial \sigma}(\sigma_n)\hat{b}_k = R\hat{b} = RF^{-1} = C_{DZ}b \qquad (4.510)$$

Similarly, consider a matrix $\mathcal{L}$ whose elements are

$$\mathcal{L}_{nk} = \int_{\sigma_n}^{1} P_k(\sigma) \, d\sigma \qquad (4.511)$$

Then the matrix $C_{INT} = \mathcal{L}F^{-1}$ will perform vertical integration, denoted in the following sections by the I_σ^1 operation since

$$I_\sigma^1 b = \int_0^1 b \, d\sigma = \sum_{k=0}^{N} \int_{\sigma_n}^{1} P_k(\sigma_n) \, d\sigma \hat{b}_k = \mathcal{L}\hat{b} = \mathcal{L}F^{-1}b = C_{INT}b. \qquad (4.512)$$

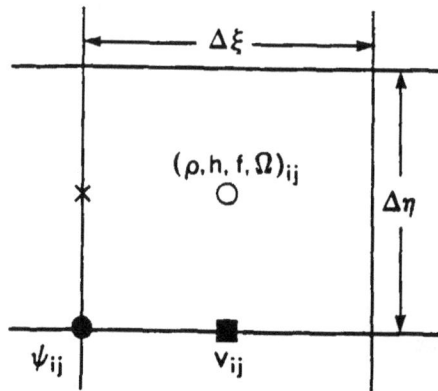

Fig. 4.20
Placement of variables on the Arakawa C grid. (from Haidvogel *et al.* (1991))

In the model simulations, Chebyshev polynomials were employed as the basis functions. The matrix operators F, R, and $\mathcal{L}$ and collocation levels σ_n for this basis set are given in the Appendix of Haidvogel *et al.* (1991).

In the horizontal coordinates (ξ, η), a traditional, centered, second-order finite difference approximation is adopted. In particular, the horizontal arrangement of variables is as shown in Fig.4.20. This is equivalent to the well-known Arakawa "C" grid, which is well suited for problems with horizontal resolution that is fine compared to the first internal radius of deformation (Arakawa and Lamb, 1977).

Using the representations described above, the semi-discrete form of the dynamic equations (4.495) to (4.999) becomes

$$
\frac{\partial}{\partial t}\left(\overline{\left(\frac{h}{mn}\right)^{\xi}} u\right) + \delta_{\xi}\left[\overline{\overline{u^{\xi}\left(u\overline{\left(\frac{h}{n}\right)^{\xi}}\right)}^{\xi}}\right] + \delta_{\eta}\left[\overline{\overline{u^{\eta}\left(v\overline{\left(\frac{h}{m}\right)^{\eta}}\right)}^{\xi}}\right] + \delta_{\sigma}\left[\overline{u\overline{\left(\frac{h\Omega}{mn}\right)^{\xi}}}\right]
$$

$$
-\overline{\left[\overline{\left(\frac{fh}{mn}\right)^{\xi\eta}} + \overline{v^{\xi}}\overline{h}^{\xi\eta}\delta_{\xi}\overline{\left(\frac{1}{n}\right)}^{\xi\eta} - \overline{u^{\eta}}\overline{h}^{\xi\eta}\delta_{\eta}\overline{\left(\frac{1}{m}\right)}^{\xi\eta}\right]\overline{v^{\xi}}}^{\eta}
$$

$$
= -\overline{\left(\frac{h}{n}\right)^{\xi}}\delta_{\xi}\phi + (1-\sigma)\overline{\left(\frac{gh\rho}{2\rho_0 n}\right)^{\xi}}\delta_{\xi}h + \mathcal{F}_{u} + D_{u} \qquad (4.513)
$$

$$
\frac{\partial}{\partial t}\left(\overline{\left(\frac{h}{mn}\right)^{\eta}} v\right) + \delta_{\xi}\left[\overline{\overline{v^{\xi}\left(u\overline{\left(\frac{h}{n}\right)^{\xi}}\right)}^{\eta}}\right] + \delta_{\eta}\left[\overline{\overline{v^{\eta}\left(v\overline{\left(\frac{h}{m}\right)^{\eta}}\right)}^{\eta}}\right] + \delta_{\sigma}\left[\overline{v\overline{\left(\frac{h\Omega}{mn}\right)^{\eta}}}\right]
$$

$$+ \overline{\left[\overline{\left(\frac{fh}{mn} \right)}^{\xi\eta} + \overline{v^\xi \overline{h}^{\xi\eta} \delta_\xi \left(\frac{1}{n} \right)}^{\xi\eta} - \overline{u^\eta \overline{h}^{\xi\eta} \delta_\eta \left(\frac{1}{m} \right)}^{\xi\eta} \right] \overline{u}^\eta}^\xi$$

$$= -\overline{\left(\frac{h}{n} \right)}^\eta \delta_\eta \phi + (1-\sigma) \overline{\left(\frac{gh\rho}{2\rho_0 m} \right)}^\eta \delta_\eta h + \mathcal{F}_v + D_v \qquad (4.514)$$

$$\frac{\partial \phi}{\partial \sigma} = -\left(\frac{gh\rho}{2\rho_0} \right) \qquad (4.515)$$

$$\frac{\partial}{\partial t} \left(\frac{h\rho}{mn} \right) + \delta_\xi \left[u \overline{\left(\frac{h\rho}{n} \right)}^\xi \right] + \delta_\eta \left[v \overline{\left(\frac{h\rho}{m} \right)}^\eta \right] + \delta_\sigma \left[\frac{h\Omega\rho}{mn} \right] = \mathcal{F}_\rho + D_\rho \qquad (4.516)$$

$$\delta_\xi \left[u \overline{\left(\frac{h}{n} \right)}^\xi \right] + \delta_\eta \left[v \overline{\left(\frac{h}{m} \right)}^\eta \right] + \delta_\sigma \left[\frac{h\Omega}{mn} \right] = 0 \qquad (4.517)$$

Here δ_ξ and δ_η denote simple centered finite-difference approximations to $\partial/\partial\xi$ and $\partial/\partial\eta$, with the differences taken over the distances $\Delta\xi$ and $\Delta\eta$, respectively; $\overline{(\ )}^\xi$ and $\overline{(\ )}^\eta$ represent averages taken over the distances $\Delta\xi$ and $\Delta\eta$; δ_σ represents a vertical derivative evaluated according to the prescription given in Arakawa and Lamb (1977); and I_σ^1 indicates the analogous vertical integral.

In their continuous form, the horizontal momentum and continuity equations (4.515), (4.516), and (4.517) can be written as

$$\frac{\partial}{\partial t} \left(\frac{hu}{mn} \right) = R_u - \left(\frac{h}{n} \right) \frac{\partial \phi}{\partial \xi} \qquad (4.518a)$$

$$\frac{\partial}{\partial t} \left(\frac{hv}{mn} \right) = R_v - \left(\frac{h}{m} \right) \frac{\partial \phi}{\partial \eta} \qquad (4.518b)$$

and

$$\frac{\partial}{\partial \xi} \left(\frac{hu}{n} \right) \frac{\partial}{\partial \eta} \left(\frac{hv}{m} \right) \frac{\partial}{\partial \sigma} \left(\frac{h\Omega}{mn} \right) = 0 \qquad (4.518c)$$

where R_u and R_v represent the sum of all other terms in the u and v equations, respectively. Performing a vertical average, the equations become

$$\frac{\partial}{\partial t} \left(\frac{h\overline{u}^\sigma}{mn} \right) = \overline{R}_u^\sigma - \left(\frac{h}{n} \right) \frac{\partial \overline{\phi}^\sigma}{\partial \xi} \qquad (4.519a)$$

$$\frac{\partial}{\partial t} \left(\frac{h\overline{v}^\sigma}{mn} \right) = \overline{R}_v^\sigma - \left(\frac{h}{m} \right) \frac{\partial \overline{\phi}^\sigma}{\partial \eta} \qquad (4.519b)$$

and

$$\frac{\partial}{\partial \xi} \left(\frac{h\overline{u}^\sigma}{n} \right) + \frac{\partial}{\partial \eta} \left(\frac{h\overline{v}^\sigma}{m} \right) = 0 \qquad (4.519c)$$

where the overbar ($\overline{}^{\sigma}$) indicates a vertically averaged quantity.

By virtue of the rigid lid boundary condition (4.500), the depth-averaged flow is horizontally non-divergent. This enables us to introduce a streamfunction, $\psi(\xi, \eta, t)$, such that

$$\overline{u}^{\sigma} = -\left(\frac{n}{h}\right)\frac{\partial\psi}{\partial\eta} \tag{4.520a}$$

and

$$\overline{v}^{\sigma} = -\left(\frac{m}{h}\right)\frac{\partial\psi}{\partial\xi} \tag{4.520b}$$

By re-arrangement of eqs.(4.519a) and (4.519b):

$$\frac{\partial}{\partial t}\left(\frac{\overline{u}^{\sigma}}{m}\right) = \left(\frac{n}{h}\right)\overline{R}_u^{\sigma} - \frac{\partial\overline{\phi}^{\sigma}}{\partial\xi}$$

and

$$\frac{\partial}{\partial t}\left(\frac{\overline{v}^{\sigma}}{n}\right) = \left(\frac{m}{h}\right)\overline{R}_v^{\sigma} - \frac{\partial\overline{\phi}^{\sigma}}{\partial\eta}$$

Finally, taking the curl of these equations yields a vorticity equation for the depth-averaged flow:

$$\frac{\partial}{\partial t}q = \frac{\partial}{\partial t}\left[\frac{\partial}{\partial\xi}\left(\frac{\overline{v}^{\sigma}}{n}\right) - \frac{\partial}{\partial\eta}\left(\frac{\overline{u}^{\sigma}}{m}\right)\right] = \frac{\partial}{\partial\xi}\left[\frac{m}{h}\overline{R}_v^{\sigma}\right] - \frac{\partial}{\partial\eta}\left[\frac{n}{h}\overline{R}_u^{\sigma}\right] \equiv R_q \tag{4.521a}$$

or, using eq.(4.520):

$$\frac{\partial}{\partial t}\left[\left(\frac{m}{hn}\right)\frac{\partial^2\psi}{\partial\xi^2} + \left(\frac{n}{hm}\right)\frac{\partial^2\psi}{\partial\eta^2} + \frac{\partial}{\partial\xi}\left(\frac{m}{hn}\right)\frac{\partial\psi}{\partial\xi} + \frac{\partial}{\partial\eta}\left(\frac{n}{hm}\right)\frac{\partial\psi}{\partial\eta}\right] = R_q \tag{4.521b}$$

The introduction of the streamfunction ψ serves two purposes. First, it automatically guarantees horizontal non-divergence, as required by eq.(4.519c). Second, it eliminates any dependence on the depth-averaged component of the pressure field ϕ which contains an unknown contribution arising due to the rigid lid.

The vorticity equation (4.521b) is solved by first obtaining an updated value of q by application of the leapfrog (second-order) time-differencing scheme (?). The associated value of the streamfunction is determined from the generalized elliptic equation:

$$\left(\frac{m}{hn}\right)\frac{\partial^2\psi}{\partial\xi^2} + \left(\frac{n}{hm}\right)\frac{\partial^2\psi}{\partial\eta^2} + \frac{\partial}{\partial\xi}\left(\frac{m}{hn}\right)\frac{\partial\psi}{\partial\xi} + \frac{\partial}{\partial\eta}\left(\frac{n}{hm}\right)\frac{\partial\psi}{\partial\eta} = q \tag{4.522}$$

A solution to eq.(4.522) is fully prescribed by specifying values for ψ on the channel walls (in the case of a periodic channel), or on all four lateral boundaries (in the case of a closed domain). In the model test problems described below, eq.(4.522) was solved using the elliptic equation solvers HWSCRT and MUD2, developed at the National Center for Atmospheric Research (NCAR).

The N internal modes of the velocity distribution $((u,v)_k$ for $1 \leq k \leq N)$ are obtained by direct time-stepping of eqs.(4.513) and (4.514), having removed their depth-averaged component. The total density eq.(4.516) can be similarly advanced for all $0 \leq k \leq N$. A leapfrog step is used in both cases. (A periodic application of an Euler or leapfrog-trapezoidal time step is used to diminish the computaional mode.)

Having obtained a complete specification of the $u, v,$ and ρ fields at the next time level, the vertical velocity and internal pressure fields can be obtained by vertical integration, in particular:

$$\phi(\sigma) = \left(\frac{gh}{2\rho_0}\right) I_\sigma^1 \rho \tag{4.523}$$

and

$$\Omega(\sigma) = \left(\frac{mn}{h}\right) I_\sigma^1 \left\{ \delta_\xi \left[u \overline{\left(\frac{h}{m}\right)}^\xi \right] + \delta_\eta \left[v \overline{\left(\frac{h}{m}\right)}^\eta \right] \right\} \tag{4.524}$$

CHAPTER V

NORMAL MODES

1. Introduction

Any water body (either completely closed or partially open) undergoes natural or free oscillations which are referred to as normal modes. Several different physical phenomena can set a water body into oscillation, i.e., excite its normal modes. The frequencies of the fundamental normal mode and its higher harmonics can be determined solely from a knowledge of the geometry of the water body and the water depths. The question of normal modes was first discussed in connection with tidal theories (LaPlace, 1775, 1776; Hough, 1898).

Consider an artificial situation in which a thin layer of water covers the Earth's surface entirely. We ask the question how this water can move freely, subject to gravity and the earth's rotation. Hough (1898) showed that the free motion can occur in either of two ways. Oscillations of the first class (OFC) are essentially gravity waves whose periods are modified by the Earth's rotation. However, OFC can exist even if the Earth does not rotate. Oscillations of the second class (OSC) owe their very existence to the Earth's rotation and have periods greater than 24 h. If the earth's rotation tends to zero, OSC will lose their periodicity and will degenerate into steady currents. An example of OSC is the Rossby wave.

If σ is the frequency of oscillation and ω is the frequency of rotation, then oscillations of the first class are those for which $\sigma \to \sigma_0 (\neq 0)$ as $\omega \to 0$ and oscillations of the second class are those for which $\sigma \to O(\omega)$ as $\omega \to 0$. Bjerknes *et al.* (1934) distinguished between these two types of oscillation by means of the ratio $\sigma/2\omega$. Gravity modes (oscillations of the first class) are those for which $\sigma/2\omega \geq 1$. Elastoid-inertia modes (oscillations of the second class) are those for which $\sigma/2\omega \leq 1$. For the gravity modes, gravity appears in the frequency equation. In the case of the rotational (elastoid-inertia) modes, the frequency for a given mode is a function mainly of the ratio of the depth of the liquid to the radius of the container and gravity does not play an important role in the frequency equation. Here this discussion is restricted to gravity modes. In the mathematical analysis this restriction is imposed by introducing the approximations of the shallow water theory called the quasi-static approximation (Bjerknes *et al.*, 1934). This means that in the vertical direction, equilibrium exists not only before the motion but also during the motion, with vertical accelerations of the liquid being considered negligible compared to that of gravity.

In studying tidal motions on a rotating Earth Kelvin (1879) considered a shallow layer of water in a circular flat-bottomed cylinder and assumed the quasi-static approximation to the pressure field. Kelvin considered small rotations and neglected the curvature of the free surface due to rotation. If σ is the frequency of the rotating mode, σ_0 is the frequency of the non-rotating mode, and ω is the rotation frequency, then the result obtained is

$$\sigma^2 = \sigma_0^2 + 4\omega^2$$

This result shows that rotation increases the frequency and thus increases the restoring tendency of the system when disturbed. However, if the curvature of the free surface is taken into account, this is not always true, especially for the higher modes.

Since one of the manifestations of normal modes in water bodies is in the form of seiches, we will start with a discussion of this phenomenon following mainly Wilson (1972).

2. Seiches

Probably the first scientific study of seiches was that of Forel (1892), and Chrystal (1905) was probably the first to put forth a hydrodynamic theory of seiches. Important contributions to the theory of seiches were reviewed by Harris (1908), Lamb (1945), Proudman (1953), Defant (1961), Wilson (1972), and Miles (1974).

Merian (1828) gave a theory for free oscillations of water in a rectangular basin of length L and uniform depth h, the period T being given by

$$T = \frac{2L}{\sqrt{gH}} \tag{5.1}$$

where g is gravity. Forel (1892) applied this formula to seiches in lakes. For real lakes with variable depth, he chose an average value of H to replace the variable depths. Lagrange (1781) showed that the velocity of c of a long wave is given by

$$c \sim \sqrt{gH} \tag{5.2}$$

From eqs.(5.1) and (5.2)

$$T = \frac{2L}{c} = \frac{\lambda}{c} \tag{5.3}$$

where λ is the wavelength of the oscillation (assuming it is in the form of a wave). Thus, the length of the wave is twice that of the water body (or basin). Forel explained this apparent paradox as being due to the superposition of two long waves whose length is twice that of the basin and travelling in opposite directions.

In the following, an attempt will be made to visualize a seiche as a special type of standing wave. For this, consider two progressive waves travelling in opposite directions in water of uniform depth. At every quarter period the crests and troughs are either in phase or out of phase. At half-wavelength intervals $(x = \lambda/4, 3\lambda/4, 5\lambda/4, \ldots)$ surface elevation is continuously zero with time. Such points are called nodes and the points that intermediate to these are the antinodes. This type of standing wave can also result if a progressive wave is reflected (without dissipation) at a vertical wall. Then, there will be an antinode of amplitude $2A$ (A being the amplitude of the progressive wave) at the wall and a first node at $x = \lambda/4$ from the wall.

A seiche is a special case of a standing wave that would result from interposing a second vertical barrier at any of the points $x = \lambda/2, 3\lambda/2, 2\lambda, \ldots$ The standing wave or seiche exists due to repeated reflections (assuming no dissipation) from the two vertical walls, where it would have its antinodes. On the other hand, if the second vertical barrier were inserted at any point other than a multiple of $\lambda/2$, the standing wave would become an irregular motion of the water surface. Thus, one can think of a seiche as a standing wave that is commensurate with the basin length L.

The seiche is uninodal for $L = \lambda/2$, binodal for $L = \lambda$, trinodal for $L = 3\lambda/2, \ldots$, n-nodal for $L = n\lambda/2$. Hence, from eq.(5.3), the period T_n of the nth mode of oscillation in a rectangular basin of length L and uniform depth H is

$$T_n = \frac{2L}{n\sqrt{gH}} \tag{5.4}$$

Here $n = 1, 2, 3, \ldots$ For an open bay of rectangular geometry (length L) and uniform depth H, the period is given by

$$T_n = \frac{4L}{n\sqrt{gH}} \tag{5.5}$$

Here $n = 1, 3, 5, \ldots$ This is a generalization of the Merian formula and is valid for a one-dimensional oscillation (no transverse motions). Note that at the nodes the motion is purely horizontal and at the antinodes it is purely vertical. The higher nodal (binodal, trinodal, etc.) seiches that may occur simultaneously with the fundamental mode (i.e., uninodal oscillation) are higher harmonics of the fundamental.

From eq.(5.4)

$$\frac{T_n}{T_1} = 1, \frac{1}{2}, \frac{1}{3}, \ldots, \frac{1}{n}, \quad n = 1, 2, \ldots, n$$

However, for irregular water bodies with variable depth (unlike in the case of a narrow rectangular basin of uniform depth), such a simple relation as above need not exist. Another point worth remembering is that neither the use of an average depth $\bar{H}$ nor a better version of this, as done by du Boys (see Defant, 1961)

$$T_n \sim \frac{2}{n} \int_0^L \frac{dx}{[gh(x)]^{1/2}} \qquad (5.6)$$

improves the Merian formula significantly.

Next, the concept of regarding seiches as a combination of free and forced oscillations will be developed. Any natural system, when displaced from its equilibrium position, will try to regain its equilibrium position (due to a restoring force) and will exhibit free oscillations once the disturbing force is removed. The nature of these oscillations depends on the system alone, the influence of the disturbing force being restricted to setting the initial amplitude of the oscillation. After some time, the free oscillations will gradually dissipate. In a water body or basin, the seiche is a type of free oscillation of the water, the restoring force being gravity. However, in nature the seiches could be of a forced nature because the disturbing force, instead of being instantaneous, can act over some period of time.

The equation of motion for a linear vibrating mass spring system subject to a displacement X due to a disturbing force $F(t)$ is, in the canonical form

$$\ddot{X} + 2\beta\omega\dot{X} + \omega^2 X = \frac{F(t)}{m} \qquad (5.7)$$

where β is a nondimensional damping coefficient, m is the mass of the vibrating body, ω is the angular frequency, and $m\omega^2$ is a spring constant for the restoring force. The solution of eq.(5.7) can be visualized as the combination of a free and forced part of a transient and steady-state part. To obtain the solution for the free oscillation, put $F(t) = 0$. Then

$$X_0 = e^{-\beta\omega t}[a\sin(\gamma t) + b\cos(\gamma t)] \qquad (5.8)$$

where a and b are amplitudes of the motion determined by the initial conditions. The natural frequency γ of the system is

$$\gamma = \omega(1 - \beta^2)^{1/2} \qquad (5.9)$$

and the natural period T is given by

$$T = \frac{2\pi}{\gamma} \qquad (5.10)$$

The frictional damping, which is given by β, makes the free oscillations decay at a rate such that the amplitude decreases in one cycle by $e^{-\delta}$ where δ is the logarithmic decrement and is given by

$$\delta = \beta\omega T \qquad (5.11)$$

For the forced solution, one must use eq.(5.7) in complete form and take a periodic disturbing force as follows:

$$\frac{F(t)}{m} = F\cos(\sigma t + \epsilon) \tag{5.12}$$

where ϵ is an arbitrary phase angle. Then the forced solution is

$$X_f = \frac{F\mu}{\omega^2}\cos(\sigma t + \epsilon - \alpha) \tag{5.13}$$

where

$$\mu = \left\{ \left[1 - \left(\frac{\sigma}{\omega}\right)^2\right]^2 + \left[2\beta\left(\frac{\sigma}{\omega}\right)\right]^2 \right\}^{-1/2} \tag{5.14}$$

$$\tan\sigma = \frac{2\beta(\sigma/\omega)}{1 - (\sigma/\omega)^2} \tag{5.15}$$

Here, μ is the dynamic amplification of the oscillation and α is a phase angle by which the forced oscillation lags the disturbing force. Thus, the total solution is

$$X = X_0 + X_f \tag{5.16}$$

Here, X_0 decays with time whereas X_f persists as long as the disturbing force is applied. One can deduce that

$$\mu = \frac{X_{\max}}{(F/\omega)^2} = \frac{X_{\max}}{X_i} \tag{5.17}$$

where X_i is the amplitude of the input displacement.

Ordinarily, μ and α are shown as ordinates versus σ/ω as abscissa. If the damping coefficient $\beta < \frac{1}{2}$, then from eq.(5.9) the natural frequency γ of the system is approximately given by ω. Hence, the ratio σ/ω in eqs.(5.14) and (5.15) is effectively the ratio of the forced to the natural frequency. The dynamic amplification μ approaches its peak value when $\sigma/\omega \sim 1$. When this happens, resonance occurs and the amplitude of motion will be several times greater than the amplitude of the disturbing force.

For small frequency ratios $\sigma/\omega \ll 1$, the magnification is small, $\mu \sim 1$, and the motion follows the excitation (i.e., $\alpha \to 0$). For $\sigma/\omega \gg 1$, the resulting motion is much smaller than that of the exciting force. Then, $\mu \to 0$ and the motion tends to become out of phase (i.e., $\alpha \to 180°$). Hence, the degree of resonance is determined by the damping factor 2β. Miles and Munk (1961) defined the degree of resonance through the factor Q, which is the maximum value of the dynamic amplification μ. From eq.(5.14) if $\sigma/\omega \sim 1$ (as occurs at resonance)

$$\mu_{\max} \equiv Q \equiv \frac{1}{2\beta} \tag{5.18}$$

In the frequency range $(1 - \beta) < \sigma/\omega < (1 + \beta)$, if damping factor 2β is small, the power amplification μ^2 has a value greater than $Q^2/2$. Hence, the frequency band width (over which the power amplification exceeds half its maximum value Q^2) is $1/Q$. Thus, the sharper is the resonance, the narrower will be the spectral energy peak. This can be quantitatively expressed by stating that near resonance

$$\frac{Q^2}{\mu^2} \sim 1 + 4Q^2 \left(1 - \frac{\sigma}{\omega}\right)^2 \tag{5.19}$$

The following results can be easily deduced from the above relations. For low Q conditions (i.e., heavy dissipation), a large rate of absorption of energy from the disturbing force to the oscillating system is necessary whereas for high Q (small damping) only a small energy absorption rate is sufficient for resonance. Miles and Munk (1961) showed that for a water body with a rather regular topography, low damping prevails. Hence, the response is of the high Q type. Hence, a relatively small amount of energy (e.g., from atmospheric pressure gradients) at the correct frequency can excite strong resonance. However, if the topography is irregular, damping is heavy and a low Q situation prevails.

Next, some theoretical aspects of free and forced seiches will be considered. With reference to a Cartesian coordinate system (x, y), let M_x and M_y be the components of the transport, ζ is the water level deviation from its equilibrium position, H is the water depth, P_a is the atmospheric pressure, τ_x^s and τ_y^s are the wind stress components, and r is the bottom friction coefficient. Omitting the Coriolis term and assuming that $\frac{\partial}{\partial y} = 0$, and $v = 0$

$$\frac{\partial M}{\partial t} + rM + g(H + \zeta)\frac{\partial \zeta}{\partial x} = F_S(x, t) \tag{5.20}$$

where subscript x on M is omitted. The external force is given by

$$F_S(x,t) = \frac{\tau_x^s}{\rho} + \frac{H + \zeta}{\rho}\frac{\partial P_a}{\partial x} \tag{5.21}$$

The continuity equation is

$$\frac{\partial \zeta}{\partial t} + \frac{\partial M}{\partial x} = 0 \tag{5.22}$$

Equations (5.20) and continuity equation can be transformed into two hyperbolic equations in the dependent variables M and ζ:

$$\frac{\partial^2 \zeta}{\partial t^2} + r\frac{\partial \zeta}{\partial t} - g\frac{\partial}{\partial x}\left[(H + \zeta)\frac{\partial \zeta}{\partial x}\right] = -\frac{\partial F_S}{\partial x} \tag{5.23a}$$

$$\frac{\partial^2 M}{\partial t^2} + r\frac{\partial M}{\partial t} - g(H + \zeta)\frac{\partial^2 M}{\partial x^2} = \frac{\partial F_S}{\partial t} \tag{5.23b}$$

The solutions of eq.(5.23) with the right-hand sides set to zero give the solutions for the free oscillation, whereas the solutions for the complete equations are the forced oscillations.

In a rectangular basin of length L and uniform depth H, in which a free oscillation is generated by equating the disturbing force F_S to zero, eqs.(5.22) and (5.23) give

$$\frac{\partial^2 \zeta}{\partial t^2} + r\frac{\partial \zeta}{\partial t} - c^2\frac{\partial^2 \zeta}{\partial x^2} = 0 \qquad (5.24)$$

$$\frac{\partial^2 M}{\partial t^2} + r\frac{\partial M}{\partial t} - c^2\frac{\partial^2 M}{\partial x^2} = 0 \qquad (5.25)$$

Since these equations have the same form in ζ and M, one can use the method of separation of variables to solve them:

$$\zeta(\text{or } M) = X(x)T(t) \qquad (5.26)$$

The solution can be shown to be

$$\zeta(\text{or } M) = e^{-rt/2}[A\cos(kx) + B\sin(kx)][C\cos(\gamma t) + D\sin(\gamma t)] \qquad (5.27)$$

where the angular frequency γ of the free oscillation is given by

$$\gamma = \omega\left(1 - \frac{k}{2\omega}\right)^{1/2} \qquad (5.28)$$

The wave number k and the angular frequency ω are related through

$$\omega = kc \qquad (5.29)$$

in which either k or ω must be determined.

To determine the constants of integration A, B, C, D, the following boundary conditions must be used. At the ends of the basin, $x = 0, L$ transport M must be zero for all time. Thus,

$$M(x = 0) = 0 \qquad \text{and} \qquad M(x = L) = 0 \qquad (5.30)$$

From the continuity equation and taking a as the amplitude of free oscillation,

$$\begin{aligned} \zeta &\sim ae^{-rt/2}\cos(kx)\cos(\gamma t + \epsilon) \\ M &\sim \frac{a\gamma}{k}e^{-rt/2}\sin(kx)\sin(\gamma t + \epsilon) \end{aligned} \qquad (5.31)$$

The wave number k can be determined from the second equation of (5.30) to give

$$kL = n\pi, \qquad n = 1, 2, 3, \ldots \tag{5.32}$$

Then, γ and ω can be determined from eqs.(5.28) and (5.29). Since eq.(5.31) represents a standing wave whose amplitude is a at $t = 0$ and decays exponentially with time, this oscillation is similar to the mechanical system discussed earlier. Thus,

$$r = 2\beta\omega \tag{5.33}$$

Next, the disturbing force will be explicitly introduced. Since the role of edge waves in storm surges and resonant coupling to the atmosphere will be considered later in this section, this disturbing force will be prescribed as an atmospheric pressure pulse moving with a uniform velocity V over the water body along the length of the water body. The atmosphere pulse is assumed to be sinusoidal with a pulse length $2l$. The amplitude of the pulse is the pressure gradient $(\partial P_a/\partial x)_{max}$ or simply ΔP.

We will refer the reader to Wilson (1972) for the mathematical details of the forced oscillations, but will summarize his results on the excitation of seiches by pressure pulses.

a) When the speed of movement V of the pressure pulse agrees with the phase velocity c of a free wave in the water body, resonance occurs and the dynamic amplification μ, defined as $\mu = 1/2\beta$, can have large values.

b) The size of the pressure pulse $2l$ relative to the distance L it traverses over the water body determines the response factor.

c) When the size of the pulse is not optimal for the maximum response of the water body, higher seiche modes also appear.

Proudman (1953) was probably among the first to invoke resonant coupling between a squall line and the gravity waves in the water on the coast of the United Kingdom to account for certain water level disturbances. However, in the hydrodynamic context, Lamb (1945) recognized such a possibility and discussed the role of the dynamic magnification μ.

3. The normal mode approach

The work of Haurwitz (1951) was referred to earlier in which he mentioned that the duration of the development of the storm surge must depend on the seiche type motion (normal modes). Following Schwab (1975), a method of calculating storm surges using the normal mode approach will be discussed.

In the equations of motion, if all the inhomogeneous terms are omitted, the solutions of the homogeneous system are the normal modes and there is, in principle, an infinite spectrum of these modes. However, when the system is forced (such as happens in the storm surge case when the forcing is from the atmosphere), the particular solutions of the inhomogeneous equations can be determined by an expansion in terms of the normal modes with time-dependent coefficients. Reid (1958) used this procedure to calculate the edge waves on the shelf. The advantage of using the normal mode approach is that the expansion automatically satisfies the space dependence. Each expansion coefficient satisfies a first-order inhomogeneous ordinary differential equation in time and these equations can be solved by numerical techniques. Since this is an uncoupled system, no expensive matrix inversion techniques are required. Although the normal mode approach is, in principle, very elegant and efficient, this approach is rarely used in storm surge studies, probably because the method has not lived up to its expectations, especially when applied to real water bodies with irregular coastlines, depth variations, and openings. Rao (1974) used this technique for certain idealized situations.

Following Schwab (1975), the equations for horizontal motion can be written in the following vector form:

$$\frac{\partial M}{\partial t} - f[M] = -gD\nabla\zeta + \tau \tag{5.34}$$

where M is the horizontal transport vector and is defined as

$$M \equiv \int_{-D}^{h} V \, dz$$

where V is the horizontal velocity vector, ∇ is the horizontal gradient operator, $\tau(x, y, t)$ is the prescribed forcing function, and [ ] is an operator that denotes rotation of a vector 90° clockwise in the horizontal plane (the other variables have been previously defined). The continuity equation is

$$\frac{\partial \zeta}{\partial t} + \nabla M = 0 \tag{5.35}$$

The boundary condition is

$$M \cdot n = DV \cdot n = 0 \tag{5.36}$$

where n is a vector normal to the shoreline. The normal modes represent the solutions of eqs.(5.34)–(5.36) when $\tau = 0$.

Following Rao and Schwab (1974), the transport vector M will be expressed as the sum of a nondivergent part and an irrotational part:

$$M = M^\phi + M^\psi \tag{5.37}$$

where

$$M^\phi \equiv D\nabla\phi$$
$$M^\psi \equiv -[\nabla\Psi]. \tag{5.38}$$

Here, ϕ and Ψ represent, respectively, the velocity potential and the stream function for the transport field. We have the following properties:

$$\nabla \cdot [D^{-1}M^\phi] = 0$$
$$\nabla \cdot M^\Psi = 0 \tag{5.39}$$

Equation (5.37) is valid as long as M is independent of depth (because it is vertically integrated) and the boundary conditions are adiabatic.

The boundary condition (5.36) becomes

$$M^\phi \cdot n = 0$$
$$M^\Psi \cdot n = 0 \tag{5.40}$$

or in terms of ϕ and Ψ this becomes

$$D\frac{\partial \phi}{\partial n} = 0$$
$$\Psi = 0 \tag{5.41}$$

on the shoreline.

The divergence of the transport field and the vorticity of the velocity field can be written as follows:

$$\nabla \cdot D\nabla\phi = -\nabla \cdot M$$
$$\nabla \cdot D\nabla\Psi = \nabla \cdot [D^{-1}M] \tag{5.42}$$

The parameters ϕ and Ψ will be expressed in terms of the spectra of the elliptic operators in eq.(5.42). That is, the following characteristic value problems are considered:

$$\nabla \cdot D\nabla\phi_\alpha = -\lambda_\alpha \phi_\alpha$$
$$D\frac{\partial \phi_\alpha}{\partial n} = 0 \tag{5.43}$$

on the boundary, and

$$\nabla \cdot D^{-1}\nabla\Psi_\alpha = -\mu_\alpha \Psi_\alpha$$
$$\Psi_\alpha = 0 \tag{5.44}$$

on the boundary. The subscript α refers to the number of the spectral components. Equations (5.43) and (5.44) are self-adjoint under the more stringent boundary condition

$$D^{-1}\Psi_\alpha = 0$$

Thus, the characteristic values λ_α and μ_α are real, and the related eigenfunctions ϕ_α and Ψ_α give rise separately to an internally orthogonal set. The orthogonality condition is

$$\int D^{-1} M^\phi{}_\alpha M^\phi{}_\beta dA = \lambda_\alpha \int \phi_\alpha \phi_\beta dA = gA\delta_{\alpha\beta}$$

$$\int D^{-1} M^\psi{}_\alpha M^\psi{}_\beta dA = \mu_\alpha \int \Psi_\alpha \Psi_\beta dA = gA\delta_{\alpha\beta}$$

(5.45)

where δ is the Kronecker delta and A is the surface area of the water body.

From eq.(5.38)

$$M^\phi{}_\alpha = -D\nabla_\alpha$$
$$M^\psi{}_\alpha = -[\nabla \Psi_\alpha]$$

(5.46)

To represent M^ϕ and M^ψ define the following nondimensional expansion coefficients,

$$P_\alpha = \frac{1}{gA} \int D^{-1} M^\phi{}_\alpha M^\phi dA = \frac{1}{gA} \int D^{-1} M^\phi{}_\alpha dA$$

$$Q_\alpha = \frac{1}{gA} \int D^{-1} M^\psi{}_\alpha M^\psi dA = \frac{1}{gA} \int D^{-1} M^\psi{}_\alpha dA$$

(5.47)

Then,

$$M^\phi = \sum_\alpha P_\alpha M^\phi{}_\alpha$$

$$M^\psi = \sum_\alpha Q_\alpha M^\psi{}_\alpha$$

(5.48)

Because of eq.(5.45), the right sides in (5.48) represent the left side in a least square sense if the summation in (5.48) covers the complete spectra of eqs.(5.43) and (5.44).

From eq.(5.45) it is evident that the divergent part of M determines H. Thus,

$$\zeta_\alpha = g^{-1} \lambda_\alpha^{1/2} \phi_{\alpha\beta}$$

(5.49)

Then, the orthogonality relation is

$$\int \zeta_\alpha \zeta_\beta dA = A\delta_{\alpha\beta}$$

(5.50)

Then, ζ can be expanded to

$$\zeta = \sum_\alpha R_\alpha \zeta_\alpha$$

(5.51)

where

$$R_\alpha \equiv \frac{1}{A} \int \zeta_\alpha \zeta \, dA$$

(5.52)

Define

$$A_{\alpha\beta} \equiv \{M^{\phi}{}_{\alpha}, [M^{\phi}{}_{\beta}]\}$$
$$B_{\alpha\beta} \equiv \{M^{\phi}{}_{\alpha}, [M^{\psi}{}_{\beta}]\}$$
$$C_{\alpha\beta} \equiv \{M^{\psi}{}_{\alpha}, [M^{\phi}{}_{\beta}]\}$$
$$E_{\alpha\beta} \equiv \{M^{\psi}{}_{\alpha}, [M^{\psi}{}_{\beta}]\}$$

(5.53)

The notation $\{A, B\}$ represents the inner product

$$\{A, B\} = \frac{1}{gA} \int f D^{-1} A \cdot B \, dA$$

From Eq.(5.53) it can be seen that

$$A_{\alpha\beta} = -A_{\beta\alpha}$$
$$B_{\alpha\beta} = -C_{\beta\alpha}$$
$$E_{\alpha\beta} = -E_{\beta\alpha}$$

(5.54)

Substituting eqs.(5.48) and (5.51) into (5.34) and (5.35) and taking $\tau = 0$ and then using the orthogonality conditions (5.47) and (5.50) gives the following spectral prediction equations:

$$\frac{dP_{\alpha}}{dt} - \sum_{\beta} A\alpha\beta P_{\beta} - \sum_{\beta} B_{\alpha\beta}Q_{\beta} - v_{\alpha}R_{\alpha} = 0$$

$$\frac{dQ_{\alpha}}{dt} - \sum_{\beta} C\alpha\beta P_{\beta} - \sum_{\beta} E_{\alpha\beta}Q_{\beta} = 0$$

(5.55)

$$\frac{dR_{\alpha}}{dt} + v_{\alpha}P_{\alpha} = 0$$

Define the following column vectors and matrices:

$$\overline{P} \equiv \text{col}\,(P_{\alpha}) \qquad \overline{Q} \equiv \text{col}\,(Q_{\alpha})$$

$$\overline{R} \equiv \text{col}\,(R_{\alpha}) \qquad \overline{S} \equiv \begin{pmatrix} P \\ Q \\ R \end{pmatrix}$$

(5.56)

and the following matrices:

$$A \equiv |A_{ij}| \qquad C \equiv |C_{ij}|$$
$$B \equiv |B_{ij}| \qquad E \equiv |E_{ij}|$$
$$\langle v \rangle \equiv \text{diagonal}\,\langle v_{\alpha} \rangle$$

(5.57)

Then eq.(5.55) can be written in the matrix form

$$\frac{d\overline{S}}{dt} + a\overline{S} = 0 \tag{5.58}$$

where a is a square matrix defined by

$$a = \begin{vmatrix} -A & -B & -\langle v \rangle \\ -C & -D & 0 \\ \langle v \rangle & 0 & 0 \end{vmatrix}$$

Making use of the property that the time dependence of $\overline{S}$ is given by $e^{i\sigma t}$ where σ is the frequency of oscillation, eq.(5.58) becomes

$$i\sigma \overline{S} + a\overline{S} = 0 \quad \text{or}$$
$$(\sigma \mathbf{I} - ia)\overline{S} = 0 \tag{5.59}$$

where $\mathbf{I}$ is the identity matrix. The eigenvalue problem (4.91) can be solved for the characteristic (real) values of σ and the eigenvectors $\overline{S}$. Once the eigenvectors are known, using eqs.(5.46), (5.48) and (5.51), the M and ζ fields can be derived.

Note that here, only gravity modes have been considered, which are also referred to as oscillations of the first class (OFC). For a real water body with irregular geometry, the characteristic value problem (eqs.(5.43), (5.44), and (5.59)) must be solved numerically. Analogous to the tidal problem one can construct cotidal and corange lines. To be able to do this, write

$$\zeta = \text{Real} \sum_\alpha R_\alpha \zeta_\alpha e^{i\sigma t} = A(x, y)\cos[\sigma t - \theta(x, y)] \tag{5.60}$$

The isolines of $A(x, y)$ give the cotidal diagram and the isolines of $\theta(x, y)$ represent the corange lines.

Let the solutions of M and ζ with $\tau = 0$ (i.e., the free solutions) be M_F and ζ_F, respectively. These can be expressed as

$$M_F = M_{j\alpha}(x, y)e^{i\sigma_{\alpha} t}$$
$$\zeta_F = \zeta_{j\alpha}(x, y)e^{i\sigma_{j\alpha} t} \tag{5.61}$$

The space-dependent normal mode functions can be found from eqs.(5.48) and (5.51). Here, α is a subscript to order the normal modes and j can take a value of either 1 or 2 to represent the normal mode or its complex conjugate. Note that the complex functions satisfy the normal mode equations

$$i\sigma_{j\alpha} M_{j\alpha} - f[M_{j\alpha}] = -gD\nabla \zeta_{j\alpha}$$
$$i\sigma_{j\alpha} \zeta_{j\alpha} + \nabla M_{j\alpha} = 0 \tag{5.62}$$

These functions are orthogonal (in a general Hilbert sense).

Let $M_{K\beta}^*$ and $\zeta_{K\beta}^*$ be the complex conjugates of the normal mode functions for $\sigma_{K\beta}$. The conjugate equations for (5.62) are

$$-i\sigma_{K\beta}^* - f[M_{K\beta}^*] = -gD\nabla\zeta_{K\beta}^*$$

$$-i\sigma_{K\beta}^* + \nabla M_{K\beta}^* = 0$$

(5.63)

The orthogonality condition is

$$\int\left(\frac{M_{j\alpha}M_{K\beta}^*}{gD} + \zeta_{j\alpha}\zeta_{K\beta}^*\right)dA = X_{j\alpha}\delta_{jK}\delta_{\alpha\beta}$$

where $X_{j\alpha}$ is the normalization for the normal mode associated with $\sigma_{j\alpha}$. The solution of the forced problem (i.e., $\tau \neq 0$) can be expressed as

$$M(x, y, t) = \sum_{j=1}^{2}\sum_{\alpha} A_{j\alpha}(t)M_{j\alpha}(x, y)$$

$$\zeta(x, y, t) = \sum_{j=1}^{2}\sum_{\alpha} A_{j\alpha}(t)h_{j\alpha}(x, y)$$

(5.64)

Here, $A_{j\alpha}$ represents the complex time-dependent amplitude factor for the normal mode with frequency $\sigma_{j\alpha}$.

Substituting eq.(5.64) into (5.34) and (5.35) gives

$$\sum_{j=1}^{2}\sum_{\alpha}\frac{dA_{j\alpha}}{dt}M_{j\alpha} + f\sum_{j=1}^{2}\sum_{\alpha} A_{j\alpha}[M_{j\alpha}] = -gD\sum_{j=1}^{2}\sum_{\alpha} A_{j\alpha}\zeta_{j\alpha} + \tau$$

$$\sum_{j=1}^{2}\sum_{\alpha}\frac{dA_{j\alpha}}{dt}\zeta_{j\alpha} + \sum_{j=1}^{2}\sum_{\alpha} A_{j\alpha}\nabla M_{j\alpha} = 0$$

(5.65)

Multiply the first equation of (5.65) by $M_{K\beta}^*$, the second equation by $\zeta_{K\beta}^*$; the first equation of (5.62) by $M/(gD)$, the second equation by ζ (here, M and ζ are given by eq.(5.64)) and add these and integrate over the area of the water body and use the orthogonality condition (5.63) to give

$$X_{j\alpha}\left(\frac{dA_{j\alpha}}{dt} + i\sigma_{j\alpha}A_{j\alpha}\right) = \int\frac{\tau M_{j\alpha}^*}{gD}dA$$

(5.66)

or

$$\frac{dA_{j\alpha}}{dt} + i\sigma_{j\alpha}A_{j\alpha} = \frac{1}{X_{j\alpha}}\int\frac{\tau M_{j\alpha}^*}{gD}dA$$

(5.67)

Since τ is a known quantity, the equation for $A_{j\alpha}(t)$ is an ordinary inhomogeneous differential equation of the first order.

Formally, the solution of eq.(5.67) can be written as

$$A_{j\alpha}(t) = \int_0^t B_{j\alpha}(\tau) e^{i\sigma_{j\alpha}^{(\tau-t)}} \, d\tau \tag{5.68}$$

where

$$B_{j\alpha} = \frac{1}{X_{j\alpha}} \int \frac{\tau M_{j\alpha}^*}{gD} \, dA \tag{5.69}$$

In real situations the integrand on the right side of eq.(5.69) will not be a simple analytical function of time, and a numerical integration must be used. Schwab (1975) used the following finite-difference scheme,

$$\frac{1}{\det}[A_{j\alpha}(t+\det) - A_{j\alpha}(t)] + \frac{i}{2}[A_{j\alpha}(t+\det) + A_{j\alpha}(t)]$$
$$= \frac{1}{2}[B_{j\alpha}(t+\det) + B_{j\alpha}(t)] \tag{5.70}$$

This can be rearranged to give

$$A_{j\alpha}(t+\det) = \left(\frac{2 - i\sigma_{j\alpha}\det}{2 + i\sigma_{j\alpha}\det}\right) A_{j\alpha}(t) + \left(\frac{\det}{2 + i\sigma_{j\alpha}\det}\right)[B_{j\alpha}(t+\det) + B_{j\alpha}(t)] \tag{5.71}$$

According to Kurihara (1965a, 1965b) this scheme is unconditionally stable for any det. However, if det is greater than one-sixth of the period of the fastest wave in the system, phase errors will then occur.

The integration procedure is as follows. Once the normal modes are determined, the normalization factors $X_{j\alpha}$ can be calculated from eq.(5.65). Since τ is given for all time, the $B_{j\alpha}$ values that enter eq.(5.71) can be determined from eq.(5.69). Since the earlier mode calculation also gives the normal mode frequencies, $\sigma_{j\alpha}$, the time-dependent expansion coefficients can be determined from eq.(5.67). Schwab (1975) claimed that although the numerical integration involved appears to be a considerable effort, the ability to limit the range of the index α in eq.(5.71) makes this approach more efficient than the ordinary finite-difference techniques.

4. Solutions for lakes and bays with uniform and variable depth

The energy imparted by a moving atmospheric disturbance to a water body depends on the degree of resonant coupling that is possible when the speed of movement of the atmospheric disturbance is approximately equal to that of the free gravity waves in the water body. Chrystal (1908) considered an atmospheric pressure jumpline moving with constant speed and intensity over infinite and semi-infinite water bodies, and he also looked into the question of resonance. Takegami (1938) studied the influence of initial conditions in the case of a pressure jump moving from the open sea towards the coast.

If the horizontal scale of the atmospheric disturbance is smaller than the horizontal scale of the water body, the water body can be considered semi-infinite. On the other hand, if the scale of the water body is comparable with or smaller than the scale of the atmospheric disturbance, then the reflection of the gravity waves at the lateral boundaries of the water body must be considered. In connection with the study of a surge on Lake Michigan, Harris (1957) gave a solution for the case of a finite canal valid for a short interval of time.

Rao (1967) studied the water level oscillations in a lake due to an atmospheric disturbance passing over the lake. He ignored the influence of the earth's rotation (his study was aimed at Lake Erie which is a narrow elongated lake) and bottom friction (somewhat justified as long as interest focuses on the initial transient motion).

Following Rao, consider a lake of uniform depth H, uniform width, and uniform density ρ. Take the origin of a Cartesian coordinate system at the undisturbed level of the lake surface and z axis upward, y axis along the width, and x axis along the length of the lake. The lake has rigid boundaries at $x = 0$ and $x = L$. In the one-dimensional case the vertically integrated equations of motion and continuity are, after ignoring the nonlinear terms and using the hydrostatic relation,

$$\frac{\partial M}{\partial t} = -c^2 \frac{\partial \zeta}{\partial x} + R \qquad (5.72)$$

$$\frac{\partial \zeta}{\partial t} = -\frac{\partial M}{\partial x} \qquad (5.73)$$

where

$$M = \int_{-h}^{0} u \, dz$$

u being the velocity component in the x direction. The external force R is expressed as

$$R = \frac{\tau}{\rho}$$

where τ is the surface wind stress. In the above equations ζ is the deviation of the water level from its equilibrium position and $c^2 = gH$ (c is the speed of free long gravity waves in the lake). Ignoring rotation gives a nondispersive system and all the gravity waves propagate with the same speed. The boundary conditions are

$$M = 0 \quad \text{at } x = 0 \text{ and } x = L. \qquad (5.74)$$

Initially, the lake is at rest, i.e., $M = 0$ and $\zeta = 0$ at $t = 0$.

Rao defined the setup as the difference in the water levels at $x = L$ and $x = 0$. Equations (5.72) and (5.73) are solved for a prescribed R as a function of x and t. For convenience, define the following dimensionless parameters:

$$\frac{x}{L} \equiv x^*$$

$$\frac{ct}{L} \equiv t^*$$

$$\frac{c^2 \zeta}{R_0 L} \equiv \zeta^* \qquad (5.75)$$

$$\frac{cM}{R_0 L} \equiv M^*$$

$$\frac{R}{R_0} \equiv R^*$$

where R_0 is a scale value of the wind stress.

Substituting eq.(5.75) into (5.73) and (5.74) and then adding and subtracting eq.(5.74) in (5.73) gives (omitting the asterisk for convenience)

$$\frac{\partial}{\partial t}(M \pm \zeta) \pm \frac{\partial}{\partial x}(M \pm \zeta) = R \qquad (5.76)$$

Essentially, this states that

$$\frac{d}{dt}(M \pm \zeta) = R \qquad \text{for } \frac{dx}{dt} = \pm 1 \qquad (5.77)$$

For a lake of uniform depth, the positive and negative characteristics are straight lines given by $x \pm t + \text{constant}$.

Following Rao (1967), consider the case of a semi-infinite stress band moving with a speed V (nondimensionalized by $\sqrt{gH}$), as shown in Fig.5.1 For this case

$$R = 0 \qquad \text{for} \quad t \le \frac{x}{V}$$

$$R = 1 \qquad \text{for} \quad t \ge \frac{x}{V} \qquad (5.78)$$

The top part of Fig.5.1 shows two successive positions of the stressband and the bottom parts show R in the $x - t$ plane. For details on the integration of eq.(5.77) using the method of characteristics, see Rao (1967) who considered the two cases $V > 1$ (atmospheric disturbances travelling faster than $\sqrt{gH}$ and $V < 1$.

The case of a finite stress band can be studied by the superposition (this is permissible in the linear case studied here) of a positive stress band and a negative stress band of the same intensity (as the positive stress band), both moving with

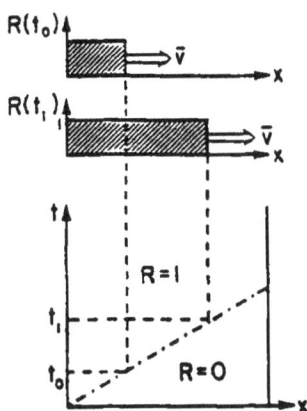

Fig. 5.1
Schematic in the (x, t) plane of a semi-infinite wind-stress band
moving across a lake of uniform depth. The dashed–dotted line shows
the position of the leading edge of the stress band. (Rao, 1967)

the same speed and in the same direction but with their jumps separated by a finite distance. The formulae for the setup in the case of the negative stress band are the same as the formulae for the positive stress band, except that the sign off the setup must be reversed and $t - T$ substituted where t appears (here, T is the dimensionless time interval between transits of the front and rear of the stress band at a fixed point). Hence, write

$$R = 0 \quad \text{for} \quad t \leq \frac{x}{V} \quad \text{and} \quad t \geq \frac{x}{V} + T$$
$$R = 1 \quad \text{for} \quad \frac{x}{V} \leq t \leq \frac{x}{V} + T \tag{5.79}$$

Again, see Rao (1967) for a detailed form of the solutions. The interesting results obtained from this study can be summarized as follows. With reference to Fig.5.2 for a given V, the maximum setup for all time and all band widths is achieved for a semi-infinite stress band. If $V = 1$, the maximum is reached for all band widths $a > 1$ (i.e., for all stress band widths $a > V$. The most important result is that for all t, all a, and all V, the maximum setup is achieved for the case of an instantaneous semi-infinite stress band. In the case of the semi-infinite stress band, the maximum setup decreases from its dimensionless upper limit of 2 at $1/V = 0$ to a lower limit of 1 at $1/V = 3$, after which the response curve shows a sawtooth

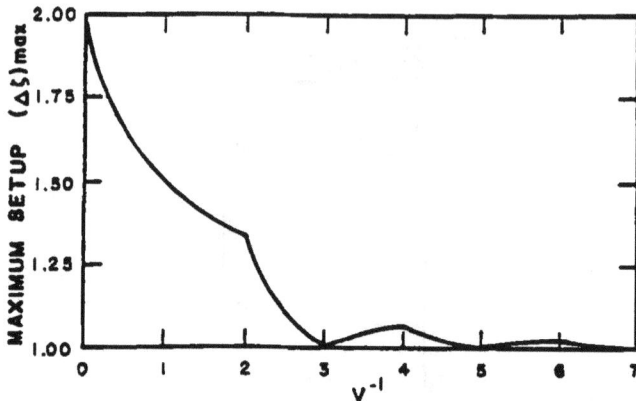

Fig. 5.2
Maximum setup (ordinate) versus V^{-1} for a semi-infinite
stress band. (Rao, 1967)

type behavior (see Fig.5.2) and the setup reaches a value of 1 asymptotically as
$1/V \to \infty$.

To study the response in the case of finite band widths, two cases must be
considered: $a > 1$ and $a < 1$. For the case $a < 1$, the maximum setup for all t
and all V is achieved at $V = 1$, which shows the importance of coupling. For the
case $a > 1$, the maximum setup is achieved for $a/V = 1$ ($a/V = 1$ means that the
time taken by the stress band to pass a fixed point in the lake is the same as the
time taken by a free gravity wave to cross the lake). In the case of a pressure jump
(i.e., stress band of zero width) due to resonant coupling, a setup of magnitude 1
is produced at $V = 1$ and the setup is 0 for all other V. The setup for different
transit times is shown in Fig.5.3.

Rao (1967) also solved the response of the fundamental mode (his Appendix
A) and showed, for the semi-infinite case, that the maximum setup of 1.62 occurs
at $1/V = 0$ and decreases to 0.81 as $1/V$ increases from 0 to 3. After this, the
setup shows essentially the same features as the total response but with a smoother
variation.

Irish and Platzman (1962) performed a statistical study of storm surges on Lake
Erie and concluded that resonant coupling is not important. Rao (1967) explained
this apparent disagreement between these results and his results by pointing out
that statistical evidence for resonance requires a classification based on band width.
Since Irish and Platzman did not sort out their data according to band width, they

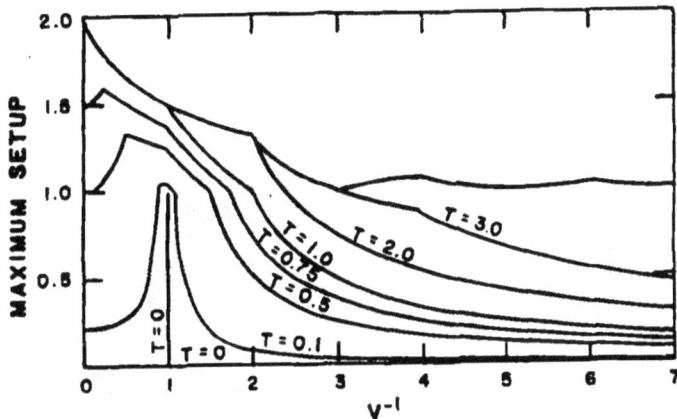

Fig. 5.3

Maximum setup versus V^{-1} for various transit times T. (Rao, 1967)

did not find significant support for resonant coupling.

Platzman (1963) found in a numerical calculation of storm surges on Lake Erie that the fundamental mode is quite well developed in some cases and quenched in others. Rao (1969) examined this by calculating the residual energy in the fundamental mode (residual energy is the energy left after the passage of the atmospheric disturbance). Rao showed in his Appendix B that there are certain adverse combinations of width and propagation speed of the stress band that quench the fundamental mode by extracting a major part of the energy supplied to the lake during the passage of the disturbance over the lake.

Rao (1969) extended his earlier study to the case of a bay (in the case of a bay, only one lateral boundary is closed). In the case of the lake, the boundary condition at both the lateral boundaries is zero volume transport whereas in the case of the bay, at the closed end a zero volume transport is invoked and at the open end the water fluctuation is prescribed as zero.

Next, a lake with variable depth is considered. Murty (1971) used the method of characteristics to study the case of a lake with a depth discontinuity. The formulation and the method of solution are similar to that of Rao (1967). For a lake of length L and uniform density ρ, let the left boundary be $x = 0$, the depth discontinuity at $x = L/2$, and the right boundary at $x = L$. The vertically integrated forms of the equations of motion and continuity in the areas to the left side of the depth discontinuity (subscript 1) and the right side of the discontinuity (subscript

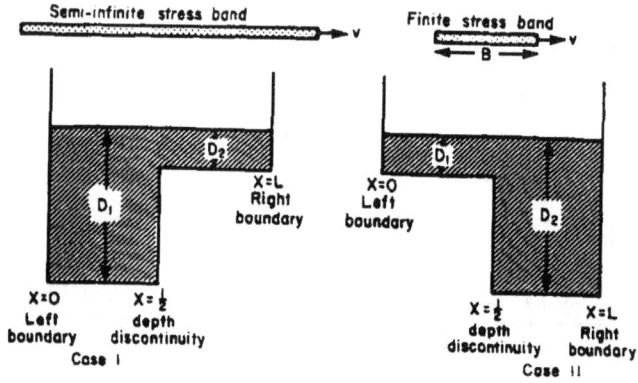

Fig. 5.4
Semi-infinite stress band moving over deeper to shallower water and
a finite stress band moving over shallower to deeper water.

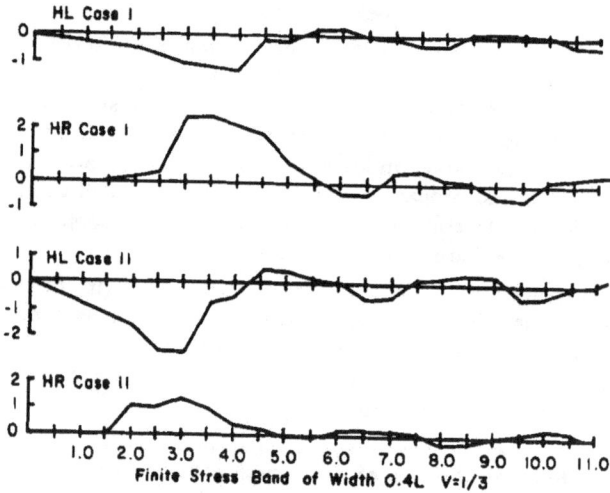

Fig. 5.5
Variation of water level at left and right boundaries for cases
I and II for a finite stress band of width O.4L.

2) are

$$\frac{\partial M_{1,2}}{\partial t} = -g H_{1,2} \frac{\partial \zeta_{1,2}}{\partial x} + R \qquad (5.80)$$

$$\frac{\partial \zeta_{1,2}}{\partial t} = -\frac{\partial M_{1,2}}{\partial x} \qquad (5.81)$$

where

$$M_{1,2} \equiv \int_{-D_{1,2}}^{h_{1,2}} u_{1,2}\, dz$$

Here, M is the volume transport through a vertical section and u is the velocity component along the length of the lake; $R = \tau/\rho$, τ is the wind stress, and ρ is water density. The boundary conditions are

$$\begin{aligned} M_1 &= 0 \quad \text{at} \quad x = 0 \\ M_2 &= 0 \quad \text{at} \quad x = L \end{aligned} \qquad (5.82)$$

At the depth discontinuity, continuity of M and ζ is invoked:

$$\begin{aligned} M_1 &= M_2 \\ \zeta_1 &= \zeta_2 \end{aligned} \quad \text{at } x = \frac{L}{2} \qquad (5.83)$$

Let $c_{1,2}$ represent the speeds of long gravity waves in regions 1 and 2. Then,

$$c_{1,2}^2 = g H_{1,2} \qquad (5.84)$$

Adding and subtracting eqs.(5.80) and (5.81) and using eq.(5.84) gives

$$\frac{d}{dt}(M_{1,2} \pm c_{1,2}\zeta_{1,2}) = R \qquad \text{for} \quad \frac{dx}{dt} = \pm c_{1,2} \qquad (5.85)$$

This means the quantity $M_{1,2}\zeta_{1,2}$ is constant along the characteristics $dx/dt = \pm c_{1,2}$. Since both regions of the lake have uniform depths individually, the characteristics in both regions are straight lines with slopes c_1 and c_2.

The calculations were performed both for semi-infinite and finite stress bands travelling over deep to shallow (case I) and shallow to deep (case II) water. The left side of Fig.5.4 shows an atmospheric disturbance of the semi-infinite stress band type moving from left to right with a constant speed V for case I. The right side shows a finite stress band for case II.

Murty (1971) performed the calculations for various combinations of parameters such as band width V, c_1, and c_2. The water levels at the left and right boundaries for cases I and II for a finite stress band of width $0.4L$ are shown in Fig.5.5. The important results follow. For semi-infinite stress bands, the water level at the left side is predominantly negative whereas it is positive on the right side and it becomes both positive and negative for finite band widths. The results for the case of a depth discontinuity differ from that without the discontinuity in that, whereas for the latter case the setup becomes periodic some time after the disturbance crosses the lake, this does not happen in the former case. Dingle and Young (1965) also used the method of characteristics to simulate storm surges in an idealized lake due to a moving squall line.

5. Systems with branches

Defant (1961) used the concept of electrical networks to describe procedures dealing with a system with branches (each assumed to have uniform depth). Occasionally the method of characteristics has been used to determine the characteristic oscillations of a basin with a branch (e.g., see Horikawa and Nishimura (1968)). However, these authors assumed uniform depth and a water body of regular shape. For this reason, although they give a rough estimate of the periods, these techniques cannot be used for accurate determination of the periods. Rao (1968) determined numerically free oscillations in the Bay of Fundy, taking into account its two branches, Chignecto Bay and Minas Basin. Henry and Murty (1972) determined the resonance modes for a system with five branches. Rao (1968) also applied corrections for rotational and frictional effects in one-dimensional systems based on the work by Platzman and Rao (1964). Murty and Boilard (1970) used essentially the same technique to determine the free oscillations of an inlet on the west coast of Canada.

5.1 Resonance in multibranched inlets

Henry and Murty (1972) determined the resonant periods for a multibranched inlet system on the west coast of Canada. Figure 5.6 shows the location of the Rivers Inlet complex treated in this study. Figure 5.7 schematically shows the five branches of this system with the grid scheme.

The finite-difference forms of the continuity and momentum equations are for a variable size grid:

$$\frac{\zeta_{j+1} - \zeta_j}{\Delta x_j} = -\frac{\sigma(M_j + M_{j+1})}{g(A_j + A_{j+1})} \tag{5.86}$$

and

$$\frac{M_{j+1} - M_j}{\Delta x_j} = \sigma B_i \frac{(\zeta_j + \zeta_{j+1})}{2} \tag{5.87}$$

where M is the volume transport through a vertical cross-section with area A and surface width B; ζ is the water-level deviation from the equilibrium position; g is gravity; Δx is the grid size at grid point j; and σ is the frequency of oscillation. Frictional and Coriolis effects are ignored.

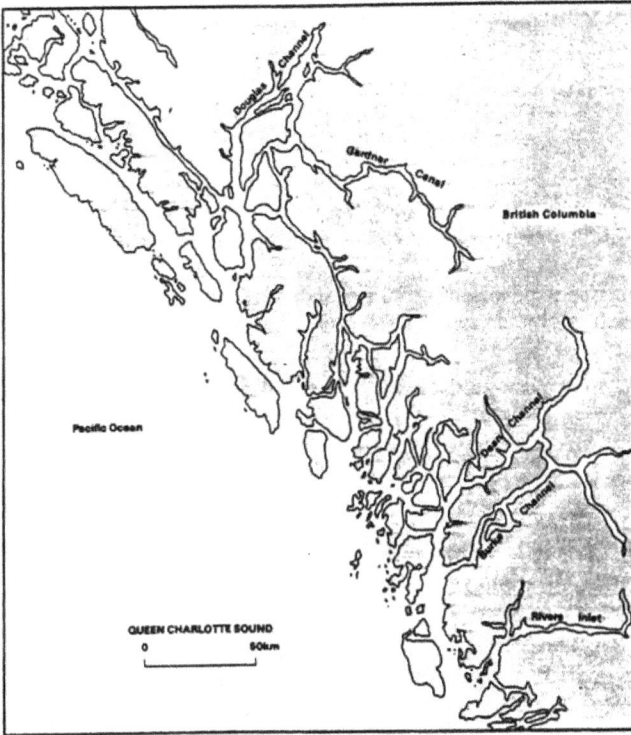

Fig. 5.6
Canadian west coast inlets.

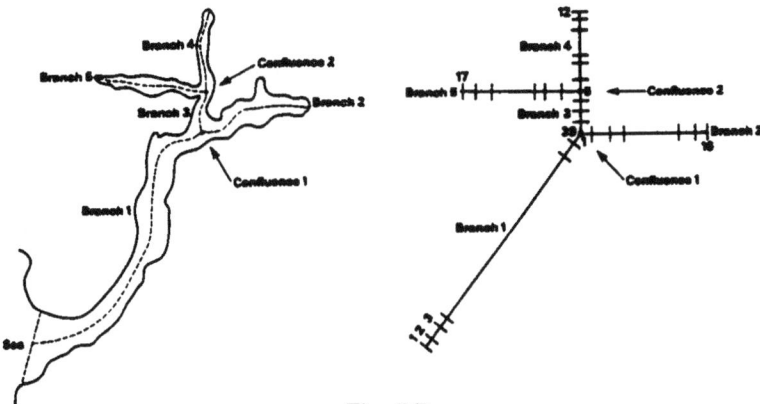

Fig. 5.7
Rivers Inlet on the west coast of Canada, map and schematic.

The conditions at confluence 1 are

$$M_{39}^{(1)} = M_1^{(2)} + M_1^{(3)}$$ (5.88)

and

$$\zeta_{39}^{(1)} = \zeta_1^{(2)} = \zeta_1^{(3)}$$ (5.89)

where superscript denotes the branch and subscript denotes the grid point in a given branch. At confluence 2, the following conditions have to be satisfied:

$$M_5^{(3)} = M_1^{(4)} + M_1^{(5)}$$ (5.90)

and

$$\zeta_5^{(3)} = \zeta_1^{(4)} = \zeta_1^{(5)}$$ (5.91)

Table 5.1 Natural periods of Rivers Inlet computed by: (A) starting at sea and monitoring at head of branch 5. (B) starting at sea and monitoring at head of branch 2. (C) starting at head of branch 4 and monitoring at head of branch 2. Numerals denote genuine modes whereas the modes identified by F are spurious.

Mode No.	(A) Period (min)	(A) No. iterations	(B) Period (min)	(B) No. iterations	(C) Period (min)	(C) No. iterations
1	88.99	2	88.99	2	88.99	3
$F_{1,2}$	–	–	37.17	36	–	–
2	33.75	2	33.75	4	33.75	3
$F_{2,3}$	–	–	–	–	27.50	30
3	25.08	3	25.08	2	25.08	5
$F_{3,4}^1$	–	–	–	–	24.52	34
$F_{3,4}^2$	22.04	52	–	–	–	–
4	21.67	5	–	–	21.67	3
$F_{4,5}$	20.10	46	–	–	–	–
5	15.27	3	15.27	2	15.27	2
$F_{5,6}$	–	–	–	–	14.33	251
6	11.36	3	11.36	2	11.36	3
$F_{6,7}$	–	–	9.95	313	–	–
7	9.86	2	9.86	7	9.86	3

Branch 1 is open to the sea and has 39 grid points, branch 2 has 16, branch 3 has 5, branch 4 has 12, and branch 5 has 17. The boundary conditions of the problem are

$$\zeta_1^{(1)} = 0 \tag{5.92}$$

$$M_1^{(1)} \text{is arbitrary} \tag{5.93}$$

and

$$M_{16}^{(2)} = M_{12}^{(4)} = M_{17}^{(5)} = 0 \tag{5.94}$$

A standard iteration technique described by Rao (1968) has been used to determine the resonance periods. (A) of Table 5.1 shows the periods determined by starting at sea and applying the final boundary condition at the head of branch 5.

Although some modes are located with a few iterations, others took considerably more iterations. Let M^1 and M^2 denote the values $M_{17}^{(5)}$ at the two values of σ bracketing the zero-crossing at the end of the fine search with frequency increment $\Delta\sigma$.

Table 5.2 shows the zero-crossing behavior for a well-behaved and an ill-behaved mode for two different values of $\Delta\sigma$. This table shows that for a well-behaved mode, $M_{17}^{(5)}$ is a smooth function of σ near the zero-crossing and the values remain small. For an ill-behaved mode, the values of M^1 and M^2 remain large. The authors ascribed this behavior to accumulation of numerical errors and called the ill-behaved modes false modes (these could be normal modes for a part of the system, however) and denoted them by F.

Table 5.2 The values $M_{17}^{(5)}$ (denoted by M^1 and M^2) at the two values of σ bracketing the zero-crossing in fine searches with two different frequency increments $\Delta\sigma$ for a well-behaved mode (number 4 for the case A in Table 5.1) and for all ill-behaved mode (number $F_{3,4}^2$ for case A in Table 5.1).

Values of $M^{(5)}{}_{17}$ $(cm^3 s^{-1})$			
Well–behaved mode $\Delta\sigma = 10^{-10}$	Well–behaved mode $\Delta\sigma = 10^{-12}$	Ill–behaved mode $\Delta\sigma = 10^{-10}$	Ill–behaved mode $\Delta\sigma = 10^{-12}$
M^1 -2.138×10^2	-3.225×10^3	-2.463×10^{11}	-1.135×10^{13}
M^2 2.850×10^{-4}	2.625×10^{-6}	1.918×10^9	1.318×10^{12}

Table 5.1 also shows the modes, starting at sea (branch 1) and applying the final boundary condition at the head of branch 2 (i.e., monitoring in branch 2). Whereas the true modes denoted by numerals stay the same (the reason for the disappearance of genuine mode 4 will be considered below) an entirely different set of false modes appears, shown as (B). (C) shows the results from starting at the head of branch 4 and monitoring at the head of branch 2.

To understand this particular behavior of the modes, the nodal structures in terms of M and ζ were examined. Figure 5.8 shows the structures of modes 1 and 2 (both well behaved) and compares the well-behaved mode 3 with the false mode $F_{3,4}^2$.

An examination of the model structures revealed the following important points. 1) When motion in the branch where the solution is begun is small compared to motion in the other branches, a false mode could occur and 2) when the motion in both the starting and monitoring branches is small, a genuine mode could disappear.

Standard iteration technique is not suitable to determine free oscillations of multibranched inlets because all permutations and combinations of starting and monitoring must be used to eliminate the false modes. For a water body with n branches, n^2 separate calculations are necessary.

Henry and Murty (1972) used two alternate methods. In the first (the eigenvalue method), the partial differential equations were replaced by a set of difference-differential equations, i.e., discrete in space but continuous in time. The set of equations can be arranged in a matrix form where eigenvalues give the frequencies of natural oscillation of the system. This method causes neither spurious or false modes, nor misses any true modes. However, it suffers from the disadvantage that complex systems with many branch matrices of very large order have to be handled and computer storage limitations might not permit the matrix evaluation.

The second method is the impulse response method. In this method the Green's function is used. Let $f_1(t)$ be the input at some point x_1 in the system. Then the response r_2 at some other point x_2 in the system is given by

$$r_2(x_1, t) = \int_{-\infty}^{\infty} K(x_1, x_2, t - \tau) f_1(\tau) d\tau \qquad (5.95)$$

where K is a Green's function. For the tsunami problem, x_1 is at the mouth of the inlet. Hence, $K(x, t)$ expresses the response at any point x to a unit impulse (a unit delta function) imposed at the mouth.

Fourier Transformation of eq.(5.95) gives

$$R_2(x, \omega) = K(x_1, x_2, \omega) F(\omega) \qquad (5.96)$$

where $R(x, \omega)$ is the circular frequency. For the tsunami problem, (5.96) simplifies to

$$R(x, \omega) = K(x, \omega) F_1(\omega) \qquad (5.97)$$

where $R(x, \omega)$ is the transform of the response at point x for input, $f_1(t)$ at the mouth, and $F_1(\omega)$ is the transform of $f_1(t)$.

If x does not coincide with a mode of any of the natural modes of the system, then $|K(x, \omega)|$ has vertical asymptotes at each of the resonant frequencies. Figure 5.9 shows this function for Rivers Inlet. Although the impulse response method is probably less accurate (Table 5.3) it is more practical than the other two methods.

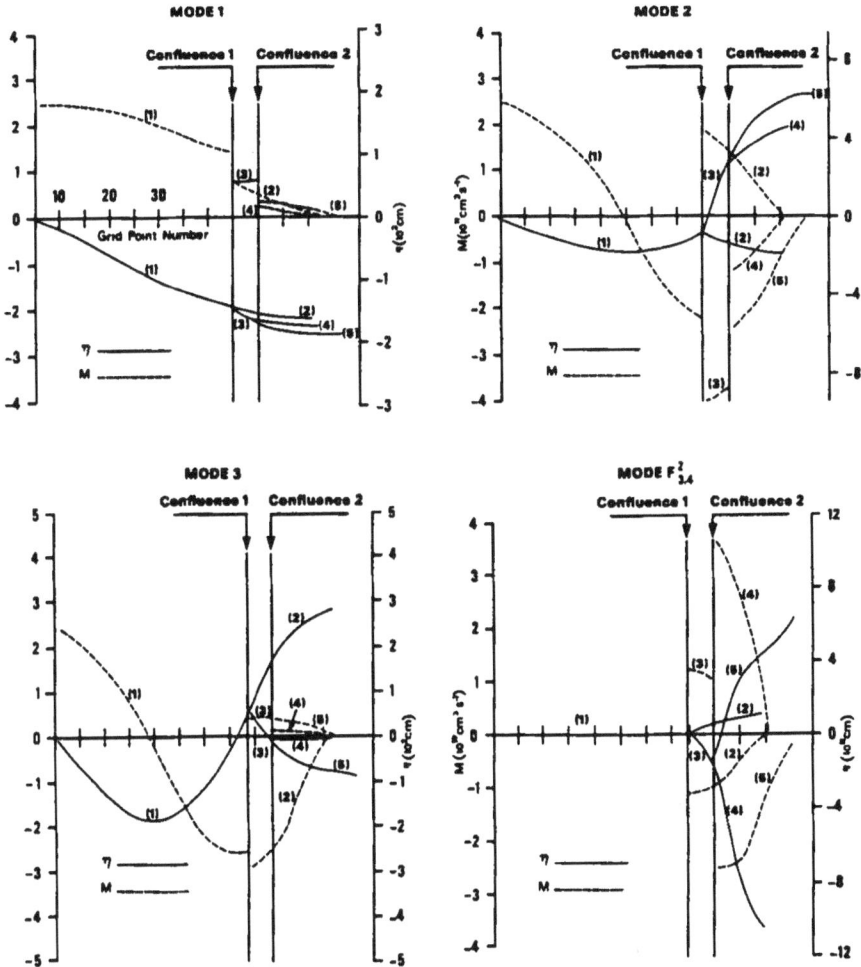

Fig. 5.8

Structure of modes 1, 2, 3 and $F_{3,4}^2$ in Rivers Inlet on the coast of
Canada in terms of volume transport (broken line) and water level (solid line).

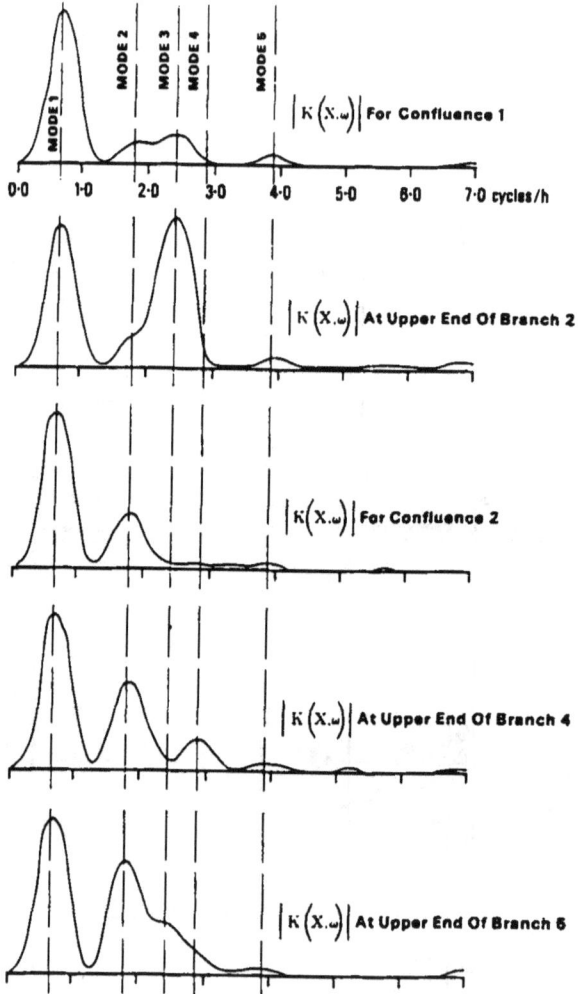

Fig. 5.9

Function $|K(X,\omega)|$ at 5 stations in Rivers Inlet
on the west coast of Canada

Table 5.3 Comparison of period (min.) for the first few genuine modes of rivers inlet computed by iterative method, direct eigenvalue method and impulse method.

Modal No.	Iterative method	Eigenvalue method	Impulse response method
1	88.99	88.97	87.38
2	33.75	33.79	36.69
3	25.08	25.06	24.82
4	21.67	21.68	21.35
5	15.27	15.25	15.38
6	11.36	11.36	
7	9.86	9.86	

6. Resonance calculation for irregular-shaped basins

Raichlen (1966) calculated natural modes of basins of arbitrary shape and depth by what he called two-dimensional models. In fact, these are the so-called one-dimensional (only one horizontal dimension) models. Raichlen's three-dimensional model corresponds to the traditional two-dimensional model, and he simply outlines his model.

Lee and Raichlen (1971) developed a theory for calculating the resonant modes in harbors with arbitrary geometry but uniform depth. The solution of the two-dimensional problem is obtained in terms of an integral equation, later approximated into a matrix form. The area of study is divided into two regions, the region outside the harbor mouth and the region inside the harbor. Boundary conditions at the harbor entrance are applied by matching the wave amplitude and its normal derivative for the exterior solutions.

These authors examined the response of the harbor to incoming waves through an "amplification factor", defined (Lee and Raichlen, 1971, p. 2169) as the ratio of the wave amplitude at any position inside the harbor to the sum of the incident and the reflected wave amplitude at the coastline (with the harbor entrance closed). They also performed some laboratory experiments to model Long Beach Harbor, California.

Clarke and Thomas (1972) calculated the normal modes of a spindle-shaped basin of uniform depth and applied the study to Port Kembla's outer harbor on the east coast of Australia. This harbor, about 80 km south of Sydney, is well known for its seiche activity. Hidaka (1931) coined the named "spindle shape" to a basin whose perimeter at the surface can be defined as the intersection of two confocal parabolae.

The perimeter of a spindle-shaped basin can be defined by

$$y^2 = \begin{cases} a(a - 2x) & \text{for } x \geq 0 \\ a(a + 2x) & \text{for } x < 0 \end{cases} \tag{5.98}$$

One interesting result from this case is that at the entrance to the harbor, instead of a node (as is usually assumed in resonance calculations) an antinode was detected. This was a result of reflection off the harbor walls of the long waves which, as they emerge from the harbor entrance, will form a resonant standing wave system outside the harbor. Only the symmetric modes will be excited when the standing wave system is such that there is an antinode at the harbor entrance.

Olsen and Hwang (1971) calculated the resonance modes for a basin with arbitrary shape and variable depth. One new feature of this theory is that the boundary condition at the entrance to the basin need not be prescribed. These authors applied this study to Keauhou Bay, Hawaii.

Let $H(x, y)$ be the water depth and ϕ the velocity potential. The linearized form of the long-wave equation is

$$\frac{i}{g}\frac{\partial^2 \Phi}{\partial t^2} = \frac{\partial}{\partial x}\left(H\frac{\partial \Phi}{\partial x}\right) + \frac{\partial}{\partial y}\left(H\frac{\partial \Phi}{\partial y}\right) \tag{5.99}$$

Assume periodic excitation from outside that ultimately causes a steady periodic motion inside the harbor. For this write

$$\Phi = \phi(x, y)\exp(-i\sigma t) \tag{5.100}$$

where σ is the frequency of periodic motion. Then from eqs.(5.99) and (5.100),

$$\frac{\partial}{\partial x}\left(H\frac{\partial \Phi}{\partial x}\right) + \frac{\partial}{\partial y}\left(H\frac{\partial \Phi}{\partial y}\right) + \frac{\sigma^2}{g}\phi = 0 \tag{5.101}$$

Figure 5.10 shows that in the outer region a uniform depth (equal to depth at the matching boundary) will be assumed and an analytical solution will be obtained, whereas in the inner region a numerical solution will be given.

Because depth is uniform in the outer region, the Helmholtz equation is

$$\nabla^2 \phi + k^2 \phi = 0 \tag{5.102}$$

where

$$k^2 \equiv \frac{\sigma^2}{gH} \tag{5.103}$$

The following boundary conditions should be satisfied. Along the shoreline

$$\frac{\partial \phi}{\partial n} = 0 \tag{5.104}$$

and at ∞

$$\phi = \phi_0 \tag{5.105}$$

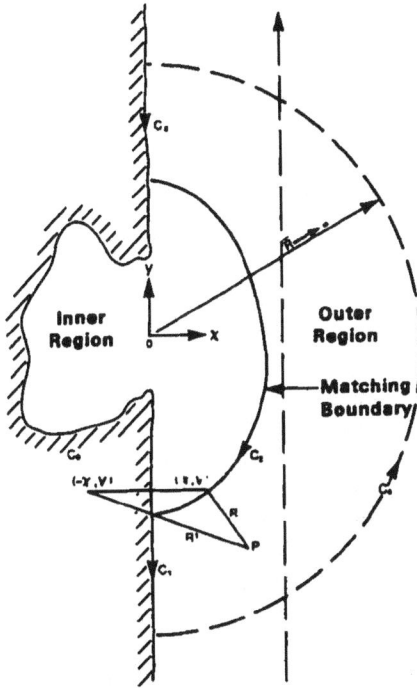

Fig. 5.10
Schematic diagram used to obtain boundary condition between inner
and outer regions. (Olsen and Hwang, 1971)

Olsen and Hwang (1971) made some simplifications. They assumed a straight
shoreline located at $x = 0$. Then, for a straight-crested standing wave with crest at
an angle β to the x axis,

$$\phi_0 = \cos(kx \cdot \cos \beta) \exp(-iky \sin \beta) \qquad \text{for} 0 < \beta < \pi \qquad (5.106)$$

However, if the wave front propagates parallel to the x axis, then β is equal to zero
and ϕ_0 reduces to

$$\phi_0 = \cos kx \qquad (5.107)$$

This condition is consistent with the assumption that the wave motion is the same
on either side of the boundary. Then

$$\phi = \phi_0 + \phi' \qquad (5.108)$$

where ϕ_0 is the contribution due to the incident waves and ϕ' is due to the presence of the harbor. Individually, ϕ_0 and ϕ' satisfy eq.(5.102). However, the boundary condition (5.104) becomes

$$\frac{\partial}{\partial n}(\phi_0 + \phi') = 0 \tag{5.109}$$

and this has to be satisfied on the shoreline C_1 and C_3.

The condition at the matching boundary is that ϕ_0 and ϕ' on one side of the boundary must be equal to ϕ_0 and ϕ' on the other side. Let G be a Green's function and S the contour of integration. Then, to determine the perturbation velocity potential Kernal ϕ', write

$$\phi' = \int_S \left(\phi' \frac{\partial G}{\partial n} - G \frac{\partial \phi'}{\partial n} \right) ds \tag{5.110}$$

The contour integration is made around C_4, C_3, C_2, and C_1, as shown in Fig.5.10. The Green's function, G, must satisfy the following conditions: 1) the Helmholtz Equation (5.102); 2) have a singularity at $R = 0$, and 3) vanish at ∞.

The Hankel function of the first kind and zero the order, $H_0^{(1)}(kR)$, could be chosen as the Green's function because

$$H_0^{(1)}(kR) \rightarrow \begin{cases} \dfrac{i}{\pi} \ell n(kR) \text{ as } R \rightarrow 0 \\[3mm] -\dfrac{i}{\pi} \left[\dfrac{2\pi}{kR} \exp\left\{ i\left(kR + \dfrac{\pi}{4} \right) \right\} \right]^{\frac{1}{2}} \text{ as } R \rightarrow \infty \end{cases} \tag{5.111}$$

If the contour C_4 is extended to ∞ then $R \rightarrow \infty$ and eq.(5.110) becomes

$$\phi' = \int_{C_4} \left(\phi' \frac{\partial G}{\partial n} - G \frac{\partial \phi'}{\partial n} \rightarrow 0 \quad \text{as } R \rightarrow \infty \right. \tag{5.112}$$

For numerical evaluation purposes, Olsen and Hwang (1971) made further simplifications. As shown in Fig.5.10, the boundaries C_1, C_2, and C_3, are taken to be straight lines. Also, the coordinate system is fixed, as shown. This means that at $d\, a\, b$ and $c\, a\, e$, a rigid wall is assumed (Fig.5.11).

Olsen and Hwang pointed out that the simplification will have little effect on the results for normal incident waves as long as a is large enough compared to the width at the mouth of the bay. They cautioned, however, that edge waves generated by obliquely incident waves will be affected. Under these simplifications, eq.(5.110) becomes

$$\phi'(x,y) = \int_{-\infty}^{\infty} \left(\phi' \frac{\partial G}{\partial x} - G \frac{\partial \phi'}{\partial x} \right) d\zeta \tag{5.113}$$

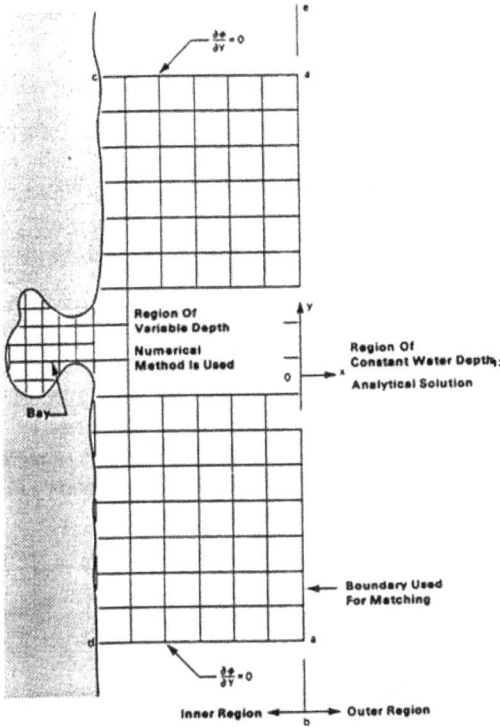

Fig.5.11
Schematic diagram of the bay and method used in analysis.
(Olsen and Hwang, 1971)

Define

$$R' \equiv \left\{ (x + \xi)^2 + (y - \zeta)^2 \right\}^{\frac{1}{2}} \tag{5.114}$$

and

$$R \equiv \left\{ (x + \xi)^2 + (y - \zeta)^2 \right\}^{\frac{1}{2}} \tag{5.115}$$

where ξ and ζ are the coordinates on the matching boundary. A choice of

$$G = -\frac{i}{4} \left\{ H_0^{(1)}(kR) + H_0^{(1)}(kR') \right\} \tag{5.116}$$

satisfies not only the three conditions listed above but also the following condition:

$$\frac{\partial G}{\partial x} = 0 \qquad (5.117)$$

along the y axis. Then eq.(5.113) becomes

$$\phi'(x, y) = -\frac{i}{2} \int_{-\infty}^{\infty} \frac{\partial \phi'(0, \zeta)}{\partial x} H_0^{(1)}(k|y - \zeta|) \, d\zeta \qquad (5.118)$$

As $a\,b$ and $a\,e$ are assumed to be rigid walls, then

$$\frac{\partial \phi'}{\partial x} = 0 \qquad (5.119)$$

Also, if the incident wave is parallel to the y axis, then eq.(5.107) holds and

$$\phi(x, y) = \cos(kx) - \frac{i}{2} \int_{-a}^{a} \frac{\partial \phi(0, \zeta)}{\partial x} H_0^{(1)}(k|y - \zeta|) \, d\zeta \qquad (5.120)$$

The prime on ϕ in the integral is suppressed for convenience.

Next, Olsen and Hwang outlined a finite-difference procedure to obtain the solution of eq.(5.107) in the inner region under the boundary condition (5.120). With reference to a rectangular grid system,

$$\begin{aligned} x_j &= j\Delta x \text{ for } 0 \le j \le je \\ y_k &= k\Delta y \text{ for } 0 \le k \le ke \end{aligned} \qquad (5.121)$$

Equation (5.101) can be written as

$$\frac{D_{j+\frac{1}{2},k}}{(\Delta x)^2}(\phi_{j+1,k} - \phi_{j,k}) - \frac{D_{j-\frac{1}{2},k}}{(\Delta x)^2}(\phi_{j,k} - \phi_{j-1,k}) + \frac{D_{j,k+\frac{1}{2}}}{(\Delta y)^2}(\phi_{j,k+1} - \phi_{j,k})$$

$$- \frac{D_{j,k-\frac{1}{2}}}{(\Delta y)^2}(\phi_{j,k} - \phi_{j,k-1}) = -\frac{\sigma^2}{g}\phi_{j,k} \qquad (5.122)$$

Defining the following quantities,

$$B_{j,k} \equiv -D_{j,k-\frac{1}{2}}\frac{\Delta x}{\Delta y} \qquad (5.123)$$

$$D_{j,k} \equiv -D_{j-\frac{1}{2},k}\frac{\Delta y}{\Delta x} \qquad (5.124)$$

$$E_{j,k} \equiv D_{j+\frac{1}{2},k}\frac{\Delta y}{\Delta x} + D_{j-\frac{1}{2},k}\frac{\Delta x}{\Delta y} + D_{j,k+\frac{1}{2}}\frac{\Delta x}{\Delta y} + D_{j,k-\frac{1}{2}}\frac{\Delta x}{\Delta y} - \frac{\sigma^2}{g}\Delta x\Delta y \qquad (5.125)$$

$$F_{j+\frac{1}{2},k} \equiv -D_{j+\frac{1}{2},k}\frac{\Delta y}{\Delta x} \qquad (5.126)$$

$$H_{j,k} \equiv -D_{j,k+\frac{1}{2}}\frac{\Delta x}{\Delta y} \qquad (5.127)$$

From eqs.(5.122) to (5.127), then

$$B_{j,k}\phi_{j,k-1} + D_{j,k}\phi - j - 1, k + E_{j,k}\phi_{j,k+1} + H_{j,k}\phi_{j,k+1} = 0 \qquad (5.128)$$

This equation can be written in the following matrix form:

$$M \phi = 0 \qquad (5.129)$$

This square matrix, M has a size $(je + 1)(ke + 1)$.
    The boundary condition (5.104) can be written as

$$\left(\frac{\partial \phi}{\partial x}\right)_{i,j} = \frac{\phi_{i+1,j} - \phi_{i-1,j}}{2\Delta x} \qquad (5.130)$$

with a similar expression for $\partial\phi/\partial y$.
    The boundary condition (5.120) becomes singular at $y = \zeta$. This is evaluated as follows:

$$\phi(x,y) = \phi_0 - \frac{i}{2}\int_{-a}^{a} H_0^{(1)}(k|y - \zeta|)\frac{\partial \phi}{\partial x}(0, \zeta)\, d\zeta$$

$$= \phi_0 - \frac{i}{2}\int_{-a}^{y-\epsilon} H_0^{(1)}(k|y - \zeta|)\frac{\partial \phi}{\partial x}\, d\zeta$$

$$- \frac{i}{2}\int_{y+\epsilon}^{a} H_0^{(1)}(k|y - \zeta|)\frac{\partial \phi}{\partial x}(0, \zeta)\, d\zeta$$

$$- \frac{i}{2}\int_{y-\epsilon}^{y+\epsilon} H_0^{(1)}(k|y - \zeta|)\frac{\partial \phi}{\partial x}(0, \zeta)\, d\zeta \qquad (5.131)$$

The integral on the right side of this equation is singular but if it is assumed that $\partial\phi/\partial x$ is uniform in the neighborhood of ϵ, then it can be evaluated analytically to give

$$I_3 = -\frac{i}{2}\int_{y-\epsilon}^{y+\epsilon} H_0^{(1)}(k|y - \zeta|)\frac{\partial \phi}{\partial x}(0, \zeta)\, d\zeta = B\frac{\partial \phi}{\partial x}$$

$$= -\frac{1}{2}\left[\begin{array}{c} -b\epsilon + \frac{ck^2}{3}\epsilon^3 + \frac{2\epsilon}{\pi}\left(\ln\frac{k\epsilon}{2} - 1\right) \\ + \frac{2ak^2\epsilon^3}{3\pi}\left(\ln\frac{k\epsilon}{2} - \frac{1}{3}\right)\frac{\partial \phi}{\partial x}(0, y) \end{array}\right] \qquad (5.132)$$

with $a = -0.2500000$, $b = 0.36746691$, $c = 0.06728818$.
    By taking $\epsilon = \Delta y$, the first two integrals on the right side of eq.(5.131) can be evaluated by the trapezoidal quadrature technique. The numerical calculations for Keauhau Bay showed that the wave amplification inside the harbor was greater for shorter wave periods.

7. Secondary undulations

Nakano (1932a) defined secondary undulations on tide gauge records as the peculiar
zigzag parts of tidal curve. Omori (1906) investigated the secondary undulations
on the Pacific coast of Japan. Honda *et. al.* (1908) made a systematic study of
the secondary undulations in 50 bays and coastal areas in Japan. These secondary
undulations are essentially normal modes of oscillations of a bay and could be excited
by several sources, such as passage of storms, squall lines, or coupling of the bay to
wave motion outside the bay. Hence, tsunamis could excite secondary undulations.

Nakano (1932a) gave a tentative classification of the secondary undulations into
types A, B, and C. Although this type of classification is not rigorous, it is included
here because some type of classification of secondary undulations is useful.

In type A, the secondary undulations appear as coherent trains of waves with
approximately the same form. In type B, the waves are not as regular and coherent
as in type A but are not completely irregular either. In type C, the secondary
undulations are more or less irregular. The secondary undulations occur and persist
after the passage of the main tsunami waves.

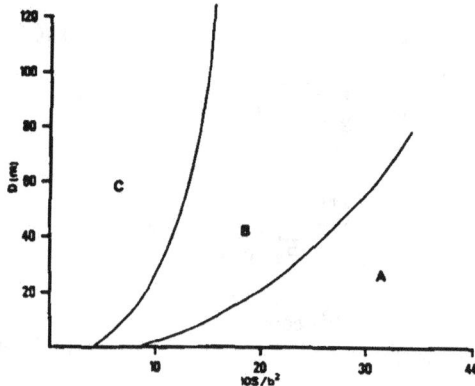

Fig. 5.12
Classification of secondary undulations. (Nakano, 1932a)

Nakano, in attempting to correlate the type of secondary undulation observed
in a bay with the form of the bay, considered 16 bays in Japan (Table 5.4). In
the final column, the type indicated is based on the observed undulations in the
bay under consideration. Nakano plotted the data of this table with the depth D
of the bay as ordinate and $10\,S/b^2$ as the abscissa where S is the area of the bay
and b is the breadth (Fig.5.12). The three types of secondary undulations appear
distinct. Another result from Nakano's study is, in an U-shaped bay, the energy of

the secondary undulations neither accumulates nor is dissipated as efficiently as in a V-shaped bay.

Nakano showed that the rate of accumulation of energy in a bay is proportional to $(S/b^2)(1/D^{3/2})$ where D is the (uniform) depth of the bay, S is the surface area, and b the width at the mouth. As the rate of accumulation of energy determines the secondary undulations, the above parameter is important in classifying bays for their resonant characteristics. The advantage of such a classification is that detailed numerical models can be constructed for only a few representative bays and the results for other bays can be deduced.

Table 5.4 Classification of secondary undulations in some Japan bays. (Nakano, 1932a).

Bays	Mean Depth (m)	Area $S(km^2)$	Breadth $b(km)$	S/b^2	Characteristic (class)
Otaru	9.1	7.19	4.84	0.367	C
Hamanaka	8.5	47.00	8.35	0.674	C
Hakodate	10.7	61.90	8.32	0.894	B
Turuga	28.5	67.62	7.22	1.297	B
Tonoura	11.6	0.54	0.83	0.784	B
Ryoisi	33.1	12.80	6.10	0.344	C
Ohunato	19.0	12.18	2.68	1.696	B
Ayukawa	12.7	1.42	1.98	0.362	C
Tokyo	52.7	1622.10	22.08	8.327	A
Moroiso (Misaki)	3.0	0.36	0.30	4.000	A
Simoda	15.0	3.44	2.12	0.765	C
Kusimoto	5.0	11.06	3.83	0.754	C
Hososima	11.7	2.34	1.65	0.852	A
Kagosima	103.0	1220.15	24.45	2.041	B
Nagasaki	17.0	13.36	2.50	2.138	A
Hutami	23.9	4.04	1.57	1.639	B

Nakano (1932b) also considered the secondary undualtions in coupled bays. In this situation, the secondary undualtions could exhibit a beat-type behavior when the amplitude of the water level increases and decreases periodically. Figure 5.13 shows a coupled bay system in Japan and the mareograms recorded.

Nakano (1932b) developed a theory for the oscillation of the coupled system. Consider a single rectangular bay and let the x axis be in the direction parallel to the bay axis and let the z axis be vertical. Let the origin be taken at the mouth of the bay at the bottom. Let L be the length, B the uniform breadth, H the uniform depth of the bay, and ξ_i and ζ_i be the horizontal displacement of the water and the

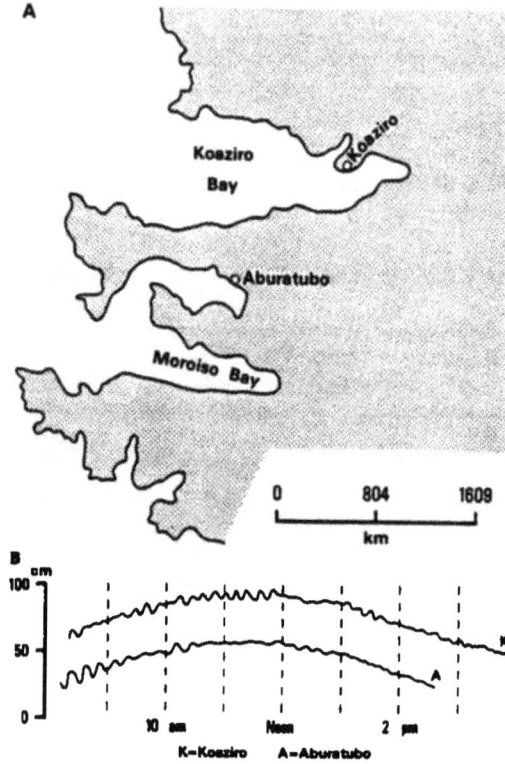

Fig. 5.13
(A) Two coupled bays, on Japan coast and
(B) secondary undulations in the bays. (Nakano, 1932b)

vertical motion of the free surface. Then

$$\xi_i = B\cos(\ell x)\sin(at + \alpha)$$
$$\zeta_i = BH\ell\sin(\ell x)\sin(at + \alpha) \tag{5.133}$$

The kinetic energy K_i and the potential energy P_i inside the bay are given by

$$K_i = \frac{\rho b H}{2}\int_0^L \dot{\xi}_i^2\, dx = \frac{\rho g H L}{4}\,\dot{\xi}_0^2$$
$$P_i = \frac{\rho b g}{2}\int_0^L \zeta_i^2\, dx = \frac{\rho b g H^2 \ell^2 L}{4}\,\xi_0^2 \tag{5.134}$$

where ξ_0 is the value of ξ_i at $x = 0$ and the dot denotes differentiation with respect to time. The potential energy outside the bay is generally small and could be neglected but the kinetic energy outside the bay is given by

$$K_a = \frac{\rho BHLQ}{4}\,\dot\xi_0^2 \qquad (5.135)$$

where Q is a dimensionless quantity.

The kinetic and potential energies of the system consisting of regions inside as well as outside the bay are given by

$$K = K_i + K_a = \frac{\rho BHL}{4}(1+Q)\dot\xi_0^2$$

$$P = P_i + P_a = \frac{\rho Bg H^2 \ell^2 L}{4}\xi_0^2 \qquad (5.136)$$

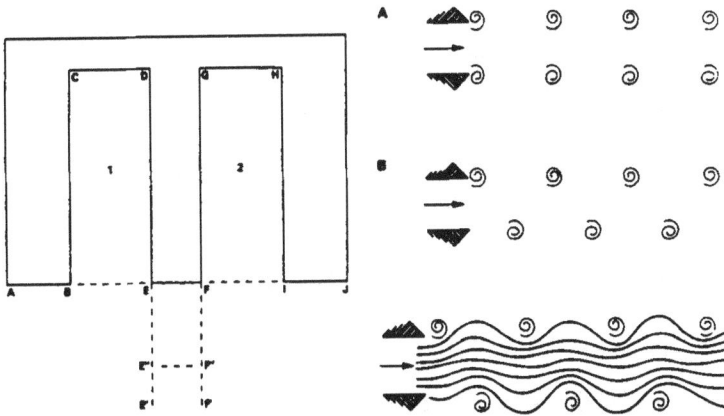

Fig. 5.14

(Left) Schematic diagram to compute secondary undulations in coupled bays, (Nakano, 1932b). (Right) Schematic diagram illustrating excitation of secondary undulations by currents, (Nakano, 1933).

Consider the situation where two bays exist side by side (Fig.5.14). Let the depth of the ocean immediately outside the bay be uniform and equal to H. Let subscripts 1 and 2 denote the two bays and let

$$X_{10} \equiv BH\xi_{10}$$

$$X_{20} \equiv BH\xi_{20} \qquad (5.137)$$

Equations (5.136) and (5.137) then give

$$K_1 = \frac{\rho L}{4BH}(1+Q)\dot{X}_{10}$$

$$K_2 = \frac{\rho L}{4BH}(1+Q)\dot{X}_{20}^2$$

$$P_1 = \frac{\rho g}{4}\frac{\ell^2 L}{B}X_{10}^2 \qquad\qquad (5.138)$$

$$P_2 = \frac{\rho g \ell^2 L}{4B}X_{20}^2$$

If there were no coupling between the two bays, they would oscillate independently and the nodes (for vertical motion) would be at BE and FI and the antinodes
at CD and GH. (The water level will be increasing in the region EFE'F' when the
water levels at CD and GH will be decreasing and vice versa.) However, the water
near the shore, EF, creates a coupling between the two bays. Consider an imaginary
barrier at E'' and FF''', both of which have no inertia, and that the vertical water
motion over EFF''E'' is uniform.

When a quantity of water X_0 flows across BE or FI, some partakes in the
coupling. Let the quantity σX_0 (where $0 < \sigma < 1$) flow through the region EFF''E'';
this water moves the pistons at EE'' and FF''. Here σ can be treated as a coupling
coefficient. Let d denote the distance between the points E and F and let z be the
magnitude of vertical motion in EFF''E''. Then

$$B\,dz = -\sigma(X_{10} + X_{20}) \qquad\qquad (5.139)$$

The potential energy associated with the vertical motion in EFF''E'' is

$$P_{12} = \int_0^z BH\,\rho g\,dz = \frac{\rho g \sigma^2}{2BH}(X_{10} + X_{20})^2 \qquad\qquad (5.140)$$

The kinetic energy of this portion of the water could be written as

$$K_{12} = \frac{\mu_1}{2}\dot{X}_{10}^2 + \frac{\mu_2}{2}\dot{X}_{20}^2 + \mu\dot{X}_{10}\dot{X}_{20} \qquad\qquad (5.141)$$

where μ_1, μ_2, and μ are functions of σ. If the two bays are identical, then

$$\mu_1 = \mu_2 = \mu'$$

Hence, the kinetic, KE, and potential, PE, energies of the whole system are given
by

$$KE = K = K_1 + K_2 + K_{12}$$

$$= \frac{\rho L}{4Bh}(1+Q)\left(\dot{X}_{10}^2 + \dot{X}_{20}^2\right) + \frac{\mu'}{2}\left(\dot{X}_{10}^2 + \dot{X}_{20}^2\right) + \mu\dot{X}_{10}\dot{X}_{20} \qquad\qquad (5.142)$$

$$PE = P = P_1 + P_2 + P_{12}$$

$$= \frac{\rho g \ell^2 L}{4B}(X_{10}^2 + X_{20}^2) + \frac{\rho g \sigma^2}{2BH}(X_{10} + X_{20})^2 \qquad (5.143)$$

From the Lagrangian equation,

$$\frac{d}{dt}\left(\frac{\partial K}{\partial \dot{X}_{k_0}}\right) - \frac{\partial K}{\partial X_{k_0}} = -\frac{\partial P}{\partial X_{k_0}}, \qquad K = 1, 2 \qquad (5.144)$$

one obtains

$$\left\{\frac{\rho L}{2Bh}(1+Q) + \mu'\right\}\ddot{X}_{10} + \mu\ddot{X}_{20} + \left(\frac{\rho g \ell^2 L}{2B} + \frac{\rho g \sigma^2}{BH}\right)X_{10} + \frac{\rho g \sigma^2}{BH}X_{20} = 0 \quad (5.145)$$

$$\left\{\frac{\rho L}{2Bh}(1+Q) + \mu'\right\}\ddot{X}_{20} + \mu\ddot{X}_{10} + \left(\frac{\rho g \ell^2 L}{2B} + \frac{\rho g \sigma^2}{BH}\right)X_{20} + \frac{\rho g \sigma^2}{BH}X_{10} = 0 \quad (5.146)$$

Nakano (1932b) solved these equations under certain assumptions. The details will be omitted but both bays execute oscillations with period,

$$T\bigg/\left(1 + \frac{\delta_1 + \delta_2}{2}\right)$$

and their amplitudes vary as

$$\frac{2T}{|\delta_2 - \delta_1|}$$

Here, T, the period of each bay (assuming that the other bay does not exist) is given by

$$T = \frac{4L}{\sqrt{gH}}(1+Q)^{1/2} \qquad (5.147)$$

and

$$\delta_1 \equiv \frac{\nu}{N} - \frac{(\mu' + \mu)}{2M}$$
$$\delta_2 \equiv \frac{(\mu - \mu')}{2M} \qquad (5.148)$$

with

$$\nu \equiv \frac{\rho g \sigma^2}{BH}$$
$$M \equiv \frac{\rho L}{2BH}(1+Q) \qquad (5.149)$$

and

$$N \equiv \frac{\rho g \ell^2 L}{2B}$$

Because the phases of X_{10} and X_{20} corresponding to the period $2T/|\delta_2 - \delta_1|$ are always opposed, one bay oscillates vigorously and the other more or less rests, and vice versa (Fig.5.14B).

Nakano (1933) considered the possibility of excitation of secondary undulations in bays by tidal or other current. He stated that any current past the mouth of a bay could be a source of excitation of secondary undulation in the bay in analogy to a jet of air into the mouth piece of an organ pipe producing a standing oscillation of the air column in the pipe. Nakano concluded that such a generation of secondary undulations was observed at the Strait of Naruto in Japan.

Nakano (1933) used the concept of rows of vortices to give an explanation for coupling of the current to the bay. When a jet of water rushes through a narrow opening such as a strait (Fig.5.14 shows the edges of the opening by the hatched areas) into a wider area such as a bay, either the symmetric system of vortices shown in (A) or the antisymmetric system is stable. The lower part of Fig.5.14 shows the streamline pattern for the stable system. The jet oscillates and let N be this frequency. Let ℓ be the distance between any two successive vortices on one side and u be the velocity of the vortices. Then

$$N = \frac{u}{\ell} \tag{5.150}$$

If this frequency N agrees with the frequency of natural oscillation of the bay, then secondary undulations could be excited in the bay.

8. Helmholtz mode

Following mechanical and acoustical analogy, the so-called Helmholtz mode will be defined and then a hydrodynamic explanation invoked. With reference to Fig.5.15, the equation of motion for the mechanical system shown can be written as (Raichlen, 1966)

$$M\ddot{x} + c\dot{x} + kx = F_0 \cos(\omega_f t) \tag{5.151}$$

where M is the mass of the oscillating body, c is a linear damping coefficient, k is a spring constant, ω_f is a (circular) forcing frequency, and dots denote differentiation with respect to t.

The following steady-state solution can be assumed:

$$x = X \cos(\omega_f t - \phi) \tag{5.152}$$

where X is the maximum displacement and ϕ is a phase angle between the input and output functions. The parameters X and ϕ can be made nondimensional as follows:

$$\frac{X}{X_{st}} = 1 \left/ \left\{ \left[1 - \left(\frac{\omega_f}{\omega_n} \right)^2 \right]^2 + (2\zeta \frac{\omega_f}{\omega_n})^2 \right\}^{1/2} \right. \tag{5.153}$$

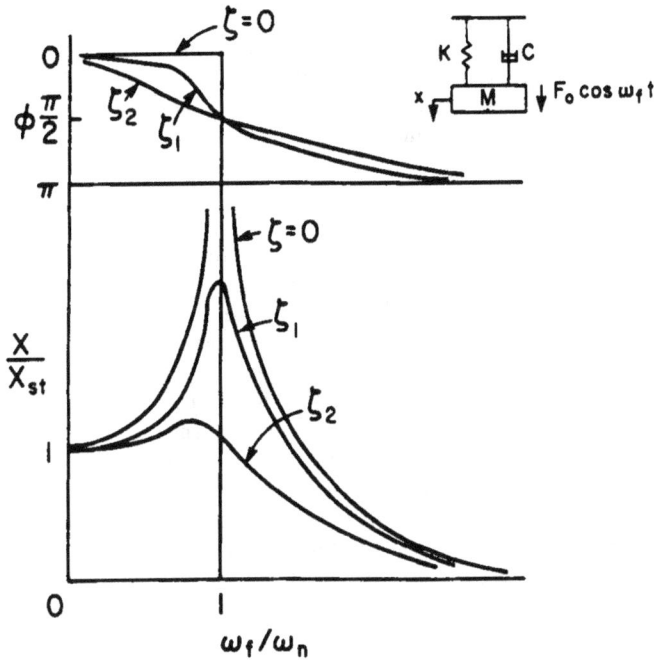

Fig. 5.15

Resonance characteristics of a system with a single degree of freedom. (Raichlen, 1966)

$$\tan \phi = \frac{2\zeta \omega_f/\omega_n}{1 - (\omega_f/\omega_n)^2} \tag{5.154}$$

where

$$X_{st} \equiv \frac{F_0}{k},$$

$$\omega_n \equiv \left(\frac{k}{M}\right)^{1/2} \tag{5.155}$$

$$\zeta \equiv \frac{c}{2M\omega_n}$$

Figure 5.15 represents graphically eqs.(5.153) and (5.154). First, consider the behavior of X/X_{st}. For the case of small frictional dissipation, when the frequency is approximately equal to the undamped natural frequency of the system, the forcing function X_{st} is greatly amplified. As the damping ζ gets bigger, the difference between the resonant frequency and ω_n increases. For low values of the frequency ratio, the amplitudes of the input and the output are approximately equal. However, for frequencies considerably above the resonant frequency, the response decreases substantially and the maximum displacement of the mass approaches zero. If the damping is zero, eq.(5.153) gives infinite amplitude at resonance. However, this result, which is obtained from the linear theory, must be modified at great amplitudes to include the influence of the nonlinear effects.

Next, consider the behavior of the phase angle with respect to the forcing frequency. For low values of the frequency ratio, the forcing function is mainly in phase with the output displacement, and at higher values they are 180° out of phase. At resonance, the phase angle becomes 90° and hence, the force $F_0 \cos(\omega_f t)$ is in phase with the velocity $\ddot{x}$. Thus, when the mass is going through its zero displacement position, a maximum force is impressed upon the system.

The number of degrees of freedom of a system is the number of independent coordinates that are required to describe the motion of the system. Raichlen (1966) cited the vibration of a clamped circular membrane as an example of a system possessing infinite degrees of freedom, whereas the spring-mass-dashpot system considered here is an example of a system with a single degree of freedom.

In acoustics, an example of a single degree of freedom system is the so-called Helmholtz resonator, which consists of a cavity of volume V connected to a tube of length l and area of cross-section A. The equation of motion for this system is

$$M\ddot{x} + r_a\dot{x} + \frac{x}{B} = P\cos(\omega_f t) \tag{5.156}$$

where x is the volume displacement, c is the wave velocity, r_a is the radiation loss coefficient, and

$$\begin{aligned} M &\equiv \frac{\rho l}{A} \\ B &\equiv \frac{V}{\rho c^2} \end{aligned} \tag{5.157}$$

The natural frequency of the Helmholtz resonator is given by

$$\omega_n \equiv c\sqrt{\frac{A}{lV}} \tag{5.158}$$

Since eqs.(5.156) and (5.151) are similar, it can be seen that when the frequency is equal to ω_n, the ratio of the volume displacement to the applied pressure will be ∞ when r_a is zero. Also, the ratio of the volume displacement to the applied pressure varies, as shown in Fig.5.15.

Next, consider the Helmholtz mode in the context of hydrodynamics. Miles (1971) used the term "Helmholtz mode", Platzman (1972) used "co-oscillating mode", and Lee and Raichlen (1972) referred to it as the "pumping mode". Basically, Helmholtz resonance represents the balance between the kinetic energy of water flowing in through a narrow connecting channel and the potential energy from the rise in the mean water level within the harbor (Freeman *et al.*, 1974). It is an additional gravitational mode of a substantially longer period than the fundamental free oscillation, as can be seen below.

To conceptualize the Helmholtz mode, Platzman (1972) presented the following argument. Suppose that at the mouth of a rectangular bay an adjustable barrier exists and that this barrier is gradually moved from the two sides of the bay to the center, completely closing off the bay. The open modes with periods initially of the form $2T/(2n-1)$, $n = 1, 2, 3, \ldots$, will be transformed continuously into the closed mode periods of T/n, $n = 0, 1, 2, \ldots$. It is obvious that the fundamental mode for the open bay transforms into the zeroth mode for the closed basin, and as the barrier closes, this period approaches ∞. For small openings, the period of the Helmholtz mode is less than ∞ but greater than the period for a completely open bay. Platzman (1972) showed that rotation changes the period of the Helmholtz mode by, at most, 3%.

The classic theory for the Helmholtz mode can be applied only to a single channel harbor. Freeman *et al.* (1974) extended this to a harbor (or basin) with multiple channels for openings). The dissipative forces (due to the eddy viscosity of the fluid and to the energy radiated from the mouth) are ignored. These forces affect the amplification factor at resonance and will shift the resonant frequency slightly. The solution developed by Freeman *et al.* (1974) for the frequency ω_0 is

$$\omega_0 = \sqrt{\frac{g}{A} \sum_{i=1}^{n} \left(\frac{S_i}{L_i}\right)^{1/2}} \ \ \text{rad} \cdot \text{s}^{-1} \tag{5.159}$$

where g is gravity, A is the surface area of the harbor, S_i is the cross-sectional area of the ith channel, and L_i is the length of the ith channel.

Miles and Munk (1961) introduced the so-called harbor paradox in which they showed that narrowing of a harbor mouth (relative to the other dimensions) diminishes the protection from seiching. For a quantitative estimation of this in terms of the sharpness or Q at resonance, the reader is referred to their paper. Miles and Lee (1975) used equivalent electric circuit analysis to study this problem. Garrett (1970) showed that the harbor paradox, as originally postulated by Miles and Munk (1961), only holds for the Helmholtz mode.

9. Open boundary conditions

Garrett (1975) considered the problem of the proper boundary conditions to be used when calculating the co-oscillating tide between an ocean and a gulf. He also showed that the traditional method (e.g., see Defant, 1961) is incorrect. The traditional method of separating the co-oscillating and independent tides in a gulf is illustrated in Fig.5.16. The independent tide is due to the direct action of the astronomical forces on the gulf and is obtained as the solution of the tidal equations, including the tide-generating forces with the boundary conditions of zero normal velocity over the gulf's boundary and with the elevation at the gulf's mouth equal to the observed tide.

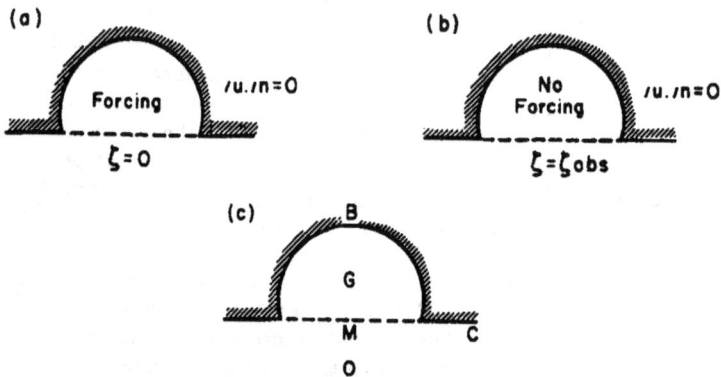

Fig. 5.16

Defant's (1961) distinction between independent and co-oscillating tides. (a) Independent tide; (b) co-oscillating tide; (c) gulf G with mouth M and coastline B set into the coastline C of ocean O. (Garrett, 1975)

With reference to an ocean and gulf, as shown in Fig.5.16, the equations of motion and continuity can be written as

$$\frac{\partial \hat{u}}{\partial t} + f\hat{u} + g\nabla(\hat{\zeta} - \hat{\zeta}_e) + \hat{F} = 0 \qquad (5.160)$$

$$\frac{\partial \hat{\zeta}}{\partial t} + \nabla \cdot (h\hat{u}) = 0 \qquad (5.161)$$

Here, $\hat{u}(x, t)$ is the current and $\hat{\zeta}(x, t)$ is the water level deviation. The depth H and the Coriolis parameter f can vary with x, and $\hat{F} = \lambda(x)\hat{u}$ is a linear bottom stress law. Let $\hat{\zeta}_e$ be the potential or equilibrium tide. Then $g\nabla\hat{\zeta}_e$ is the tide-generating force. On the coastlines B and C of the gulf and ocean (Fig.5.16), the normal velocity $\hat{u} \cdot n = 0$. Also, the normal mass flux $H\hat{u} \cdot n$ and elevation $\hat{\zeta}$ must be continuous across the mouth M of the gulf. Assume periodic solutions of the following form:

$$\hat{u} = R_e \hat{u}(x) e^{i\omega t}$$

$$\hat{\zeta} = R_e \zeta(x) e^{i\omega t} \qquad (5.162)$$

The periodic forcing is represented by

$$\hat{\zeta}_e = R_e \zeta_e(x) e^{i\omega t} \qquad (5.163)$$

Then, from eqs.(5.160) to (5.163),

$$i\omega u + fu + g\nabla(\zeta - \zeta_e) + \lambda u = 0 \qquad (5.164)$$

$$i\omega\zeta + \nabla \cdot (Hu) = 0 \qquad (5.165)$$

Next, the solutions for u and ζ will be split into two parts: $u^{(1)}$, $\zeta^{(1)}$ and $u^{(2)}$, $\zeta^{(2)}$. Let subscripts G and O represent gulf and ocean, respectively. The solutions to eqs.(5.164) and (5.165) with the boundary condition $u^{(1)} \cdot n = 0$ on M, B, and C are $u^{(1)}$ and $\zeta^{(1)}$. These solutions (with subscripts G and O) refer to the tide that would be generated in the gulf and ocean when they are assumed to be separated by a thin rigid wall. Here, n is the outward-pointing normal to the gulf and let S be the distance measured long M. The solutions of eqs.(5.164) and (5.165) without the forcing term $g\nabla\zeta_e$ and with the boundary conditions $u^{(2)} \cdot n = 0$ on B and C and $Hu^{(2)} \cdot n = F(S)$ on M are $u^{(2)}$ and $\zeta^{(2)}$. Here, $F(S)$ is a function to be determined. Since $Hu \cdot n$ is continuous, this function is the same for the gulf and the ocean.

The total solution of eqs.(5.164) and (5.165) is given then, by $u^{(1)} + u^{(2)}$ and $\zeta^{(2)}$. The condition that the elevation is continuous across the mouth of the gulf can be written as

$$\zeta_G^{(1)} + \zeta_G^{(2)} = \zeta_O^{(1)} + \zeta_O^{(2)} \quad \text{on M} \qquad (5.166)$$

Note that $\zeta_G^{(1)}$ and $\zeta_O^{(1)}$ are determined entirely from the geometry and the tide-generating force but $\zeta_G^{(2)}$ and $\zeta_O^{(2)}$ are dependent on the flux $F(S)$ on M.

The next problem is the determination of $F(S)$. For this, assume that $Hu^{(2)}\dot{n} = \delta(S - \sigma)$ produces water levels $\zeta_G(S) = K_G(S, \sigma)$ and $\zeta_O(S) = -K_O(s\sigma)$ on the gulf side and ocean side of the mouth, $\delta(S)$ being the Dirac delta function. Then eq.(5.166) becomes an integral equation for the current $F(\sigma)$,

$$\zeta_G^{(1)}(S) + \int_M K_G(S, \sigma) + F(\sigma) d\sigma = \zeta_O^{(1)}(S) - \int_M K_O(S, \sigma) F(\sigma) d\sigma$$

Write

$$F(\sigma) = F_G(\sigma) + F_O(\sigma)$$

where

$$\int_M [K_O(S,\sigma) + K_G(S,\sigma)]F_G(\sigma)d\sigma = -\zeta_G^{(1)}(S) \tag{5.167}$$

$$\int_M [K_O(S,\sigma) + K_G(S,\sigma)]F_G(\sigma)d\sigma = \zeta_O^{(1)}(S) \tag{5.168}$$

The mass flux $F(\sigma)$ produces a response $\mathbf{u}^{(2)}$ and $\zeta^{(2)}$. For the water level, write

$$\zeta_G^{(2)}(x) = \zeta_{GG}^{(2)}(x) + \zeta_{GO}^{(2)}(x) \tag{5.169}$$

where $\zeta_{GG}^{(2)}$ is the elevation produced in the gulf due to mass flux $F_G(S)$ at the mouth and $\zeta_{GO}^{(2)}$ is due to $F_O(S)$.

Garrett (1975) developed the formal solutions using the impedance approach. These solutions will not be discussed here. To summarize, the tidal response in the gulf can be interpreted as follows, in terms of the forcing functions. 1) Direct astronomical forcing produces separate responses in the gulf and the ocean when it is assumed that the mouth of the gulf is closed. 2) The water levels for the gulf and the ocean will be different at the mouth of the gulf. To balance this difference, a mass flux $F(S)$ is needed and this is obtained as the solution of eq.(5.167). 3) This mass flux $F(S)$ produces a response over the gulf and the ocean.

In terms of the response, the tide in the gulf can be assumed to be made up of the following parts: i) response $\zeta_G^{(1)}$ of the enclosed gulf to direct forcing, ii) response $\zeta_{GG}^{(2)}$ of the gulf associated with the mass flux $F_G(S)$ in the mouth required to balance $\zeta_G^{(1)}(S)$, and iii) response $\zeta(2)_{GO}$ of the gulf associated with the mass flux in the mouth required to balance $\zeta_O^{(1)}(S)$.

Although Garrett's work is of interest, like many other analytical works, it is not easy to use in practical situations. More comprehensible and amenable to practical application is the paper by Heaps (1975) which will be considered in detail below. Heaps applied his results to resonant oscillations between Lake Michigan and Green Bay. Observations showed that long wave disturbances in Lake Michigan acting at the mouth of Green Bay set up resonant co-oscillations in Green Bay with periods of 9.0 and 12.4 h. Heaps attributed these to the fundamental mode of Lake Michigan and the M_2 tide, respectively. The period of the fundamental mode of Green Bay is 10.8 h and since this lies between 9.0 and 12.4 h, the conditions for near-resonance exist in Green Bay.

Let ζ be the actual (observed) water level at the mouth of Green Bay and let a portion of it, namely $\hat{\zeta}$, be attributed to disturbances from Lake Michigan. Then $\zeta - \hat{\zeta}$ is due to the response of Green Bay. Here, the assumption is made that the presence of Green Bay has no effect on $\hat{\zeta}$. Two distinct cases can be visualized here. In the first case, the water level is prescribed at the mouth of Green Bay (i.e., assume a node at the mouth, as is traditionally done) and its resonance periods are calculated. In the second case, the total system is considered, i.e., Lake Michigan together with Green Bay, and the resonance periods of the total system are considered. The periods calculated for both cases will agree only when each mode of the total system has a node at the mouth (a node is a location where the vertical water level deviation from the equilibrium position is zero).

Observations show that the period of the fundamental mode of Green Bay (connected freely with Lake Michigan) is 10.8 h. Numerical models for the period of the fundamental mode of Green Bay (assuming a node at the mouth) gave a period of 9.75 h. These results will now be interpreted mathematically.

The equation of motion and continuity equation for free oscillations in a narrow rectangular gulf of length l and uniform depth H are, after ignoring the friction and nonlinear terms,

$$\frac{\partial u}{\partial t} = -g\frac{\partial \zeta}{\partial x} \tag{5.170}$$

$$H\frac{\partial u}{\partial x} = -\frac{\partial \zeta}{\partial t} \tag{5.171}$$

The condition of perfect reflection at the head requires

$$u = 0 \quad \text{at } x = 0 \tag{5.172}$$

The solutions for the rth mode of free oscillation can be written as

$$\begin{aligned} u &= U(x)e^{i\sigma_r t} \\ \zeta &= Z(x)e^{i\sigma_r t} \end{aligned} \tag{5.173}$$

where

$$Z \equiv H\cos(\alpha_r \xi) \tag{5.174}$$

$$U \equiv -\left(\frac{ig}{c}\right)D\sin(\alpha_r \xi) \tag{5.175}$$

$$\alpha_r \equiv \frac{\sigma_r l}{c} \tag{5.176}$$

$$c \equiv \sqrt{gH} \tag{5.177}$$

$$\xi \equiv \frac{x}{l} \tag{5.178}$$

The period of the rth mode is given by $2\pi/\sigma_r$. Here, D is a constant that denotes the amplitude of ζ at the closed end.

Taking the real parts in eqs.(5.173) and (5.175), it can be seen that high water occurs at $x = 0$ at $t = 0$. From eqs.(5.174) and (5.175),

$$\frac{\zeta}{u} = \frac{ihB_r}{c} \quad x = l \tag{5.179}$$

with

$$B_r = \cot \sigma_r \qquad (5.180)$$

Here, B_r is proportional to the impedance of the gulf evaluated at σ_r. From eq.(5.179) it can be seen that equating B_r to zero is the same as

$$\zeta = 0 \quad \text{at} \quad x = l \qquad (5.181)$$

(i.e., there is a node at the mouth). From eq.(5.180) the frequencies $\sigma_r(r = 1, 2, 3, \ldots)$ are given by $\cdot\sigma_r = 0$. Thus,

$$\alpha_r = (2r - 1)\frac{\pi}{2} \qquad (5.182)$$

From this,

$$\sigma_r = (2r - 1)\frac{\pi c}{2l} \qquad (5.183)$$

The periods are given by

$$T_r = \frac{2\pi}{\sigma_r} = \frac{4l}{c}\frac{1}{(2r - 1)} \qquad (5.184)$$

There is no *a priori* reason to assume that the free oscillations of Green Bay have a node at the mouth. Thus, in general, B_r is not zero. Assume that it is positive. Then from eq.(5.179) it can be seen that ζ leads u by a quarter period at $x = l$ and from eq.(5.180)

$$\alpha_r = (2r - 1)\frac{\pi}{2} - \tan^{-1} B_r \qquad (5.185)$$

Then,

$$\sigma_r = (2r - 1)\frac{\pi c}{2l} - \frac{c}{l}\tan^{-1} B_r, \qquad (5.186)$$

or

$$T_r = \frac{4l/c}{2r - 1 - (2/\pi)}\tan^{-1} B_r \qquad (5.187)$$

These results tend to the earlier ones if $B_r \to 0$.

Comparison of eqs.(5.184) and (5.187) shows that the periods T_r are greater when B_r is positive. In other words, when a node is assumed at the mouth, the periods of free oscillation are smaller than those when a node is not assumed at the mouth. For Green Bay, the corresponding periods for the fundamental mode are 9.75 and 10.8 h, respectively. For the Bay of Fundy, for example, the corresponding periods are 9.05 and 13.3 h, respectively.

Thus, the condition (5.179) incorporates the influence of Lake Michigan on Green Bay. Here, one assumption that was made is that B_r is real (which is a necessary condition for free oscillations). Complex B_r represents damped oscillations, the damping agent being the external water body (Lake Michigan in this example).

Heaps (1975) also considered the forced response of a gulf from the outer sea and showed that the surface elevation at the mouth may be equated to the elevation of the incident wave only when $B_r = 0$. Since, in general, $B_r \neq 0$, prescribing the water level at the mouth cannot represent a totally external input since such a prescribed water level contains a contribution from the gulf's response.

10. Numerical models for resonance calculations

10.1 Eigenvalues of the symmetrical matrix

Murty (1977) discussed numerical methods for resonance calculations in complex one-dimensional systems, such as coastal inlets and rivers, and semi-analytic techniques for two-dimensional systems, such as bays and gulfs. The discussion below will be confined to numerical techniques for two-dimensional systems. First, smaller water bodies such as harbors will be considered where the Earth's rotational effects are not significant. Then, how to include rotation will be considered. Finally, certain examples will be given.

Typical examples of resonance calculations in two-dimensional basins (assuming a node at the mouth) without including the effects of rotation are given in Loomis (1970, 1972, 1973). In this technique, one begins with the shallow-water wave equation, see Ramming and Kowalik (1980)

$$\nabla \cdot (H\nabla\zeta) = \frac{1}{g}\frac{\partial^2\zeta}{\partial t^2} \tag{5.188}$$

where ζ is the velocity potential and $H(x,y)$ is the water depth. The boundary of the water body is taken in a general way so that two types of boundaries can exist:

$$\frac{\partial\zeta}{\partial n} = 0 \quad \text{on } B_1 - \text{shoreline} \tag{5.189a}$$

and

$$\zeta = 0 \quad \text{on } B_2 - \text{open boundary} \tag{5.189b}$$

Substituting

$$\zeta(x,y,t) = e^{i\omega t}\zeta_1(x,y) \tag{5.190}$$

into eq.(5.188) gives

$$\nabla \cdot (H\nabla\zeta_1)e^{i\omega t} = -\frac{\omega^2}{g}e^{i\omega t} \tag{5.191}$$

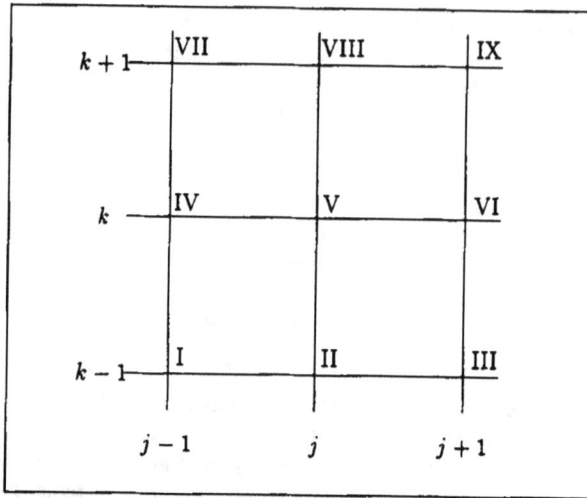

Fig. 5.17
One-dimensional enumeration in two-dimensional space.

For convenience, omit the subscript on ζ_1, divide by $e^{i\omega t}$, and define $\lambda \equiv \omega^2/g$ to give

$$\nabla \cdot (H\nabla\zeta) = \lambda\zeta \tag{5.192}$$

Note that a value λ for which there is a nonzero solution $\zeta(x,y)$ is an eigenvalue and the corresponding $\zeta(x,y)$ is an eigenfunction. The grid is shown in Fig.5.17.

In writing the finite-difference form of eq.(5.188), the following form is used,

$$\frac{\partial}{\partial x}(H\frac{\partial\zeta}{\partial x}) = \frac{1}{h}[\frac{H_{j+1,k} + H_{j,k}}{2}\frac{\zeta_{j+1,k} - \zeta_{j,k}}{h}$$

$$-\frac{H_{j-1,k} + H_{j,k}}{2}\frac{\zeta_{j-1,k} - \zeta_{j,k}}{h} \tag{5.193}$$

A similar approach is used to derive a numerical expression along the y coordinate. In the above scheme the derivative to the right from the j,k point is

$$H\frac{\partial\zeta}{\partial x} = \frac{H_{j+1,k} + H_{j,k}}{2}\frac{\zeta_{j+1,k} - \zeta_{j,k}}{h} \tag{5.194}$$

Therefore, the boundary condition $\partial\zeta/\partial x = 0$ can be fulfilled by taking the $H_{j,k} + H_{j+1,k} = 0$ so that eq. (5.194) is zero. Equating H to zero on the land gives the same difference equations as though $\partial \zeta / \partial x$ and/or $\partial \zeta / \partial y$ were set to zero. This arrangement takes care of the B_1-type boundaries. Boundaries of B_2 type $\zeta(i,j) = 0$ are handled differently.

The finite-difference form of eq.(5.192) is

$$(H_{j+1,k} + H_{j,k})\zeta_{j+1,k} + (H_{j-1,k} + H_{j,k})\zeta_{j-1,k} + (H_{j,k+1} + H_{j,k})\zeta_{j,k+1}$$

$$+(H_{j,k-1} + H_{j,k})\zeta_{j,k-1} + (\lambda 2h^2 - 4H_{j,k} - H_{j+1,k} - H_{j-1,k} - H_{j,k+1} - H_{j,k-1})\zeta_{j,k} = 0 \tag{5.195}$$

This can be rearranged in the form

$$a_{j+1,k}\zeta_{j+1,k} + a_{j-1,k}\zeta_{j-1,k} + a_{j,k+1}\zeta_{j,k+1} + a_{j,k-1}\zeta_{j,k-1}$$

$$+a_{j,k}\zeta_{j,k} = -\lambda\zeta_{j,k} \tag{5.196}$$

Let $\vec{\zeta}$ represent a vector with components $\zeta(i,j)$ and let A represent the matrix of coefficients. Then eq.(5.196) becomes

$$A\vec{\zeta} = \lambda\vec{\zeta} \tag{5.197}$$

Although eq.(5.196) gives the impression of double indexing for ζ and A, a single index $I(j,k) = 1, 2, \ldots, n$ can be defined so that a single distinct integer could be assigned to each (j,k) using capital letters for the new indices. Thus, if $(j,k) \leftrightarrow J$ and $(j,k+1) \leftrightarrow K$, then

$$a_{j,k+1} = a_{J,K} \tag{5.198}$$

which will be the coefficient of Φ_K in the Jth equation. The coefficient of Φ_J in the Kth equation will be $a_{K,J}$. It can be shown that

$$a_{J,K} = a_{K,J}$$

so that A is symmetric, which can be deduced from eq.(5.195).

To take care of a boundary condition of the type $\Phi_J = 0$, an index $J > n$ can be assigned to those boundary grid points so that when the nn matrix A is constructed, those $A_{I,J}$ values will lie outside the matrix, which in essence makes

$$\Phi_J = 0$$

Since A is symmetric, the eigenvalues of eq.(5.197) will be real. Thus, the solutions are $\lambda_1, \vec{\zeta}_1; \lambda_2\vec{\zeta}_2; \ldots; \lambda_n, \vec{\zeta}_n$ and the normal mode frequencies are given by

$$f_i = \frac{\omega_i}{2\pi} = \frac{\sqrt{g\lambda_i}}{2\pi} \qquad i = 1, 2, \ldots, n \tag{5.199}$$

The vectors $\vec{\zeta_i}$, $i = 1, 2, \ldots, n$ describe the envelope of the normal mode.

Now we shall return to the Fig.5.17 and consider a small domain of 3 grid points along the x direction and the same number of grid points along the y direction. The only internal point is $j = 2$, $k = 2$, the remaining points are the boundary points. The purpose of this exercise is to learn how to implement the boundary condition and how to change double indexing into a single index. Accounting for the boundary condition is easy; it is sufficient to set depth over the land equal to zero and the B_1 boundary condition will be fulfilled. The method of changing double indexing into a single index is demonstrated in Fig.5.17 by the Latin numerals. Let us write the numerical equations for all the points. In the four corners i.e., at the points I, III, VII and IX the sea level is zero. At the remaining points eq.(5.195) yields,

$$H_V(\zeta_V - \zeta_{II}) + 2h^2\lambda\zeta_{II} = 0 \tag{5.200a}$$

$$H_V(\zeta_V - \zeta_{IV}) + 2h^2\lambda\zeta_{IV} = 0 \tag{5.200b}$$

$$H_V(\zeta_{VI} - \zeta_V) - H_V(\zeta_V - \zeta_{IV}) + H_V(\zeta_{VIII} - \zeta_V) - H_V(\zeta_V - \zeta_{II})$$
$$+ 2h^2\lambda\zeta_V = 0 \tag{5.200c}$$

$$-H_V(\zeta_{VI} - \zeta_V) + 2h^2\lambda\zeta_{VI} = 0 \tag{5.200d}$$

$$-H_V(\zeta_{VIII} - \zeta_V) + 2h^2\lambda\zeta_{VIII} = 0 \tag{5.200e}$$

Thus we have five equations and five unknowns to define the sea level. The solution to this set exists if the coefficient determinant is equal to zero. This will also define the set of eigenfrequencies. The matrix of coefficients is symmetrical (eq.(5.201)) because the coefficients $a_{j,k} = a_{k,j}$ and so one can apply standard procedures for solving the symmetrical determinant

$$\begin{pmatrix} \alpha - H_V & 0 & H_V & 0 & 0 \\ 0 & \alpha - H_V & H_V & 0 & 0 \\ H_V & H_V & \alpha - 4H_V & H_V & H_V \\ 0 & 0 & H_V & \alpha - H_V & 0 \\ 0 & 0 & H_V & 0 & \alpha - H_V \end{pmatrix} \begin{pmatrix} \zeta_{II} \\ \zeta_{IV} \\ \zeta_V \\ \zeta_{VI} \\ \zeta_{VIII} \end{pmatrix} = \begin{pmatrix} 0 \\ 0 \\ 0 \\ 0 \\ 0 \end{pmatrix} \tag{5.201}$$

Here $\alpha = 2h^2\lambda$.

10.2 Investigation of the free oscillations by forcing function

Motion generated in the ocean by wind or atmospheric pressure can be considered as a superposition of forced and free oscillations. Whenever the period of an external force coincides with the free oscillation mode, the resulting motion is amplified. We shall describe an approach which utilizes this phenomena by studying response of the water body to forcing. The method was conceived by Radach (1971) and used by Papa (1977) to compute the free oscillations in the Ligurian Sea, by Wüber and Krauss (1979) for the Baltic Sea, and by Gotlib and Kagan (1981) for the World Ocean. The set of equations (3.1)–(3.3) is linearized and frictional effects neglected,

$$\frac{\partial u}{\partial t} - fv = -g\frac{\partial \zeta}{\partial x} \qquad (5.202)$$

$$\frac{\partial v}{\partial t} + fu = -g\frac{\partial \zeta}{\partial y} \qquad (5.203)$$

$$\frac{\partial \zeta}{\partial t} = -\frac{\partial}{\partial x}uD - \frac{\partial}{\partial y}vD \qquad (5.204)$$

To force oscillation through this system of equations two methods are used: a) application of the forcing function similar to atmospheric pressure over the whole integration domain, and b) forcing through the open boundary. For the first method to be implemented the forcing function may be chosen similar to the atmospheric pressure term and in eqs.(5.202) and (5.203) gradient of pressure p_a should be introduced. Wüber and Krauss (1979) experimented with the pressure p_a making it both function of time and space.

$$p_a(x,y,t) = \begin{cases} (1 - \cos \pi t/T)F(x,y), & \text{for } \leq 0 \leq T; \\ 0, & \text{otherwise} \end{cases} \qquad (5.205)$$

Function $F(x,y)$ is chosen so that the distinct space modes will be forced and cos or sin dependence will usually do a good job. Our experience with this method indicates that it is very effective in studying the longest periods modes. Usually, a pressure field was applied initially for about 30 hrs, and then it was set to zero. Because, the model is inviscid and conserves energy the system was allowed to oscillate for the long enough time so that the oscillations were redistributed to the various portions of the free wave spectra. The time series of sea level and velocity were chosen and Fourier analysis performed to find out in which portion of the spectra the oscillations were amplified.

When part of the domain has an open boundary the system can be forced through this boundary by changing in time sea level or velocity. Setting at the open boundary oscillation having unit amplitude and having periods, e.g., of 1 to 30 hrs in hourly increment one can investigate response of the system by using eqs.(5.202)–(5.204). Again, time series of the sea level and velocity are used for the Fourier analysis and amplitude of the signal versus frequency will show the behavior of the system. Such a preliminary investigation if repeated in the vicinity of observed peak with a shorter time increment, will give the exact location of the maximum (Kowalik, 1993).

Computations were carried out by numerical equation (3.95)–(3.97) written on the staggered grid given in Fig.3.10.

$$\frac{u_{j,k}^{m+1} - u_{j,k}^{m}}{T} - f\bar{v}^{u,m} = -\frac{g}{h}(\zeta_{j,k}^{m} - \zeta_{j-1,k}^{m}) \qquad (5.206)$$

$$\frac{v_{j,k}^{m+1} - v_{j,k}^{m}}{T} + f\bar{u}^{v,m+1} - \frac{g}{h}(\zeta_{j,k+1}^{m} - \zeta_{j,k}^{m}) \qquad (5.207)$$

$$\frac{\zeta_{j,k}^{m+1} - \zeta_{j,k}^{m}}{T} = -\frac{1}{h}(u_{j+1,k}^{m+1}D_{u,j+1,k}^{m} - u_{j,k}^{m+1}D_{u,j,k}^{m}) - \frac{1}{h}(v_{j,k}^{m+1}D_{v,j,k}^{m} - v_{j,k-1}^{m+1}D_{v,j,k-1}^{m})$$
$$(5.208)$$

We shall demonstrate an application of the method through two examples: the longest mode of the free gravitational oscillations in a rectangular water body of a constant depth, and vorticity waves trapped around an oceanic plateau. In the first experiment, the rectangular water body with a constant depth equal to 139.83 m and elongated in the north direction with a length equal to 800 km and width equal to 400 km, is open at the northern boundary. This domain is covered by a numerical grid of 40 km size. The forcing function is taken to generate basic open–mouth mode of oscillation.

$$p_a = 10^4(1 - \cos\frac{\pi(m - 1)T}{30\text{hr}}) \times (1 + \cos\frac{\pi(k - 1)}{20}) \qquad (5.209)$$

Here coefficient 10^4 is chosen to obtain reasonable values of the sea level. The time portion of the forcing function grows from zero at the initial moment (time step $m = 1$) to 2 when time of the process equals to 30 hrs ($(m - 1)T = 30$ hrs). The space portion is a function of the y coordinate (k index). At the southern boundary ($k = 1$) it attains a maximum, and at the northern (open) boundary ($k = 21$) it is zero. At the land boundary the velocity normal to the boundary is zero and at the northern open boundary the sea level is set to zero. Integration is performed in time for 30 hrs of the process and then the forcing function is set to zero, but integration continues for 1024 hrs. Time series data are recorded and analyzed by FFT routine. First, the case without Coriolis force is calculated.

In Fig.5.18 the sea level amplitude as a function frequency is depicted. The maximum energy in this plot is at the 24 hr period, and the second maximum is close to 8 hr period. These are the basic periods of the open mouth mode of the free oscillations. Periods are easily obtained from the formula (5.5). For the longest period, introducing depth and length into this formula, we arrive at

$$T_1 = \frac{4l}{\sqrt{gH}} = \frac{4 \times 800 \times 10^3}{\sqrt{9.81 \times 139.83}} = 24\text{hrs} \qquad (5.210)$$

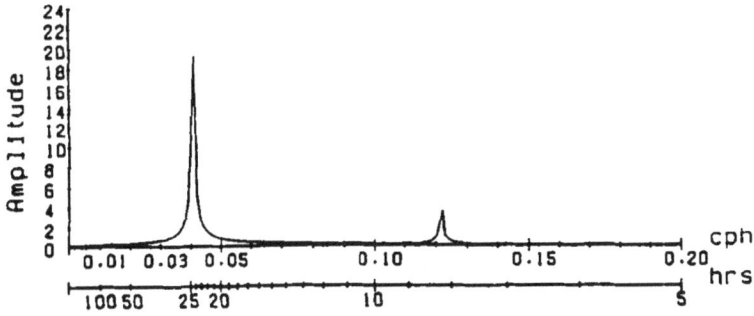

Fig.5.18

Spectra of the sea level amplitude in cm.

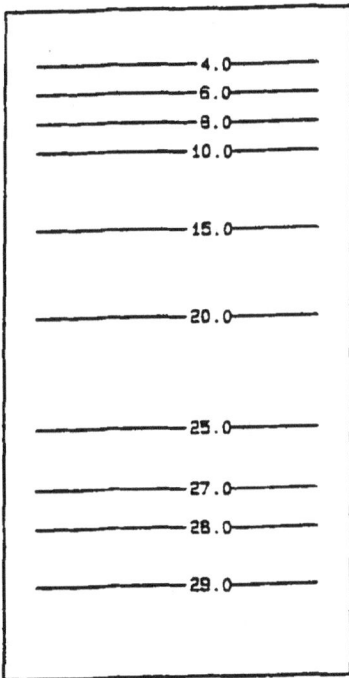

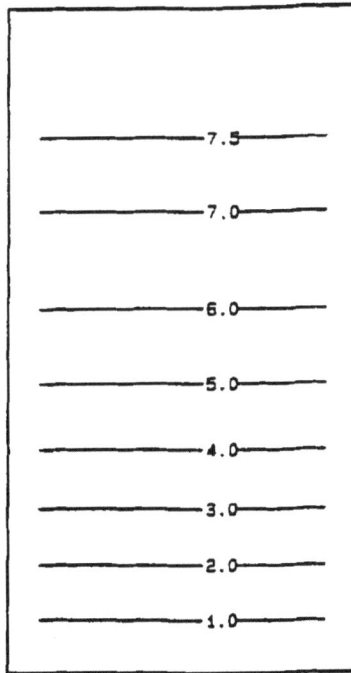

Fig. 5.19a   Fig. 5.19b

Amplitude of the sea level in cm (a), and amplitude of velocity in cm/s (b),
of the free oscillation. Northern (upper) boundary is open.

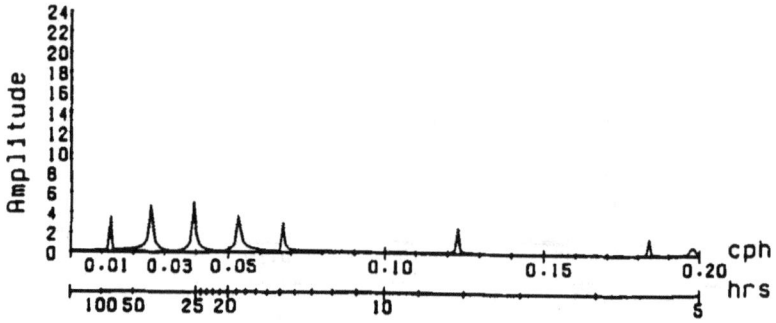

Fig. 5.20

Spectra of the sea level amplitude in cm.

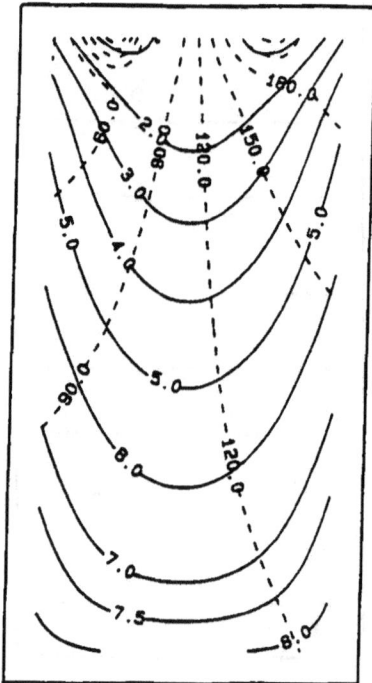

Fig. 5.21a

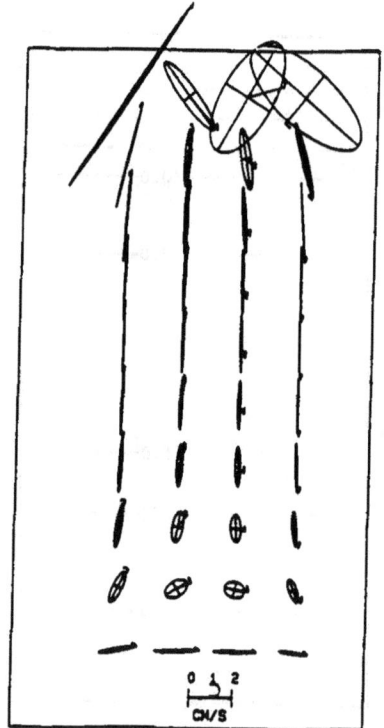

Fig. 5.21b

Amplitude of the sea level in cm – continuous line, and phase – dash line (a); velocity ellipses in cm/s (b), for the 25 hrs period.

The sea level amplitude and velocity are given in Fig.5.19a and Fig.5.19b respectively. At the open end, the sea level is diminishing to zero, while velocity has the largest value. Inclusion of the Coriolis force complicates the free oscillation picture. Amplitude spectra show the maximum at about 25 hrs, but it is not dominating over the other peaks – Fig.5.20. The sea level amplitude and velocity ellipses for the 25 hr period are plotted in Fig.5.21a and Fig.5.21b. Sea level increases from the open boundary towards the south, but the motion is no longer rectilinear; due to Coriolis force the sea level isolines describe a semi-ellipse, with an amphidromic point at the open boundary. The phase, or the time of occurrence of a given amplitude at a given location, is depicted by broken line and the amplitudes are shown by continuous line. Here, $360^0 = 26$ hrs, and the phase lines show that the wave rotates around the amphidromic point in a counter-clockwise direction. Actually, the boundary condition at the open boundary ($\zeta = 0$) is fulfilled exactly only at the amphidromic point where sea level diminishes to zero. Therefore, to the right and to the left from this point the phase distortion is observed. The velocity ellipse demonstrates, that the presence of the Coriolis force has changed the rectilinear N–S motion to the elliptical motion. Close to southern boundary the motion changed from N–S direction to the E–W direction. Small arrows attached to the ellipses are indicators of the rotation.

Hunkins (1986) investigated resonant forcing of the topographic wave by the diurnal tide at the Yermak Plateau. For this purpose he considered an idealized submarine plateau – Fig.5.22 and Fig.5.23. This Plateau's inner diameter is 140 km and the depth is equal to 800 m. From the inner diameter to the outer diameter of 280 km the depth increases parabolically up to 3200 m, and outside outer diameter the depth is constant. Hunkins solved analytically an equation for the vorticity waves and obtained a resonant period of about 24 hrs. To investigate this mode of the free oscillations we shall apply the system of eqs.(5.202)–(5.204). The system is forced through the southern open boundary – Fig.5.23. At the southern boundary of the domain, the sea level having unit amplitude and having periods, of 1 to 30 hrs in hourly increment is applied. The remaining boundaries are closed and the reflection boundary condition is used $\zeta = 0$. After initial period of 30 hrs the model is run for 1024 hrs (the wave at the southern boundary are applied continuously), and this series of hourly data is used for the Fourier analysis. The horizontal grid distance is 10 km, therefore to diminish the amount of data, sea level and velocity are recorded at every second grid point. The response of the system is investigated by the amplitude spectra and by the velocity spectra. The spectra are first constructed at every point and afterwards an average spectra for the whole integration domain is constructed. The average spectra are depicted in Fig.5.24 for the sea level and in Fig.5.25 for the velocity. The maxima in the both spectra are somewhat differently distributed in the frequency domain. Most of the peaks in Fig.5.24 are due to the forcing function, which clearly has periods in hourly increment.

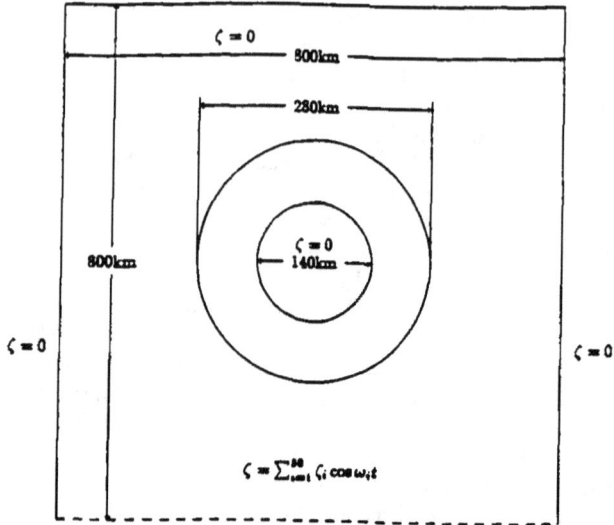

Fig. 5.22
Domain and boundary conditions for the undersea plateau experiment.

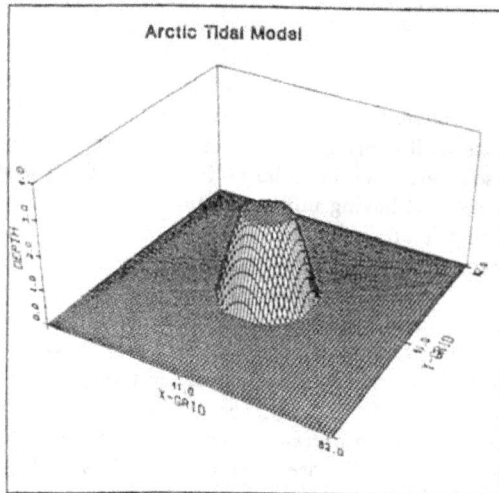

Fig. 5.23
Three-dimensional view of the submarine plateau.

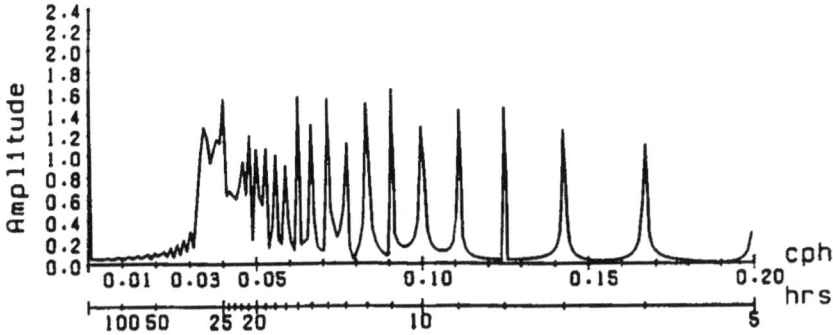

Fig. 5.24

Sea level (in cm) spectra. Peaks located at hourly distance are due to forcing function.

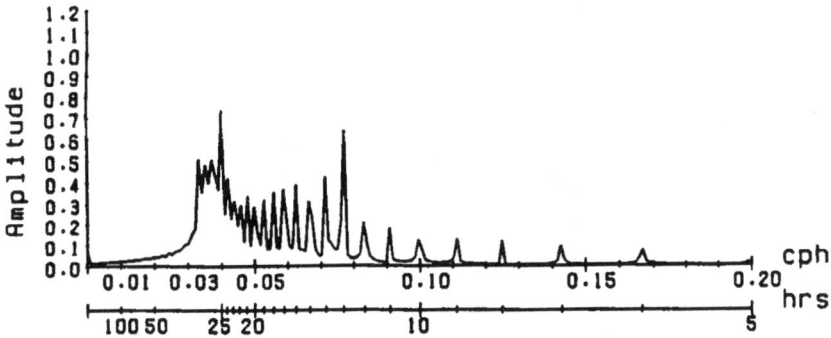

Fig. 5.25

Velocity (cm/s) spectra. Peaks located at hourly distance are due to forcing function.

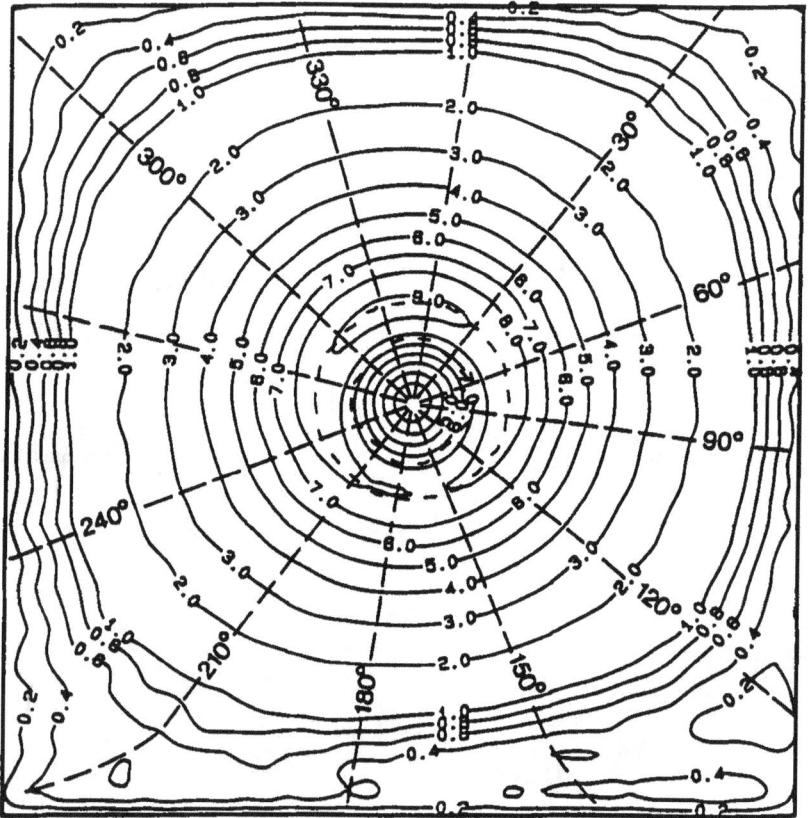

Fig. 5.26
Topographically trapped wave by submarine plateau.
Sea level amplitude – continuous lines. Phase – broken lines.
Two broken circular lines denote inner and outer diameter of the submarine plateau.

Both in the sea level and in velocity spectra the periods of about 25 hrs is clearly seen, and this period dominates the currents. The sea level amplitude in the spectra peaks is close to 1 cm, hence the amplification is not seen, but these are average spectra. In Fig.5.26 we have depicted the contours of the sea level amplitude and phase for this period. This wave travels around plateau in 25 hrs. It has amphidromic point at the plateau's center and the maximum of sea level amplitude is located at the outer diameter of the plateau. This is obviously topographically trapped wave propagating clockwise around the submarine plateau.

10.3 Platzman's resonance–iteration approach

Next, numerical models for resonance calculations when the Coriolis terms are included will be considered. To this author's knowledge, such calculations were originally made by Platzman (1972), and some modifications and alternate techniques were suggested by Hamblin (1972) and Rao (1974). Here, the so-called resonance-iteration method introduced by Platzman will be briefly discussed.

The equations of motion and continuity for an inviscid, homogeneous fluid are in the vector form

$$\frac{\partial \mathbf{V}}{\partial t} = -g\nabla(\zeta - \bar{\zeta}) - 2\Omega X \mathbf{V} \tag{5.211}$$

$$\frac{\partial \zeta}{\partial t} = -\nabla \cdot h\mathbf{V} \tag{5.212}$$

In the above equation, $g\nabla\bar{\zeta}$ is the tidal force.

For convenience, a matrix column $a \equiv (\zeta, \mathbf{V})$ is defined. Then eq.(5.211) and (5.212) become

$$\frac{\partial a}{\partial t} = \mathcal{L}(a - \bar{a})$$

$$\mathcal{L} \equiv i \begin{bmatrix} 0 & \nabla H \\ g\nabla & 2\Omega X \end{bmatrix} \tag{5.213}$$

where

$$\bar{a} \equiv (\bar{\zeta}, 0).$$

The boundary conditions considered here are of the adiabatic type, i.e., normal component of

$$HV = 0 \tag{5.214}$$

or

$$\zeta = 0. \tag{5.215}$$

These boundary conditions do not permit flux of energy into or out of the region at any time. Equation (5.214) applies to solid boundaries and (5.215) is used for the openings.

The properties of the operator $\mathcal{L}$ are determined by the scalar product of two arbitrary vectors a' and a,

$$\{a', a\} = \int (g\zeta'\zeta + H\mathbf{V}'\mathbf{V}) \, dS = 0 \tag{5.216}$$

where dS is an area element and the integration is carried out around the boundary. It can be shown that the system (5.213–5.215) is self-adjoint and hence, the eigenvalues of $\mathcal{L}$ are real. If these eigenvalues are denoted by σ and the corresponding eigenvectors by $A \equiv (z, \mathbf{V})$, then

$$\mathcal{L}A = \sigma A. \qquad (5.217)$$

Then $Ae^{i\sigma t}$ is a normal mode solution of eq.(5.214) with $\bar{a} = 0$.

Noting that $\mathcal{L}$ is purely imaginary, real representation of the normal mode with frequency $|\sigma|$ is

$$Ae^{i\sigma t} \pm A^* e^{-i\sigma t}$$

where A^* is a complex conjugate of A.

The scalar product of A^* with each side of eq.(5.217) gives the primitive Rayleigh quotient

$$\sigma = \frac{\{A^*, \mathcal{L}A\}}{\{A^*, A\}} \qquad (5.218)$$

The numerator and denominator are

$$\{A^*, \mathcal{L}A\} = i \int [gz^* \nabla \cdot H\mathbf{V} + h\mathbf{V}^* \cdot (g\nabla z + 2\Omega X \mathbf{V})] \, dS$$

$$\{A^*, A\} = \int (g|z|^2 + H|\mathbf{V}|^2) \, dS$$

In using eq.(5.218) in estimating σ from the approximation to $A \equiv (z, \mathbf{V})$, one can take in principle any z and $\mathbf{V}$ that satisfies the boundary conditions. Other constraints are the kinematic and dynamic parts of eq.(5.217) which are, respectively,

$$\sigma z = i\nabla \cdot H\mathbf{V} \qquad (5.219)$$

$$(\sigma - 2\Omega i X)\mathbf{V} = ig\nabla z \qquad (5.220)$$

The parameter z can be explicitly expressed in terms of $\mathbf{V}$ in the kinematic part; the dynamic part can be manipulated to express $\mathbf{V}$ explicitly in terms of z:

$$\sigma \mathbf{V} = iFg\nabla z$$

$$F = \left(1 - \frac{4\Omega^2}{\sigma^2}\right)^{-1} \left[1 + \left(\frac{2\Omega i}{i}\right)X\right] \qquad (5.221)$$

The operator $\sigma - 2\Omega X$ is nonsingular except for the trivial case when σ and Ω are zero. Hence, in the operator F, the latitudes (if any) where $4\Omega^2 = \sigma^2$ are only apparent singularities.

Irish and Platzman (1972) pointed out that three alternatives are possible for the use of the Rayleigh quotient. 1) The kinematic and dynamic constraints are not used. Then the relevant operator is $\mathcal{L}$ in eq.(5.213). The normal mode frequencies σ are identical to the eigenvalues of this operator and hence are equal to the stationary values of the corresponding Rayleigh quotient (5.218). 2) The dynamic constraint is imposed and the velocities are eliminated in the traditional way. Then the relevant operator is the second order Laplacian tidal operator and its spectrum is parametrically dependent on σ. In this case, the Rayleigh quotient becomes an explicit equation quadratic in σ with coefficients that are homogeneous quadratic functions of the particle velocities or displacements. Platzman (1972) used the Richardson lattice for space discretization and a leapfrog scheme for time discretization. Murty and Taylor (1975) computed two-dimensional free oscillations of Georgian Bay (on Lake Huron), Green Bay (on Lake Michigan), and Chedabucto Bay (in Nova Scotia). However, the method used here is not as elegant as that of Platzman (1972) but is more straightforward and can be used only when the frequencies of the gravitational modes and those of the rotational modes are well separated.

The free oscillations (normal modes) of any water body can be conveniently classified into gravitational and rotational modes. The gravitational modes owe their existence to disturbances of the mass field whereas the rotational modes primarily depend on deformations of the potential vorticity (Platzman, 1974). A spectrum of gravitational modes would exist in the absence of the Coriolis force and, similarly, a spectrum of rotational modes would exist in the absence of gravity. Gravitational modes cluster at the short-period end of the spectrum and rotational modes cluster at the long-period end.

Identification of a given mode (whether it is of gravitational or rotational type) can be made on the basis of several factors (Platzman, 1975). The distribution of the periods provides some basis for identification. As a typical length of the disturbances (that excites these modes) increases, the period of a rotational mode will decrease and the period of a gravitational mode will increase. Since the length scale of the water body is an upper limit of the length scale of the disturbance, there is a limiting minimum value for the rotational mode periods and a limiting maximum value for the gravitational mode periods.

Murty and Taylor (1975) used the time-dependent two-dimensional equations of motion and continuity and solved them numerically. For the basic condition for calculating the free oscillations, they invoked the standard (but not strictly correct) condition of zero surface elevation at the mouth. The free oscillations were excited by applying a wind stress for some length of time over the bay. The resonant frequencies were abstracted from the numerical model output by spectrally analyzing the computed water level at selected locations in the bay. One disadvantage of this technique is that all the modes may not systematically be identified. However, the results for Green Bay and Georgian Bay agree with the results of other calculations.

Tronson (1975) calculated the normal modes of the South Australian Gulf system.

Heath (1975) calculated the normal modes of Lake Wakatipu in New Zealand. One very interesting feature is that the most energetic mode is not the fundamental but the second one. Quoting from Heath (1975, p. 235),

> "One reason generally given for most of the seiche energy in a lake being in the fundamental mode is that this mode has the longest wavelength and is therefore less affected by scattering than the higher order modes. It can be shown by Fourier decomposition of the initial shape of the lake level that in a unidirectional lake most of the seiche energy might also be expected to be initially in the fundamental mode. However, further examination of this method of generation indicates that this dog-leg shaped lake might respond like three uni-directional lakes acting separately with the elevation distribution of the fundamental mode in the two meridional legs being similar to that of the second mode of the entire lake."

11. Kelvin waves, Sverdrup waves, and Poincaré waves

There are classes of normal mode solutions with special properties that have been referred to as Kelvin waves Sverdrup waves, and Poincaré waves. These wave types have been frequently invoked to explain the tidal phenomena in water bodies. For an excellent review on this topic, see Platzman (1971). Other relevant works are Defant (1961), Proudman (1953), Voyt (1974), and LeBlond and Mysak (1978). Simons (1980) has given a rather concise summary, and this discussion will essentially follow his line of argument.

Earlier, the gravitational and rotational modes were introduced, also referred to as oscillations of the first class (OFC) and oscillations of the second class (OSC). It was also pointed out that OFC are motions with large divergence whereas OSC are essentially nondivergent. For introducing the concepts of different types of wave motion mentioned here, discussion begins with the linearized version of the vertically integrated equations, and these will be applied to a rectangular basin of uniform depth:

$$\frac{\partial U}{\partial t} - fV = -c^2 \frac{\partial \zeta}{\partial x}$$
$$\frac{\partial V}{\partial t} + fU = -c^2 \frac{\partial \zeta}{\partial y}$$
$$\frac{\partial \zeta}{\partial t} + \frac{\partial U}{\partial x} + \frac{\partial V}{\partial y} = 0$$

$$(5.222)$$

where U and V are the x and y components of the volume transport, ζ is the deviation of the water level from its equilibrium position, f if the Coriolis parameter, and $c^2 = gH$ where H is the uniform water depth.

Assuming a time factor $e^{i\sigma t}$ where σ is the frequency, and eliminating U and V from eq.(5.222) gives the wave equation

$$(\sigma^2 - f^2)\zeta + c^2\nabla^2\zeta = 0 \tag{5.223}$$

The boundary condition of zero normal transport to a boundary can be stated as

$$f\frac{\partial\zeta}{\partial S} + i\sigma\frac{\partial\zeta}{\partial n} = 0 \tag{5.224}$$

where S and n are the coordinates along and perpendicular, respectively, to the wall. Equation (5.223) can be satisfied by

$$\exp[i(kx + ly)]$$

where the frequency σ is given by

$$\sigma^2 = f^2 + c^2(k^2 + l^2) \tag{5.225}$$

For the nonrotating case, eq.(5.224) is easily satisfied by standing waves with wave numbers $k = m\pi/L$ and $l = n\pi/B$ where m and n are integers and L and B are the length and breadth of the basin, respectively. However, in the rotating case, because of the compacted nature of the boundary condition (5.224) it is difficult to determine the normal modes. However, even in the rotating case, for an infinitely long channel, there are some elementary wave solutions that do not satisfy the boundary conditions.

In eq.(5.222), if $V = 0$, then the solutions to the resulting equations are

$$\zeta = \zeta_0 e^{-fy/c}e^{ik(x-ct)}$$
$$U = -\frac{c^2}{f}\frac{\partial\zeta}{\partial y} \tag{5.226}$$
$$V = 0.$$

The wave speed is the same for the nonrotating and the rotating cases, but in the latter, the wave amplitude decreases exponentially from right to left for an observer looking in the direction of wave propagation. The rate of decrease of the amplitude from right to left is proportional to c/f, the Rossby radius of deformation. These waves are known as Kelvin waves.

Another elementary solution can be obtained by setting $\partial/\partial x$ and $\partial/\partial y$ to zero. This will result in inertial oscillations with frequency f (note that ζ is zero for these and there is only horizontal motion). If only the gradients in one horizontal direction are ignored e.g., $\partial/\partial x = 0$, then the solutions are

$$\zeta = \zeta_0 e^{i(ly-\sigma t)}$$
$$\sigma = f^2 + c^2 l^2 \tag{5.227}$$

These waves are referred to as Sverdrup waves and have horizontal crests. For a straight coast parallel to the x axis, two Sverdrup waves travelling in opposite directions may be combined to form a standing wave that satisfies the boundary condition (5.224). Thus, standing Sverdrup waves with wave numbers $l = n\pi/\beta$ are the normal modes of an infinitely long rotating channel.

The more general solutions of eq.(5.223) are known as Poincaré waves. Pairs of progressive Poincaré waves can be combined into standing waves that display cellular patterns. For an infinite channel, in analogy with Sverdrup waves, Poincaré waves can be made to satisfy the boundary conditions by properly choosing the transverse wave numbers.

At a transverse barrier in the channel, none of these waves could be made to satisfy the boundary condition (5.224). Hence, the analytical determination of the normal modes of a rotating rectangular bay is more difficult, and it is convenient to resort to numerical techniques.

12. Influence of ice cover on normal modes

There are several water bodies in the world in which the ice cover alters their topography (i.e., shape of the coastline, water depths, number of openings that provide coupling to the external ocean) and thus the frequencies of the normal modes. Here, we will consider the modification of the normal modes (and tides) in Tuktoyaktuk Harbour, Northwest Territories, by ice cover. Tuktoyaktuk Harbour is located on the southern part of the Beaufort Sea (Fig.5.27) The 200 m depth contour is assumed to be the seaward limit of the Herschel–Bathurst Shelf which has an average depth of about 50 m and an average width of 108 km. It is assumed that the Amundsen Gulf extends from the narrowest portion of the Dolphin and Union Strait northwestward to the far edge of the Herschel–Bathurst Shelf, i.e., about 556 km, with an average depth of 200 m. Tuktoyaktuk Harbour is connected to the Beaufort Sea through the shallow Kugmallit Bay. For the computation of the free oscillations (normal modes) of Kugmallit Bay, the relevant dimensions are not certain (a length of 25 km and an average depth of 2 m was assumed). Note that this average depth of 2 m could be reduced to 1 m by ice cover, or increased to 3 m by a positive storm surge. Bilelo and Bates (1971) showed that the thickness of the ice cover could average 1.8 m in Kugmallit Bay by the end of winter, and this cover usually exists seaward of the shelf. The configuration of Kugmallit Bay may be altered by ice cover (Reimnitz and Barnes, 1974) because parts of the bay are very shallow, and especially if a feature like 'bottom fast ice' is presented landward of the 2 m depth contour. An ice cover can affect the topography, not only by changing the water depth but also by changing the surface area, cross-sectional area, and width of the openings. There is a difference in the ice cover between Kugmallit Bay and Tuktoyaktuk Harbour. For a brief period in early summer, the ice in Tuktoyaktuk Harbour remains landfast whereas the landfast ice of the Bay undergoes breakup.

Tuktoyaktuk Harbour is located on the eastern edge of the Mackenzie River

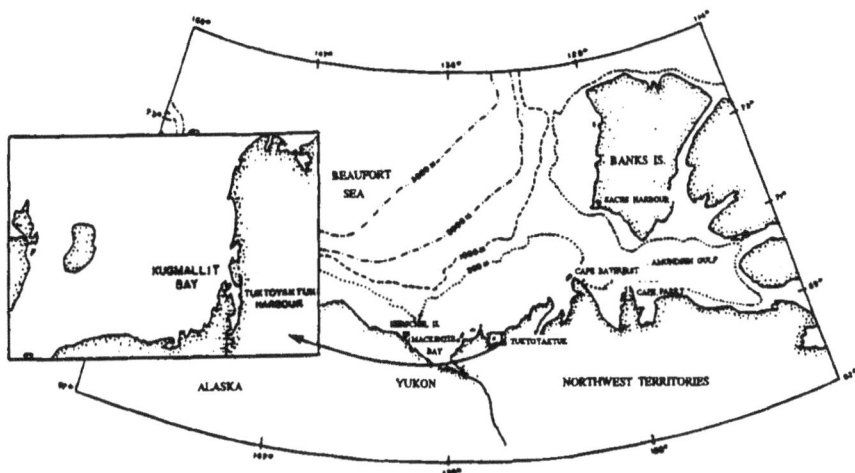

Fig. 5.27
Geography of the southern part of the Beaufort Sea area.
Inset shows the Kugmallit Bay and Tuktoyaktuk Harbour.

delta and is connected through the eastern and western entrances to Kugmallit Bay (Fig.5.27). The harbor, which is not yet completely surveyed, is at least 6.5 km long (and perhaps as long as 11.5 km), is 1.8 km wide, and attains a maximum depth of 30 m. The limiting depth of about 4 m occurs just seaward of the entrances to the harbor. The maximum depth in the western entrance is only 40 m wide. For the computation of the normal modes, the length of the harbor was taken as 11 km and the average depth as 5 m. Ice cover can alter the shape, depth, and number of ports of the harbor that provide coupling to Kugmallit Bay. The wetted shape of the harbor and the number and dimensions of the ports are influenced by the ice cover. Note that the western entrance to the harbor might be closed entirely by ice. The depth of the harbor increases after the disappearance of the ice cover, and also during storm surges.

Estimates will be provided for the following resonant oscillations: one-quarter resonant wavelength of the Herschel–Bathurst Shelf, the fundamental mode edge wave on the shelf, the longitudinal and transverse modes of Kugmallit Bay, the modes of the Amundsen Gulf, and the fundamental and Helmholtz modes of Tuk-toyaktuk Harbour.

For the shelf, the one-quarter resonant wavelength frequency is equal to 0.18 cph as computed from $\sqrt{gH}/4B$ where H=average depth of water=50 m, B=average width of the shelf=108 km, and g=acceleration due to gravity=980cm/s^2.

An atmospheric disturbance travelling with a speed U would generate a fundamental edge wave mode whose frequency ν is given by

$$\nu = \left(\frac{gs}{U} - g \right) 2\pi$$

where the bottom slope s is of the order of 10^{-3} and the Coriolis parameter f is 1.3710^{-4}s^{-1}. The computed edge wave mode frequency is 0.45 cph. For four disturbances in 1961, Hunkins (1964) gave values of U ranging from 6 to 15 m·s^{-1}. These values give an edge wave frequency of 0.64 to 0.45 cph. The 15 m·s^{-1} value for U is based on the observation by Schafer (1966) that the storm of October 1963 travelled 150 km in three hours near Point Barrow.

The open (longitudinal) mode of Kugmallit Bay during ice cover may have a frequency which lies in the semidiurnal band. The closed (transverse) mode of this bay appears to have a frequency of about 0.3 cph. The frequency of the open mode during the absence of ice cover is about 0.12 cph after applying a mouth correction (Defant, 1961).

For the Amundsen Gulf, the length was taken as 556 km and an average depth of 200 m. From the Merian formula, the frequency for the non-rotating case was estimated to be 0.073 cph (13.6-h period). Rotation (Rao, 1966) would increase the frequency to 0.07 cph, which is quite close to the semidiurnal period.

For the harbor, the frequency of the fundamental mode for the closed case is estimated to be about 1.0 cph in the absence of ice cover and in the absence of a surge. As noted previously, ice cover reduces the water depth and a positive surge increases the water depth. With ice cover and no surge, the frequency of the closed mode is estimated to be 0.9 cph; without ice cover but with surge, the frequency is 1.1 cph.

Helmholtz resonance represents the balance between the kinetic energy of the water flowing in through a narrow connecting channel and the potential energy from the rise in the mean water level within the harbor. It is an additional gravitational mode of substantially longer period than the fundamental free oscillation. Let L_1 and L_2 be the lengths of the western and eastern entrances and s_1 and s_2 their areas of cross-section. Then the frequency of the Helmholtz (ν_H) mode is given by

$$\nu_H = \sqrt{\frac{g}{A} \left(\frac{s_1}{L_1} + \frac{s_2}{L_2} \right)^{1/2}} \ \text{s}^{-1}$$

where A is the surface area of the water body. Table 5.5 lists the frequencies of the various modes estimated in this section.

The following results were deduced from the various spectra as shown in Figures 3 to 11 of Murty (1985). There is considerable noise in the tidal data in summer and the noise is considerably reduced during ice cover. Periods of reduced noise in summer appear to be correlated with easterly and offshore winds. It is concluded that the input of noise mainly occurs over and inshore of the shelf. It seems likely that the reduction of noise in winter is characteristic of fast ice areas.

Table 5.5 Frequencies (cph) of the various modes.

Mode	Ice-free	Ice cover
Herschel–Bathurst Shelf		
1/4 resonant wave length	0.18	0.18
fundamental mode edge wave	0.45	0.45
Kugmallit Bay		
open, longitudinal	0.12	0.087
closed, transverse	0.3	not seen
Tuktoyaktuk Harbour		
closed case (1st mode)	1.0	0.9
Helmholtz mode	0.6 to 0.7	unknown

CHAPTER VI

STEADY STATE PROCESSES

1. Oceanographic examples

We shall give here three examples of the problems to be solved by the steady state numerical methods. The first one is a diagnostic modeling which is truly a steady state problem. The second one arises from the solution for the time dependent stream function. This stream function at each new time step is described by an elliptical problem and therefore, can be solved through the steady state approach. We also show that to every steady state problem there is an associated unsteady problem and solution of the steady state is derived in the limit when $t \to \infty$. The third problem to be solved is an oscillatory motion when the time dependence is eliminated by introducing solution in the form $\exp(i\omega t)$.

1.1 Diagnostic modeling

Traditionally oceanographers relied on large data sets of observed temperature and salinity fields to define the field of motion in the ocean. A dynamical method developed many years ago served this purpose. For calculation of velocity distribution using this method one has to rely on finding the level of no motion in the ocean. A diagnostic model assumes a knowledge of the density field in the ocean and the wind field at the ocean surface, and solves for distribution of the currents as a boundary value problem, thus avoiding complications related to the level of no motion. The method was stated by Friedrich (1966), but developed and widely used by Sarkisyan (1977) and his colleagues. The method spurred development of numerical methods for the steady state problems. The so-called Il'in's solution has been found and is often referred to as Il'in's problem (Gupta, 1983). This solution we have outlined in Ch.II, Sec.9.1. The construction of the diagnostic model is described in Ch.I, Sec 3.1 under the heading of "steady state Ekman problem". In fact this model is very closely linked to the Ekman problem, especially to the vertically integrated equation, introduced by Ekman in 1923. Starting from the Ekman type of equation (with the density and wind given from the measurements) we find only one unknown in the set of equations that is the sea level distribution. To construct an additional equation for the sea level, the set of equations is vertically integrated and a stream function is introduced. The introduction of the stream function in the steady state equation is possible through the vertically integrated equation of continuity. The method is so simple and the physics is so clear that it still finds frequent application. A very good description of this method is given by Galt (1984). The core of the method is the stream function equation (1.115).

$$\frac{\partial}{\partial \lambda} k_\lambda \frac{\partial \Psi}{\partial \lambda} + \frac{\partial}{\partial \phi} k_\phi \frac{\partial \Psi}{\partial \phi} + \frac{\partial \alpha}{\partial \phi} \frac{\partial \Psi}{\partial \lambda} - \frac{\partial \alpha}{\partial \lambda} \frac{\partial \Psi}{\partial \phi}$$

$$= \frac{\partial}{\partial \lambda} (R\tau_\phi^s / H) - \frac{\partial}{\partial \phi} (R\tau_\lambda^s \cos \phi / H) + \frac{\partial B_\lambda}{\partial \phi} - \frac{\partial B_\phi}{\partial \lambda} \qquad (6.1)$$

A unique solution to this equation can be obtained if the stream function is defined along the boundary. At the shore line the normal component of the transport equals to zero and the boundary condition for the stream function states

$$\Psi = 0 \qquad \text{along the boundary} \qquad (6.2)$$

Along the open boundaries the inflow or outflow should be specified. If there are islands located in the integration domain, it is no longer a simply connected region but becomes a multiply connected domain and the method to construct the boundary conditions and solution for such domains are outlined in Ch.IV, Sec.5.4, and in Ramming and Kowalik (1980).

The coefficients at the highest derivatives k_λ, k_ϕ are function of the bottom friction. They are usually very small. Therefore, the terms containing friction in eq.(6.1) can be neglected or moved to the right-hand side in eq.(6.1). Solution to such a problem is searched by an approximation method. Initially, only the first order equation is solved,

$$\frac{\partial \alpha}{\partial \phi} \frac{\partial \Psi}{\partial \lambda} - \frac{\partial \alpha}{\partial \lambda} \frac{\partial \Psi}{\partial \phi}$$

$$= \frac{\partial}{\partial \lambda} (R\tau_\phi^s / H) - \frac{\partial}{\partial \phi} (R\tau_\lambda^s \cos \phi / H) + \frac{\partial B_\lambda}{\partial \phi} - \frac{\partial B_\phi}{\partial \lambda} \qquad (6.3)$$

Denoting solution of eq.(6.3) as Ψ_1 one can add at the next step, the frictional terms to the right hand side of the above equation as a small correction, thus

$$\frac{\partial \alpha}{\partial \phi} \frac{\partial \Psi}{\partial \lambda} - \frac{\partial \alpha}{\partial \lambda} \frac{\partial \Psi}{\partial \phi}$$

$$= \frac{\partial}{\partial \lambda} (R\tau_\phi^s / H) - \frac{\partial}{\partial \phi} (R\tau_\lambda^s \cos \phi / H) + \frac{\partial B_\lambda}{\partial \phi} - \frac{\partial B_\phi}{\partial \lambda} - \frac{\partial}{\partial \lambda} k_\lambda \frac{\partial \Psi_1}{\partial \lambda} - \frac{\partial}{\partial \phi} k_\phi \frac{\partial \Psi_1}{\partial \phi} \qquad (6.4)$$

The above approach lowers the order of equation from second to first. The first-order equation does not require the boundary conditions over all the integration domain, and the domain is not required to be closed by the boundaries. Consider, the velocity field measured across the shelf. Taking this velocity as the boundary data we can extend the stream line pattern in both direction (i.e., upstream and downstream) along the shore from the "source line" by application of eq.(6.4). Eq.(6.3) has been applied by Kantha *et al.*(1982) to describe the circulation in the Atlantic Ocean.

1.2 Relationship between the methods used for the solution of steady and unsteady problems

The general circulation model introduced by Bryan (1969) is based on the surface lead condition. With this condition the vertically integrated equation of continuity does not include sea level changes in time and as in the steady state problem it allows to introduce the stream function. Equation for the stream function was given in the Ch.I, Sec.3.2 and Ch.IV, Sec.5.2, and it actually describes the change of the stream function in time, nevertheless the methods applied to derive solution are related to the steady state solution. To simplify the notation we shall consider barotropic motion in the rectangular system of coordinates due only to the wind stress.

Starting with the vertically averaged and linearized equations of motion (3.1) and (3.2),

$$\frac{\partial u}{\partial t} - fv = -g\frac{\partial \zeta}{\partial x} + \frac{\tau_x^s}{\rho_o H} \tag{6.5}$$

$$\frac{\partial v}{\partial t} + fu = -g\frac{\partial \zeta}{\partial y} + \frac{\tau_y^s}{\rho_o H} \tag{6.6}$$

and equation of continuity with $\frac{\partial \zeta}{\partial t} = 0$,

$$\frac{\partial}{\partial x}(uH) + \frac{\partial}{\partial y}(vH) = 0 \tag{6.7}$$

The stream function is introduced through eq.(6.7) as

$$uH = -\frac{\partial \psi}{\partial y} \quad \text{and} \quad vH = \frac{\partial \psi}{\partial x} \tag{6.8}$$

Cross differentiating (6.5) and (6.6) and using (6.7) we arrive at

$$\frac{\partial}{\partial t}[(\frac{\partial^2}{\partial x^2} + \frac{\partial^2}{\partial y^2})\psi - \frac{1}{H}(\frac{\partial H}{\partial x}\frac{\partial \psi}{\partial x} + \frac{\partial H}{\partial y}\frac{\partial \psi}{\partial y})] - \frac{f}{H}(\frac{\partial \psi}{\partial x}\frac{\partial H}{\partial y} - \frac{\partial \psi}{\partial y}\frac{\partial H}{\partial x}) + \beta\frac{\partial \psi}{\partial x}$$

$$= H\left[\frac{\partial}{\partial x}\left(\frac{\tau_y^s}{H}\right) - \frac{\partial}{\partial y}\left(\frac{\tau_x^s}{H}\right)\right] \tag{6.9a}$$

This can be rewritten in a compact form as

$$\frac{\partial}{\partial x}\left(\frac{1}{H}\frac{\partial^2\psi}{\partial x\partial t}\right) + \frac{\partial}{\partial y}\left(\frac{1}{H}\frac{\partial^2\psi}{\partial y\partial t}\right) + \frac{\partial}{\partial y}\left(\frac{f}{H}\frac{\partial\psi}{\partial x}\right) - \frac{\partial}{\partial x}\left(\frac{f}{H}\frac{\partial\psi}{\partial y}\right) = \frac{\partial}{\partial x}\left(\frac{\tau_y^s}{H}\right) - \frac{\partial}{\partial y}\left(\frac{\tau_x^s}{H}\right) \tag{6.9b}$$

Applying vector and Jacobian notation the above equation can be again rewritten as

$$\frac{\partial}{\partial t}\left(\Delta\psi - \frac{1}{H}\nabla H\nabla\psi\right) - \frac{f}{H}J(\psi, H) + \beta\frac{\partial\psi}{\partial x} = Hcurl_z\left(\frac{\vec{\tau^s}}{H}\right) \tag{6.9c}$$

To solve in time eq.(6.9c) let us apply a two-level formula,

$$\frac{(\Delta\psi - \frac{1}{H}\nabla H\nabla\psi)^{m+1} - (\Delta\psi - \frac{1}{H}\nabla H\nabla\psi)^m}{T}$$

$$- \frac{f}{H}J(\psi, H)^m + \beta\frac{\partial\psi}{\partial x}^m = Hcurl_z\left(\frac{\vec{\tau^s}}{H}\right)^m \tag{6.10}$$

In the above equation the stream function at the old time step is known and can be moved to the right-hand side. Equation for an unknown stream function at the $m+1$ time step

$$\left(\Delta\psi - \frac{1}{H}\nabla H\nabla\psi\right)^{m+1}$$

$$= T\left(\Delta\psi - \frac{1}{H}\nabla H\nabla\psi\right)^m + \frac{f}{H}J(\psi, H)^m - \beta\frac{\partial\psi}{\partial x}^m + Hcurl_z\left(\frac{\vec{\tau^s}}{H}\right)^m \tag{6.11}$$

is of elliptical type and as such can be solved by the methods applied for the steady state problems. This is an example of the unsteady problem solved by an application of the steady state methods. Often the reverse situation is encountered. That is, instead of seeking solutions to the steady state problem, an associated unsteady problem is considered which is solved with constant forcing and therefore in the limit will yield solution to the steady state problem.

Consider, the steady state problem described by an equation

$$\mathcal{L}\psi = f(x, y) \tag{6.12}$$

and an associated unsteady problem

$$\frac{\partial\psi}{\partial t} = \mathcal{L}\psi - f(x, y) \tag{6.13}$$

Here $\mathcal{L}$ denotes elliptical operator. Solution of eq.(6.13) when $t \to \infty$ should agree in the limit with the solution of eq.(6.12). As an example consider the one dimensional problem

$$\frac{\partial^2 \psi}{\partial x^2} = 0 \quad \text{and associated unsteady problem} \quad \frac{\partial \psi}{\partial t} = \frac{\partial^2 \psi}{\partial x^2} \qquad (6.14)$$

Assuming an explicit two–level numerical formula for the time stepping

$$\frac{\psi_j^{m+1} - \psi_j^m}{T} = \frac{\psi_{j+1}^m + \psi_{j-1}^m - 2\psi_j^m}{h^2} \qquad (6.15)$$

a stable (in time) numerical formula is derived under the following stability condition (see Ch.2, Sec.8.3).

$$0 \le \frac{T}{h^2} \le \frac{1}{2} \qquad (6.16)$$

1.3 Oscillatory motion – linear equations

Tidal motion expressed by a linear set of equations can be transformed into steady state problem as well. Consider the system of vertically integrated equations (3.1)–(3.3) by neglecting the nonlinear (advective) terms and simplifying the frictional terms so that the bottom friction only is included, we arrive at following equations of motion,

$$\frac{\partial u}{\partial t} - fv = -g\frac{\partial \zeta}{\partial x} - \frac{\tau_x^b}{\rho_o H} \qquad (6.17)$$

$$\frac{\partial v}{\partial t} + fu = -g\frac{\partial \zeta}{\partial y} - \frac{\tau_y^b}{\rho_o H} \qquad (6.18)$$

and the equation of continuity,

$$\frac{\partial \zeta}{\partial t} + \frac{\partial}{\partial x}(uH) + \frac{\partial}{\partial y}(vH) = 0 \qquad (6.19)$$

Assuming a linear form for the bottom stress given by eq.(1.130a) we shall rewrite bottom friction as

$$\frac{\tau_x^b}{\rho_o H} = \frac{r_1 u}{\rho_o H} = R_1 u \qquad (6.20)$$

A similar expression for the y component holds as well. Tide wave, being of an oscillatory character in time produces oscillatory changes of velocity and sea level, therefore the variables in eqs.(6.17)–(6.19) can be represented as

$$u = u^* \exp(i\omega t); \quad v = v^* \exp(i\omega t); \quad \zeta = \zeta^* \exp(i\omega t) \qquad (6.21)$$

Introducing (6.21) into the system (6.17)–(6.19) a new set of equation describing space variation only is derived:

$$-i\omega u^* - fv^* = -g\frac{\partial \zeta^*}{\partial x} - R_1 u^* \tag{6.22a}$$

$$-i\omega v^* + fu^* = -g\frac{\partial \zeta^*}{\partial y} - R_1 v^* \tag{6.22b}$$

$$-i\omega \zeta^* + \frac{\partial(Hu^*)}{\partial x} + \frac{\partial(Hv^*)}{\partial x} = 0 \tag{6.22c}$$

Introducing u^* and v^* from the first and second equations into the third one, an equation of the second order for the sea level is obtained:

$$-\Delta \zeta^* + \alpha\frac{\partial \zeta^*}{\partial x} + \beta\frac{\partial \zeta^*}{\partial y} + \gamma \zeta^* = 0 \tag{6.23}$$

where

$$\alpha = \frac{1}{H}\left[\frac{1}{R_1 - i\omega}\frac{\partial H}{\partial y} - \frac{\partial H}{\partial x}\right]$$

$$\beta = \frac{1}{H}\left[\frac{1}{R_1 - i\omega}\frac{\partial H}{\partial x} - \frac{\partial H}{\partial y}\right]$$

$$\gamma = \frac{i\omega}{gH}[(R_1 - i\omega)^2 + f^2]/(R_1 - i\omega) \tag{6.24}$$

Eq.(6.23) is of the elliptical type and the steady state methods can be applied to derive its solution. The full derivation of equations and boundary conditions for this problem is given by Ramming and Kowalik (1980).

2. Numerical approximation

A prototype equation to be used in ensuing considerations is the diagnostic equation (6.1), but written in the Descart's system of coordinates

$$r_1\Delta\psi + \frac{f}{H}J(H,\psi) - \frac{r_1}{H}\nabla H \nabla \psi + \beta\frac{\partial \psi}{\partial x} = F(x,y) \tag{6.25}$$

We shall also use the general elliptic equation

$$a\frac{\partial^2 \psi}{\partial x^2} + b\frac{\partial^2 \psi}{\partial y^2} + c\frac{\partial \psi}{\partial x} + d\frac{\partial \psi}{\partial y} + e\psi = F(x,y) \tag{6.26}$$

defined in the domain $\mathcal{D}$ with boundary conditions prescribed along the boundary Γ. Behavior of the physical coefficients in eq.(6.25) is important when numerical or analytical solutions are searched. The coefficient at the highest derivative (r_1) is positive but, very small in the deep ocean. The coefficients at the first derivatives (like $\frac{\partial H}{\partial x}$) change due to the depth distribution. They can be both positive and negative reaching the largest values at the continental slope. From this cursory analysis one can see that eq.(6.25) will tend to change its order when $r_1 \rightarrow 0$.

Let us consider numerical approximation to the various terms in eq.(6.26). Fig.6.1 shows the grid point distribution and will be helpful in locating the derivatives. Construction of the second order derivative is given in Ch.II, Sec.1. One can start by writing down the first derivatives, and consider the second derivatives as a combination of the first derivatives; thus along the x direction

$$\text{der}_{1/2} = \frac{\psi_{j+1,k} - \psi_{j,k}}{h} \quad \text{and} \quad \text{der}_{-1/2} = \frac{\psi_{j,k} - \psi_{j-1,k}}{h} \qquad (6.27a)$$

$$\frac{\partial^2 \psi}{\partial x^2} \simeq \frac{\text{der}_{1/2} - \text{der}_{-1/2}}{h} = \frac{\psi_{j+1,k} + \psi_{j-1,k} - 2\psi_{j,k}}{h^2} \qquad (6.27b)$$

In a similar fashion the equation for the second derivative along the y coordinate is constructed. Here we can also use the undefined coefficients method elucidated in Ch.II, Sec 9.8. Thus representing the second derivative as a function of unknown coefficients α_1, α_0 and α_3 at the grid points 1, 0 and 3 in Fig.6.1

$$\frac{\partial^2 \psi}{\partial x^2} \simeq \alpha_1 \psi_{j+1,k} + \alpha_0 \psi_{j,k} + \alpha_3 \psi_{j-1,k} \qquad (6.28)$$

Developing ψ into Taylor series at the j, k point and comparing terms at the both sides of eq.(6.28), three equations for the coefficients, follow

$$\alpha_1 + \alpha_0 + \alpha_3 = 0; \quad \alpha_1 - \alpha_3 = 0; \quad \frac{h^2}{2}(\alpha_1 + \alpha_3) = 1 \qquad (6.29)$$

Introducing the values of the coefficients into eq.(6.28) we arrive again at the eq.(6.27b). The methods of undefined coefficients although often requires long algebraic calculations is a good general approach for obtaining the finite difference formulas.

We can reduce errors and improve the accuracy by considering the higher order approximation. This of course may have some disadvantages as the number of points represented in the finite-difference formula is increasing. Denoting the Laplacian

$$\Delta \psi = \frac{\partial^2 \psi}{\partial x^2} + \frac{\partial^2 \psi}{\partial y^2} \quad \text{by} \quad \nabla^2 \psi \qquad (6.30)$$

and introducing the notation used by Fox (1962) and Smith (1978)

$$S_1 = \psi_1 + \psi_2 + \psi_3 + \psi_4 \qquad S_2 = \psi_5 + \psi_6 + \psi_7 + \psi_8$$

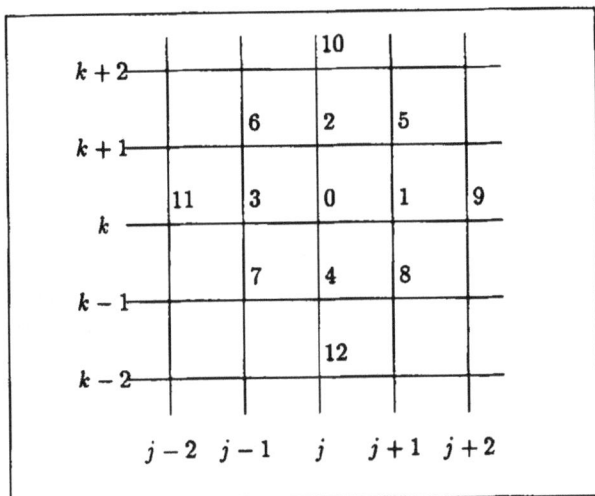

Fig. 6.1

$$S_3 = \psi_9 + \psi_{10} + \psi_{11} + \psi_{12} \tag{6.31}$$

On expanding all these function around the j, k or 0 (see Fig.6.1) point we arrive at

$$S_1 = 4\psi_0 + h^2\nabla^2\psi_0 + \frac{1}{12}h^4(\nabla^4 - 2\mathcal{D}^4)\psi_0 + \frac{1}{360}h^6(\nabla^6 - 3\mathcal{D}^2\nabla^2)\psi_0 + \ldots \tag{6.32a}$$

$$S_2 = 4\psi_0 + 2h^2\nabla^2\psi_0 + \frac{1}{6}h^4(\nabla^4 + 4\mathcal{D}^4)\psi_0 + \frac{1}{180}h^6(\nabla^6 + 12\mathcal{D}^4\nabla^2)\psi_0 + \ldots \tag{6.32b}$$

$$S_3 = 4\psi_0 + 4h^2\nabla^2\psi_0 + \frac{4}{3}h^4(\nabla^4 - 2\mathcal{D}^4)\psi_0 + \frac{8}{45}h^6(\nabla^6 - 3\mathcal{D}^4\nabla^2)\psi_0 + \ldots \tag{6.32c}$$

Here $\mathcal{D}^2 = \dfrac{\partial^2}{\partial x \partial y}$. The above expressions can be combined in a few ways to derive the Laplace operator on the nine points or biharmonic operator on thirteen points. Elimination of $\mathcal{D}^4\psi$ from eq.(6.32a) and eq.(6.32b) provides

$$\nabla^2\psi_0 = \frac{4S_1 + S_2 - 20\psi_0}{6h^2} - \frac{1}{12}h^2\nabla^4\psi_0 + O(h^4) \tag{6.33}$$

For Laplace's equation the second term on the right of eq.(6.33) equals zero. For the Poisson's equation $\nabla^2\psi = f$, the second term on the right is known as $\nabla^4\psi = \nabla^2 f$.

In conclusion, we have derived fourth order approximation formula constructed on a nine point grid. Same formula can be found through the undefined coefficients method. The biharmonic operator plays presently quite an important role in defining the horizontal friction processes (see Ch.I, Sec.4.3). Its numerical counterpart can be found from eqs.(6.32) by eliminating ∇^2 and $\mathcal{D}^4$,

$$\nabla^4 \psi \simeq \frac{1}{h^4}(20u_0 - 8S_1 + 2S_2 + S_3) + O(h^2) \tag{6.34}$$

We now, return to eq.(6.25) and show that the formal application of the finite difference approximation for the equations with the variable coefficients should be made cautiously. Consider one-dimensional motion given by an equation

$$\mathcal{L}\psi = r_1 \frac{\partial^2 \psi}{\partial x^2} - \frac{f}{H}\frac{\partial H}{\partial y}\frac{\partial \psi}{\partial x} = 0 \tag{6.35a}$$

and simplifying the notation let us rewrite it as

$$\mathcal{L}\psi = a\frac{\partial^2 \psi}{\partial x^2} + c\frac{\partial \psi}{\partial x} = 0 \tag{6.35b}$$

These are two first terms along the x direction from eq.(6.25). For numerical solution we take the following finite-difference equation

$$\mathcal{L}_h \psi = \frac{1}{h^2}[(\sqrt{a} + \frac{ch}{4\sqrt{a}})^2(\psi_{j+1,k} - \psi_{j,k}) - (\sqrt{a} - \frac{ch}{4\sqrt{a}})^2(\psi_{j,k} - \psi_{j-1,k})]$$

$$= \frac{a}{h^2}(\psi_{j+1,k} + \psi_{j-1,k} - 2\psi_{j,k}) + \frac{c}{2h}(\psi_{j+1,k} - \psi_{j-1,k}) + \frac{h^2 c^2}{16a}\frac{\partial^2 \psi}{\partial x^2} \tag{6.36}$$

Thus we constructed the finite-difference equation which approximates eq.(6.35) up to the second order. Unfortunately this statement is not true. The coefficients in the last term of eq.(6.36) i.e., $\frac{c^2}{16a}$ should be close to unity for the second order approximation. At the following two regions such approximation may fail: a) in the deep ocean, since $r_1 \to 0$, and b) at shelf break, since the depth variation can be large there. This situation exists not only for the large scale steady motion but for the diffusion–advection processes, as well. Consider for example the diffusion–advection process described by

$$u\frac{\partial C}{\partial x} = D_z \frac{\partial^2 C}{\partial x^2} \tag{6.37}$$

Here we denoted concentration as capital C so it will differ from coefficient c. In the above equation $a = D_z$ and $c = -u$ thus for the second order of approximation

$$\frac{c^2}{16a} = \frac{u^2}{16D_z} \simeq 1 \tag{6.38}$$

Again for the advection dominated process it will be difficult to obtain second order of approximation. More examples of such approximation are given in Ch.II, Sec.9.1. The actual application of the numerical methods requires that the finite-difference formulas will satisfy certain properties. In Ch.II for the time integration we learned to apply upstream–downstream approximation for the first derivative so that the numerical scheme for the diffusion equation will remain stable. A similar situation occurs for the steady state when an iterative method of solution is applied. Let us consider eq.(6.35b) with the coefficients given by eq.(6.35a) and use the following numerical formula to solve the equation,

$$a\frac{\partial^2 \psi}{\partial x^2} + c\frac{\partial \psi}{\partial x} \simeq$$

$$\frac{a}{h^2}(\psi_{j+1,k} + \psi_{j-1,k} - 2\psi_{j,k}) + \frac{c+|c|}{2}\frac{\psi_{j+1,k} - \psi_{j,k}}{h} + \frac{c-|c|}{2}\frac{\psi_{j,k} - \psi_{j-1,k}}{h} \quad (6.39)$$

It is useful to notice that the signs of the coefficients of the function $\psi_{j,k}$ are in all terms the same. This is done so that the coefficients at this central term will dominate over all the other terms. Developing the above numerical formula into a Taylor series at the j, k point we find the following differential equation

$$a\frac{\partial^2 \psi}{\partial x^2} + \frac{|c|h}{2}\frac{\partial^2 \psi}{\partial x^2} + c\frac{\partial \psi}{\partial x} + O(h^2) = 0 \quad (6.40)$$

which is somewhat different from the original equation given at the left-hand side of eq.(6.39). Equation (6.40) is of second order of approximation but the directional derivatives used in eq.(6.39) are only of first order approximation. The difference between the first and the second order is given by the term $\dfrac{|c|h}{2}\dfrac{\partial^2 \psi}{\partial x^2}$. This term is similar to the original second order term when the coefficient a is changed to $|c|h/2$. Because the coefficient a denotes the bottom friction coefficient we shall call $|c|h/2$ a numerical friction coefficient. It is useful to note that this coefficient depends on the space step h, and therefore with diminishing space step, the error (the numerical friction) is diminishing as well. The numerical friction coefficient contains also the physical coefficient c. Since c depends on the depth, abrupt changes in the depth over short horizontal distances will influence the computational results. Finally, it is useful to notice that the numerical friction coefficient may be larger than the physical bottom friction coefficient and thus it can distort the physical processes as well. This is a very important point because one has to be sure that a poorly chosen approximation is not changing the physical processes.

3. Boundary conditions

Consider the boundary condition $\dfrac{\partial \psi}{\partial x} = 0$ for the stream function. This derivative is usually expressed on the right boundary by the forward difference formula and on the left boundary as the backward difference. Such derivatives have only the first-order accuracy of approximation. We shall present here two methods for increasing the order of approximation. First method, given by Samarski (1971), uses for this purpose, information from the differential equation for which a boundary condition is written. The second method introduces a ficticious grid point located beyond the boundary.

Starting from eq.(6.35b) with a forcing function $f(x)$.

$$a\frac{\partial^2 \psi}{\partial x^2} + c\frac{\partial \psi}{\partial x} = f(x)$$

we shall use the following form of the above equation

$$\frac{\partial^2 \psi}{\partial x^2} + c_1\frac{\partial \psi}{\partial x} = f_1(x) \tag{6.41}$$

Here $c_1 = c/a$ and $f_1(x) = f(x)/a$.

Solution to this equation is searched subject to the boundary conditions:

$$\frac{\partial \psi}{\partial x}\Big|_{x=0} = 0; \quad \text{and} \quad \frac{\partial \psi}{\partial x}\Big|_{x=l} = 0 \tag{6.42}$$

The finite-difference form of the boundary condition at the left boundary is taken as a backward difference

$$\frac{\psi_1 - \psi_0}{h} = \frac{\partial \psi_0}{\partial x} + \frac{h}{2}\frac{\partial^2 \psi_0}{\partial x^2} + O(h^2) \tag{6.43}$$

In order to derive the second order approximation for eq.(6.43), the error $\dfrac{h}{2}\dfrac{\partial^2 \psi_0}{\partial x^2}$ is expressed by eq.(6.41)

$$\frac{\partial^2 \psi_0}{\partial x^2} = f_1(x) - c_1\frac{\partial \psi_0}{\partial x}$$

Introducing this formula into eq.(6.43) gives

$$\frac{\psi_1 - \psi_0}{h} - \frac{h}{2}\left(f_1(x) - c_1\frac{\partial \psi_0}{\partial x}\right) = \frac{\partial \psi_0}{\partial x} + O(h^2) \tag{6.44a}$$

This second-order scheme for the first derivative, can now be applied to satisfy the boundary condition (6.42),

$$\frac{\psi_1 - \psi_0}{h} - \frac{h}{2}\left(f_1(x) - c_1\frac{\partial \psi_0}{\partial x}\right) = 0$$

Changing the differential form of $\dfrac{\partial \psi_0}{\partial x}$ to a difference one

$$\frac{\psi_1 - \psi_0}{h} - \frac{h}{2}(f_1(x) - c_1\frac{\psi_1 - \psi_0}{h}) = 0 \tag{6.44b}$$

it results in the following two-point formula for the boundary condition at the left boundary

$$\frac{\psi_1 - \psi_0}{h} = \frac{hf_1}{2 + hc_1} \tag{6.45}$$

A slightly different approach for the construction of the higher order approximation to the boundary condition is the introduction of the central difference formula with a ficticious grid point located beyond the boundary. The central difference in the boundary point x_0 is

$$\frac{\partial \psi}{\partial x} = \frac{\psi_1 - \psi_{-1}}{2h} + O(h^2) \tag{6.46}$$

To find the unknown ψ_{-1} we again bring into consideration eq.(6.41) and eliminate ψ_{-1} between eq.(6.41) and eq.(6.46). Rewriting eq.(6.41) in the finite-difference form at the point x_0

$$\frac{\psi_1 + \psi_{-1} - 2\psi_0}{h^2} + c_1\frac{\psi_1 - \psi_{-1}}{2h} = f_1 \tag{6.47}$$

we find

$$\psi_{-1} = [-\psi_1(2 + hc_1) + 4\psi_0 + 2f_1h^2]/(2 - hc_1) \tag{6.48}$$

From the boundary condition (6.42), $\psi_1 = \psi_{-1}$, and eq.(6.48) simplifies to

$$\psi_1 = \psi_0 + \frac{h^2 f_1}{2} \tag{6.49}$$

This equation again allows to express the two-point boundary condition as the second order approximation formula. Why then do the two formulas given by eq.(6.49) and eq.(6.45) differ? The answer is hidden in the manner both formulas were approximated. While eq.(6.49) is based on a symmetrical finite-difference approach; at some point in the construction of the first formula, namely, in eq.(6.44b) we used the first order approximation to obtain this finite-difference formula. Thus the first formula seems to have less than the second order of approximation.

4. Numerical methods

Consider again eq.(6.26)

$$a\frac{\partial^2 \psi}{\partial x^2} + b\frac{\partial^2 \psi}{\partial y^2} + c\frac{\partial \psi}{\partial x} + d\frac{\partial \psi}{\partial y} + e\psi = F(x, y) \tag{6.50}$$

in the rectangular region. With the coefficients $a > 0$, $b > 0$, $e < 0$, and all the other coefficients are both positive and negative but bounded. Employing the finite-difference second-order formulas written on the grid given in Fig.6.1 we arrive at the following approximation for eq.(6.50).

$$a_{j,k} \frac{\psi_{j+1,k} + \psi_{j-1,k} - 2\psi_{j,k}}{h^2} + b_{j,k} \frac{\psi_{j,k+1} + \psi_{j,k-1} - 2\psi_{j,k}}{h^2}$$

$$+ c_{j,k} \frac{\psi_{j+1,k} - \psi_{j-1,k}}{2h} + d_{j,k} \frac{\psi_{j,k+1} - \psi_{j,k-1}}{2h} + e_{j,k}\psi_{j,k} = F_{j,k} \qquad (6.51)$$

This can be rewritten in a more compact way as

$$\beta_1 \psi_{j+1,k} + \beta_2 \psi_{j-1,k} + \beta_3 \psi_{j,k+1} + \beta_4 \psi_{j,k-1} - \beta_0 \psi_{j,k} = h^2 F_{j,k} \qquad (6.52)$$

Here the coefficients are

$$\beta_1 = a_{j+1,k} + 0.5hc_{j,k}; \quad \beta_2 = a_{j-1,k} - 0.5hc_{j,k}$$

$$\beta_3 = b_{j,k+1} + 0.5hd_{j,k}; \quad \beta_4 = b_{j,k-1} - 0.5hd_{j,k}; \quad \beta_0 = 2(a_{j,k} + b_{j,k} - 0.5h^2 e_{j,k})$$
$$(6.53)$$

For all the β_i to be positive the space step has to be chosen so small that

$$0 < h < min\{\frac{2a_{j,k}}{|c_{j,k}|}; \quad \frac{2b_{j,k}}{|d_{j,k}|}\} \qquad (6.54)$$

Assuming that the number of grid points in the integration domain is equal to N we obtain N linear equations which can be written in the matrix–vector notation as

$$\mathcal{A}\Phi = \mathcal{B} \qquad (6.55)$$

The matrix $\mathcal{A}$ contains only real coefficients. On its diagonal the elements β_0 are located. Off–diagonal elements are the coefficients β_i. Column vector Φ_i contains $i = 1, \ldots, N$ elements and these elements are actually equal to $\psi_{j,k}$. One can better understand this type of notation by assuming that the one dimensional enumeration has been used and elements of the matrix are now related to each grid point enumerated from 1 to N. The methods of solution of this system can be of direct or iterative nature. The direct methods, such as the Gauss elimination procedure, require big computer storage and are time-consuming. The line inversion method used for solution of the one dimensional problems (Appendix 1) is also a direct method. It factorizes $\mathcal{A}$ into triangular matrix. The line inversion method can be applied effectively for the sparse matrices where coefficients are clustered around the main diagonal, while the Gauss elimination is inefficient for such matrices because it "eliminates" both zero and non–zero elements of the matrix.

The iterative methods are constructed on the simple algorithms which are easily repeated. They can be also relatively easily applied to the nonlinear set of equations. Convergence criteria for the iterative methods are expressed through the matrix notation given in the one-dimensional enumeration. Actual computation for the two dimensional problem by the iterative methods is done with the help of two-dimensional notation given in eq.(6.53), so we are compelled to keep both notations.

4.1 Simple iterative method (Jacobi method)

For application of an iterative method eq.(6.52) is recast as

$$\psi_{j,k}^{m+1} = \left[\beta_1\psi_{j+1,k}^m + \beta_2\psi_{j-1,k}^m + \beta_3\psi_{j,k+1}^m + \beta_4\psi_{j,k-1}^m - h^2 F_{j,k}\right]/\beta_0 \qquad (6.56)$$

Here m denotes successive iteration step, m and $m+1$ specify respectively the old and the new iterations. This notation and procedure reminds us very much of time stepping; an analogy which will be used in the ensuing considerations. In the matrix notation the same procedure is

$$\Phi^{m+1} = A\Phi^m + B \qquad (6.57)$$

The elements of matrix A and B differ from eq.(6.55). Basing on the above notation one can write the following iterative process starting from an arbitrary initial approximation Φ^0:

$$\Phi^1 = A\Phi^0 + B$$
$$\Phi^2 = A\Phi^1 + B$$
$$\vdots \qquad \vdots \qquad\qquad (6.58)$$
$$\Phi^m = A\Phi^{m-1} + B$$

If an analytical solution to eq.(6.50) is known as Φ then the question is whether the above iterative process converges to this analytical solution. The convergence depends on the properties of the matrix A (Faddeeva, 1959). Introducing the norm of the matrix

$$|A| = \text{Max}_j \sum_{k=1}^{N} |\alpha_{j,k}| = \mu \qquad (6.59)$$

Here $\alpha_{j,k}$ are the matrix coefficients, j stands for the rows, k for the columns, and N denotes number of columns in the matrix A. Thus by means of eq.(6.59) the sum of the absolute values of the elements in the first row, next in the second row and so on, is compared and the maximum is chosen to be a norm. The iterative process is convergent if $\mu < 1$. For equation (6.56) this condition states

$$(|\beta_1| + |\beta_2|)/|\beta_0| < 1 \qquad (6.60)$$

Similar condition holds for the columns as well, therefore the general condition is

$$|\beta_1| + |\beta_2| + |\beta_3| + |\beta_4| < |\beta_0| \qquad (6.61)$$

This convergence condition simply states that the diagonal elements in the matrix should dominate over the off-diagonal elements.

As we mentioned before an iterative process reminds us of the time integration procedure. We shall use here this analogy to find a convergence condition of the simple iterative scheme by relating it to the stability condition of the time dependent process (Roache, 1976). Let us take for this purpose eq.(6.51) along the x direction only

$$a_{j,k}\frac{\psi_{j+1,k} + \psi_{j-1,k} - 2\psi_{j,k}}{h^2} + c_{j,k}\frac{\psi_{j+1,k} - \psi_{j-1,k}}{2h} = 0 \qquad (6.62)$$

Applying the above iterative process to eq.(6.62) we obtain

$$\psi_{j,k}^{m+1} = \frac{\psi_{j+1,k}^m + \psi_{j-1,k}^m}{2} + \frac{c_{j,k}h^2}{2a_{j,k}}\frac{\psi_{j+1,k}^m - \psi_{j-1,k}^m}{2h} \qquad (6.63)$$

To demonstrate what kind of time dependent process the above equation represents we subtract from the both sides of eq.(6.63) the value at the central grid point $\psi_{j,k}^m$,

$$\psi_{j,k}^{m+1} - \psi_{j,k}^m = \frac{h^2}{2}\frac{\psi_{j+1,k}^m + \psi_{j-1,k}^m - 2\psi_{j,k}^m}{h^2} + \frac{c_{j,k}h^2}{2a_{j,k}}\frac{\psi_{j+1,k}^m - \psi_{j-1,k}^m}{2h} \qquad (6.64)$$

First term on the left-hand side denotes the time stepping with the time step $T = 1$, and on the right-hand side we have the Laplace operator and the first derivative. Applying to this equation the standard procedure for checking stability (Ch.II, Sec.3.2) we find the stability coefficient

$$|\lambda| = [\cos^2 \kappa h + (\frac{ch}{2a})^2 \sin^2 \kappa h]^{1/2} \qquad (6.65)$$

For stability $|\lambda| \leq 1$. To fulfill this requirement

$$\frac{ch}{2a} < 1 \qquad (6.66)$$

This is exactly the condition given in eq.(6.54), it will insure that the off-diagonal elements in the matrix are positive and it will result in the dominance of the diagonal elements. Unfortunately this dominance does not hold always. Condition (6.66) has an additional and important information for the stability. The process is only conditionally stable; when $a \to 0$ or c grows large the process is unstable. For the steady oceanic flow it means that when the coefficient at the first derivative is large (for the steady flow a is bottom friction coefficient, and c is related to the depth change along x direction), the iteration process fails. For the advection–diffusion equation $a = D_h$ and $c = -u$ the iteration process fails when advection dominates over diffusion. Condition (6.66) for the latter equation is

$$\frac{uh}{2D_h} \leq 1 \qquad (6.67)$$

Since we know from the Ch.II that the advective scheme for the central space differ-ence is unstable. We might suspect that the conditional convergence of the simple iterative scheme is caused by application of the central–difference approximation for the first derivatives. Stability condition given by eq.(6.67) is strictly related to the presence of the diffusive (Laplacian) term. To construct an unconditionally convergent iterative process we shall apply an approach used for the time dependent processes, i.e., we introduce a directional derivative. The finite-difference scheme which includes the directional derivatives is given by eq.(6.39)

$$\frac{a}{h^2}(\psi_{j+1,k}+\psi_{j-1,k}-2\psi_{j,k})+\frac{c+|c|}{2}\frac{\psi_{j+1,k}-\psi_{j,k}}{h}+\frac{c-|c|}{2}\frac{\psi_{j,k}-\psi_{j-1,k}}{h}=0 \quad (6.68)$$

Taking the central term from the above equation to serve as the new iterated value, the iterative process is given by

$$\psi_{j,k}^{m+1}(\frac{2a}{h^2}+\frac{|c|}{h})=\frac{a}{h^2}(\psi_{j+1,k}^m+\psi_{j-1,k}^m)+\frac{c+|c|}{2}\frac{\psi_{j+1,k}^m}{h}-\frac{c-|c|}{2}\frac{\psi_{j-1,k}^m}{h} \quad (6.69)$$

Since the result we intend to derive is quite interesting we carefully perform stability analysis. Expressing the error (Ch.II, Sec.3.2), as $\delta\psi=\lambda^m e^{i\kappa h}$ and introducing this expression into eq.(6.69) we arrive at

$$\lambda(\frac{2a}{h}+|c|)=\frac{2a}{h}\cos\kappa h+ic\sin\kappa h+|c|\cos\kappa h=0 \quad (6.70)$$

Denoting

$$q_1=\frac{2a}{h}\qquad q_2=|c|$$

and multiplying λ by complex conjugate, we arrive at

$$\lambda^2=\frac{(q_1+q_2)^2\cos^2\kappa h+q_2^2\sin^2\kappa h}{(q_1+q_2)^2} \quad (6.71)$$

From eq.(6.71) it follows that $|\lambda|\leq 1$, and the numerical scheme (6.69) is uncon-ditionally stable, therefore the iterative process always converges. For the uncon-ditional convergence we have paid a price; eq.(6.69) is of the first order of approx-imation only and it includes a numerical friction coefficient which may distort the physical processes (see eq.(6.40) in this chapter). In this finite difference scheme the coefficients on the main diagonal (i.e. coefficients of the term $\psi_{j,k}$) always dominate over the nondiagonal terms. We can conclude that the application of the stability notion to the iterative process is an effective method in finding convergence of the iterative process without any need for resorting to the tedious investigations of the matrix properties. Finally, we shall write a simple program in FORTRAN for Jacobi iteration process.

```
                    nq=100
122                 ni=ni+1
C Counter ni counts the number of iterations, it runs up to nq=100.
                    tt=0
C tt will be maximum relative error estimated between old and new value
                    do 120 j=2,49
                    do 120 k=2,24
C computation are done in rectangular domain with stream function equal
C to zero along boundaries
             psn(j,k)=a(j,k)*pso(j+1,k)+b(j,k)*pso(j-1,k)+
                    c(j,k)*pso(j,k+1)+ d(j,k)* pso(j,k-1)+f(j,k)
C Here psn-new value, pso-old value, a,b,c,d - coefficients
             if(abs(psn(j,k).lt.1.oe-26) go to 120
C so we do not divide by very small number
             se=abs(psn(j,k)-pso(j,k))/abs(psn(j,k))
C se is a relative error between old and new value
             if(se.gt.tt)then
             tt=se
C but we are interested only in the maximum of the error which is tt
             end if
120                 continue
             if(tt.lt.0.01) go to 124
C if tt=0.01 we assume that scheme have converged to the solution
C put tt equal to 0.001 and see difference
             if(ni.eq.100)go to 124
C possibly scheme does not converge if it requires so many iterations
                    do j=2,49
                    do k=2,24
             pso(j,k)=psn(j,k)
             end do
             end do
C this loop is to change variables before starting new iteration
```

go to 122

C go back to 122 to start new iteration

124   **continue**

C jump out from computational loop when solution has converged

4.2 Gauss–Seidel method

In the simple iterative method the new value is calculated based on the old values. The Gauss–Seidel method accelerates this process by using the values recently obtained during the same iteration step. Such an approach is possible because the grid points and calculations performed in these grid points are in a fixed ordering – see Fig.6.2. One usually starts calculations from the low values of the j, k index and proceeds towards the larger index number (see direction given in Fig.6.2). If the computation is just to be done at the j, k point, the new values at the points with the index less than j, k are already available (these are values below the dash line in Fig.6.2).

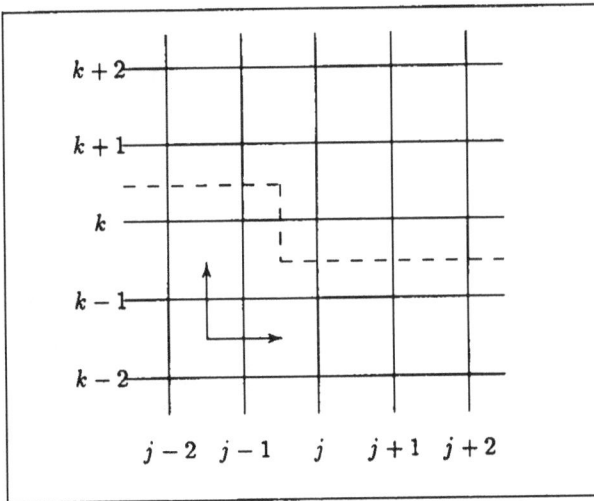

Fig. 6.2

To illustrate this approach let us take eq.(6.52)

$$\beta_1 \psi_{j+1,k} + \beta_2 \psi_{j-1,k} + \beta_3 \psi_{j,k+1} + \beta_4 \psi_{j,k-1} - \beta_0 \psi_{j,k} = h^2 F_{j,k} \qquad (6.72)$$

or rearranging it for the iterative process as

$$\psi_{j,k}^{m+1} = \frac{1}{\beta_0}[\beta_1\psi_{j+1,k}^m + \beta_2\psi_{j-1,k}^{m+1} + \beta_3\psi_{j,k+1}^m + \beta_4\psi_{j,k-1}^{m+1} - h^2F_{j,k}] \qquad (6.73)$$

This equation shows how the computation at the j, k grid point utilizes recently derived values at the $j - 1, k$ and $j, k - 1$ points. The convergence of this iterative process is again subject to the dominance of the diagonal term over the off–diagonal terms given by eq.(6.61). Now, we can use the analogy to the time stepping schemes and deduce a condition on the coefficients to preserve stability or in the language of the iterative processes to achieve dominance of the diagonal terms. We return to the one-dimensional eq.(6.62)

$$a_{j,k}\frac{\psi_{j+1,k} + \psi_{j-1,k} - 2\psi_{j,k}}{h^2} + c_{j,k}\frac{\psi_{j+1,k} - \psi_{j-1,k}}{2h} = 0 \qquad (6.74)$$

Applying the above iterative process, eq.(6.74) becomes

$$\psi_{j,k}^{m+1} = \frac{\psi_{j+1,k}^m + \psi_{j-1,k}^{m+1}}{2} + \frac{c_{j,k}h^2}{2a_{j,k}}\frac{\psi_{j+1,k}^m - \psi_{j-1,k}^{m+1}}{2h} \qquad (6.75)$$

Now, searching for the stability parameter λ it follows that

$$\lambda[2 - e^{-i\kappa h}(1 - Rc)] = (1 + Rc)e^{i\kappa h} \qquad (6.76a)$$

and

$$\lambda = \frac{(1 + Rc)e^{i\kappa h}}{2 - e^{-i\kappa h}(1 - Rc)} \qquad (6.76b)$$

Here $Rc = \dfrac{c_{j,k}h}{2a_{j,k}}$. We can conclude that this scheme is only stable if $Rc \le 1$. This conclusion is similar to the one we derived for the simple iterative scheme. The application of the upwind–downwind numerical scheme for the first derivative will eliminate this condition and the scheme will be unconditionally stable

The Gauss–Seidel scheme from the point of view of time dependent schemes may be called a semi-implicit scheme. This observation allows us to bring into the scope of the iterative processes the various semi-implicit time dependent schemes. One scheme which has second order approximation in space and uses semi–implicit scheme for the first derivative is the angular–derivative (Ch.II, Sec.5.2), similar properties has for the second derivative – DuFort–Frankel scheme (Ch.II, Sec.8.2).

Considering again eq.(6.74) but with the first derivative approximated by the angular-derivative,

$$a_{j,k}\frac{\psi_{j+1,k}^m + \psi_{j-1,k}^{m+1} - 2\psi_{j,k}^{m+1}}{h^2} + \frac{c_{j,k}}{h}\left(\frac{\psi_{j,k}^{m+1} + \psi_{j+1,k}^m}{2} - \frac{\psi_{j-1,k}^{m+1} + \psi_{j,k}^m}{2}\right) = 0 \quad (6.77)$$

Introducing as before the nondimensional Reynolds cell number $Rc = \dfrac{ch}{2a}$ the iterative scheme takes the form,

$$(2 - Rc)\psi_{j,k}^{m+1} = \psi_{j+1,k}^{m} + \psi_{j-1}^{m+1} + Rc(\psi_{j+1,k}^{m} - \psi_{j-1,k}^{m+1} - \psi_{j,k}^{m}) \qquad (6.78)$$

This scheme seems to be absolutely stable. It was never tested and is constructed to demonstrate the strength of analogy of the iterative and time dependent processes. One interesting item in this scheme is that the central grid point is applied both to the new function and to the old function, while in the previous iterative schemes the central grid point served always for calculation of a new value.

The Gauss–Seidel method is easy to program in FORTRAN. Below a short subroutine for the calculation is given.

```
              nq=100
122            ni=ni+1
C Counter ni counts the number of iterations, it runs up to nq=100.
              tt=0
C tt will be maximum relative error estimated between old and new value
              do 120 j=2,49
              do 120 k=2,24
C computation are done in rectangular domain with stream function equal
C to zero along boundaries
              pso=ps(j,k)
C The old value of the function is put into temporary storage
called pso for calculating relative error.
              ps(j,k)=a(j,k)*ps(j+1,k)+b(j,k)*ps(j-1,k)+
              c(j,k)*ps(j,k+1)+ d(j,k)* ps(j,k-1)+f(j,k)
C Here there is no difference between old and new values as soon as new
value is available it is used in the right hand side of the above equation;
a,b,c,d – coefficients
              if(abs(ps(j,k).lt.1.oe-26) go to 120
C so we do not divide by very small number
              se=abs(pso-ps(j,k))/abs(ps(j,k))
C se is a relative error between old and new value
              if(se.gt.tt)then
```

```
                              tt=se
C but we are interested only in the maximum of the error which is tt
                              end if
120                           continue
                    if(tt.lt.0.01) go to 124
C if tt=0.01 we consider that scheme have converged to the solution
C put tt equal to 0.001 and see difference
                    if(ni.eq.100)go to 124
C possibly scheme does not converge if it requires so many iterations
                        go to 122
C go back to 122 to start new iteration
124                           continue
C jump out from computational loop when solution has converged
```

4.3 Successive over-relaxation method

Successive over-relaxation method (SOR) is an accelerating method which extends the Gauss–Seidel process. Let us turn to eq.(6.73):

$$\psi_{j,k}^{m+1} = \frac{1}{\beta_0}[\beta_1\psi_{j+1,k}^m + \beta_2\psi_{j-1,k}^{m+1} + \beta_3\psi_{j,k+1}^m + \beta_4\psi_{j,k-1}^{m+1} - h^2 F_{j,k}] \qquad (6.79)$$

and add to both sides the term $\psi_{j,k}^m$. Next multiplying the right-hand side of eq.(6.79) by factor ω we obtain an equation for the SOR method

$$\psi_{j,k}^{m+1} = \psi_{j,k}^m + \omega\frac{1}{\beta_0}[\beta_1\psi_{j+1,k}^m + \beta_2\psi_{j-1,k}^{m+1} + \beta_3\psi_{j,k+1}^m + \beta_4\psi_{j,k-1}^{m+1} - \beta_0\psi_{j,k}^m - h^2 F_{j,k}]$$
$$(6.80)$$

The ω is called an acceleration or relaxation factor. It usually changes in the range $1 < \omega < 2$. If $\omega = 1$ the method changes into that of Gauss–Seidel. The choice of the optimum of the relaxation factor is important for the SOR, because simply taking let's say value 1.3 is not going to accelerate the process. Here the various approaches are possible as was shown by Carré (1961) and O'Brien (1968), Young (1971). Time stability analysis for the simplest SOR involving one-dimensional Laplace operator

$$\psi_j^{m+1} - \psi_j^m = \omega(\psi_{j+1}^m + \psi_{j-1}^{m+1} - 2\psi_j^m) \qquad (6.81)$$

gives

$$\lambda - 1 = \omega(e^{i\kappa h} + \lambda e^{-i\kappa h} - 2) \qquad (6.82)$$

and defines that stability holds for $0 < \omega < 2$.

5. Direct methods

5.1 Line inversion method

The line inversion method has been delineated in Appendix 1. It is best suited for the one-dimensional problems. Multidimensional problems can be tackled by splitting them into a system of one dimensional problems (see Ch.II, Sec.8.3). The method may be used both in the direct approach or in conjunction with the iterative processes, since it has built in strong convergence (or stability) requirements (see Appendix 1).

Let us turn to eq.(6.26) and consider only the one-dimensional case

$$a^* \frac{\partial^2 \psi}{\partial x^2} + c^* \frac{\partial \psi}{\partial x} + e^* \psi = F(x) \tag{6.83a}$$

Introducing a computational mesh with the grid distance h the finite-difference form follows,

$$-a_j^* \frac{-2\psi_j + \psi_{j+1} + \psi_{j-1}}{h^2} - c_j^* \frac{\psi_{j+1} - \psi_{j-1}}{2h} - e_j^* \psi_j = -F_j \tag{6.83b}$$

We have denoted the coefficients in the above equations by star, because lower case a, b, c letter we shall use in a three-point formula. The change of sign made in this equation is to comply with the form of equation given in Appendix 1. The j index runs from 1 on the left boundary to J on the right one. It is important to understand that the approximation of the first derivative in the above equation should be changed from central to forward or backward derivatives accordingly to the conditions set on the coefficients. These conditions will be stated below and they are also given by eq.(A11). The above equation is subject to the following boundary conditions,

on the left boundary,

$$\alpha_1 \psi_1 + \alpha_2 \frac{\psi_2 - \psi_1}{h} = A \tag{6.84a}$$

and, on the right boundary,

$$\beta_1 \psi_J + \beta_2 \frac{\psi_J - \psi_{J-1}}{h} = B \tag{6.84b}$$

Here A and B are given functions. Coefficients $\alpha_1, \alpha_2, \beta_1, \beta_2$ are equal to zero or unity depending on what kind of boundary condition is considered, i.e., the Dirichlet condition (given function), the Neumann condition (given first derivative), or the mixed condition. Rearranging eq.(6.83b) to the three-point formula

$$-a_j \psi_{j-1} + b_j \psi_j - c_j \psi_{j+1} = d_j \tag{6.85}$$

one can proceed with the algorithm given in Appendix 1. In this equation to apply
the line inversion method the coefficients should be:

$$a_j > 0; b_j > 0; c_j > 0 \quad \text{and} \quad b_j \geq a_j + c_j \tag{6.86}$$

The main part of the solution consists of a transformation of the three-point formula
eq.(6.85) into the two-point formula

$$\psi_j = s_j \psi_{j+1} + e_j \tag{6.87}$$

This expression allows starting computation from the boundary and carry it out
into the domain. Through the above transformation the matrix of coefficients has
been factorized from a band matrix with the non-zero main diagonal and two sub-
diagonals symmetrical to the main diagonal, into an upper triangular matrix with
the main diagonal and one sub-diagonal above the main diagonal. To illustrate this
transformation a few equations based on eq.(6.85), in the close proximity to the left
and right boundaries are given:

$$\psi_1 = 0$$
$$0\psi_1 + b_2\psi_2 - c_2\psi_3 = d_2$$
$$-a_3\psi_2 + b_3\psi_3 - c_3\psi_4 = d_3$$
$$\vdots \quad \vdots \tag{6.88}$$
$$-a_{J-2}\psi_{J-3} + b_{J-2}\psi_{J-2} - c_{J-2}\psi_{J-1} = d_{J-2}$$
$$-a_{J-1}\psi_{J-2} + b_{J-1}\psi_{J-1} - c_{J-1}\psi_J = d_{J-1}$$
$$\psi_J = 0$$

This system of equations can be written in the matrix form as

$$\begin{pmatrix} b_2 & -c_2 & 0 & 0 & 0 & 0 \\ -a_3 & b_3 & -c_3 & 0 & 0 & 0 \\ \vdots & \vdots & \vdots & \vdots & \vdots & \vdots \\ 0 & 0 & 0 & -a_{J-2} & b_{J-2} & -c_{J-2} \\ 0 & 0 & 0 & 0 & -a_{J-1} & b_{J-1} \end{pmatrix} \begin{pmatrix} \psi_2 \\ \psi_3 \\ \vdots \\ \psi_{J-2} \\ \psi_{J-1} \end{pmatrix} = \begin{pmatrix} d_2 \\ d_3 \\ \vdots \\ d_{J-2} \\ d_{J-1} \end{pmatrix} \tag{6.89}$$

Main diagonal, in the coefficient matrix, consists of b_j, the coefficients a_j and c_j are
located off-diagonal. Transformation given by eq.(6.87) changes this matrix into

$$\begin{pmatrix} 0 & -s_1 & 0 & 0 & 0 & 0 \\ 0 & 1 & -s_2 & 0 & 0 & 0 \\ \vdots & \vdots & \vdots & \vdots & \vdots & \vdots \\ 0 & 0 & 0 & 1 & s_{J-2} & 0 \\ 0 & 0 & 0 & 0 & 1 & 0 \end{pmatrix} \begin{pmatrix} \psi_1 \\ \psi_2 \\ \vdots \\ \psi_{J-1} \\ \psi_J \end{pmatrix} = \begin{pmatrix} e_1 \\ e_2 \\ \vdots \\ e_{J-1} \\ e_J \end{pmatrix} \tag{6.90}$$

The line inversion method (often called tridiagonal algorithm) has a very desirable computational properties. An iterative process based on this method shows the tendency for self-correction (Appendix 1) and converges quite fast. The self-correction is also related to stability, since any error introduced in the course of computation is not amplified. The method of line inversion was extended by Lindzen and Kuo (1969) to a wide class of partial differential equations, so that not only triadiagonal line inversion is considered but pentadiagonal (and so on) as well.

5.2 Application of the line inversion method to two-dimensional problems

An iterative process can be "organized" so that it can be solved by the line inversion method. Consider, the general two-dimensional elliptic problem given by eq.(6.50). We shall write an iterative algorithm in such a manner that the line inversion method will be applied first along the x axis and afterwards along the y axis. For implementing this approach we devise an iteration consisting of two sub-iterations. Equation (6.52), which is numerical form of eq.(6.50), is used on the first sub-step for the line inversion along the x direction:

$$\psi_{j,k}^{m+1/2} = \frac{1}{\beta_0}(\beta_1\psi_{j+1,k}^{m+1/2} + \beta_2\psi_{j-1,k}^{m+1/2} + \beta_3\psi_{j,k+1}^{m} + \beta_4\psi_{j,k-1}^{m} - h^2F_{j,k}) \quad (6.91a)$$

Unknown $\psi^{m+1/2}$ is computed from the known values of ψ^m. On the second sub-step, the line inversion is applied along the y direction

$$\psi_{j,k}^{m+1} = \frac{1}{\beta_0}(\beta_1\psi_{j+1,k}^{m+1/2} + \beta_2\psi_{j-1,k}^{m+1/2} + \beta_3\psi_{j,k+1}^{m+1} + \beta_4\psi_{j,k-1}^{m+1} - h^2F_{j,k}) \quad (6.91b)$$

and function ψ^{m+1} is calculated while $\psi^{m+1/2}$ is given. To speed up the computations one can use a relaxation parameter similar to ω used in the SOR.

Since the above iterative process, has an analog in the time domain we can resort to the time dependent equation and solve it for the steady state when $t \to \infty$. Adding to eq.(6.50) time derivative of the stream function

$$\frac{\partial\psi}{\partial t} = a\frac{\partial^2\psi}{\partial x^2} + b\frac{\partial^2\psi}{\partial y^2} + c\frac{\partial\psi}{\partial x} + d\frac{\partial\psi}{\partial y} + e\psi - F(x,y), \quad (6.92)$$

solution to eq.(6.92) is searched by the split method described in Ch.II, Sec.8.3. To apply this method eq.(6.92) is split into a system of equations:

$$\frac{1}{2}\frac{\partial\psi}{\partial t} = \frac{a}{2}\frac{\partial^2\psi}{\partial x^2} + \frac{b}{2}\frac{\partial^2\psi}{\partial y^2} + \frac{c}{2}\frac{\partial\psi}{\partial x} + \frac{d}{2}\frac{\partial\psi}{\partial y} + \frac{e}{2}\psi - \frac{F(x,y)}{2} \quad (6.93a)$$

$$\frac{1}{2}\frac{\partial\psi}{\partial t} = \frac{a}{2}\frac{\partial^2\psi}{\partial x^2} + \frac{b}{2}\frac{\partial^2\psi}{\partial y^2} + \frac{c}{2}\frac{\partial\psi}{\partial x} + \frac{d}{2}\frac{\partial\psi}{\partial y} + \frac{e}{2}\psi - \frac{F(x,y)}{2} \quad (6.93b)$$

Time stepping in the above equations is done in the two substeps, first along the x direction and afterwards along the y direction.

$$\frac{\psi^{m+1/2} - \psi^m}{T} = \frac{a}{2}\frac{\partial^2\psi^{m+1/2}}{\partial x^2} + \frac{b}{2}\frac{\partial^2\psi^m}{\partial y^2}$$

$$+\frac{c}{2}\frac{\partial\psi^{m+1/2}}{\partial x} + \frac{d}{2}\frac{\partial\psi^m}{\partial y} + \frac{e}{2}\psi^{m+1/2} - \frac{F^m(x,y)}{2} \tag{6.94a}$$

$$\frac{\psi^{m+1} - \psi^{m+1/2}}{T} = \frac{a}{2}\frac{\partial^2\psi^{m+1/2}}{\partial x^2} + \frac{b}{2}\frac{\partial^2\psi^{m+1}}{\partial y^2}$$

$$+\frac{c}{2}\frac{\partial\psi^{m+1/2}}{\partial x} + \frac{d}{2}\frac{\partial\psi^{m+1}}{\partial y} + \frac{e}{2}\psi^{m+1} - \frac{F^{m+1}(x,y)}{2} \tag{6.94b}$$

This implicit scheme should, after long time of integration, provide a steady state solution, which ought to be identical to the solution obtained by the iterative process given in eq.(6.91).

6. Higher order accuracy schemes for convection-dominated flow

One of the obstacles faced by the ocean dynamics modeler is to balance various requirements related to computer and computational scheme, such as, economy of numerical computations (usually one deals with very large arrays), stability of the numerical algorithms, and accuracy. Before all the other requirements we have to put accuracy of the numerical schemes, because the way a numerical scheme reproduces the physical process has to be well understood. The elliptical problems, we have considered, are difficult to solve numerically because the coefficient at the highest derivative (friction coefficient or eddy diffusivity) are always positive but often are very small, and the coefficients at the first derivatives are highly variable. An interaction of the two processes e.g., in transport equation, an interaction of diffusion and advection, in large scale flow, an interaction of friction and bottom slopes, leads to the dependence of numerical accuracy on the physical process. In eq.(6.36) we have observed, that the second order approximation is achieved only if the velocity and eddy diffusion coefficient are related in a specific way. We have encountered these problems while constructing the various iterative schemes. In advection dominated transport, application of the numerical schemes derived from the central difference (for the first derivative) to an iterative process often failed to provide a convergent process. For convergent iterative scheme we had to apply the upwind–downwind approximation which in turn caused the effect of numerical viscosity (eq.6.40). The convergence of the process constructed with the second-order central difference approximation was strictly related to the condition on the cell Reynolds number

$$\frac{uh}{2D_h} \leq 1 \tag{6.95}$$

This number for the diffusive processes is called Peclet number, for the friction processes the number is called Reynolds number.

Let us turn back to the transport equation we investigated in Ch.II for the time dependent processes, and consider now a steady state process,

$$u\frac{\partial c}{\partial x} = D\frac{\partial^2 c}{\partial x^2} \qquad (6.96)$$

In Ch.II, Sec.9.1 an analytical solution to this equation between the grid point $j-1$ and j was given as,

$$c = c_{j-1} + (c_j - c_{j-1})\frac{\exp[\frac{u}{D}(x - x_{j-1})] - 1}{\exp[\frac{u}{D}(x_j - x_{j-1})] - 1} \qquad (6.97)$$

Multiplying and dividing the argument of the exponential function by h, and introducing cell Peclet number as $Pc = \dfrac{uh}{D}$, the above solution takes form

$$c = c_{j-1} + (c_j - c_{j-1})\frac{\exp[Pc\frac{(x-x_{j-1})}{h}] - 1}{\exp[Pc\frac{(x_j-x_{j-1})}{h}] - 1} \qquad (6.98)$$

For small Peclet numbers $Pc \le 1$ the above solution varies smoothly from $j - 1$ to j point. If $Pc > 1$ the solution is changing abruptly in the proximity to the right boundary. This change occurs in a very thin layer and the space step used for numerical solution is too large to resolve this layer. To keep Pc number in the range given by eq.(6.95) one has to diminish space step, i.e., the space resolution should be increased.

Now, we shall consider merits of the various approximations for the first derivative and the avenues for the higher accuracy will be delineated. To facilitate discussion we shall use Fig.6.3 which elucidates the grid distribution; here both velocity and concentration are given in the same grid points.

We start again by considering the first order upwind–downwind scheme, given in Sec.4.1 of this chapter.

$$u\frac{\partial c}{\partial x} \simeq \frac{u^+}{h}(c_j - c_{j-1}) + \frac{u^-}{h}(c_{j+1} - c_j) \qquad (6.99)$$

Here $u^+ = 0.5(u + |u|)$, $u^- = 0.5(u - |u|)$. Expanding the right-hand side of eq.(6.99) into Taylor series one will obtain an original equation in j point and approximation errors.

$$\frac{u^+}{h}(c_j - c_{j-1}) + \frac{u^-}{h}(c_{j+1} - c_j) = (u\frac{\partial c}{\partial x})_j - (\frac{|u|h}{2}\frac{\partial^2 c}{\partial x^2})_j \qquad (6.100)$$

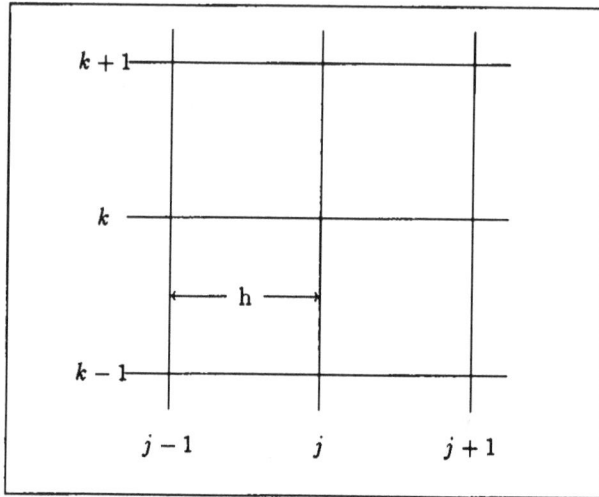

Fig. 6.3

The approximation error is here of the first order and the term containing the error is analogous to physical diffusion/friction. Moreover, we have found that this scheme applied in the iterative process always leads to the unconditional convergence of the process, because in the coefficient matrix the diagonal terms always dominate over the off-diagonal terms. This approximation reproduces the concentration field in a smooth monotonic fashion without any spikes, unfortunately due to the high numerical diffusion the sharp gradients are usually smoothed and therefore, significant error can be observed in the calculated results.

The second approximation we applied to the first derivative is the central differencing scheme

$$u\frac{\partial c}{\partial x} \simeq u\frac{(c_{j+1} - c_{j-1})}{2h} \tag{6.101}$$

Expanding again the right-hand side into a Taylor series, the original equation is obtained and the approximation error is proportional to h^2

$$u\frac{(c_{j+1} - c_{j-1})}{2h} = (u\frac{\partial c}{\partial x})_j + \frac{uh^2}{6}\frac{\partial^3 c}{\partial x^3} \tag{6.102}$$

As is obvious from the above equation for the second order of approximation both velocity u and the third derivative ought to be bounded. In Sec.4.1 of this chapter it was shown that the central difference scheme applied to the iterative process, together with the diffusive term, leads to the conditionally convergent (stable) process. For the convergence of the process the cell Peclet number $\frac{uh}{2D} \leq 1$. With this condition the diagonal terms in coefficient matrix dominates over the off–diagonal terms. This approximation reproduces the concentration field with good precision, unfortunately the sharp gradients are often reproduced with wiggles. Thus the order of approximation alone is not sufficient for reproducing the processes taking place in the very thin, boundary layer.

Very often much better results can be derived by the second, third or higher order upwind–downwind approximation schemes. Among various schemes we choose a second order upwind–downwind difference scheme,

$$u\frac{\partial c}{\partial x} \simeq \frac{u^+}{2h}(3c_j - 4c_{j-1} + c_{j-2}) + \frac{u^-}{2h}(-c_{j+2} + 4c_{j+1} - 3c_j) \tag{6.103}$$

Using Taylor series to check the order of approximation it follows,

$$\frac{u^+}{2h}(3c_j - 4c_{j-1} + c_{j-2}) + \frac{u^-}{2h}(-c_{j+2} + 4c_{j+1} - 3c_j) = (u\frac{\partial c}{\partial x})_j - (\frac{uh^2}{3}\frac{\partial^3 c}{\partial x^3})_j \tag{6.104}$$

Therefore, this derivative has the second order of approximation. An iterative scheme constructed for the transport equation with advection given by eq.(6.104) can be written as

$$c_j^{m+1}(\frac{2D}{h} + \frac{3|u|}{2h}) = \frac{D}{h}(c_{j+1}^m + c_{j-1}^m)$$

$$-u(c_{j+1}^m - c_{j-1}^m) + |u|(c_{j+1}^m + c_{j-1}^m) + \frac{u}{4}(c_{j+2}^m - c_{j-2}^m) - \frac{|u|}{4}(c_{j+2}^m + c_{j-2}^m) \tag{6.105}$$

We investigate the convergence of this process through the stability parameter. Thus expressing c as $\lambda^m e^{i\kappa h}$ and introducing it into eq.(6.105),

$$\lambda = \frac{(\frac{2D}{h} + 2|u|)\cos \kappa h - \frac{|u|}{2}\cos 2\kappa h + iu(-2\sin \kappa h + \frac{u}{2}\sin 2\kappa h)}{(\frac{2D}{h} + \frac{3|u|}{2})} \tag{6.106}$$

The absolute value of this parameter can be calculated numerically. Comparison of the stabilities for the first order upwind scheme (6.71) and for the central difference scheme (6.65) indicates that the second order upwind scheme has better stability than the central difference scheme. Approximation given by eq.(6.103) carry over better properties from the both mentioned schemes: it reproduces sharp concentration gradients practically without any wiggles. The solution to eq.(6.105) can be obtained by SOR method and the relaxation factor is smaller or larger than 1, depending on the numerical scheme and cell Peclet number (Shyy, 1985).

In this manner, we have opened a host of methods which have been delineated in Ch.II, nonetheless the problems still remain, because even with the present day computers it is still difficult to make very fine meshes so that the cell Peclet number will be small in the whole computation domain. A few avenues can be delineated for further improvement. For the steady state processes, before starting each iteration, the field can be checked for the non–monotonic behavior, by searching for large gradients. One solution proposed by Zhu and Rodi (1991) is to apply a composite procedure. Such a procedure combines the second-order upwind, central differencing and first-order upwind schemes, with the switch from one to the other scheme being automatically controlled by an advection boundedness criterion. Leonard (1988) starting from the same premises, proposed to improve QUICK algorithm (Ch.II, Sec.9.4) in the areas of sharp changes. The application of the better (and more complicated) approach is again governed by boundedness criterion. Another procedure which may find some future application is to use an automatic generation of the fine grid in the portion of the domain with the high gradients.

Appendix 1

1. Line inversion (factorization) method

Line inversion method, also called tridiagonal method, is usually applied to search solution of the three-point equation (Ramming and Kowalik, 1980),

$$-a_k \Psi_{k-1} + b_k \Psi_k - c_k \Psi_{k+1} = d_k \tag{A1}$$

subject to the boundary condition at the left end of the line

$$\Psi_1 = \chi_1 \Psi_2 + \mu_1 \quad \text{at} \quad k = 1 \tag{A2a}$$

and at the right end of the line

$$\Psi_K = \chi_2 \Psi_{K-1} + \mu_2 \quad \text{at} \quad k = K \tag{A2b}$$

Here $k = 1, 2, \ldots, K - 1, K$ denote indices of the grid points located on the line; $a_k, b_k, c_k, \chi_1, \mu_1, \chi_2, \mu_2$ are given coefficients.

The boundary conditions (A2a) and (A2b) define the value of Ψ at the boundaries through the first internal grid point, but to advance the solution by the three-point formulae (A1) we need to specify the values at the two points (i.e. we have to know the value at the boundary point and at the point located next to the boundary). Therefore we shall search a solution to (A1) via the two-point expression,

$$\Psi_k = s_k \Psi_{k+1} + e_k \tag{A3}$$

Here s_k and e_k are coefficients to be defined later by the known coefficients a_k, b_k, c_k, d_k. For this purpose (A3) can be rewritten at the point $k - 1$ and afterwards the new formula is introduced into (A1),

$$-a_k(s_{k-1} \Psi_k + e_{k-1}) + b_k \Psi_k - c_k \Psi_{k+1} = d_k,$$

which leads to the two-point relation

$$\Psi_k = \frac{c_k \Psi_{k+1}}{b_k - a_k s_{k-1}} + \frac{d_k + a_k e_{k-1}}{b_k - a_k s_{k-1}} \tag{A4}$$

Comparing (A4) and (A3) we arrive at the expressions for the new coefficients,

$$s_k = \frac{c_k}{b_k - a_k s_{k-1}} \tag{A5}$$

and

$$e_k = \frac{d_k + a_k e_{k-1}}{b_k - a_k s_{k-1}} \tag{A6}$$

These are recurrence formulas to calculate s_k and e_k as a function of s_{k-1} and e_{k-1} and the coefficients of (A1).

In summary the process of computations contains two successive sweeps. At the first sweep, from $k = 2$ to $k = K - 1$ the coefficients are calculated via (A5) and (A6), and at the second sweep, from $k = K - 1$ to $k = 2$ the dependent variable Ψ is derived from (A3).

To start the calculation by (A5) and (A6) one has to know s_1 and e_1. We obtain these coefficients by considering (A3) at the point $k = 1$ and comparing it to (A2a), thus

$$s_1 = \chi_1 \quad \text{and} \quad e_1 = \mu_1 \tag{A7}$$

To start the second sweep we need to define Ψ_K. This time we compare (A2b) and (A3) at the point $K - 1$. For (A3) written at the point $K - 1$

$$\Psi_{K-1} = s_{K-1}\Psi_K + e_{K-1} \tag{A8}$$

the value of Ψ_{K-1} from (A2b) is introduced

$$\frac{\Psi_K - \mu_2}{\chi_2} = s_{K-1}\Psi_K + e_{K-1}$$

and from above the boundary value at $k = K$ follows

$$\Psi_K = \frac{\mu_2 + \chi_2 e_{K-1}}{1 - \chi_2 s_{K-1}} \tag{A9}$$

It serves to begin successive computation by (A3).

Let us notice that the above delineated method can be called a right-hand sweep, since the unknown function is calculated starting from the right-hand side and proceeds towards the left-hand side. One can, as well, devise a symmetrical procedure which will start from the left-hand side. For this (A3) can be redefined as

$$\Psi_{k+1} = s_{k+1}\Psi_k + e_{k+1} \tag{A10}$$

and s_{k+1} and e_{k+1} will be defined by introducing (A10) into Ψ_{k+1} term of (A1).

We do not intend to proceed along this line, rather we shall discuss stability and the range of the coefficients which will define stability of the above delineated process. General stability condition requires that

$$a_k > 0, \quad c_k > 0, \quad b_k \geq a_k + c_k$$

$$0 \leq \chi_1 \leq 1, \quad 0 \leq \chi_2 \leq 1, \quad \text{and} \quad \chi_1 + \chi_2 \leq 2 \tag{A11}$$

The above conditions hold when matrix of the coefficients for the three-point equation is positive-definite.

Conditions (A11) assures that the denominator in (A5) and (A6) will never vanish. Let us check this statement. Consider (A5)

$$s_k = \frac{c_k}{b_k - a_k s_{k-1}}$$

and since from (A11) $b_k = a_k + c_k + \epsilon_k$, where $\epsilon_k \geq 0$, we rewrite the above as

$$s_k = \frac{c_k}{c_k + \epsilon_k + a_k(1 - s_{k-1})}$$

For $k = 2$, $s_1 = \chi_1 \leq 1$, hence $\delta_1 = 1 - s_1 \geq 0$, and

$$s_2 = \frac{c_2}{c_2 + \epsilon_2 + a_2 \delta_1} \leq 1$$

Continuing this process we deduce that:

$$s_k \leq 1 \qquad \text{for} \quad k = 1, 2, \ldots, K \tag{A12}$$

and obviously the denominator of (A5) or (A6) never equals zero. If condition (A12) holds it is easy to prove that an error in Ψ_{k+1} does not increase when Ψ_k is calculated. To show this tendency of the line inversion method to self-correction let us investigate the behavior of an error from (A3)

$$\delta \Psi_k = s_k \delta \Psi_{k+1} \quad \text{or} \quad |\delta \Psi_k| \leq |s_k| \cdot |\delta \Psi_{k+1}|$$

and since $|s_k| \leq 1$, we deduce that the error introduced at $k + 1$ step diminishes when k step is calculated or,

$$|\delta \Psi_k| \leq |\delta \Psi_{k+1}| \tag{A13}$$

The line inversion method is also quite economical when the number of the basic operations is considered. Let us assume that the method has been used to integrate in time by an implicit algorithm, therefore at every time step we have to perform two sweeps through the domain. An explicit algorithm used for the same purpose "sweeps " through the same domain only once, and is usually at least two time faster than the line inversion method.

Actually the line inversion method serves to solve a system of equations and as such should be compared to another method for example to the Gauss method. Here a simple approach is to count the number of the basic operations required to derive a solution defined by (A1), (A2a) and (A2b). Via expressions (A5) and (A6) six operations (addition, subtraction, multiplication and division) are performed at $K - 1$ points, expression (A3) requires 2 operation at K points and finally (A9) is calculated by 5 operation. Thus line the inversion method requires,

$$Q = 6(K - 1) + 2K + 5 = 8K - 1 \tag{A14}$$

operations. The number Q is proportional to the K - total number of unknowns. The Gauss method will require $Q \simeq K^3$ basic operations to derive a solution.

The line inversion method is usually applied to a few coordinates, therefore the equations and the process of solution are often quite similar, to avoid repetition of calculations the method was generalized for the vectorized variables and is called matrix line inversion method (Yanenko, 1971). Let us consider a classical Ekman (1905) problem, for the time dependent motion in the rotating frame with vertical friction, and let us construct the following numerical scheme

$$\frac{u^{m+1} - u^m}{T} - f\frac{(v^{m+1} + v^m)}{2} = \frac{N_z}{2}(\frac{\partial^2 u^{m+1}}{\partial z^2} + \frac{\partial^2 u^m}{\partial z^2}) \tag{A15a}$$

$$\frac{v^{m+1} - v^m}{T} + f\frac{(u^{m+1} + u^m)}{2} = \frac{N_z}{2}(\frac{\partial^2 v^{m+1}}{\partial z^2} + \frac{\partial^2 v^m}{\partial z^2}) \tag{A15b}$$

Subject to the boundary condition at the free surface $(z = 0)$

$$\tau_x^w = N_z\frac{u_1 - u_2}{h_z} \quad \text{or} \quad u_1 = u_2 + \frac{\tau_x^w h_z}{N_z} \tag{A16a}$$

and

$$\tau_y^w = 0 \quad \text{or} \quad v_1 = v_2 \tag{A16b}$$

At the bottom both velocity components are zero,

$$u_L = 0 \quad \text{and} \quad v_L = 0 \tag{A17}$$

From (A15) by introducing the finite-difference formulas along the z direction, we arrive at

$$-\frac{N_z T}{2h_z^2}u_{l-1}^{m+1} + (1 + \frac{N_z T}{h_z^2})u_l^{m+1} - 0.5fv_l^{m+1} - \frac{N_z T}{2h_z^2}u_{l+1}^{m+1} = d1^m \tag{A18a}$$

$$-\frac{N_z T}{2h_z^2}v_{l-1}^{m+1} + (1 + \frac{N_z T}{h_z^2})v_l^{m+1} + 0.5fu_l^{m+1} - \frac{N_z T}{2h_z^2}v_{l+1}^{m+1} = d2^m \tag{A18b}$$

Let us generalize the above by introducing a matrix and vector notation. The vector for the unknown variable

$$\Psi_l = \begin{pmatrix} u_l \\ v_l \end{pmatrix} \tag{A19}$$

has as the components u_l and v_l velocity components at the l grid point. A three-point equation similar to (A1) can be introduced as well,

$$-A_l\Psi_{l-1} + B_l\Psi_l - C_l\Psi_{l+1} = D_l \tag{A20}$$

subject to the boundary condition at $l = 1$

$$\Psi_1 = P_1 \Psi_2 + R_1 \tag{A21a}$$

and at $l = L$

$$\Psi_L = P_2 \Psi_{L-1} + R_2 \tag{A21b}$$

Here $A_l, B_l, C_l, P_1, P_2, R_1, R_2$ are square matrices.

For the example considered above (i.e., for eqs.(A18)) the general equation (A20) becomes

$$\begin{pmatrix} -\frac{N_z T}{2h_z{}^2} & 0 \\ 0 & -\frac{N_z T}{2h_z{}^2} \end{pmatrix} \begin{pmatrix} u_{l-1}^{m+1} \\ v_{l-1}^{m+1} \end{pmatrix} + \begin{pmatrix} 1 + \frac{N_z T}{h_z{}^2} & -f/2 \\ f/2 & 1 + \frac{N_z T}{h_z{}^2} \end{pmatrix} \begin{pmatrix} u_l^{m+1} \\ v_l^{m+1} \end{pmatrix}$$
$$+ \begin{pmatrix} -\frac{N_z T}{2h_z{}^2} & 0 \\ 0 & -\frac{N_z T}{2h_z{}^2} \end{pmatrix} \begin{pmatrix} u_{l+1}^{m+1} \\ v_{l+1}^{m+1} \end{pmatrix} = \begin{pmatrix} d1^m \\ d2^m \end{pmatrix} \tag{A22}$$

Subject to the boundary condition:

at the surface

$$\begin{pmatrix} u_1^{m+1} \\ v_1^{m+1} \end{pmatrix} = \begin{pmatrix} 1 & 0 \\ 0 & 1 \end{pmatrix} \begin{pmatrix} u_2^{m+1} \\ v_2^{m+1} \end{pmatrix} + \begin{pmatrix} \frac{\tau_z^w h_z}{2N_z} \\ 0 \end{pmatrix} \tag{A23a}$$

(Notice division by 2 in the last term, this is because boundary condition (A16a) has to be split between step m and $m + 1$).

and at the bottom

$$\begin{pmatrix} u_L^{m+1} \\ v_L^{m+1} \end{pmatrix} = \begin{pmatrix} 0 \\ 0 \end{pmatrix} \tag{A23b}$$

Returning to the general equation (A20), we shall search a solution via the two-point expression

$$\Psi_l = S_l \Psi_{l+1} + E_l \tag{A24}$$

similar to the previously introduced (A3). Again by introducing (A24) into (A20) we arrive at the matrix of coefficients

$$S_l = (B_l - A_l S_{l-1})^{-1} C_l \tag{A25a}$$

and

$$E_l = (B_l - A_l S_{l-1})^{-1} (D_l + A_l E_{l-1}) \tag{A25b}$$

Stability conditions are defined in terms of the matrix norm (Faddeeva, 1959)

$$\|B_l^{-1} A_l\| + \|B_l^{-1} C_l\| < 1, \quad \|P_1\| < 1, \quad \|P_2\| < 1 \tag{A26}$$

Matrix line inversion method is well suited for any number of variables but actually for the two variables one can apply more economical method by introducing a complex number $\Psi = u + iv$. Here $i = \sqrt{-1}$. For this case the whole process of calculation will be done through the ordinary line inversion. Equations (A15) can be recast into one equation by multiplying (A15b) by i and adding on either side (A15a) and (A15b), hence

$$\frac{Q^{m+1} - Q^m}{T} - if\left(\frac{Q^{m+1} + Q^m}{2}\right) = \frac{N_z}{2} \frac{(\partial^2 Q^{m+1} + Q^m)}{\partial z^2} \tag{A27}$$

Here $Q = u + iv$.

Equation (A27) can be easily rendered into a finite difference form along z coordinate and afterwards the line inversion method given by (A3), (A5) and (A6) is used to derive the solution. The velocity components u and v are obtained as real and imaginery part of Q.

2. FORTRAN program for the line inversion algorithm

```
            PARAMETER(KE=20)

            REAL PS(KE),A(KE),B(KE),C(KE),D(KE),S(KE),E(KE)

            REAL KAI1,MU1,KAI2,MU2
C KE is the maximum value of counter (index) K. We have chosen KE=20,
C it may be changed accordingly to desired resolution.
C PS denotes function Ψ, and A,B,C are coefficients in eq.(A1).
C S and E are coefficients of (A3). KAI1 and MU1 are coefficients
C of (A2a). KAI2 and MU2 are coefficients of (A2b).
C To start calculation define S(1) and E(1)

            S(1)=KAI1

            E(1)=MU1
C KAI1=χ1, and MU1=μ1 in boundary condition (A2a).
C continue to calculate coefficients S(K) and E(K) from (A5) and (A6).

            DO 10 K=2,KE-1

            DENOM=B(K)-S(K-1)*A(K)

            S(K)=C(K)/DENOM

            E(K)=(D(K)+A(K)*E(K-1))/DENOM

      10          CONTINUE
C Define unknown PS staring from boundary condition at K=KE (A9)

            A1=MU2+KAI2*E(KE-1)

            A2=1.-KAI2*S(KE-1)

            PS(KE)=A1/A2
```

```
C Notice that to find PS(KE) the coefficients E and S are
C taken at the point KE-1, therefore integration in the loop 10
C can be extended only to KE-1.
C Applying above condition we can continue to calculate PS from (A3)
              DO 20 K=KE-1,1,-1
              PS(K)=PS(K+1)*S(K)+E(K)
20            CONTINUE
              END
```

References

Abbot, M.B. 1966. *An Introduction to the Method of Characteristics.* Thames & Hudson, London, 243pp.

Abbott, M.B., A. Danesgaard, and G.S. Rodenhuis. 1973. System 21, Jupiter: a design system for two-dimensional nearly-horizontal flow. *J. Hydraul. Res.*, 11, 1–28.

de G.Allen D.N. and R.V. Southwell. 1955. Relaxation method applied to determine the motion, in two dimensions, of a viscous fluid past a fixed cylinder. *Q.J. Mech. Appl. Math.*, 8, 129–145.

Alvarez, J.A. 1973. Numerical Prediction of Storm Surges in the Rio de la Plata Area. Ph.D. Thesis, University of Buenos Aires, Argentina. 70 pp.

Arakawa, A. 1966. Computational design for long–term numerical integration of the equations of fluid motion: two–dimensional incompressible flow. Part 1, *J. Comput. Physics*, 1, 119–1423.

Arakawa, A. 1972. Design of the U.C.L.A. general circulation model, numerical simulation of weather and climate. Dept. Meteor., University of California, Los Angeles.

Arakawa, A. and V.R. Lamb. 1977. Computational design of the basic dynamical processes if the UCLA general circulation model. In: *Methods of Computational Physics*, Vol. 17, Academic Press.

Arakawa, A. and V.R. Lamb. 1981. A potential enstrophy and energy conserving scheme for the shallow water equation. *Mon. Wea. Rev.*, 109, 18–36.

Asselin, R. 1972. Frequency filter for time integrations. *Mon. Wea. Rev.*, 100(6), 487–490.

Backhaus, J. 1976. Zur hydrodynamik im flachwassergebiet, ein numerisches modell. *Dtsch. Hydrogr. Z.* 29, 222–238.

Backhaus, J.O. 1985. A three-dimensional model for the simulation of shelf-sea dynamics. *Dtsch. Hydrogr. Z.*, **38**, 165–187.

Batteen, M.L. and Y.-J. Han. 1981. On the computational noise of the finite-difference schemes used in ocean models. *Tellus*, **33**, 387–396.

Berntsen, H., Z. Kowalik, S. Saelid and K. Soerli, 1981. Efficient numerical simulation of ocean hydrodynamics by splitting procedure. *Modeling, Identification and Control*, **4(2)**, 181–199.

Bilelo, M.A. and R.E. Bates. 1971. Ice thickness observations: North American Arctic and Subarctic 1966-1967, 1967–1968. Rept. 43, Part 4, Hanover, N.H., Cold Regions Res. Envir. Lab.

Birchfield, G.E. and T.S. Murty. 1974. A numerical model for wind-driven circulation in Lakes Michigan and Huron. *Mon. Wea. Rev.*, **102(2)**, 157–165.

Bjerknes, V., J. Bjerknes, H. Solberg and T. Bergeron. 1934. *Hydrodynamic Physique.* Vol. II, Chap XI. Les Presses Universitaires de France, Paris. 457–491.

Blackadar, A.K. 1962. The vertical distribution of wind and turbulent exchanges in a neutral atmosphere. *J. Geophys. Res.*, **67**, 3095–3120.

Blumberg, A.F. 1975. A numerical investigation into dynamics of estuarine circulation. Chesapeake Bay Institute. Tech. Rept. 91.

Blumberg, A.F. 1978. The influence of density variations on estuarine tides and circulation. *Est. Coastal Mar. Sci.*, **6**, 209–215.

Blumberg, F.A. and G.L. Mellor, 1983. Diagnostic and prognostic numerical circulation studies of the South Atlantic Bight. *J. Geophys. Res.*, **88**, 4579–4592.

Book, D.L., J.P. Boris and K. Hain. 1975. Flux-corrected transport, II: Generalizations of the method. *J. Comput. Physics*, **18**, 248–283.

Boris, J.P. and D.L. Book. 1973. Flux-corrected transport, I: SHASTA, a fluid transport algorithm that works. *J. Comput. Physics*, **11**, 38–69.

Boussinesq, J. 1903. *Théorie analytique de chaleur.* Vol. 2. Gauthier - Villars, Paris.

Bowden, K.F. and P. Hamilton. 1975. Some experiments with a numerical model of circulation and mixing in a tidal estuary. *Est. Coastal Mar. Sci.*, **3**, 281–301.

Brown, P.S. and J.P. Pandolfo. 1978. Merging finite difference schemes having dissimilar time-differencing operators. *Mon. Wea. Rev.*, **106**, 268–270.

Bryan, K.A. 1963. A numerical investigation of a nonlinear model of a wind-driven ocean. *J. Atmos. Sci.*, **20**, 594–606.

Bryan, K. 1969. A numerical method for the study of the circulation of the World Ocean. *J. Comput. Physics*, **4**, 347–376.

Bryan, K. and M.D. Cox. 1972. An approximate equation of state for numerical models of ocean circulation. *J. Phys. Oceanogr.*, **2**, 510–514.

Brettschneider, G. 1967. Anwendung des hydrodynamish–numerischen Verfahrens zur Ermittlung der M_2– Mitschwingungsgezeit in der Nordsee. *Mitteilungen des Instituts für Meereskunde der Universität Hamburg*, VII.

Burden, R.L., J.D. Faires and A.C. Reynolds. 1981. *Numerical Analysis*. Prindle, Weber and Schmidt, Boston, Mass. 598pp.

Butler, H.L. 1979. Coastal flood simulation in stretched coordinates, 1030–1048. Proc. 16th Coastal Eng. Conf., 27 Aug.–3 Sept., 1978, Hamburg. ASCE, N.Y.

Camerlengo, A.L. and J.J. O'Brien. 1980. Open boundary condition in rotating fluids. *J. Comput. Physics*, **35**, 12–35.

Carré, B.A. 1961. The determination of the optimum accelerating factor for successive over-relaxation. *Computer. J.*, **4(1)**.

Carrier, G.F. and H.P. Greenspan. 1958. Water waves of finite amplitudes on a sloping beach. *J. Fluid Mech.*, **4**, 97–109.

Carton, J.A. 1984. Coastal circulation caused by an isolated storm. *J. Phys. Oceanogr.*, **14**, 114–124.

Chapman, D.C. 1985. Numerical treatment of cross-shelf open boundaries in a barotropic coastal ocean model. *J. Phys. Oceanogr.*, **15**, 1060–1075.

Charney, J.G. 1949. On a physical basis for numerical prediction of large-scale motions in the atmosphere. *J. Meteor.*, **6**, 371–385.

Charnock, H. and J. Crease. 1957. Recent advances in science. North Sea surges. *Sci. Prog.*, **45**, 494–511.

Chilicka, Z, Z. Kowalik and Z. Wierzbicki. 1983. Construction of a numerical model of storm surges with a refined grid. *Oceanologia*, 16, 5–14.

Chrystal, G. 1905. On the hydrodynamical theory of seiches (with bibliography on seiches). *Trans. Roy. Soc. Edin.*, 41(3), 599–649.

Chrystal, G. 1908. An investigation of seiches of Loch Earn by the Scottish Loch Survey. 5. Mathematical appendix on the effect of pressure disturbances upon the seiches in a symmetric parabolic lake. *Trans. Roy. Soc. Edin.*, 46(5), 455–517.

Clarke, D.J. and N.D. Thomas. 1972. The two-dimensional flow oscillations of a fluid in a spindle-shaped basin; application to Port Kembla outer harbour N.S.W. Australia. *Aust. J. Mar. Freshwater Res.*, 23,1–9.

Cox, M.D., 1984. A primitive equation three-dimensional model of the world ocean. GFDL Ocean Group Tech. Rept. No. 1, GFDL Princeton Univ., 250 pp.

Cox, M.D., 1985. An eddy resolving numerical model of the ventilated thermocline. *J. Phys. Oceanogr.*, 15, 1312–1324.

Courant, R. and Hilbert D. 1962. *Methods of Mathematical Physics.* Interscience Publ. Inc., New York.

Crandall, S.H. 1956. *Engineering Analysis — a Survey of Numerical Procedures.* McGraw-Hill Publ., N.Y. 417 pp.

Crank, J. and P. Nicholson. 1947. A practical method for numerical evaluation of solutions of partial differential equations of the heat-conduction type. *Proc. Cambridge Phil. Soc.*, 43(50), 50–67.

Crean, P.B. 1978. A numerical model of barotropic mixed tides between Vancouver Island and the mainland and its relation to studies of the estuarine circulation, 283–313. In: J.C.J. Nihoul (ed.), Hydrodynamic of Estuaries and Fjords. Proc. 9th Int. Liège Colloq. Ocean Hydrodyn. Scientific Publ. Co., Amsterdam.

Crean, P.B., T.S. Murty, and J.A. Stronach. 1988. *Mathematical Modelling of Tides and Estuarine Circulation.* Springer-Verlag, N.Y. 471 pp.

Crowley, W.P. 1970. A numerical model for viscous, free surface, barotropic wind driven ocean circulations. *J. Comput. Physics*, 5, 139–168.

Cummins, P.F. and L.A. Mysak. 1988. A quasi-geostrophic circulation model of the northeast Pacific. Part I: A preliminary numerical experiment. *J. Phys. Oceanogr.*, **18**, 1261–1286.

Danard, M.B. and T.S. Murty. 1992. Storm surge mitigation through vegetation canopies. *Nat. Hazards*, in press.

Davies, A.M. 1981. Three-dimensional modelling of surges. 45–74. In: D.H. Peregrine (ed.) *Floods due to High Winds and Tides.* Academic Press, London. 109 pp.

Davies, A.M. 1985. Application of a sigma coordinate sea model to the calculation of wind-induced curents. *Cont. Shelf Res.*, **4**(4), 389–423.

Defant, A. 1961. *Physical Oceanography.* Pergamon Press, N.Y. 598 pp.

Dingle, A.N. and C. Young. 1965. Computer applications in the atmospheric sciences. Dept. Meteor. Ocean., University of Michigan, Ann Arbor, Mich. 35 pp.

DuFort, E.C. and S.P. Frankel. 1953. Stability conditions in the numerical treatment of parabolic difference equations. *Math. Tables and Other Aides to Computation*, **7**, 135–152.

Eckart, C.H. 1960. *Hydrodynamics of Ocean and Atmospheres.* Pergamon Pr., 290pp.

Egan, B.A and J.R. Mahoney. 1972. Numerical modeling of advection and diffusion of urban area source pollutants. *J. Appl. Meteor.*, **11**, 312–322.

Ekman, V.W. 1905. On the influence of the Earth's rotation on ocean currents. *Arkiv. Mat. Astron. Fys.*, **2**, 11.

Ekman, V.W. 1923. Veber horizontal cirkulation bei winderzeuøgten Meeres Strömungen. *Ark. Mat. Astron. Fys.* **17**, 26.

Eliassen, A. 1956. A Procedure for Numerical Integration of the Primitive Equations of the Two-Parameter Model of the Atmosphere. Sci. Rept. No. 4, Dept. Meteor., U.C.L.A., Los Angeles. 53 pp.

Elliot, A.J. 1976. A numerical model of the internal circulation in a branching tidal estuary. Chesapeake Bay Institute. Special Rept. 54.

Endoh, M. 1977. Formation of thermocline front by cooling of the sea and inflows of the fresh water. *J. Oceanogr. Soc. Jpn.*, **33**, 6–15.

Faddeeva, V.N. 1959. *Computational Methods of Linear Algebra*, Dover Publ. New York, 252 pp.

Falconer, R.A. 1980. Numerical modelling of tidal circulation in harbors, pp. 31–48. *J. Waterway Port Coastal Ocean Div. Proc.*, ASCE, Feb. 1980.

Farmer, D.M. 1976. The influence of wind on the surface layer of a stratified inlet: Part II. Analysis. *J. Phys. Oceanogr.*, **6**, 941–952.

Fiadeiro M.E. and G. Veronis. 1977. On the weighted-mean schemes for the finite-difference approximation to the advection-diffusion equation. *Tellus*, **29**, 512–522.

Fischer, G. 1959. Ein numerisches Verfahren zur Errechnung von Windstau und Gezeiten in Randmeeren. *Tellus*, **2(1)**, 60–76.

Fischer, G. 1965a. On a finite-difference scheme for solving the nonlinear primitive equations for barotropic fluid with application to the boundary current problem. *Tellus*, **17(4)**, 405–412.

Fischer, G. 1965b. Comments on some problems involved in numerical solutions of tidal hydraulics equations. *Mon. Wea. Rev.*, **93**, 110–111.

Fischer, G. 1965c. On a finite difference scheme for solving the primitive equations for a barotropic fluid with application to the boundary curent problem. *Tellus*, **17**, 405–412.

Flather, R.A. and N.S. Heaps. 1975. Tidal computations for Morecambe Bay. *Geophys. J. Roy. Astr. Soc.*, **42**, 489–517.

Fofonoff, N.P. and F. Webster. 1971. Current measurements in the western Atlantic. *Phil. Trans. Roy. Soc. Lond.*, **A210**, 423–436.

Forel, F.A. 1892. Le Leman (collected papers). Two volumes. Rouge, Lausanne, Switzerland.

Foreman, M.G.G. 1986. An accuracy analysis of boundary conditions for the forced shallow water equations. *J. Comput. Physics*, **64(2)**, 334–367.

Fox, L. 1962. *Numerical Solution of Ordinary and Partial Differential Equations*, Addison–Wesley Publ. Co., Reading, Mass.

Freeman, N.G., P.F. Hamblin and T.S. Murty. 1974. Helmholtz resonance in harbours of the Great Lakes. Proc. 17th Conf. Great Lakes Res. Int. Assoc. Great Lakes Res. Proc. **15**: 399–411.

Friedrich, H.J. 1966. Numerische berechnungen der allgemeinen zirculation in meere nach einem differenzen verfahren, vornehmlich für den Atlantischen Ozean. Ph.D. diss. Iniv. Hamburg, 60 pp.

Friedrich, H.J. and S. Levitus. 1972. An approximation to the equation of state for sea water, suitable for numerical ocean models. *J. Phys. Oceanogr.*, **2**, 514-517.

Fromm, J. 1964. The time dependent flow of an incompressible viscous fluid. *Methods Comput. Physics*, **3**, 346–382.

Galt, J.A. 1984. Derivation of simplified diagnostic model for the interpretation of oceanic data. OCSEAP, Final Report, **26**, 251–300.

Garabedian, P.R. 1964. *Partial Differential Equations*. John Wiley & Sons, N.Y. 672 pp.

Garrett, C.J.R. 1970. Bottomless harbors. *J. Fluid Mech.* **43**, 433–449.

Garrett, C.J.R. 1975. Tides in gulfs. *Deep-Sea Res.* **22**, 23–25.

Garratt, J.R. 1977. Review of drag coefficients over oceans and continents. *Mon. Weather Rev.*, **7(105)**, 915–929.

Gary, J.M. 1973. Estimate of truncation error in transformed coordinate, primitive equation atmospheric models. *J. Atmos. Sci.*, **30**, 223–233.

Gent, P.R. and M.A. Cane. 1989. A reduced gravity, primitive equation model of the upper equatorial ocean. *J. Compt. Phys.*, **81**, 444–480.

Gill, A.L. 1982. *Atmosphere-Ocean Dynamics*. Academic Press, Intl. Geophysical Series, Vol.30, 662 pp.

Gjevik, B. and T. Straume. 1989. Model simulations of the M_2 and K_1 tide in the Nordic Seas and the Arctic Ocean. *Tellus*, **41A**, 73–96.

Godunov, S.K. and V.S. Riabenki. 1977. *Difference Schemes*, (in Russian), Publ.

by Nauka, Moscow, 439pp.

Gotlib, V.Yu. and B.A. Kagan. 1981. On resonance generation of semi–diurnal tides in the World Ocean. *Physics Atm. Ocean*, **17**, 502–512.

Grammelvedt, A. 1969. A survey of finite difference schemes for the primitive equations for a barotropic fluid. *Mon. Wea. Rev.*, **97**, 384–404.

Greatbatch, R.J. and T. Otterson. 1991. On the formulation of open boundary conditions at the mouth of a bay. *J. Geophys. Res.*, **96(C10)**, 18,431–18,445.

Greenberg, D.A. 1975. Mathematical Studies of Tidal Behaviour in the Bay of Fundy. Ph.D. Thesis, Univ. of Liverpool, Liverpool, U.K. 139 pp.

Greenberg, D.A. 1976. Mathematical Description of the Bay of Fundy–Gulf of Maine Numerical Model. Tech. Note 16, Mar. Envir. Data Service, Ottawa.

Greenberg, D.A. 1977. Mathematical Studies of Tidal Behaviour in the Bay of Fundy. Rept. No. 46, Mar. Sci. Directorate, Ottawa.

Greenberg, D.A. 1979. A numerical model investigation of tidal phenomena in the Bay of Fundy and Gulf of Maine. *Mar. Geod.*, **2(2)**, 161–187.

Groen, P. and G.W. Groves. 1962. Surges. 611–646. **In:** M.N. Hill (ed.), *The Sea*. Vol. 1. John Wiley & Sons, New York.

Gupta, M.M. 1983. A survey of some second order difference schemes for the steady state convection-diffusion equations. *Intl. J. Num. Methods Fluid*, **3**, 319–331.

Haidvogel, D.B. and W.R. Holland. 1978. The stability of ocean currents in eddy-resolving general circulation models. *J. Phys. Oceanogr.*, **8**, 393–413.

Haidvogel, D.B., A.R. Robinson, and F.F. Schulman. 1980. The accuracy, efficiency of stability of three numerical models with application to open ocean problems. *J. Comp. Sci.*, **34**, 1–53.

Haidvogel, D.B., J.L. Wilkin, and R. Young. 1991. A semi-spectral primitive equation ocean circulation model using vertical sigma and orthogonal curvilinear horizontal coordinates. *J. Comp. Sci.*, **94**, 151–185.

Haltiner, G.J. and R.T. Williams. 1980. *Numerical Prediction and Dynamic Meteorology*, 2nd ed. Wiley & Sons, N.Y.

Hamblin, P.F. 1972. Some free oscillation of a rotating natural basin. Ph. D. thesis, Univ. of Washington, Seattle, 97pp.

Hamilton, P. 1975. A numerical model of the vertical circulation of tidal estuaries and its application to the Rotterdam Waterway. *Geophys. J. Roy. Astr. Soc.,* **40,** 1–21.

Hamilton, J. 1978. The quarter-diurnal tide in the English Channel. *Geophys. J. Roy. Astr. Soc.,* **53,** 541–552.

Haney, R.L., 1971. Surface thermal boundary condition for ocean circulation models. *J. Phys. Oceanogr.,* **1,** 241–248.

Haney, R.L., 1974. A numerical study of the response of an idealized ocean to large-scale surface heat and momentum flux. *J. Phys. Oceanogr.,* **4,** 145–167.

Haney, R.L. 1991. On the pressure gradient force over steep topography in sigma coordinate ocen models. *J. Phys. Oceanogr.,* **21,** 610–619.

Hansen, W. 1949. Die halbtägigen Gezeiten im Nordatlantisepen Ozean. *Dtsch. Hydr. Zeit.,* **2,** 44–61.

Hansen, W. 1962. Hydrodynamical methods applied to oceanographic problems. Proc. Symp. Mathem.–Hydrodyn. Methods of Phys. Oceanogr., Inst. Meereskunde Univ. Hamburg, 25–34.

Harper, B.A. and R.J. Sobey. 1983. Open boundary conditions for open-coast hurricane storm surge. *Coast. Eng.,* **7,** 41–60.

Harris, D.L. 1957. The effect of a moving pressure disturbance on the water level in a lake. *Meteor. Mongr.,* **2(10),** 46–57.

Harris, D.L. 1967. A critical survey of the storm surge prediction problem, pp. 47–64. Proc. Symp. Tsunamis Storm Surges. The 11th Pac. Sci. Congr., Tokyo, 1966. Committee for the Pacific Science Congress, March 1967, Tokyo.

Harris, D.L. and C.P. Jelesnianski. 1964. Some problems involved in the numerical solutions of tidal hydraulic equations. *Mon. Wea. Rev.,* **92,** 409–422.

Harris, R.A. 1908. Seiches in lakes, bays, etc. *Manual of Tides* Part V, Chap. IX, Appendix 6. U.S. Coast and Geodetic Survey, Washington, D.C.

Haurwitz, B. 1951. The slope of lake surfaces under variable wind stresses. AD699402, Corps Eng., Washington, D.C.

Heaps, N.S. 1969. A two-dimensional numerical sea model. *Phil. Trans. Roy. Soc. London* Ser. A, **275**, 93–137.

Heaps, N.S. 1972. On the numerical solution of the three-dimensional hydrodynamical equations for tides and storm surges. *Mem. Soc. Roy. Soc. Sci. Liège Coll. Huit*, **2**, 143–180.

Heaps, N.S. 1974. Development of a three-dimensional numerical model of the Irish Sea. *Rapp. P.-V. Reun. Cons. Int. Explor. Mer.*, **167**, 147–162.

Heaps, N.S. 1975. Resonant tidal cooscillations in a narrow gulf. *Arch. Meteor. Geophys. Bioklimat. Ser. A*, **24**, 361–384.

Heath, R.A. 1975. Surface oscillations of Lake Wakatipu, New Zealand. *N.Z. J. Mar. Freshwater Res.*, **9**, 223–238.

Hebenstreit, G.T., F.I. Gonzalez and A.F. Morris. 1985. Near–source tsunami simulation of Valparaiso Harbor, Chile. Proc. Int. Tsunami Symp., Murty T.S. and W.J. Rapatz (eds.), Inst. Ocean Sciences, Sidney, B.C.

Henry, R.F. 1975. Storm Surges. Tech. Rept. No. 19, Beaufort Sea Project. Dept. Envir. Victoria, B.C. 41 pp.

Henry, R.F. 1981. Richardson–Sielecki schemes for the shallow–water equations, with applications to Kelvin waves. *J. Comput. Physics*, **41**, 389–406.

Henry, R.F. 1982. Automated Programming of Explicit Shallow-Water Models. Part I. Linearized Models with Linear or Quadratic Friction. Pac. Mar. Sci. Rept. No. 3. Inst. Ocean Sciences, Victoria, B.C. 71 pp.

Henry, R.F. and N.S. Heaps. 1976. Storm surge in the southern Beaufort Sea. *J. Fish. Board Can.*, **33(10)**, 2362-2376.

Henry, R.F. and T.S. Murty. 1972. Resonance periods of multi-branched inlets with tsunami amplication. Dept. Envir. Mar. Sci. Rept. 28, 47–79.

Hess, K.W., 1985. Assessment model for estuarine circulation and salinity. NOAA Tech. Memo. NESDIS AISC 3, Mar. Envir. Ass. Div. Washington D.C.

Hibberd, S and D.H. Peregrine. 1979. Surf and run–up on a beach: a uniform bore. *J. Fluid Mech.*, **95**, 323–345.

Hidaka, K. 1931. The oscillations of water in spindle-shaped and elliptic basins as well as the associated problems. *Mem. Mar. Obs. Kobe.*, **4**, 99–219.

Hinze, J.O. 1975. *Turbulence.* McGraw–Hill Book Co. 790pp.

Holland, W.R. 1977. Ocean general circulation models. In: E. Goldberg (ed.), *The Sea.* Vol. 6. Wiley & Sons, N.Y.

Holland, W.R. 1978. The role of mesoscale eddies in the general circulation of the ocean-numerical experiments using a wind-driven quasi-geostrophic model. *J. Phys. Oceanogr.*, **8**, 363–392.

Holland, W.R. 1986. Quasi-geostrophic modelling of eddy-resolving ocean circulation. 203–231. *Advanced Physical Oceanographic Numerical Modelling*, J.J. O'Brien (ed.), D. Reidel.

Holland, W.R. and L.B. Lin. 1975a. On the generation of mesoscale eddies and their contribution to the oceanic general circulation. I. A preliminary numerical experiment. *J. Phys. Oceanogr.*, **5**, 642–657.

Holland, W.R. and L.B. Lin. 1975b. On the generation of mesoscale eddies and their contribution to the oceanic general circulation. II. A parameter study. *J. Phys. Oceanogr.*, **5**, 658–669.

Holloway, G. 1987. Systematic forcing of large–scale geophysical flow by eddy-topography interaction. *J. Fluid Mech.*, **184**, 463–476.

Honda, K., T. Terada, Y. Yoshida and D. Isitani. 1908. Secondary undulations of oceanic tides. Sci. Rept. Tohoku Imp. Univ. Sendai. **1(1)**, 61–66.

Horikawa, K and H. Nishimura. 1968. Characteristic oscillation in a basin with a branch. *Coast. Eng. Jpn.*, **11**, 59–68.

Hough, S.S. 1898. On the application of harmonic analysis to the dynamical theory of tides. Part II. On the general integration of Laplace's dynamical equations. *Phil. Trans. Roy. Soc. A*, **191**, 138–185.

Hunkins, K. 1964. Tide and storm surge observations on the Chukchi Sea. *Limnol. Oceanogr.*, **10**, 29–39.

Hunkins, K. 1986. Anomalous diurnal tidal currents on the Yermak Plateau. *J. Mar. Res.*, **44**, 51–69.

Huppert, H.E. 1980. Topographic effects in stratified fluids. 117–140. **In:** H.J. Freeland *et al.* (eds.), *Fjord Oceanography*, Plenum Press, N.Y.

Il'in, A.M. 1969. Differencing scheme for a differential equation with a small parameter affecting the highest derivative. *Math. Notes Acad. Sci. USSR*, 6, 596–602.

Irish, S.M. and G.W. Platzman. 1961. An investigation of the meteorological conditions associated with extreme wind tides on Lake Erie. U.S. Weather Bur. Tech. Rept. No. 4. Dept. Meteor., Univ. of Chicago. 35 pp.

Irish, S.M. and G.W. Platzman. 1962. An investigation of meteorological conditions associated with extreme wind tides on Lake Erie. *Mon. Wea. Rev.*, **90**, 39–47.

Ivchenko, V.O. 1977. Fluid dynamics on a rotating plane: its averaged equations. *LAES, Intern. J.*, Princeton, **5(6)**, 465–472.

Jamart, B.M. and J. Ozer. 1986. Numerical boundary layers and spurious residual flows. *J. Geophys. Res.*, **91**, 10,621–10631.

Janjic, Z.I. 1974. A stable centered difference scheme free of two-grid interval noise. *Mon. Wea. Rev.*, **102**, 319–323.

Jelesnianski, C.P. 1965. A numerical calculation of storm tides induced by a tropical storm impinging on a continental shelf. *Mon. Wea. Rev.*, **93**, 343–360.

Jelesnianski, C.P. 1967. Numerical computations of storm surges with bottom stress. *Mon. Wea. Rev.*, **95**, 740–756.

Jelesnianski, C.P. 1976. A sheared coordinate system for storm surge equations of motion with a mildly curved coast. NOAA Tech. Memo., NWS–TDL–GI, NOAA, Washington, D.C., 52pp.

Johns, B. 1982. Numerical integration of the shallow water equations over a sloping beach. *Intl. J. Num. Methods Fluids*, 2, 253-261.

Johns, B. and A. Ali. 1980. The numerical modelling of storm surges in the Bay of Bengal. *Q. J. R. Meteor. Soc.*, **106(447)**, 1–18.

Kagan, B.A. 1970. On the features of some finite-difference schemes used in numerical integration of tidal dynamics equations. *Atmos. Oceanic Physics*, **6**, 704–717.

Kamenkovitch, V.M., 1961. On the integration of the sea current equations in multiply connected space. *Dokl. A.N. SSSR*, **138**, 5 (in Russian).

Kantha, L.H., G.L. Mellor and A.F. Blumberg. 1982. A diagnostic calculation of the general circulation in the South Atlantic Bight. *J. Phys. Ocean.*, **12**, 805–819.

Kasahara, A. 1969. Simulation of the earth's atmosphere. N.C.A.R. Manuscr., **69–27**, 42pp.

Kelvin, Lord W., 1879. On gravitational oscillations of rotating water. *Proc. Roy. Soc. Edin.*, **10**, 92–109, Papers, **4**, 141–148.

Kielmann, J and Z. Kowalik. 1980. A bottom stress formulation for storm surge problems. *Oceanol. Acta*, **3**, 51–58.

Kim, S.K. and Y. Shimazu. 1982. Simulation of tsunami runup and estimate of tsunami disaster by the expected great earthquake in the Tokai District, Central Japan. *J. Earth Sci. Nagoya Univ.*, **30**, 1–30.

Klinck, J.M., J.J. O'Brien and H. Svendsen. 1981. A simple model of fjord and coastal circulation interaction. *J. Phys. Oceanogr.*, **11**, 1612–1626.

Komar, P.D. 1976. Boundary layer flow under steady unidirectional currents. 91–106. In: D.J. Stanley and D.J.P. Swift (eds.), *Marine Sediment Transport*, John Wiley & Sons, New York.

Kondo, H., K. Saito, Y. Mamiya, and M. Hara. 1982. On the conservation of the energy of low-frequency waves in iterative time integration schemes. *J. Meteor. Soc. Jpn.*, **60(2)**, 824–829.

Koss, W. 1971. Numerical integration experiments with variable resolution two-dimensional Cartesian grids using the box method. *Mon. Wea. Rev.*, **99**, 727–738.

Kowalik, Z. 1969. Wind–driven circulation in a shallow sea with application to the Baltic Sea. *Acta Geophysica Polonica*, **XVII(1)**, 13–38.

Kowalik, Z. 1975. On the numerical solution of steady circulation in the sea. *Acta Geophysica Polonica*, **XXIII(4)**, 327–336.

Kowalik, Z. and I. Bang. 1987. Numerical computation of tsunami run-up by the upstream derivative method. *Sci. Tsunami Hazards*, **5**(2), 77–84.

Kowalik, Z. 1993. Diurnal Tides in the Nordic Seas. In press.

Kowalik, Z. and N. Bich Hung. 1977. On a system of hydrodynamic equations for certain oceanographical problems in the region of the earth's pole and stability of its solution. *Oceanologia*, **7**, 5–20.

Kowalik, Z. and J.B. Matthews. 1983. Numerical study of the water movement driven by brine rejection from nearshore arctic ice. *J. Geophys. Res.*, **88**(C5), 2953–2958.

Kowalik, Z and N. Untersteiner. 1978. A study of the M_2 tide in the Arctic Ocean. *Dtsch. Hydrog. Z.*, **31**, H.6, 216–229.

Kowalik, Z. and P.M. Whitmore. 1991. An investigation of two tsunamis recorded at Adak, Alaska. *Sci. Tsunami Hazards*, **9**(2), 67–83.

Krishnamurti, T.N., M. Kanamitsu, B. Ceselski and M.B. Mathur. 1973. Florida State University's tropical prediction model. *Tellus*, **25**, 523–535.

Kuipers, H. and C.B. Vreugdenhil. 1973. Calculations of two-dimensional horizontal flow. Report of Basic Research S 163. Part 1. Delft Hydraulics Laboratory, The Netherlands. 64 pp.

Kurihara, Y. 1965a. On the use of implicit and iterative methods for the time integration of the wave equation. *Mon. Wea. Rev.*, **93**, 33–46.

Kurihara, Y. 1965b. Budget analysis of a tropical cyclone simulated in an axisymmetric numerical model. *J. Atmos. Sci.*, **22**, 25–59.

Kwizak, M. and A.J. Robert. 1971. A semi-implicit scheme for grid point atmospheric models of the primitive equations. *Mon. Wea. Rev.* **99**, 32–36.

Lagrange, J.L. 1781. Memoire sur la théorie du mouvement des fluides. Nouv. Mem. Acad. R. Berli, Oeuvres 4.

Lam, D.C.L. 1977. Comparison of finite element and finite difference methods for nearshore advection-diffusion transport models, 115–129. **In:** W.G. Gray, G.F. Pinder, and C.A. Brebbia (eds.), *Finite Elements in Water Resources*. Pentech Press, London.

Lamb, H. 1945. *Hydrodynamics*, 6th ed. Dover Publ. N.Y. 738 pp.

Landau, L.D. and E.M. Lifshitz. 1959. *Fluid Dynamics*. Pergamon Press, 536 pp.

LaPlace, P.S. 1775, 1776. Reserches sur plusiers points du système du monde. *Mem. Acad. Roy. Sci.*, **88**, 75–182; **89**, 177–267.

L'atkher, V.M., Militev, A.N. and Shkol'nikov, S.Ya. 1978. Calculation of the tsunami run-up. In: *Tsunami Research in the Open Ocean*. 48–55. Published by NAUKA, Moscow.

Launder, B.E and Spalding D.B. 1972. *Mathematical Models of Turbulence*. Academic Press, London, 170 pp.

Lauwerier, H.A. 1962. Some recent work of the Amsterdam Mathematical Center on the hydrodynamics of the North Sea, pp. 13–24. Proc. Symp. Math. Hydrody. Math. Phys. Ocean. University of Hamburg, Hamburg.

Lax, P.D. and B. Wendroff. 1960. Systems of conservation laws. *Comm. Pure Appl. Math.*, **13**, 217–237.

LeBlond, P.H. and L.A. Mysak. 1978. *Waves in the Ocean*. Elsevier Oceanographic Series 20, Amsterdam. 602 pp.

Lee, J.J. and F. Raichlen. 1971. Resonance in harbors of arbitrary shape. Proc. 12th Conf. Coast. Eng., Washington, D.C., Am. Soc. Civ. Eng. **3**, 2163–2180.

Lee, J.J. and F. Raichlen. 1972. Oscillations in harbors with connected basins. *J. Waterway, Harbors Coast. Eng. Div.* Proc. Am. Civ. Eng. **98**, 311–332.

Leendertse, J.J. 1967. Aspects of a computational model for long-period water-wave propagation. RAND Corporation, Memo. RM–5294–PR.

Leendertse, J.J. 1970. A Water Quality Simulation Model for Well Mixed Estuaries and Coastal Seas. Vol. I. *Principles of Computation*. RM-6230-RC. The Rand Corporation, Santa Monica, Calif.

Leendertse, J.J., R.C. Alexander and S.K. Liu. 1973. A three-dimensional model for estuaries and coastal seas: Volume I, *Principles of Computation*. RAND Corporation. R–1417–OWRR.

Leendertse, J.J. and E.C. Gritton. 1971. A Water Quality Simulation Model for

Well Mixed Estuaries and Coastal Seas. Vol. II. *Principles of Computation.* RM-6230-RC. The Rand Corporation, New York.

Leendertse, J.J. and S.K. Liu. 1975. A three-dimensional model for the estuaries and coastal seas: Volume II, *Aspects of Computation.* RAND Corporation. R-1764-OWRT.

Leith, C.E. 1965. Numerical simulation of the Earth's atmosphere. *Methods Comput. Physics*, 4, 1–28.

Leonard, A. 1974. Energy cascade in large-eddy simulations of turbulent fluid flows. *Adv. Geophys.*, 18A, 237–248.

Leonard, B.P. 1979a. A survey of finite differences of opinion on numerical muddling of the incomprehensible defective confusion equation. 1–17. In: T.J.R. Hughes (ed.) *Finite Element Methods for Convection Dominated Flows*, Am. Soc. Mech. Eng. New York.

Leonard, B.P. 1979b. A stable and accurate convective modelling procedure based on quadratic upstream interpolation. *Computer Methods in Applied Mech. and Eng.*, 19, 59–98. North-Holland Publ. Co.

Leonard, B.P. 1988. Simple high-accuracy resolution program for convective modelling of discontinuities. *Int. J. Num. Methods in Fluids*, 8, 1291–1318.

Lewis, C.H. III and Adams, W.M. 1983. Development of a tsunami-flooding model having versatile formulation of moving boundary conditions. The Tsunami Society MONOGRAPH SERIES, No. 1, 128 pp.

Lilly, D.K. 1961. A proposed staggered grid system for numerical integration of dynamic equations. *Com. Pure Applied Math.*, 9, 267–293.

Lilly, D.K. 1965. On the computational stability of numerical solutions of time-dependent nonlinear geophysical fluid dynamics problems. *Mon. Wea. Rev.*, 93, 11–26.

Lindzen, R.S. and H.L. Kuo. 1969. A reliable method of numerical integration of a large class of ordinary and partial differential equations. *Mon. Wea. Rev.*, 97(10), 732–734.

Liu, S.K. and J.J. Leendertse. 1978. Multidimensional Numerical Modeling of Estuaries and Coastal Seas. 95–164. In: Ven Te Chow (ed.), *Advances in Hydro-science*, Vol. 11.

Loomis, H.G. 1970. A method of setting up the eigenvalue problem for the linear, shallow water wave equation for irregular bodies of water with variable depth and application to bays and harbors in Hawaii. *Hawaii Ins. Geoph.*, HIG–70–32, NOAA–JTRE–16, 9pp.

Loomis, H.G. 1972. A package program for time–stepping long waves into coastal region with application to Haleiwa Harbor, Oahu. *Hawaii Ins. Geoph.*, HIG–72–21, NOAA–JTRE–79, 33pp.

Loomis, H.G. 1973. A new method for determining normal modes of irregular bodies of water with variable depth. *Hawaii Ins. Geop.*, HIG–73–10, NOAA–JTRE–86, 27pp.

Lynch, D.R. 1980. Moving boundary numerical surge model. *J. Waterway Port Coastal Ocean Div. Proc.* **106(WW3)**, 425–428.

Madala, R.V. and S.A. Piacsek, 1977. A semi-implicit numerical model for baroclinic oceans. *J. Comp. Phys.*, **23**, 167–178.

Mader, L.Ch. 1988. *Numerical Modeling of Water Waves*, Univ. of California Press, Berkeley, Calif.

Mamayev, O.I. 1958. The influence of stratification on vertical turbulent mixing in the sea. *Izv. Geophys. Ser.*, 870–875.

Mamayev, O.I. 1975. *Temperature-Salinity Analysis of World Ocean Waters*. Elsevier, Amsterdam, 374 pp.

Marchuk, G.I. 1976. *Numerical Methods in Weather Prediction*. Academic Press, New York, 277 pp.

Marchuk, A.G., L.B. Chubarov and Iu.I. Shokin. 1983. *Numerical Modeling of Tsunami Waves*, Izd. Nauka. Novosibirsk. (Los Alamos), 1985, 282 pp.

Marchuk, G.I., B.A. Kagan and R.E. Tamsalu. 1969. Numerical methods to compute tidal motion in adjacent seas. *Atmos. Oceanic Physics*, **5**, 7.

Marchuk, G.I., B.A. Kagan. 1984. *Ocean Tides*, Pergamon Press, New York, 292 pp.

Matsutomi, H. 1983. Numerical analysis of the run-up of tsunamis on dry bed. 479–493. In: K. Iida and T. Iwasaki (eds.) *Tsunamis - Their Science and Engineering*. Terra Sci. Publ. Co., Tokyo.

McPhee, M.G. 1980. An analysis of pack ice drift in summer. 62–75. Proc. AIDJEX Symp., R. Pritchard (ed.), Univ. of Washington Press, Seattle.

McWilliams, J.C. 1976. Maps from the mid-ocean dynamics experiment. Part II: Potential vorticity and its conservation. *J. Phys. Oceanogr.*, **6**, 828–846.

McWilliams, J.C. 1977. On a class of stable, slightly geostrophic mean gyres. *Dyn. Atmos. Oceans* **2**, 19–28.

Mellor, G.L. and T. Yamada. 1974. A hierarchy of turbulent closure models for planetary boundary layers. *J. Atmos. Sci.*, **31**, 1791-1806.

Mellor, G.L. and T. Yamada. 1982. Development of a turbulent closure model for geophysical fluid problems. *Rev. Geophys. Space Phys.*, **20**, 851–875.

Merian, J.R. 1828. *Über die Bewegung*. Tropfbarer Flussigkeiten in Gebässen (Basle).

Mesinger, F. 1973. A method for construction of second order accuracy difference schemes permitting no false two grid interval wave in the height field. *Tellus*, **25**, 444–458.

Mesinger, F. and A. Arakawa. 1976. *Numerical Methods Used in Atmospheric Models*. GARP Publications Series, No.17. 64pp.

Miles, J.W. 1971. Resonant response of harbors: an equivalent circuit analysis. *J. Fluid Mech.*, **46**, 241–265.

Miles, J.W. 1974. Harbor seiching. *Ann. Rev. Fluid Mech.*, **6**, 17–35.

Miles, J.W. and Y.K. Lee. 1975. Helmholtz resonance of harbors. *J. Fluid Mech.*, **67**, 445–464.

Miles, J.W. and W.H. Munk. 1961. Harbor paradox. *J. Waterway, Harbors Coast. Eng. Div.*, Proc. Am. Soc. Civ. Eng. **87**, 111–130.

Miller, R.N., A.R. Robinson and D.B. Haidvogel. 1983. A baroclinic quasi-geostrophic open ocean model. *J. Comp. Phys.*, **50**, 38–70.

Mitchell, A.R. and D.F. Griffiths. 1980. *The Finite Difference Method in Partial Differential Equations*, John Wiley & Sons, Chichester–Toronto, 272pp.

Miyakoda, K. 1962. Contribution to the numerical weather prediction computation with finite difference. *Jpn. J. Geophys.*, **3**(1), 75–190.

Monin, A.S. and A.M. Yaglom. 1971. *Statistical Fluid Dynamics*. Vol 1. The MIT Press, Cambridge, 769 pp.

Moore, D.W. and S.G.H. Philander. 1977. Modeling the tropical oceanic circulation. 319–361. In: E. Goldberg (ed.), *The Sea*. Vol. 6. Wiley & Sons, N.Y.

Munk, W.H. 1950. On the wind-driven ocean circulation. *J. Meteor.*, **7**, 79–93.

Munk, W.H. and E.R. Anderson. 1948. Notes on a theory of the thermocline. *J. Mar. Res.*, **3**, 276–295.

Murty, T.S. 1971. The response of a lake with a depth-discontinuity to a time-dependent wind-stress. *Arch. Meteor. Geophys. Bioklamat. Ser.*, A **20**, 55–66.

Murty, T.S. 1977. *Seismic Sea Waves - Tsunamis*. Bull. Fish. Res. Board Canada 198, 337pp.

Murty, T.S. 1984. *Storm Surges - Meteorological Ocean Tides*, Bull. No.212, Dept. Fish. Oceans, Ottawa, 897 pp.

Murty, T.S. 1985. Modification of hydrographic characteristics, tides, and normal modes by ice cover. *Mar. Geod.* **9**(4), 451–468.

Murty, T.S. and L. Boilard. 1970. The tsunami in Alberni Inlet caused by the Alaska earthquake of March, 1964, 165–187. **In:** W.M. Adams (ed.), *Tsunamis in the Pacific Ocean*, East West Center Press, Honolulu.

Murty, T.S. and M.I. El-Sabh. 1986. Treatment of free boundaries in numerical computation of long gravity wave problems, 261–267. **In:** R.P. Shaw et al. (eds.), *Innovative Numerical Methods in Engineering*. Springer-Verlag, New York.

Murty, T.S. and J.D. Taylor. 1975. Free ocsillations in bays and harbors. *Proc. Symp. Modelling Tech.*, Sept. 3–5, 1975, San Francisco, CA.

Nakano, M. 1932a. Preliminary note on the accumulation and dissipation of energy of the secondary undulations in a bay. *Proc. Phys. Math. Soc. Jpn. Ser.*, **3**, 44–56.

Nakano, M. 1932b. The secondary undulations in bays forming a coupled system. *Proc. Phys. Math. Soc. Jpn. Ser.*, **3**, 372–380.

Nakano, M. 1933. Possibility of excitation of secondary undulations in bays by tidal or oceanic currents. *Proc. Imp. Acad. Jpn.*, **9**, 152–155.

Niebauer, H.J. 1980. A numerical model of circulation in a continental shelf-silled fjord coupled system. *Est. Coast. Mar. Sci.*, **10**, 507–521.

Nihoul, C.J., F.C. Ronday, J.J. Peters and A. Sterling. 1978. Hydrodynamics of the Scheldt Estuary. 27–53. **In:** J.C. Nihoul (ed.), *Hydrodynamics of Estuaries and Fjords*, Elsevier, Amsterdam.

Niiler, P.P. 1976. Observations of low frequency currents on the west Florida continental shelf. *Mem. Soc. Roy. Sci. Liège*, **6**, 331–358.

Nikolayevski, V.N. 1970. Asymetric mechanics of turbulent flows. *Prikladnaya Matem. i Mechanika*, **34(3)**, 514–525.

Noye, B.J. 1977. Wind-induced circulation and water level changes in lakes, 135–145. **In:** C.A. Brebbia (ed.), *Applied Numerical Modelling*. Halstead Press, London.

O'Brien, J.J. 1968. Comments on "The over–relaxation factor in the numerical solution of the omega equation". *Mon. Wea. Rev.*, **96(2)**.

O'Brien, J.J. 1986. The diffusive problem. 127–144. In: *Advanced Physical Oceanographic Numerical Modelling*, J.J. O'Brien (ed.), D. Reidel.

O'Brien, J.J. and H.E. Hurlburt. 1972. A numerical model of coastal upwelling. *J. Phys. Oceanogr.*, **2**, 14–26.

Obukhov, A.M. 1957. O tochnosti predvychislenia advektionykh izmenenii polei pri chislenom prognoze pogody. *Izv. Akad. Nauk SSSR Ser. Geofiz.*, **9**, 1133–1141.

Olsen, K. and L.-S. Hwang. 1971. Oscillations in a bay of arbitrary shape and variable depth. *J. Geophys. Res.*, **76**, 5048–5064.

Omori, F. 1906. Note on the tidal waves caused by the Great Krakatoa eruption of 1883. *Proc. Tokyo Math. Phys. Soc.*, **2**, 455–457.

Orlanski, I. 1976. A simple boundary condition for unbounded hyperbolic flows. *J. Comp. Physics*, **21**, 251–269.

Orszag, S.A. 1971. On the elimination of aliasing in finite differencing schemes by filtering high wave number components. *J. Atmos. Sci.*, **28**, 1074.

Ozmidov, R.V. 1986. *Diffusion of Contaminants in the Ocean*. Gidrometeoizdat, Leningrad, 280pp.

Papa, L. 1977. The free oscillations of the Ligurian Sea computed by the H–N Method, *Dtsch. Hydrogr. Z.*, **30**, 81–90.

Pedersen, F.B. 1978. A brief review of present theories of fjord dynamics. 407–422. In: J.C. Nihoul (ed.), *Hydrodynamics of Estuaries and Fjords*, Elsevier, Amsterdam.

Pedersen, L.B. and L.P. Prahm. 1974. A method for numerical solution of the advection equation. *Tellus*, **XXVI**, 594–602.

Pedlosky, J. 1982. *Geophysical Fluid Dynamics*. Springer-Verlag, New York. 624 pp.

Perrels, P.A.J. and M. Karelse. 1978. A two-dimensional numerical model for salt intrusion in estuaries. 107–125. In: J.C. Nihoul (ed.), *Hydrodynamics of Estuaries and Fjords*, Elsevier, Amsterdam.

Phillips, N.A. 1951. A simple three-dimensional model for the study of large-scale extratropical flow patterns. *J. Meteor.*, **8**, 381–394.

Phillips, N.A. 1959. An example of nonlinear computational instability. *The Atmosphere and Sea in Motion*, Rockefeller Ins. Press, New York.

Phillips, O.M. 1966. *The Dynamics of the Upper Ocean*. Cambridge Univ. Press, London, 261 pp.

Platzman, G.W. 1958. The lattice structure of the finite-difference primitive and vorticity equations. *Mon. Wea. Rev.*, **86**, 285–292.

Platzman, G.W. 1963. The dynamic prediction of wind tides on Lake Erie. *Meteor. Mongr.*, **4(26)**, 44 pp.

Platzman, G.W. 1964. An exact integral of complete spectral equations for unsteady one dimensional flow. Tech. Rept. 16, Dept. Geoph. Sci., Univ. Chicago, 28pp.

Platzman, G.W. 1971. Ocean tides and related waves. 239–291,In: W.H. Reid (ed.), *Mathematical Probems in the Geophysical Sciences*, Am. Math. Soc., Providence, Rhode Island.

Platzman, G.W. 1972. Two-dimensional free oscillations in natural basins. *J. Phys. Oceanogr.* **2(2)**, 117–138.

Platzman, G.W. 1974. Normal modes of the Atlantic and Indian oceans. *Tech. Rep.*, 25, Dept. of Geophys. Sciences, Univ. of Chicago, Chicago, IL, 83pp.

Platzman, G.W. 1975. Normal modes of the Atlantic and Indian oceans. *J. Phys. Oceanogr.*, **5**, 201–222.

Platzman, G.W. and D.B. Rao. 1964. The free oscillations of Lake Erie. 359–382, In: K. Yoshida (ed.), *Studies of Oceanography.* (Hidaka volume) University of Washington Press, Seattle,

Proudman, J. 1953. *Dynamical Oceanography.* Methuen, J. Willey, London, 409 pp.

Radach, G. 1971. Ermittlung zufallsangeregter Bewegungsvorgänge für zwei Modellmeere mittels des hydrodymamischnumerischen Verfahrens, *Mitt. Inst. Meeresk. Univ. Hamburg*, Nr. 20.

Raichlen, F. 1966. Long period oscillations in basins of arbitrary shape. Coast. Eng. Spec. Conf. Santa Barbara, Calif., 1965. Am. Soc. Civ. Eng. Part I. 115–145.

Raithby, G.D. and K.E. Torrance. 1974. Upstream-weighted differencing schemes and their application to elliptic problems involving fluid flow. *Computers and Fluids*, **2**, 191–206.

Ramming, H.G. 1972. Reproduction of physical processes in coastal areas, 2197–2216. Proc. 13th Conf. Coastal Eng., Vancouver, British Columbia. ASCE, New York.

Ramming, H.G. 1976. A nested North Sea model with fine resolution in shalow coastal areas. *Mem. Soc. Roy. Sci. Liège* **10**, 9–26.

Ramming, H.G. and Z. Kowalik. 1980. *Numerical Modeling of Marine Hydrodynamics.* Elsevier, Amsterdam.

Rao, D.B. 1966. Free gravitational oscillations in rotating rectangular basins. *J. Fluid Mech.*, **25**, 523–555.

Rao, D.B. 1967. Response of a lake to a time dependent wind stress. *J. Geophys. Res.*, **72**, 1697–1708.

Rao, D.B. 1968. Natural oscillations of the Bay of Fundy. *J. Fish. Res. Bd. Can.*, **25**, 1097–1114.

Rao, D.B. 1969. Effect of travelling disturbances on a rectangular bay of uniform depth. *Arch. Meteor. Geophys. Bioklimat. Ser. A*, **18**, 171–190.

Rao, D.B. 1974. Transient response of shallow enclosed basins by the method of normal modes. Sci. Ser. No. 38, CCIW. 30 pp.

Rao, D.B. and D.J. Schwab. 1974. Two-dimensional normal modes in arbitrarily enclosed basins on a rotating earth: Application to Lakes Ontario and Superior. Spec. Rept. No. 19. Center for Great Lakes Studies, Univ. of Milwaukee, Milwaukee, Wisc. 69 pp.

Raymond, W.H. and H.L. Kuo. 1984. A radiation boundary condition for multidimensional flows. *Quart. J. Roy. Met. Soc.*, **110**, 535–551.

Reid, R.O. 1958. Effect of Coriolis force on edge waves. Part I. Investigation of the normal modes. *J. Mar. Res.*, **16**, 109–144.

Reid, R.O. 1975. Comment on "Three-dimensional structure of storm-generated currents" by G.Z. Forristall. *J. Geophys. Res.*, **80(9)**, 1184–1187.

Reid, R.O. and R.O. Bodine. 1968. Numerical model for storm surges in Galveston Bay. *J. Waterway Harbour Div.*, **94(WWI)**, 33–57.

Reid, R.O. and A.C. Vastano. 1966. Orthogonal coordinates for analysis of long gravity waves near islands. 1–20. Santa Barbara Specialty Conf. Ocean Eng. Proc. Soc. Civ. Eng.

Reid, R.O., A.C. Vastano and T.J. Reid. 1977a. Development of Surge II Program with application to the Sabine–Calcasieu area for Hurricane Carla and design hurricanes. Tech. Paper No. 77-13, Nov. 1977. U.S. Army Cor. Eng., Coastal Eng. Res. Cent., Fort Belvoir, VA., 218pp.

Reid, R.O., A.C. Vastano, R.E. Whitaker, and J.J. Wanstroth. 1977b. Experiments in storm surge simulation, 145–168. In: E.D. Goldberg, I.N. McCave, J.J. O'Brien, and J.H. Steele (eds.), *The Sea*. Vol. 6. Wiley Interscience Publ., New York, 1048pp.

Reid, R.O. and R.E. Whitaker. 1976. Wind driven flow of water influenced by a canopy. *J. Waterway Port Coast. Ocean Div. Proc.* **102(WW1)**, 61–77.

Reimnitz, E. and P.W. Barnes. 1974. Sea ice as a geologic agent on the Beaufort Sea Shelf off Alaska. In: J.C. Reid and J.E. Slater (eds.), *The Coast and Shelf off the Beaufort Sea*, Washington, D.C., Arctic Inst. North America, 301–353.

Reynolds. W.C. 1976. Computation of turbulent flow. *Annual Review of Fluid Mech.*, **8**, 183–208.

Richtmeyer, R.D. 1963. A Survey of Difference Methods for Non-Steady Fluid Dynamics. NCAR Tech. Notes 63-2. 25 pp.

Richtmeyer, R.D. and Morton K.W. 1967. *Difference Methods for Initial-Value Problems.* Interscience Publishers, 405pp.

Roache, P.J. 1976. *Computational Fluid Dynamics*, Hermosa Publisher, Albuquerque, N.M., 446pp.

Robert, A.J. 1966. The integration of a low order spectral form of the primitive meteorological equations. *J. Meteor. Soc. Jpn.*, Ser. 2 44, 237–245.

Roberts, K.V. and N.O. Weiss. 1966. Convective difference schemes. *Math. Comp.* **20**, 272–299.

Røed, L.P. and C.K. Cooper. 1986. Open boundary conditions in numerical ocean models. 411–436. In: J.J. O'Brien (ed.), *Advanced Physical Oceanographic Numerical Modeling*, Reidel.

Røed, L.P. and O.M. Smedstad. 1984. Open boundary conditions for forced waves in a rotating fluid. *SIAM, J. Sci. Stat. Comp.*, **5**, 414–426.

Rood, R.B. 1987. Numerical advection algorithms and their role in atmospheric transport and chemistry models. *Rev. Geophys.*, **25(1)**, 71–100.

Runchal, A.K. 1975. Numerical model for storm surge and tidal run-up studies, 1516–1534. Symp. Modelling Techniques, Vol. 11. 3–5 Sept., 1975, San Francisco. ASCE, New York.

Russell, G.L. and J.L. Lerner. 1981. A new finite-differencing scheme for the tracer transport equation. *Mon. Wea. Rev.*, **20**, 1483–1498.

Samarski, A.A. 1971. *Introduction to the Theory of Difference Schemes*, (in Russian), Publ. by Nauka, Moscow.

Sarkisyan, A.S. 1977. *The Diagnostic Calculations of a Large-Scale Oceanic Circulation*, 363–458. In: *The Sea*, Vol.6, E.D. Goldberg et al. (eds.), A Willey-Interscience Publ., N.Y.

Saul'ev, V.K. 1957. On a method of numerical integration of the equation of diffusion. *Doklady Akad. Nauk USSR*, 115, 1077.

Schafer, P.J. 1966. Computation of a storm surge at Barrow, Alaska. *Arch. Meteor. Geophys. Bioklimat. Ser. A*, 15(3-4), 372–393.

Schwab, D.J. 1975. A normal mode method for predicting storm surges on a lake. Spec. Rept. No. 20. Center for Great Lakes Studies, Univ. of Milwaukee, Milwaukee, Wisc. 50 pp.

Semtner, A.J., Jr. 1973. A numerical investigation of Arctic Ocean circulation. Princeton Univ., Ph.D., Oceanography, 251 pp.

Semtner, A.J., Jr. 1974. An oceanic general circulation model with bottom topography. Dept. Meteor., Univ. California, L.A., Tech. Rept. No. 9.

Shapiro, R. 1970. Smoothing, filtering and boundary effects. *Rev. Geophys. Space Physics*, 8, 359–387.

Shuman, F.G. 1957. Numerical methods in weather prediction: II – smoothing and filtering. *Mon. Wea. Rev.*, 85, 357–361.

Shuman, F.G. 1974. Analysis and experiment in nonlinear computational stability. 51–81. In: Difference and spectral methods for atmospheric and ocean dynamics problems, Proc. Symp. Novosibirsk, Part 1.

Shuman, F.G., Brown, J.A., and Campana, K. 1972. A new explicit differencing system for primitive equations. Paper presented at the AGU Spring Meeting, Washington, D.C. Trans. Am. Geophys. Union 33, 386. (Abstr.)

Shyy, W. 1985. A study of finite difference approximations to steady-state, convection-dominated flow problems. *J. Comp. Phys.*, 57, 415–438.

Sielecki, A. 1968. An energy-conserving difference scheme for the storm surge equations. *Mon. Wea. Rev.*, 96, 150–156.

Sielecki, A. and Wurtele, M.G. 1970. The numerical integration of the nonlinear shallow-water equations with sloping boundaries. *J. Comput. Physics*, **6**, 219–236.

Simons, T.J., 1973. Development of three-dimensional numerical models of the Great Lakes. Sci. Ser. No. 12. Inland Waters Directorate, Canada Centre for Inland Waters, Burlington, Ont.

Simons, T.J., 1974. Verification of numerical models of Lake Ontario: Part I. Circulation in spring and early summer. *J. Phys. Oceanogr.*, 4, 507–523.

Simons, T.J. 1978. Wind driven circulations in the Southwest Baltic. *Tellus*, **30**, 272–283.

Simons, T.J. 1980. Circulation models of lakes and inland seas. Can. Bull. Fish. Aquat. Sci., Bull. 203, Ottawa, 146 pp.

Smagorinsky, J. 1963. General circulation experiments with the primitive equations. *Mon. Wea. Rev.*, **91(3)**, 99–120.

Smith, G.D. 1978. *Numerical Solution of Partial Differential Equations: Finite Difference Methods.* Clarendon Press, Oxford, 304pp.

Smolarkiewicz, P.K. 1983. A simple positive definite advection scheme with small implicit diffusion. *Mon. Wea. Rev.*, 111, 479–486.

Smolarkiewicz, P.K. 1984. A fully multidimensional positive definite advection transport algorithm with small implicit diffusion. *J. Comput. Physics*, 54, 325–362.

Sod, G.R. 1985. *Numerical methods in fluid dynamics.* Cambridge University Press, 446 pp.

Sommerfeld, A. 1949. *Partial Differential Equations.* Academic Press, New York, 333 pp.

Spall, M.A. and W.R. Holland. 1991. A nested primitive equation model for oceanic applications. *J. Phys. Oceanogr.*, **21(2)**, 205–220.

Stommel, H. 1948. The western intensification of wind-driven ocean currents. *Trans. Am. Geophys. Union*, 29, 202–206.

Stommel, H. and H.G. Farmer. 1952. Abrupt change in width in two-layer open channel flow. *J. Mar. Res.*, 11, 205–214.

Stommel, H. and H.G. Farmer. 1953. Control of salinity in a estuary by a transition. *J. Mar. Res.*, 12, 13–20.

Stommel, H. and A. Leetmaa. 1972. Circulation on the continental shelf. Proc. U.S. Nat. Acad. Sci. 69, 3380–3384.

Takeda, H. 1984. Numerical simulation of run-up by the variable transformation. *J. Oceanogr. Soc. Jpn.*, 40, 271–278.

Takegami, T. 1938. A study of the effect of a travelling atmospheric disturbance upon a sea surface. *Mem. Coll. Sci.*, Kyoto Imp. Univ. 21, 55–74.

Tetra Tech Inc. 1978. Coastal Flooding Storm Surge Model. Part I: Methodology. Prepared by Tetra Tech Inc. for the U.S. Dept. of Insurance Administration, Washington, D.C.

Thacker, W. C. 1981. Some exact solutions to the nonlinear shallow-water wave equations, *J. Fluid Mech.*, 107, 499–508.

Trefethen, L.N. 1983. Group velocity interpretation of the stability theory of Gustafsson, Kreiss, and Sundström. *J. Comput. Physics*, 49, 199–217.

Tronson, K. 1975. The hydraulics of South Australian Gulf System II. Resonances. *Aust. J. Mar. Freshwater Res.*, 26, 363–374.

Turner, J.S. 1973. *Buoyancy Effects in Fluids*. Cambridge Univ. Press, London, 368 pp.

Varga, R.S. 1962. *Matrix Iterative Analysis*. Prentice Hall, Englewood Cliffs, N.J.

Vastano, A.C. and R.O. Reid. 1967. Tsunami response for islands: verification of a numerical procedure. *J. Mar. Res.*, 25(2), 129–139.

Veronis, G. 1966. Wind-driven ocean circulation – Part 2. Numerical solutions of the non-linear problem. *Deep-Sea Res.*, 13, 30–35.

Voltzinger, N.E., K.A. Klevanny and E.N. Pelinovsky. 1989. *Long-Wave Dynamics of the Coastal Zone*. Gidrometizdat. Leningrad, 271pp.

Voronov, V.A. and V.O. Ivchenko. 1978. On calculating the influence of mesoscale motion on the large-scale ocean circulation. *Oceanology*, **18(6)**, 637–639.

Voyt, S.S. 1974. Long waves and tides. In: P.S. Lineykin (ed.), ITOGI. *Summaries of Scientific Progress. Oceanology*. Vol. 2. G.K. Hall and Co., Boston. 33–51.

Walton, R. and B.A. Christensen. 1980. Friction factors in storm surges over inland areas. *J. Waterway Port Coast. Ocean Div. Proc.*, **106(WW2)**, 261–271.

Wang, D.P. and D.W. Kravitz. 1980. A semi-implicit two-dimensional model of estuarine circulation. *J. Phys. Oceanogr.*, **10**, 441–454.

Wanstrath, J.J. 1976. Storm Surge Simulation in Transformed Coordinates. Vol. 11: Program Documentation. Tech. Rept. No. 76-3. Coastal Eng. Res. Center, U.S. Army Corps of Engineers, Fort Belvoir, VA, 176 pp.

Wanstrath, J.J. 1977a. An open coast storm surge model with inland flooding, 676–679. Proc. 11th Tech. Conf. Hurricanes Trop. Meteor., 13–16 Dec., 1977, Miami Beach, Fla. Amer. Meteor. Soc., Boston.

Wanstrath, J.J. 1977b. Near Shore Numerical Storm Surge and Tidal Simulation. Tech. Rept. H-77-17. U.S. Army Engineer Waterways Experiment Station, CE, Vicksburg, Miss.

Warren, B.A. 1963. Topographic influences on the path of the Gulf Stream. *Tellus*, **15**, 167–183.

Weenink, M.P.H. 1956. The twin storm surges during December 21–24, 1954: a case of resonance. *Dtsch. Hydrogr. Z.*, **9**, 240–249.

Weenink, M.P.H. and P. Groen. 1958. *Proc. K. ned. Akad. Wet. B*, **61**, 198–213.

Welander, P. 1961. Numerical prediction of storm surges, 315–317. In: H.E. Landsberg and J. Van Miegham (eds.), *Advances in Geophysics*. Vol. 8. Academic Press, New York.

Whitaker, R.E., R.O. Reid, and A.C. Vastano. 1973. Drag Coefficient at Hurricane Wind Speeds as Deduced from Numericl Simulation of Dynamical Water Level Changes in Lake Okeechobee, Fla. Tech. Rept. 73-13-T. Dept. Oceanogr., Texas A&M Univ., College Station, Texas.

Wilson, B.W. 1972. Seiches. *Adv. Hydrosci.*, **8**, 1–94.

Wüber, Ch. and W. Krauss. 1979. The two–dimensional seiches of the Baltic Sea. *Ocean. Acta*, **2**, 435–446.

Wurtele, M.G., J. Paegle, and A. Sielecki. 1971. The use of open boundary conditions with the storm surge equations. *Mon. Wea. Rev.*, **99(6)**, 537–544.

Yanenko, N.N. 1971. *The Methods of Fractional Steps*, Springer-Verlag, New York.

Yeh, H. 1987. Tsunami action on a beach – summary. 206–211. Proc. Inter. Tsunami Symp., Vancouver, B.C., E.N. Bernard (ed.), NOAA/PMEL, Seattle.

Yeh, G.T. and F.K. Chou. 1979. Moving boundary numerical surge model. *J. Waterway Port Coast. Ocean Div. Proc.*, **105(WW3)**, 247–263.

Yeh, G.T. and F.F. Yeh. 1976. A generalized model for storm surges, pp. 921–933, Vol. 1. Proc. 15th Int. Conf. Coast. Eng., 7–11 July, 1976, Honolulu, Hawaii.

Yin, F.L. and I.Y. Fung. 1991. Net diffusivity in ocean general circulation models with nonuniform grids. *J. Geophys. Res.*, **96(C6)**, 10,773–10,776.

Young, J.A. 1968. Comparative properties of some time differencing schemes for linear and nonlinear oscillations. *Mon. Wea. Rev.* **96**, 357–364.

Young, D.M. 1971. *Iterative Solution of Large Linear Systems*, Acad. Press, London.

Zalesak, S.T. 1979. Fully multidimensional flux-corrected transport algorithm for fluids. *J. Comput. Physics*, **31**, 335–362.

Zhu, J. and W. Rodi. 1991. Zonal finite-volume computations of incompressible flows. *Computers Fluid*, **20(4)**, 411–420

www.ingramcontent.com/pod-product-compliance
Lightning Source LLC
Chambersburg PA
CBHW070743220326
41598CB00026B/3730